Climatology: Concepts and Applications

Climatology: Concepts and Applications

Editor: Dale Sullivan

R CALLISTO
REFERENCE

www.callistoreference.com

Callisto Reference,
118-35 Queens Blvd., Suite 400,
Forest Hills, NY 11375, USA

Visit us on the World Wide Web at:
www.callistoreference.com

This book contains information obtained from authentic and highly regarded sources. Copyright for all individual chapters remain with the respective authors as indicated. All chapters are published with permission under the Creative Commons Attribution License or equivalent. A wide variety of references are listed. Permission and sources are indicated; for detailed attributions, please refer to the permissions page and list of contributors. Reasonable efforts have been made to publish reliable data and information, but the authors, editors and publisher cannot assume any responsibility for the validity of all materials or the consequences of their use.

ISBN: 978-1-64116-165-7 (Hardback)

Trademark Notice: Registered trademark of products or corporate names are used only for explanation and identification without intent to infringe.

Cataloging-in-Publication Data

Climatology : concepts and applications / edited by Dale Sullivan.
    p. cm.
Includes bibliographical references and index.
ISBN 978-1-64116-165-7
1. Climatology. 2. Climatic changes. 3. Meteorology. I. Sullivan, Dale.
QC981 .C54 2019
551.6--dc23

Table of Contents

Preface...IX

Chapter 1   **Climate change multi-model projections for temperature extremes in Portugal** 1
C. Andrade, H. Fraga and J. A. Santos

Chapter 2   **How do carbon cycle uncertainties affect IPCC temperature projections?**.............................. 9
Roger W. Bodman, Peter J. Rayner and Roger N. Jones

Chapter 3   **Using feedback from summer subtropical highs to evaluate climate models** 16
Carmen Sánchez de Cos, Jose M. Sánchez-Laulhé, Carlos Jiménez-Alonso and
Ernesto Rodríguez-Camino

Chapter 4   **Impacts of land use and land cover changes on the climate over Northeast Brazil** 22
Ana Paula M. A. Cunha, Regina C. S. Alvalá, Paulo Y. Kubota and
Rita M. S. P. Vieira

Chapter 5   **Surface air temperature variability in global climate models** 31
Richard Davy and Igor Esau

Chapter 6   **Long-term changes in Australian tropical cyclone numbers** 39
Andrew J. Dowdy

Chapter 7   **A new perspective on Australian snow** 46
Sonya Louise Fiddes, Alexandre Bernardes Pezza and Vaughan Barras

Chapter 8   **Observed and projected precipitation variability in Athens over a 2.5 century
period**... 53
D. Founda, C. Giannakopoulos, F. Pierros, A. Kalimeris and M. Petrakis

Chapter 9   **Clausius–Clapeyron-like relationship in multidecadal changes of extreme
short-term precipitation and temperature in Japan**... 60
Fumiaki Fujibe

Chapter 10   **Decadal change of the connection between summer western North Pacific
Subtropical High and tropical SST in the early 1990s** 66
Chao He and Tianjun Zhou

Chapter 11   **Variability of weather regimes in the North Atlantic-European area: past and
future**... 73
Elke Hertig and Jucundus Jacobeit

Chapter 12   **Regional response of annual-mean tropical rainfall to global warming** 80
Ping Huang

Chapter 13   **Multi-GCM by multi-RAM experiments for dynamical downscaling on summertime climate change in Hokkaido**.. 87
Masaru Inatsu, Tomonori Sato, Tomohito J. Yamada, Ryusuke Kuno,
Shiori Sugimoto, Murad A. Farukh, Yadu N. Pokhrel and Shuichi Kure

Chapter 14   **Can marine cloud brightening reduce coral bleaching?**.. 95
John Latham, Joan Kleypas, Rachel Hauser, Ben Parkes and Alan Gadian

Chapter 15   **Climatology of convective available potential energy (CAPE) in ERA-Interim reanalysis over West Africa**.. 101
Cyrille Meukaleuni, André Lenouo and David Monkam

Chapter 16   **Tibetan ice core evidence for an intensification of the East Asian jet stream since the 1870s**... 107
G. W. K. Moore

Chapter 17   **Impact of Congo Basin deforestation on the African monsoon**............................ 115
R. Nogherotto, E. Coppola, F. Giorgi and L. Mariotti

Chapter 18   **Validation of CMIP5 models for the contiguous United States**.............................. 122
Victor Privalsky and Vladislav Yushkov

Chapter 19   **Land surface–atmosphere interaction in future South American climate using a multi-model ensemble**.. 126
R. C. Ruscica, C. G. Menéndez and A. A. Sörensson

Chapter 20   **On the link between cold fronts and hail in Switzerland**.. 133
Sebastian Schemm, Luca Nisi, Andrey Martinov, Daniel Leuenberger and
Olivia Martius

Chapter 21   **Multi-parameter multi-physics ensemble (MPMPE): a new approach exploring the uncertainties of climate sensitivity**.. 144
Hideo Shiogama, Masahiro Watanabe, Tomoo Ogura, Tokuta Yokohata and
Masahide Kimoto

Chapter 22   **An evaluation of CORDEX regional climate models in simulating precipitation over Southern Africa**.. 150
Mxolisi E. Shongwe, Chris Lennard, Brant Liebmann, Evangelia-Anna Kalognomou,
Lucky Ntsangwane and Izidine Pinto

Chapter 23   **Assessing the contribution of different factors in regional climate model projections using the factor separation method**.. 159
Csaba Torma and Filippo Giorgi

Chapter 24   **Trend and pattern classification of surface air temperature change in the Arctic region** 165
Wandee Wanishsakpong, Nittaya McNeil and Khairil A. Notodiputro

Chapter 25   **Long-term changes in the relationship between stratospheric circulation and East Asian winter monsoon**... 171
Ke Wei, Masaaki Takahashi and Wen Chen

Chapter 26   **Variations in temperature-related extreme events (1975–2014) in Ny-Ålesund, Svalbard**.. 178
Ting Wei, Minghu Ding, Bingyi Wu, Changgui Lu and Shujie Wang

Chapter 27   **Strong subsurface soil temperature feedbacks on summer climate variability over the arid/semi-arid regions of East Asia** 185
Lingyun Wu and Jingyong Zhang

Chapter 28   **How does model development affect climate projections?** 192
Jussi S. Ylhäisi, Jouni Räisänen, David Masson, Olle Räty and Heikki Järvinen

Chapter 29   **Pathways between soil moisture and precipitation in southeastern South America**............................... 198
Romina C. Ruscica, Anna A. Sörensson and Claudio G. Menéndez

Permissions

List of Contributors

Index

Preface

The study of climate and weather conditions averaged over a period of time is within the domain of climatology. It studies a number of analog techniques for weather forecasting and various climate models for the future projections of weather and climate systems. Climatology studies are approached from different dimensions like paleoclimatology, paleotempestology and historical climatology. Paleoclimatology strives to understand past climates by studying records of ice cores and tree rings. Paleotempestology determines the frequency of hurricane over millennia using similar records. Climate indices are used for the characterization of the different climatic mechanisms that contribute to daily weather. They are based on the twin objectives of simplicity and completeness. Some of such indices are El Niño–Southern Oscillation, North Atlantic oscillation, Madden-Julian oscillation, etc. This book covers in detail some existing theories and innovative concepts revolving around climatology. Different approaches, evaluations, methodologies and advanced studies on climatology have been included herein. Climatologists, meteorologists, researchers and students associated with this field will benefit alike from this book.

The information contained in this book is the result of intensive hard work done by researchers in this field. All due efforts have been made to make this book serve as a complete guiding source for students and researchers. The topics in this book have been comprehensively explained to help readers understand the growing trends in the field.

I would like to thank the entire group of writers who made sincere efforts in this book and my family who supported me in my efforts of working on this book. I take this opportunity to thank all those who have been a guiding force throughout my life.

Editor

Climate change multi-model projections for temperature extremes in Portugal

C. Andrade,[1,2]* H. Fraga[2] and J. A. Santos[2]

[1]Mathematics and Physics Department, Polytechnic Institute of Tomar, Tomar 2300-313, Portugal
[2]Centre for the Research and Technology of Agro-Environmental and Biological Sciences, Universidade de Trás-os-Montes e Alto Douro, Vila Real 5001-801, Portugal

*Correspondence to:
C. Andrade, Mathematics and Physics Department, Polytechnic Institute of Tomar, Quinta do Contador, Estrada da Serra, Tomar 2300-313, Portugal.
E-mail: c.andrade@ipt.pt

Abstract

Climate change projections for spatial-temporal distributions of temperatures in Portugal are analysed using a 13-member ensemble of regional climate model simulations (A1B scenario for 2041–2070). Bias corrections are carried out using an observational gridded dataset (E-OBS) and equally-weighted ensemble statistics are discussed. Clear shifts toward higher future seasonal mean temperatures, in central tendency and also in both tails of the distributions, are found (2–4 °C), particularly for summer and autumn maximum temperatures. Furthermore, frequencies of occurrence of daily extremes are projected to increase, particularly in summer maximum temperatures over inland Portugal. Wintertime changes are weaker than in other seasons.

Keywords: temperature extremes; climate change projections; multi-model ensemble; indices of extremes; Portugal

1. Introduction

Temperature variability and its extremes play a key role in many human and economic activities. The recent warming and the increasing number of heat waves have had deep impacts on Europe (Kuglitsch et al., 2010; Zhang et al., 2011), particularly in the Mediterranean, a prominent climate change hotspot (Diffenbaugh and Giorgi, 2012). Portugal, due to its location in Southern/Mediterranean Europe, is already highly vulnerable to the occurrence of temperature extremes, particularly to extremely high temperatures and heat waves. Furthermore, in conjunction with a gradual warming throughout the next decades, these extremes are expected to increase, in both frequency and strength (Beniston et al., 2007; Fischer and Schär, 2010; Nikulin et al., 2011). In this context, the assessment of climate change projections for these extremes is of foremost relevance to promote suitable and timely measures for adapting and mitigating their impacts. These projections are also an opportunity to potentiate new cost-effective and eco-innovative practices that may warrant a more sustainable development in the future.

Many previous studies have been devoted to the analysis of temperature extremes in Europe and in Portugal (Klein Tank and Können, 2003; Santos et al., 2007; Andrade et al., 2012; Frías et al., 2012), including their climate change projections for the next decades (Ramos et al., 2011). However, there is a lack of studies using multi-model ensembles. Furthermore, new observational gridded datasets have been recently developed, being more suitable for model validation and calibration than weather station data (models inherently simulate area-mean fields). Hence, this study aims at developing climate change projections for temperature in Portugal using a state-of-the-art ensemble of model simulations, validated using a state-of-the-art gridded observational dataset.

2. Data and methodology

Simulated daily gridded maximum (TX) and minimum temperature (TN) data from a 13-member ensemble of regional climate model (RCM) experiments, run within the ENSEMBLES Project (Christensen et al., 1996; Jacob, 2001; Lenderink et al., 2003; Böhm et al., 2006; Elguindi et al., 2007; Jaeger et al., 2008; Collins et al., 2011; Samuelsson et al., 2011), are used herein (Table I). These RCMs are nested in two widely used global climate models (GCM): ECHAM5 and HadCM3. Data for a recent past period (1961–2000; 40 years) are used as a baseline climate. Further, a future period (2041–2070) under anthropogenic greenhouse gas forcing is selected for assessing climate change projections. The A1B SRES scenario (Nakićenović et al., 2000), which is an intermediate scenario for future carbon dioxide equivalent concentrations, is considered for this purpose. Nonetheless, as the selected future period ends in 2070, the deviations amongst the different scenario pathways are not yet very expressive (IPCC, 2007; van der Linden and Mitchell, 2009). Data within a geographical sector covering mainland Portugal (36.625°N–42.375°N; 6.125°W–9.875°W; Figure S1, Supporting Information) are extracted. Due

Table I. GCM-RCM model chains used in this study, along with their developing institutions, relevant citations and original RCM grids.

GCM	RCM	Institution	Resolution
ECHAM5	HIRHAM	Danish Meteorological Institute (Christensen et al., 1996)	0.22° × 0.22° rotated
	RACMO	Koninklijk Nederlands Meteorologisch Instituut (Lenderink et al., 2003)	0.22° × 0.22° rotated
	CLM1	Max Planck Institute (Böhm et al., 2006)	0.20° × 0.20° regular
	CLM2	Max Planck Institute	0.20° × 0.20° regular
	RCA	Swedish Meteorological and Hydrological Institute (Samuelsson et al., 2011)	0.22° × 0.22° rotated
	RegCM	International Centre for Theoretical Physics (Elguindi et al., 2007)	0.22° × 0.22° rotated
	REMO	Max Planck Institute (Jacob, 2001)	0.22° × 0.22° rotated
	CLM	Eidgenössische Technische Hochschule Zürich (Jaeger et al., 2008)	0.22° × 0.22° rotated
	RCA	Swedish Meteorological and Hydrological Institute (Samuelsson et al., 2011)	0.22° × 0.22° rotated
HadCM3	RCA3	C4I Center (Samuelsson et al., 2011)	0.22° × 0.22° rotated
	HadRM3Q0	Hadley Centre (Collins et al., 2011)	0.22° × 0.22° rotated
	HadRM3Q16	Hadley Centre	0.22° × 0.22° rotated
	HadRM3Q3	Hadley Centre	0.22° × 0.22° rotated

to the climate contrasts between northern Portugal, more influenced by the North Atlantic, and its southern part, with a more typical Mediterranean climate, the target area is divided into two sectors: Region 1 (36.625°N–39.375°N) and Region 2 (39.625°N–42.375°N) – southern/northern half of Portugal (Figure S1). An observational dataset of daily gridded TX and TN (E-OBS, version 8.0), produced by the EU-FP6 project ENSEMBLES (http://ensembles-eu.metoffice.com) and provided by the European Climate Assessment & Dataset (ECA&D) project (Haylock et al., 2008), is used for validating/calibrating the simulated datasets. Model datasets are bi-linearly interpolated to the E-OBS grid. As such, both observational and simulated datasets are defined on a common 0.25° latitude × 0.25° longitude grid (Figure S1) and on a daily basis. As the E-OBS dataset is only available for land grid boxes, all subsequent analysis is conducted for this subset of grid boxes (grid boxes over the North Atlantic are white cells in all maps).

The differences in seasonal mean TX/TN between the ensemble and the E-OBS (1961–2000) show that TX tends to be less skilfully replicated than TN (Figure S2). Further, positive differences prevail in the northernmost regions for all seasons. The patterns display geographical asymmetries that can either be explained by model biases (e.g. smoothed orography and deficient representation of dynamical processes) or by limitations inherent to the E-OBS dataset (Hofstra et al., 2009), also considering the relatively low density of the available weather stations in Portugal (Figure S1). These shortcomings tend to be enhanced in regions with complex topography, such as the northern half of Portugal (Figure S1). Most of the simulations underestimate (overestimate) the mean monthly TX (TN) when compared to E-OBS (Figure S3), i.e. models underestimate thermal amplitudes.

In this study, however, these biases are corrected prior to the analysis and for each model separately. The bias, i.e. simulated minus observed means averaged over 1961–2000, are computed on the monthly basis and are then interpolated to the daily

basis by regression splines. The biases in each pair model/variable and at each grid point are then subtracted from the simulated data so that null patterns replace Figure S2. The same bias corrections are applied to the future period. These calibrations have no other effect on the statistical distributions of each model than a mere translation of its full distribution (their shape remains invariant), but have obvious implications in the ensemble probability density functions (PDFs). After bias corrections, the different models still have different skills in reproducing the spatial patterns of TN and TX from E-OBS. These discrepancies are documented by the corresponding Taylor diagrams (Figure S4). All models show root mean squared deviations (RMSDs) below 2.0 and correlation coefficients above 0.8. The empirical PDFs of each model/variable pair, after bias calibration, are plotted in Figure S5 (1961–2000) and Figure S6 (2041–2070); PDFs are estimated by kernel density estimation with normal kernel-smoothing windows. Despite the different model skills, no further corrections are carried out and all forthcoming analyses are based on equally-weighted ensemble statistics (cf. ensemble PDFs in Figures S5 and S6), which are commonly considered more reliable than single-model or weighted statistics (Christensen et al., 2010).

Besides the analysis of changes in the central tendency of temperature distributions, percentile-based extreme temperature indices are computed for both the recent past and future periods. Owing to the strong seasonality of the temperature regime in Portugal, winter [December to February (DJF)], spring [March to May, (MAM)], summer [June to August (JJA)] and autumn [September to November (SON)] are also individually analysed. In order to quantify the changes in the inter-annual variability of the seasonal mean temperatures, their 90th and 10th percentiles are selected (TX90p/TX10p and TN90p/TN10p for TX and TN, respectively). These percentiles correspond to thresholds for defining moderate to extreme anomalies in the seasonal mean temperatures: anomalously

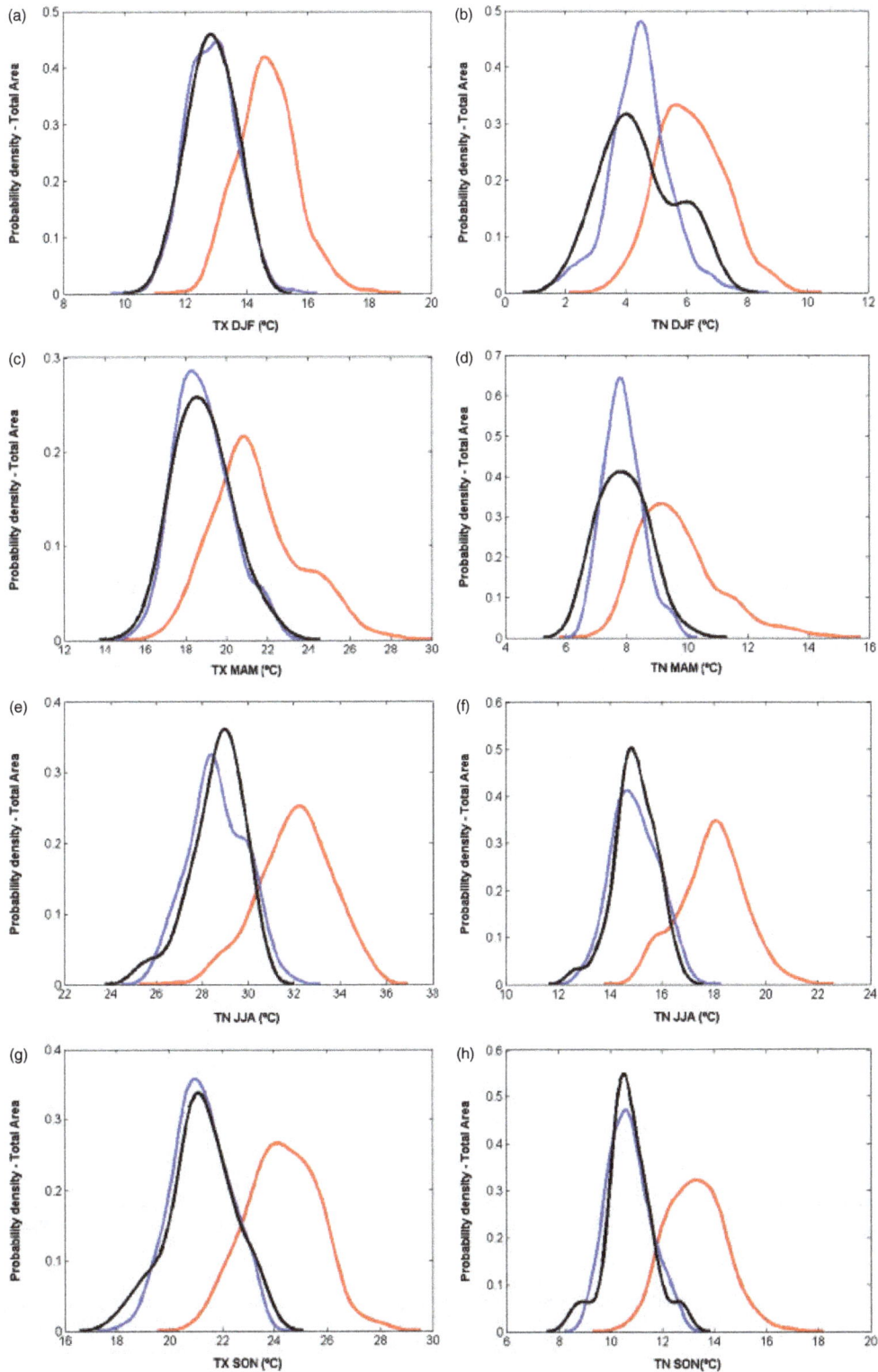

Figure 1. PDFs of seasonal mean (left panels) TX and (right panels) TN, spatially averaged over Portugal, from E-OBS (black curves), the ensembles in 1961–2000 (blue curves) and 2041–2070 (red curves) and for: (a, b) winter (DJF), (c, d) spring (MAM), (e, f) summer (JJA) and (g, h) autumn (SON). Area-means are only over land grid boxes within the Portuguese sector in Figure S1.

cold (warm) seasons below (above) TX10p/TN10p (TX90p/TN90p).

Extreme temperature indices are also computed on a daily basis (Peterson, 2005; Zhang *et al.*, 2011). For this purpose, the number of days per season (ND) above TX75p/TX90p/TX99p (warm days) or

TN75p/TN90p/TN99p (warm nights) are chosen. By definition, for the baseline (1961–2000), their mean values are of approximately 23, 9 and 1 days per season for NDTX75p/NDTN75p, NDTX90p/ NDTN90p and NDTX99p/NDTN99p, respectively. The past period percentiles are then used in the

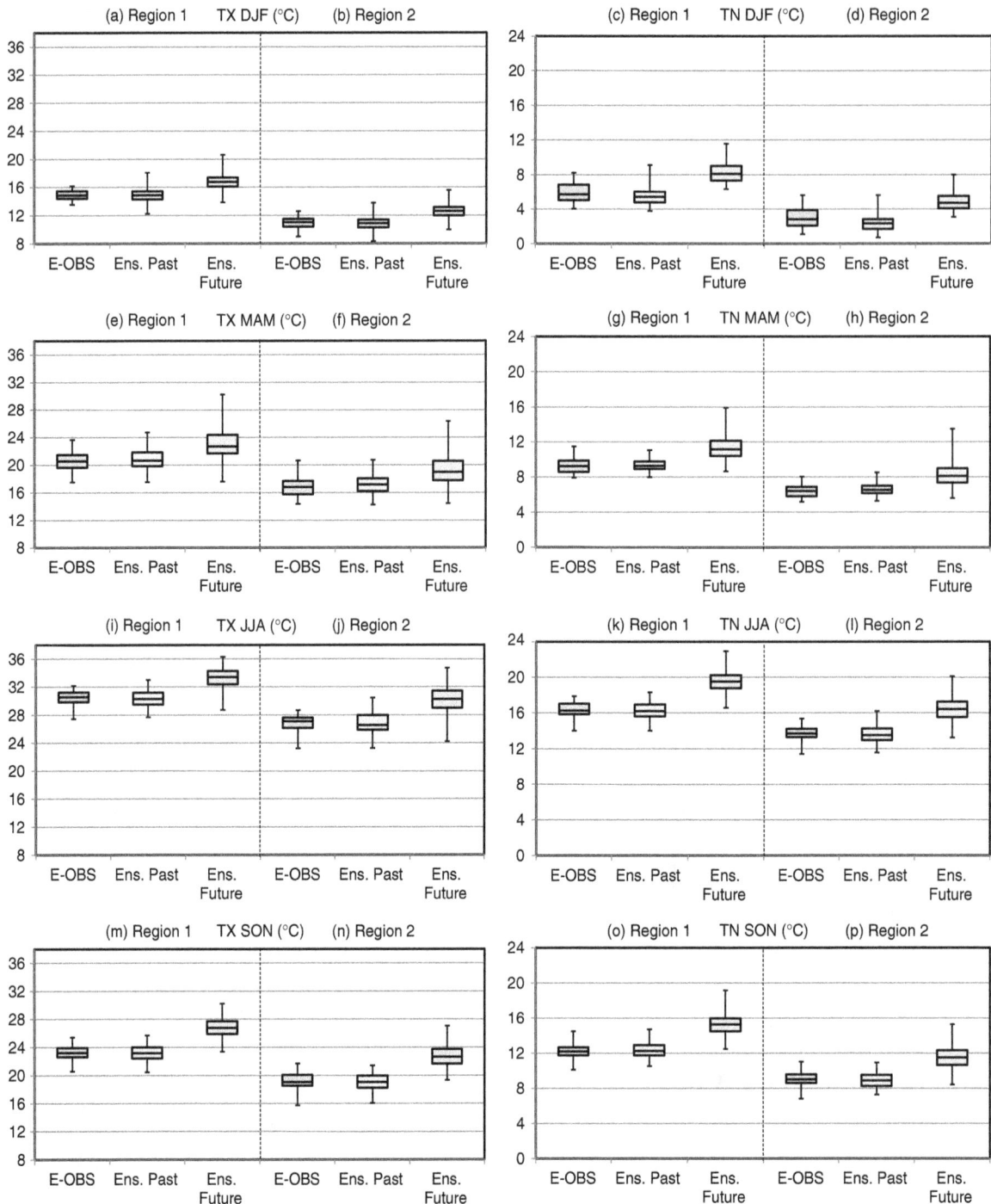

Figure 2. Boxplots of the ensemble distributions of Region 1- and Region 2-mean TX/TN in 2041–2070 (Ens. Future), 1961–2000 (Ens. Past) and E-OBS for: (a–d) winter (DJF), (e–h) spring (MAM), (i–l) summer (JJA) and (m–p) autumn (SON).

future (2041–2070) and corresponding changes in the indices are discussed. The respective cold day/night indices (Figure S7) are of secondary relevance under a warming climate and will not be discussed herein for the sake of succinctness.

3. Results

In order to identify changes in climate extremes from climate model simulations, the entire empirical statistical distribution of each variable/parameter has to be first validated by observational data (simulated vs observed). The PDFs of TN and TX (Figure 1), for E-OBS (black curves) and the ensemble (blue curves) in the past period (1961–2000), highlight the ability of the ensemble to reproduce the baseline climate. Overall, there is a good agreement between the PDFs. Furthermore, some small discrepancies may be rather attributed to limitations inherent to the E-OBS dataset itself than to the ensemble (Hofstra *et al.*, 2009).

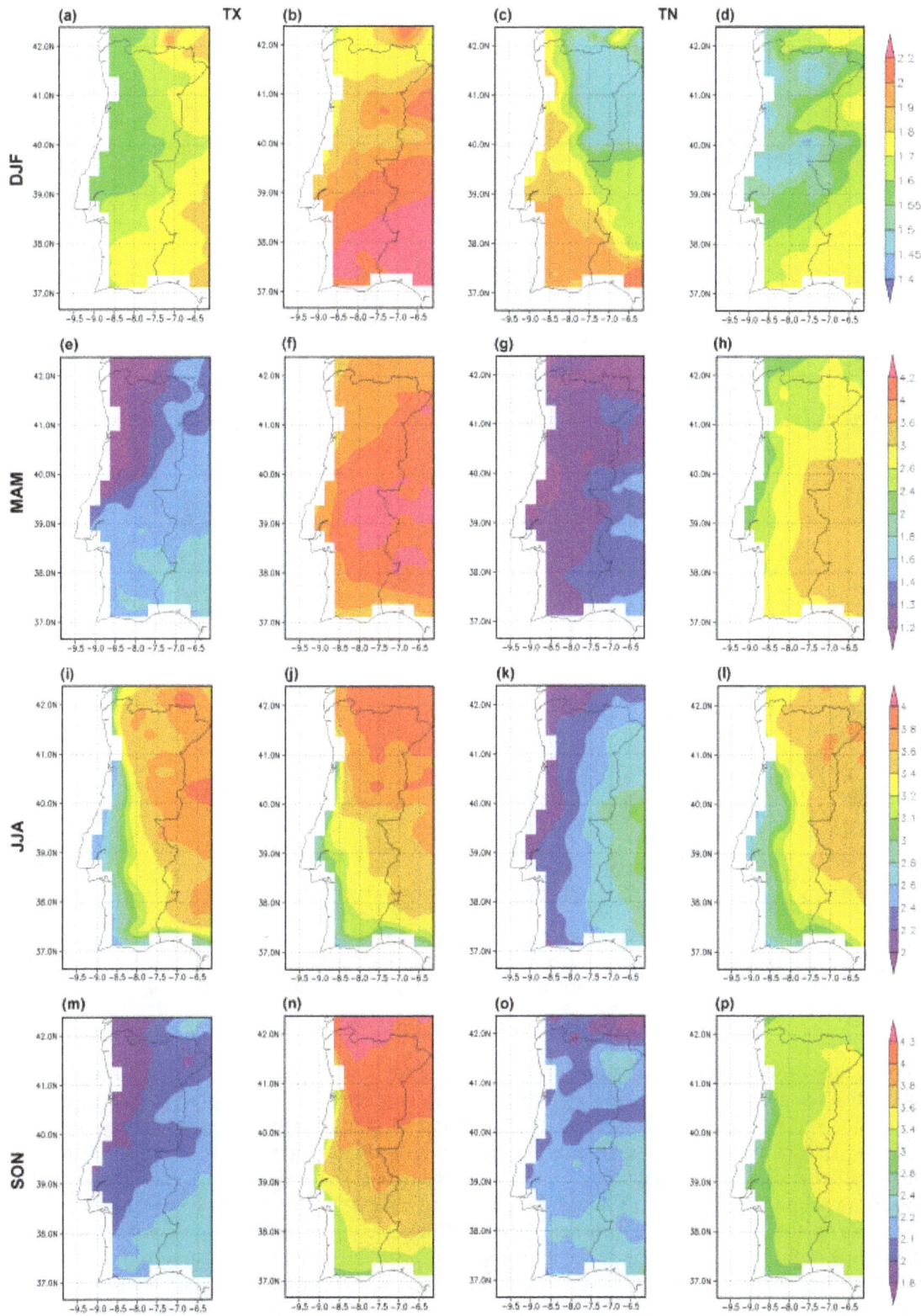

Figure 3. Differences between future (2041–2070) and recent past (1961–2000) ensemble means of TX10p, TX90p, TN10p and TN90p (°C) for: (a–d) winter (DJF), (e–h) spring (MAM), (i–l) summer (JJA) and (m–p) autumn (SON).

The climate change signal for both TX and TN (2041–2070) reveals changes in mean, median and variance. For all seasons, significant increases in both temperatures are projected (red curves in Figure 1 shifted to the right). This warming is more pronounced for TX (for spring, summer and autumn 3–4 °C) than for TN (about 2 °C). Overall, the PDFs for the future have higher kurtosis and variability, mainly for spring (Figure 1(c) and (d)), when strong positive skewness is depicted. For autumn TX there is also important positive skewness, but less pronounced. These results hint at the strengthening of the temperature extremes,

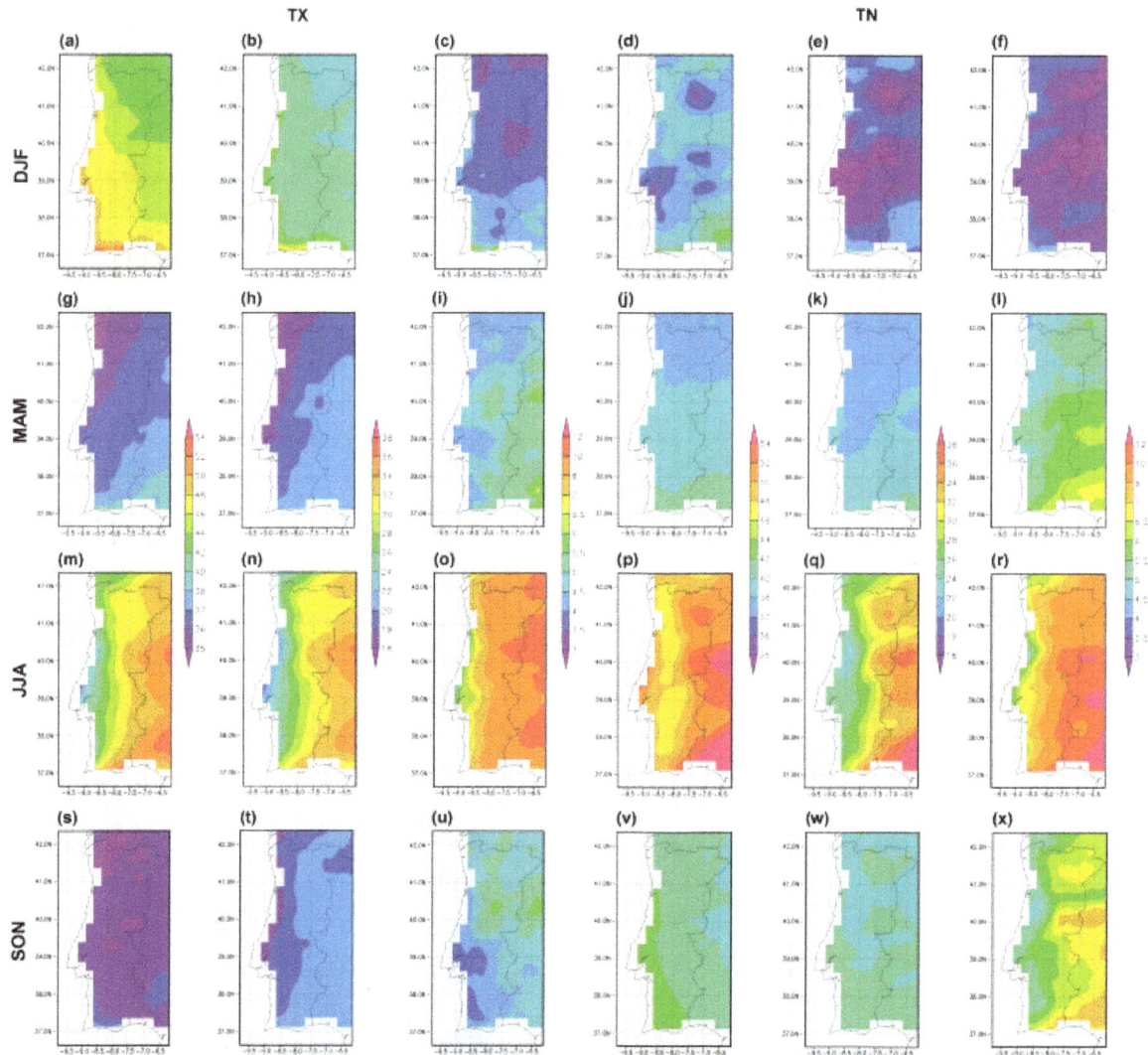

Figure 4. NDTX75p, NDTX90p, NDTX99p, NDTN75p, NDTN90p and NDTN99p (days per season) for: (a–f) winter (DJF), (g–l) spring (MAM), (m–r), summer (JJA) and (s–x) autumn (SON) in the future (2041–2070). Note that, by definition, the values of the indices for the recent past (1961–2000) are of approximately 23, 9 and 1 days per season for NDTX75p/NDTN75p, NDTX90p/NDTN90p and NDTX99p/NDTN99p, respectively.

not only through a translation of the entire PDF, but also owing to enhanced asymmetry towards higher temperatures.

For conciseness purposes, PDFs are replaced by boxplots in the analysis by sector (Figures S1 and S2). A clear warming is projected for both TX and TN and for all seasons in the two regions. Furthermore, the boxplots also reveal an overall increase in the projected TX and TN variability, mainly in summer and autumn (Figure 2(i)–(p)), without a clear distinction between the two regions.

The upward trends in seasonal mean temperatures are also clearly depicted in the ensemble-mean patterns of TX10p/TN10p and TX90p/TN90p (Figure 3). In fact, for every season and percentile, all differences between future and past values are positive, indicating a general warming. Nevertheless, the projected changes tend to be more pronounced for extremely high temperatures (TX90p/TN90p), mainly in spring (Figure 3(f) and (h)), summer (Figure 3(j) and (l))

and autumn (Figure 3(n) and (p)), which is also in clear agreement with the preceding analysis (Figures 1 and 2). The summer patterns, mainly for TN10p and TX10p (Figure 3(i) and (k)), display a strong gradient orthogonal to the coastline, denoting the climate moderation effect by the North Atlantic thermal inertia. Conversely, the inner pasts of Portugal are project to undergo much higher temperature increases, strengthening the coastal-inner climate contrasts in the future.

A number of extreme indices for TX and TN were computed on a daily basis and for each season separately. As the previous projections reveal more substantial changes in the extremely high temperatures (right/positive tail of the PDFs) than in the extremely low temperatures (left/negative tail of the PDFs), only the number of days per season above the 75th, 90th and 99th percentiles are discussed herein: NDTX75p/90p/99p and NDTN75p/90p/99p (Figure 4). On the whole, changes in TX are more pronounced than in TN, which is also in accordance with the previous outcomes for their seasonal means (Figures 1–3). The spatial heterogeneity of the climate change signal is still noteworthy, with strong contrasts between northern and southern Portugal (Figures 2 and 4). The projected increase in the number of extremely hot days in summer (Figure 4(o) and (r)) suggests an increase in both the number of heat waves and in their intensity (e.g. length and maximum temperature).

4. Conclusions

A state-of-the-art 13-member ensemble, generated by GCM–RCM chains and run within the ENSEMBLES project, is used to establish climate change projections for temperature in Portugal, more specifically for TX and TN. The use of equally weighted multi-model ensembles in this study is currently the best approach to take into account the uncertainties in climate change projections (Räisänen, 2007; Christensen et al., 2010), as is clearly depicted in Figures S3 and S4. The E-OBS observational dataset is used to validate/calibrate the simulated datasets. After calibrating the models individually (correcting bias in their seasonal means of TX and TN), the full PDFs of the simulated seasonal means of TX and TN for the recent past (1961–2000) are validated with the corresponding PDFs from E-OBS. Taylor diagrams document the skill of each model to reproduce the spatial variability of TX and TN. The identified model biases can not only be attributed to deficient representation of dynamical processes in the models, such as the upper-level polar jet stream (Delcambre et al., 2013), atmospheric blocking (Sillmann and Croci-Maspoli, 2009) and storm tracks (Woollings et al., 2012), but also to limitations in the E-OBS dataset (Hofstra et al., 2009).

For the future period (2041–2070), significant warming trends are projected for TX and TN in both seasonal and daily scales. The full PDFs of the seasonal mean TX and TN are positively shifted (2–4 °C), primarily for TX in summer and autumn (3–4 °C). Additionally, daily extremes are projected to become more frequent, particularly in summer TX over the innermost parts of Portugal. Overall, winter-time changes are much weaker than in other seasons. However, the increase in the number of extremely hot days in spring and summer, particularly in the inner part of the country, is quite remarkable. These changes are associated with alterations in the large-scale atmospheric circulation over the North Atlantic and Western Europe (e.g. Pinto et al., 2007; Sillmann and Croci-Maspoli, 2009; Woollings et al., 2012). Further, our results are in general agreement with Ramos et al. (2011), which considered only a single model (HadRM3), different scenarios (B2 and A2) and time period (2071–2100).

This climate change footprint might be particularly challenging for many socio-economic sectors in Portugal. Many agro-forestry systems in Portugal, already under heat and water stresses (e.g. Fraga et al., 2012; 2013), also threatened by forest fires (Marques et al., 2011), may experience detrimental impacts from a changing climate. Energy supply and management may be also severely affected. Further impacts deal with human health, such as mortality in recent heat waves in Portugal (Nogueira and Paixão, 2008). Therefore, integrating the implications of climate change on these systems is a key issue for developing suitable, cost-effective, environmentally sustainable and eco-innovative adaptation and mitigation strategies in Portugal.

Supporting information

The following supporting information is available:

Figure S1. Orographic map showing the Portuguese sector (36.625°N–42.375°N; 6.125°W–9.875°W) and Regions 1 (36.625°N–39.375°N) and 2 (39.625°N–42.375°N). E-OBS grid boxes are also plotted.

Figure S2. Differences between the ensemble mean and the E-OBS mean (left panel) TX and (right panel) TN (°C) in 1961–2000 for: (a, b) winter (DJF), (c, d) spring (MAM), (e, f) summer (JJA) and (g, h) autumn (SON).

Figure S3. Climate-mean seasonal cycles (baseline 1961–2000) of monthly (a) TX, and (b) TN for all simulations (Table I) and E-OBS (solid black curves).

Figure S4. Taylor diagrams showing the correlation coefficients, root mean squared differences (RMSD) and standard deviations of TX and TN relative to E-OBS for all simulations in 1961–2000 (Table I) and for: (a, b) winter (DJF), (c, d) spring (MAM), (e, f) summer (JJA) and (g, h) autumn (SON).

Figure S5. PDFs of the area-means (only over land grid boxes within the Portuguese sector in Figure S1) of (left panels) TX and (right panels) TN from all individual simulations (Table I), for the corresponding ensemble (dashed black curves) and from E-OBS (solid black curves) in 1961–2000 and for: (a, b) winter (DJF), (c, d) spring (MAM), (e, f) summer (JJA) and (g, h) autumn (SON).

Figure S6. PDFs of the area-means (only over land grid boxes within the Portuguese sector in Figure S1) of (left panels) TX and (right panels) TN from all individual simulations (Table I), for the corresponding ensemble (dashed black curves) and from E-OBS (solid black curves) in 2041–2070 and for: (a, b) winter (DJF), (c, d) spring (MAM), (e, f) summer (JJA) and (g, h) autumn (SON).

Figure S7. NDTX10p and NDTN10p (days/season) for: (a–b) winter (DJF), (c–d) spring (MAM), (e–f), summer (JJA) and (g–h) autumn (SON) in the future (2041–2070). Note that, by definition, the values of the indices for the recent past (1961–2000) are of approximately 9 days per season for NDTX10p and NDTN10p.

Acknowledgements

We acknowledge the E-OBS dataset from EUFP6 project ENSEMBLES (http://ensembles-eu.metoffice.com) and the data providers in the ECA&D project (http://eca.knmi.nl). We thank the German Federal Environment Agency and the COSMO-CLM consortium for providing COSMO-CLM data.

We also acknowledge E-OBS and the data providers in the ECA&D project (http://eca.knmi.nl). We thank Joaquim G. Pinto for discussions. This work is supported by European Union Funds (FEDER/COMPETE – Operational Competitiveness Programme) and by national funds (FCT – Portuguese Foundation for Science and Technology) under the project FCOMP-01-0124-FEDER-022692.

References

Andrade C, Leite SM, Santos JA. 2012. Temperature extremes in Europe: overview of their driving atmospheric patterns. *Natural Hazards and Earth System Sciences* **12**: 1671–1691.

Beniston M, Stephenson D, Christensen O, Ferro CT, Frei C, Goyette S, Halsnaes K, Holt T, Jylhä K, Koffi B, Palutikof J, Schöll R, Semmler T, Woth K. 2007. Future extreme events in European climate: an exploration of regional climate model projections. *Climatic Change* **81**: 71–95.

Böhm U, Kücken M, Ahrens W, Block A, Hauffe D, Keuler K, Rockel B, Will A. 2006. CLM-the climate version of LM: brief description and long-term applications. *COSMO Newsletter* **6**: 225–235.

Christensen JH, Christensen OB, Lopez P, van Meijgaard E, Botzet M. 1996. The HIRHAM 4 regional atmospheric climate model. DMI Scientific Report 96–4 pp.

Christensen JH, Kjellstrom E, Giorgi F, Lenderink G, Rummukainen M. 2010. Weight assignment in regional climate models. *Climate Research* **44**: 179–194.

Collins M, Booth BB, Bhaskaran B, Harris GR, Murphy JM, Sexton DMH, Webb MJ. 2011. Climate model errors, feedbacks and forcings: a comparison of perturbed physics and multi-model ensembles. *Climate Dynamics* **36**: 1737–1766.

Delcambre SC, Lorenz DJ, Vimont DJ, Martin JE. 2013. Diagnosing northern hemisphere jet portrayal in 17 CMIP3 global climate models: twentieth-century intermodel variability. *Journal of Climate* **26**: 4910–4929.

Diffenbaugh NS, Giorgi F. 2012. Climate change hotspots in the CMIP5 global climate model ensemble. *Climate Change* **114**: 813–822.

Elguindi N, Bi X, Giorgi F, Nagarajan B, Pal J, Solmon F, Rauscher S, Zakey A. 2007. *RegCM version 3.1 user's guide*. PWCG Abdus Salam ICTP.

Fischer EM, Schär C. 2010. Consistent geographical patterns of changes in high-impact European heatwaves. *Nature Geoscience* **3**: 398–403.

Fraga H, Santos JA, Malheiro AC, Moutinho-Pereira J. 2012. Climate Change Projections for the Portuguese Viticulture Using a Multi-Model Ensemble. *Ciência Téc Vitiv* **27**(1): 39–48.

Fraga H, Malheiro AC, Moutinho-Pereira J, Jones GV, Alves F, Pinto JG, Santos JA. 2013. Very high resolution bioclimatic zoning of Portuguese wine regions: present and future scenarios. *Regional Environmental Change*, DOI: 10.1007/s10113-013-0490-y.

Frías M, Mínguez R, Gutiérrez J, Méndez F. 2012. Future regional projections of extreme temperatures in Europe: a nonstationary seasonal approach. *Climatic Change* **113**: 371–392.

Haylock MR, Hofstra N, Tank KAMG, Klok EJ, Jones PD, New M. 2008. A European daily high-resolution gridded data set of surface temperature and precipitation for 1950–2006. *Journal of Geophysical Research* **113**: D20119.

Hofstra N, Haylock M, New M, Jones PD. 2009. Testing E-OBS European high-resolution gridded data set of daily precipitation and surface temperature. *Journal of Geophysical Research* **114**(D21), DOI: 10.1029/2009jd011799.

IPCC. 2007. *Climate Change 2007: Synthesis Report. Contribution of Working Groups I, II and III to the Fourth Assessment Report of the Intergovernmental Panel on Climate Change*, Core Writing Team: RK Pachauri and A Reisinger (eds). IPCC: Geneva, Switzerland, 104 pp.

Jacob D. 2001. A note to the simulation of the annual and inter-annual variability of the water budget over the Baltic Sea drainage basin. *Meteorology and Atmospheric Physics* **77**: 61–73.

Jaeger EB, Anders I, Luthi D, Rockel B, Schar C, Seneviratne SI. 2008. Analysis of ERA40-driven CLM simulations for Europe. *Meteorologische Zeitschrift* **17**: 349–367.

Klein Tank AMG, Können GP. 2003. Trends in indices of daily temperature and precipitation extremes in Europe, 1946–99. *Journal of Climate* **16**: 3665–3680.

Kuglitsch FG, Toreti A, Xoplaki E, Della-Marta PM, Zerefos CS, Türkeş M, Luterbacher J. 2010. Heat wave changes in the eastern Mediterranean since 1960. *Geophysical Research Letters* **37**: L04802.

Lenderink G, van den Hurk B, van Meijgaard E, van Ulden A, Cuijpers H. 2003. *Simulation of Present-Day Climate in RACMO2: First Results and Model Developments*. Ministerie van Verkeer en Waterstaat, Koninklijk Nederlands Meteorologisch Instituut.

van der Linden P, Mitchell JFB. 2009. *ENSEMBLES: Climate Change and its Impacts: Summary of Research and Results From the ENSEMBLES Project*. Met Office Hadley Centre: Exeter; 160.

Marques S, Borges JG, Garcia-Gonzalo J, Moreira F, Carreiras JMB, Oliveira MM, Cantarinha A, Botequim B, Pereira JMC. 2011. Characterization of wildfires in Portugal. *European Journal of Forest Research* **130**: 775–784.

Nakićenović N, Alcamo J, Davis G, de Vries HJM, Fenhann J, Gaffin S, Gregory K, Grubler A, Jung TY, Kram T, La Rovere EL, Michaelis L, Mori S, Morita T, Papper W, Pitcher H, Price L, Riahi K, Roehrl A, Rogner H-H, Sankovski A, Schlesinger M, Shukla P, Smith S, Swart R, van Rooijen S, Victor N, Dadi Z. 2000. Emissions scenarios. *A special report of Working Group III of the Intergovernmental Panel on Climate Change*. Cambridge University Press: Cambridge, UK and New York, NY.

Nikulin G, Kjellström E, Hansson U, Strandberg G, Ullerstig A. 2011. Evaluation and future projections of temperature, precipitation and wind extremes over Europe in an ensemble of regional climate simulations. *Tellus A* **63**.

Nogueira P, Paixão E. 2008. Models for mortality associated with heatwaves: update of the Portuguese heat health warning system. *International Journal of Climatology* **28**: 545–562.

Peterson TC. 2005. Climate change indices. *WMO Bulletin* **54**(2).

Pinto JG, Ulbrich U, Leckebusch GC, Spangehl T, Reyers M, Zacharias S. 2007. Changes in storm track and cyclone activity in three SRES ensemble experiments with the ECHAM5/MPI-OM1 GCM. *Climate Dynamics* **29**: 195–210.

Räisänen J. 2007. How reliable are climate models? *Tellus Series A-Dynamic Meteorology and Oceanography* **59**: 2–29.

Ramos A, Trigo R, Santo F. 2011. Evolution of extreme temperatures over Portugal: recent changes and future scenarios. *Climate Research* **48**: 177–192.

Samuelsson P, Jones CG, Willen U, Ullerstig A, Gollvik S, Hansson U, Jansson C, Kjellstrom E, Nikulin G, Wyser K. 2011. The Rossby Centre Regional Climate model RCA3: model description and performance. *Tellus Series a-Dynamic Meteorology and Oceanography* **63**: 4–23.

Santos JA, Corte-Real J, Ulbrich U, Palutikof J. 2007. European winter precipitation extremes and large-scale circulation: a coupled model and its scenarios. *Theoretical and Applied Climatology* **87**: 85–102.

Sillmann J, Croci-Maspoli M. 2009. Present and future atmospheric blocking and its impact on European mean and extreme climate. *Geophysical Research Letters* **36**: L10702.

Woollings T, Gregory JM, Pinto JG, Reyers M, Brayshaw DJ. 2012. Response of the North Atlantic storm track to climate change shaped by ocean–atmosphere coupling. *Nature Geoscience* **5**: 313–317.

Zhang X, Alexander L, Hegerl GC, Jones P, Tank AK, Peterson TC, Trewin B, Zwiers FW. 2011. Indices for monitoring changes in extremes based on daily temperature and precipitation data. *Wiley Interdisciplinary Reviews: Climate Change* **2**: 851–870.

How do carbon cycle uncertainties affect IPCC temperature projections?

Roger W. Bodman,[1]* Peter J. Rayner[2] and Roger N. Jones[1]

[1] Victoria Institute of Strategic Economic Studies, Victoria University, Melbourne, Australia
[2] School of Earth Sciences, The University of Melbourne, Melbourne, Australia

*Correspondence to:
R. W. Bodman, Victoria Institute
of Strategic Economic Studies,
Victoria University, 300 Flinders
St, Melbourne, Victoria 3000,
Australia.
E-mail: roger.bodman@vu.edu.au

Abstract

Carbon cycle uncertainties associated with the Intergovernmental Panel on Climate Change temperature-change projections were treated differently between the Fourth and Fifth Assessment Reports as the latter focused on concentration- rather than emission-driven experiments. Carbon cycle feedbacks then relate to the emissions consistent with a particular concentration. A valuable alternative is to include all uncertainties in a single step from emissions to temperatures. We use a simple climate model with an observationally constrained parameter distribution to explore the carbon cycle and temperature-change projections, simulating the emission-driven Representative Concentration Pathways. The resulting range of uncertainty is a somewhat wider and asymmetric *likely* range (biased high).

Keywords: carbon cycle uncertainties; temperature-change projections; simple climate model

1. Introduction

The latest global-mean surface temperature-change (ΔGMST) projections from the Intergovernmental Panel on Climate Change (IPCC) Fifth Assessment Report Working Group I (AR5 WGI; IPCC, 2013a) are based on the Representative Concentration Pathways (RCPs; Moss *et al.*, 2010). The results span a *likely* range of 0.3–4.8 °C in 2081–2100 (relative to 1986–2005) across the four RCPs, where *likely* is an assessment with a greater than 66% probability (Stocker *et al.*, 2013). The spread in the temperature-change projections stems from the set of greenhouse-gas emission trajectories used in the analysis together with model uncertainties connected to each set of emissions. The emission trajectories span a broad range of uncertainties in the future growth of anthropogenic greenhouse gases (GHGs) and the underlying socio-economic drivers (population growth, economic growth, energy intensity etc.). Model uncertainty arises from simulations of physically plausible responses by the Earth's climate system to increasing GHG emissions.

The two main differences between ΔGMST projections in the Fourth (IPCC, 2007) and Fifth Assessment Reports (IPCC, 2013a) are the changeover from the Special Report on Emission Scenarios (SRES; Nakicenovic *et al.*, 2000) to the RCPs and the different presentation of carbon cycle uncertainties in the AR5 results. Most of the AR5 projections were concentration-based rather than emission-based results and so it is difficult to assess how large the spread in uncertainty would have been if emission-driven scenarios were used and the carbon cycle uncertainties also

accounted for. Based on previous studies (Huntingford *et al.*, 2009; Bodman *et al.*, 2013; Booth *et al.*, 2013), we would expect the carbon cycle feedbacks to lead to a wider range of future feedbacks, and therefore, uncertainties in ΔGMST projections than that due to climate sensitivity alone. This is partly because the majority of complex climate models are atmosphere–ocean general circulation models (AOGCMs) and not Earth system models (ESMs), so do not include the carbon cycle. They therefore need to be supplied with CO_2 concentrations (Hibbard *et al.*, 2007). Accordingly, by applying atmospheric CO_2 inputs, the Coupled Model Intercomparison Phase 5 (CMIP5) experiments were designed to allow AOGCMs to participate (Friedlingstein *et al.*, 2014).

These CO_2 inputs include a best estimate for carbon cycle feedbacks established using the simplified climate model MAGICC (Wigley and Raper, 2001; Meinshausen *et al.*, 2011a). The carbon cycle parameter settings used for this purpose were based on the Bern-CC model in order to preserve consistency between CMIP3 and CMIP5 (Meinshausen *et al.*, 2011b) and not observationally constrained parameters as in this study (refer Section 2). Note that the Bern-CC model has a below average carbon cycle feedback (Friedlingstein *et al.*, 2006), which may contribute toward underestimating the temperature response.

The CMIP5 concentration-driven projections do not directly allow for carbon cycle feedbacks, although the RCP concentrations include an implicit carbon cycle feedback based on the Bern-CC model (Meinshausen *et al.*, 2011b). However, this means that the uncertainties associated with temperature feedbacks and other physical processes that affect atmospheric

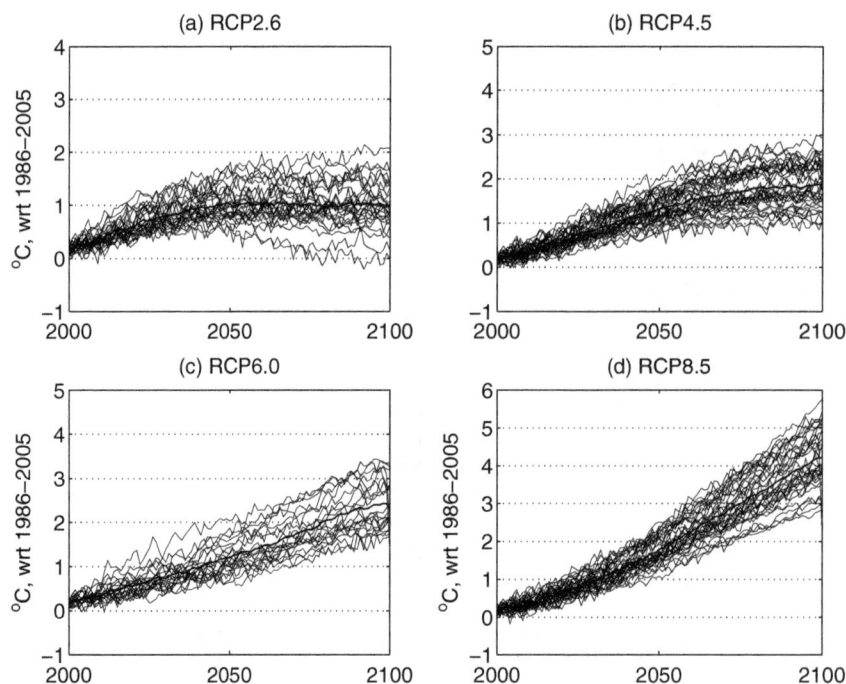

Figure 1. RCP CMIP5 ensembles for ΔGMST (relative to 1986–2005) for the four RCPs (concentration-driven experiments). Red lines are a single realization for each model, black line the ensemble mean: (a) RCP2.6, 27 models, (b) RCP4.5, 37 models, (c) RCP6.0, 20 models and (d) RCP8.5, 36 models (Source: Greg Kociuba, Australian Bureau of Meteorology).

CO_2 concentrations were not explored in the CMIP5 concentration-driven projections.

The global-mean surface air temperature anomalies derived from CMIP5 are illustrated here in Figure 1. This is the data used to calculate the likely range for the AR5 projections for each of the RCPs shown in Table 1. The calculation for the 5–95% interval is set around the mean and based on the standard deviation from the sample of models in the multi-model ensemble, assuming a normal distribution (Collins *et al.*, 2013). These 'ensembles of opportunity' do not necessarily represent the full range of potential model uncertainty, as there remain issues of model independence, limited validation and so forth (for a more detailed discussion refer Tebaldi and Knutti, 2007).

In this article, we explore how carbon cycle uncertainties affect the AR5 temperature-change projections had they been emissions-driven rather than concentration-driven experiments. The carbon cycle uncertainties referred to here include the partitioning of CO_2 between different sinks, temperature feedbacks related to plant and soil respiration and CO_2 solubility in the ocean. They do not include those associated with tipping points such as permafrost melting or extensive vegetation changes. We tested these differences using CMIP5 (Taylor *et al.*, 2012) ensembles and our implementation of the reduced complexity upwelling-diffusion energy-balance model MAGICC that explores parameter uncertainties in the carbon cycle (Wigley and Raper, 2001; Meinshausen *et al.*, 2011a; Bodman *et al.*, 2013). Parameter uncertainties are constrained by the use of observations with a Bayesian data assimilation process.

Table 1. Estimates of ΔGMST change for the RCP scenarios as per the AR5 CMIP5 results (IPCC, 2013a, 2013b) and our MAGICC simulations for two emissions-driven cases, one with carbon cycle temperature feedbacks off (MAGICC CC-off) and one with them on (MAGICC CC-on).

ΔGMST, °C at 2081–2100 relative to 1986–2005		
Scenario IPCC AR5	**Mean**	**Likely range[a]**
RCP2.6	1.0	0.3–1.7
RCP4.5	1.8	1.1–2.6
RCP6.0	2.2	1.4–3.1
RCP8.5	3.7	2.6–4.8
MAGICC CC-off	**Median**	**67% range[b]**
RCP2.6	0.7	0.2–1.3
RCP4.5	1.8	1.1–2.6
RCP6.0	2.4	1.6–3.4
RCP8.5	3.7	2.5–4.9
MAGICC CC-on	**Median**	**67% range**
RCP2.6	0.9	0.4–1.8
RCP4.5	2.1	1.2–3.2
RCP6.0	2.7	1.7–3.9
RCP8.5	4.0	2.7–5.5

[a]CMIP5 *likely range* is a 5–95% model range calculated from the ensemble of projections and assessed as *likely* (IPCC, 2013a, 2013b).
[b]Where the 67% interval is just that.

2. Method

To explore the effect of including carbon cycle uncertainties on ΔGMST change projections, we applied our version of the MAGICC version 6 model that samples plausible ranges of key parameters based on historical data (Bodman *et al.*, 2013). MAGICC is a simple climate model that has been developed and maintained for nearly 30 years (Wigley and Raper, 1987; Wigley and Raper, 2001; Meinshausen *et al.*, 2009; Meinshausen *et al.*, 2011a). It is an energy-balance model with an upwelling-diffusive ocean for the climate in conjunction with land and ocean carbon cycle components. Although substantially less complex than fully coupled three-dimensional ESMs, it has been shown to perform well in terms of emulating complex models, such as those used for CMIP3 (Meinshausen *et al.*, 2011c). In addition to calibrating MAGICC's parameters for emulating complex models, its climate parameters have been calibrated against historical observations using Monte Carlo-based Bayesian statistical techniques that run the model thousands of time, testing the model's results to arrive at probability distributions for the main parameters rather than single point estimates (Meinshausen *et al.*, 2009; Rogelj *et al.*, 2011; Rogelj *et al.*, 2012). The latter research work calibrated MAGICC's carbon cycle parameter against the C4MIP models (Friedlingstein *et al.*, 2006); whereas, here, we use a method that includes observed CO_2 concentrations (1960–2010) to estimate the key carbon cycle parameters, along with other observations to constrain key climate parameters (Bodman *et al.*, 2013). Results from complex models were not used in this process, although they do help in guiding the selection of parameter prior distributions. This calibration technique results in a posterior parameter distribution that can then be used to run MAGICC, iterating through the parameter sets to generate temperature-change results for a given greenhouse-gas concentration pathway or emission trajectory.

One important difference between the use of MAGICC to estimate ΔGMST and the process used in AR5 is that here, MAGICC uses emissions as inputs, whereas CMIP5 used concentrations selected to emulate a specific set of emissions for which the carbon cycle feedbacks were tuned to the C4MIP Bern-CC carbon cycle model (Meinshausen *et al.*, 2011b). We generated ΔGMST distributions for the four emissions-driven RCPs using the aforementioned parameter distribution for two cases, with and without the carbon cycle temperature feedbacks. The case with the carbon cycle temperature feedbacks off means that the CO_2 fertilsation effect and oceanic CO_2 uptake contribute to the temperature-change uncertainty while the carbon cycle temperature feedbacks do not (they are simply not applied as adjusting factors in the calculation of ΔGMST). The case with carbon cycle feedbacks on tested the fuller range of uncertainty.

3. Results

We calculated a set of emissions-driven RCP projections to compare with the concentration-driven AR5 results, using parameter uncertainties constrained by

the 20th century observations. The second part of Table 1 presents our results. One set is with the carbon cycle temperature feedbacks switched off (MAGICC CC-off), then a set with the carbon cycle temperature feedbacks on (MAGICC CC-on). These results use the same reference periods as the AR5, with shaded time series plots given in Figure 2.

Using the emissions associated with each RCP and no carbon cycle temperature feedbacks, we obtain results similar to the AR5 mean and breadth across the *likely range* with our 67% confidence interval. This similarity suggests that the MAGICC uncertainty range is comparable with the uncertainty range given for the complex models even though they are not equivalent. We are using parameter uncertainty combined with historical constraints, whereas the AR5 uses a climate model ensemble.

With carbon cycle temperature feedbacks included, the median, lower and upper bounds for ΔGMST all increase, the upper bound most of all. For example, for RCP8.5, the AR5 projections span 2.6–4.8 °C and the MAGICC results with the carbon cycle temperature feedbacks off span a similar range of 2.5–4.9 °C. Including carbon cycle temperature feedbacks increases the range to 2.7–5.5 °C, mostly at the upper bound. The emission-driven CC-on results produce a more asymmetric distribution biased to higher temperature increases, implying higher levels of risk if the uncertainties stemming from the carbon cycle are included.

The lower temperature-change results in the AR5 projections may be partly due to the choice of carbon cycle settings used to generate the RCP concentrations for the CMIP5 simulations (MAGICC with carbon cycle parameters based on the Bern-CC model). If the RCP concentrations were based on a median of the C4MIP models or the observationally constrained parameter set of Bodman *et al.* (2013) then the range of temperature-change outcomes would be expected to shift upwards. This could potentially be explored further using the C4MIP or Bodman *et al.* (2013) distributions, deriving upper and lower bounds for the CO_2 concentrations which could then be modeled by the AOGCMs.

Table 2 shows that the our MAGICC CC-on median CO_2 concentration values at 2100 correspond closely to the RCP-specified amounts with very small differences of 2–4 ppm for RCP2.6, RCP4.5 and RCP6.0, while RCP8.5 differs by 20 ppm or 2%. The CO_2 concentration amounts and likely uncertainty ranges are greater for the CC-on cases as compared with CC-off (Table 2). The CC-off runs have lower median values than those specified for the RCP concentration-driven runs as these are derived from MAGICC with feedbacks turned on (refer also Meinshausen *et al.*, 2011b). The MAGICC CC-off runs still have carbon cycle uncertainty stemming from the parameterisation of the CO_2 fertilization effect and carbon uptake across the land and ocean sinks (which is the reason for the uncertainty range in the MAGICC CC-off CO_2 concentration results, Table 2), whereas the concentrations for the

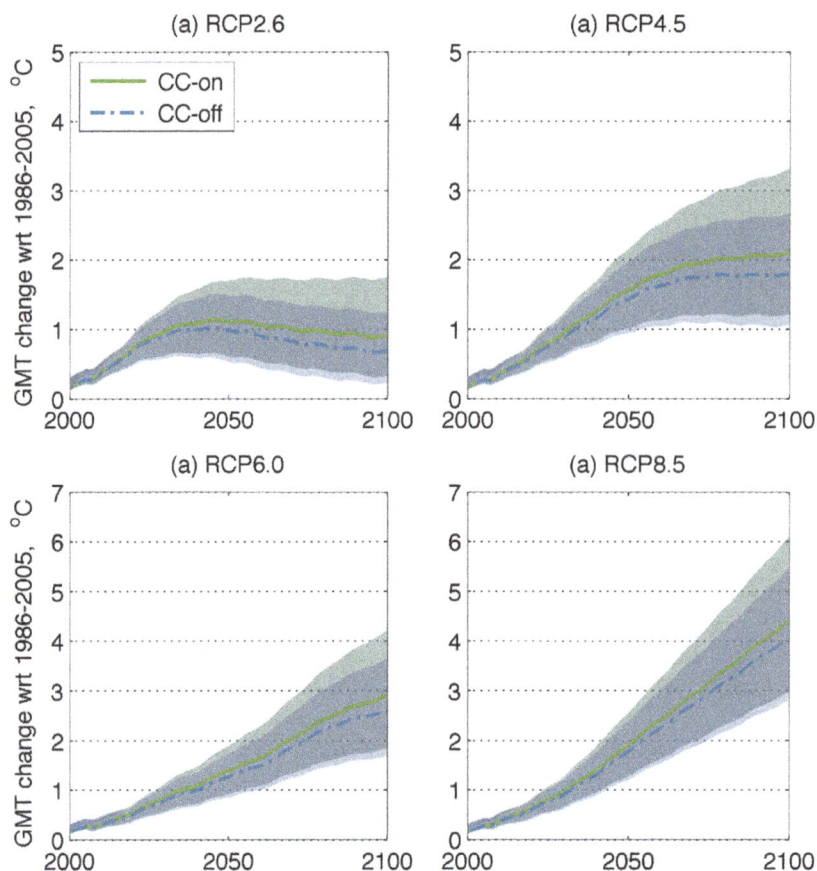

Figure 2. Plume plots for ΔGMT change projections 2000–2100, °C relative to 1986–2005. MAGICC results with carbon cycle temperature feedbacks on (CC-on) and switched off (CC-off) (a) RCP2.6, (b) RCP4.5, (c) RCP6.0 and (d) RCP8.5. Shaded regions indicate the 67% confidence interval for CC-on (green) and CC-off (blue), with median results as solid green and dashed blue lines, respectively.

CMIP5 RCP concentration-driven runs have a single pathway.

4. Discussion

We have tested the effect of allowing for carbon cycle uncertainties by deriving probabilistic ΔGMST projections using a simple ESM with a historically constrained parameter distribution. This increases the range of uncertainty as compared with concentration-driven RCP scenarios, with an asymmetric distribution for the uncertainty ranges that is biased toward higher values. The *likely range* of temperature change is also warmer [except for RCP2.6, where the MAGICC CC-on results are only slightly different to the AR5 multi-model ensemble (MME)].

The increased asymmetry is largely a result of the stronger carbon cycle temperature feedbacks being more evident in the higher forcing scenarios with greater ΔGMST (Table 1). The associated MAGICC carbon cycle parameters are only weakly constrained by the calibration process and remain a significant source of uncertainty in the forward projections.

Our MAGICC results are consistent with other studies that have examined the emission-driven RCP8.5

scenario, the only experiment in the CMIP5 protocol that includes both concentration- and emission- driven simulations (Friedlingstein *et al.*, 2014; Shao *et al.*, 2014). This is achieved with our Bayesian historical calibration method for the key climate and carbon cycle parameters – not through calibrating the simple model to emulate the complex models. Note that uncertainties in the IPCC AR5 and those presented here are not equivalent. The AR5 uncertainty range is estimated from the spread of results from an ensemble of complex climate models and assessed as *likely* by expert judgment. We utilize a joint parameter distribution derived from calibration against historical data with a single simple climate model. The likely range is then the 67% range confidence interval obtained from the spread in model results.

For the CMIP5 emission-driven RCP8.5 temperature-change results, Friedlingstein *et al.* (2014) reported a temperature range of 2.5–5.6 °C, with a MME mean of 3.9 ± 0.9 °C (1σ). Converting this to a *likely* uncertainty range, with ±1.64 times the standard deviation of the MME mean (Collins *et al.*, 2013), yields a range of 2.4–5.4 °C. This is similar to the MAGICC CC-on results reported here (Table 1; median 4.0 °C, *likely* range 2.7–5.5 °C). Likewise, for their concentration-driven RCP8.5 results, MME mean 3.7

Table 2. CO_2 concentrations at 2100 for the CMIP5 RCP pathways and our MAGICC simulations for two emissions-driven cases, one with carbon cycle temperature feedbacks off (MAGICC CC-off) and one with them on (MAGICC CC-on).

CO$_2$ concentrations at 2100 (ppm)		
Scenario	**CO$_2$ concentration**	
RCP2.6	421	
RCP4.5	538	
RCP6.0	670	
RCP8.5	936	
MAGICC CC-off	**Median**	**67% range**[a]
RCP2.6	393	372–421
RCP4.5	490	449–540
RCP6.0	611	553–678
RCP8.5	831	739–933
MAGICC CC-on	**Median**	**67% range**
RCP2.6	419	389–475
RCP4.5	536	475–643
RCP6.0	666	583–798
RCP8.5	916	775–1129

[a]Where the 67% interval is just that.

$\pm 0.7\,°C$, the equivalent *likely* range is 2.6–4.9 °C, very close to our MAGICC CC-off median 3.7 °C, *likely* range 2.5–4.9 °C.

The CO_2 concentration ranges differ from our results, with most of the CMIP5 ESMs having higher estimates. The emission-driven average CO_2 concentration at 2100 was 985 ± 97 ppm, whereas the MAGICC CC-on median was 916 ppm. However, most of the ESMs overestimate the historical CO_2 concentrations (Friedlingstein *et al.*, 2014) and therefore may be overestimating CO_2 concentrations by 2100. The carbon cycle is not the only factor affecting the temperature outcomes, with other forcing components such as aerosols being important contributors.

The HadCM3C ESM has been run using a perturbed parameter approach to sample projected temperature-change uncertainties for emission-driven RCP2.6 and 8.5 scenarios (Booth *et al.*, 2013). A similar conclusion to ours was reached, but the different model structures between their ESM and the SCM used in this study suggest further areas for investigation. They do however suggest using historical constraints on carbon cycle uncertainties, which we have done. The study by Booth *et al.* (2013) has a wider range of temperature results across their full Earth system ensemble than we find, although this is reduced when they subsample the ensemble to exclude climate sensitivities above the CMIP5 range. For example, their RCP8.5 10th–90th percentile range for the full ensemble is 4.2–8.1 °C, but 4.2–6.8 °C when subsampled. The corresponding MAGICC CC-on range is 3.0–6.5 °C when set to the matching reference periods (see also Table S1, Supporting Information).

Differences in the Booth *et al.* (2013) ESM results arise from their ensemble having many members with climate sensitivity over 3 °C (refer Figure 3(e) in Booth

et al., 2013) as well as members with stronger carbon cycle feedbacks. Some of the ensemble members also have CO_2 concentration results that diverge from present-day observations. The HadCM3C model was not explicitly calibrated against historical data, and therefore, the sampled parameter space is not constrained in the same way as our SCM.

The similarities between our MAGICC ΔGMST projections and those of Friedlingstein *et al.* (2014) and Booth *et al.* (2013) for the fewer emission-driven scenarios that have been modeled suggest that the MAGICC setup and calibration technique used here is able to generate meaningful results, albeit with its own limitations (noted below), just as the complex models have their own issues.

The role of the carbon cycle is clearly understood by the authors of the IPCC's AR5 WGI Chapter 12, who explain that uncertainties in carbon cycle feedbacks are addressed in a different way to the AR4 because temperature projections are based on the concentration-driven RCPs (Collins *et al.*, 2013). Instead of uncertainty stemming from the causal sequence emissions to concentrations to radiative forcing to temperature change, concentrations are fixed for each RCP and then the uncertainty moves to either side in the causal chain (Hibbard *et al.*, 2007). Therefore, the ΔGMST projections in the AR5 presented in the WGI Summary for Policymakers Table SPM.2 (IPCC, 2013b; IPCC, 2014), may underestimate the likelihood of reaching higher temperatures given current knowledge. Temperature uncertainties stemming from the carbon cycle are related to cumulative emissions in, for example Figure SPM.10 (IPCC, 2013b; IPCC, 2014). If the uncertainties for the whole causal chain are to be included, the uncertainties prior to obtaining concentrations then need to be added in later. What we have done is to present a range of uncertainty for ΔGMST across the four emission-driven RCPs following the more intuitive and physically natural causal chain. These different ways of presenting temperature-change projections have implications for climate policy makers and climate risk management, especially if temperature or related positive feedbacks have a marked effect on the carbon cycle.

The AR5 affirms that there is no fundamental difference between the behavior of the CMIP5 ensemble in comparison with the CMIP3. Instead, the differences are largely due to the scenarios, choice of reference periods and treatment of uncertainty. With respect to the main differences between the AR4 and AR5 ΔGMST projections, the change over from the SRES emission scenarios in the AR4 to the RCP concentration pathways in the AR5, meant that the effect of carbon cycle uncertainties on atmospheric CO_2 concentrations were not considered in the concentration-driven CMIP5 simulations (IPCC, 2013a: SPM, p20). This implies that the AR4 results were based on emission-driven modeling, while the AR5 results were based on concentration-driven modeling; this is not the case. In practice, both the CMIP3 and CMIP5 results that were used to inform the SRES and RCP projections respectively were concentration-driven. Global-mean surface temperature-change (ΔGMST) projections in the AR4 were based on multi-model ensemble results using the SRES scenarios and the CMIP3 archive. The SRES emission scenarios were converted to concentrations using either the MAGICC or Bern-CC reduced complexity ESMs (Collins *et al.*, 2013), effectively producing concentration pathways akin to the RCPs. Carbon cycle uncertainties were then added afterwards, with the upper and lower bounds of the MME mean extended by +60% and −40% using expert judgment informed by the less complex models. As the AR4 explains 'The AOGCMs cannot sample the full range of possible warming, in particular because they do not include uncertainties in the carbon cycle' (IPCC, 2007). The range derived from the AR4 MME has 'additional uncertainty estimates obtained from published probabilistic methods using different types of models and observational constraints: the MAGICC SCM and the BERN2.5CC coupled climate-carbon cycle EMIC tuned to different climate sensitivities and carbon cycle settings, and the C4MIP coupled climate-carbon cycle models' (Meehl *et al.*, 2007).

These details are re-confirmed in the AR5 (Collins *et al.*, 2013), where the limitation of using a constant fractional uncertainty is noted, particularly for a scenario such as RCP2.6. The AR5 WGI ΔGMST projections were derived similar to the AR4, from multi-model ensembles drawn from the CMIP5 concentration-driven experiments. The likely uncertainty range was then assessed as ±1.64 times the standard deviation of the MME, i.e. as a 5–95% confidence interval (Collins *et al.*, 2013). This approach was used to indicate the spread in model results but is not a formal measurement for the uncertainty. A check on our results (Table 1) found that the +60% and −40% calculation does not adequately characterize the AR5 *likely* range, especially for the very low- and very high-emission scenarios.

The different quantification of key climatic variables between one report to the next and its traceability in assessing risk is an important issue. These differences may lead to risks being either under- or over-estimated, as well as changing their significance for policymakers. All known contributions to uncertainty should be quantified where possible or at least identified. Assessments, especially Summaries for Policymakers, need to address the issue of traceability between assessments otherwise policymakers will not have a clear guide to changing climate risks.

As preparations for the next round of climate model intercomparisons (CMIP6) are being made and thoughts turn toward a Sixth Assessment Report, the choice of model experiments and long-term climate change projections should include a review of how model uncertainties are being managed and presented. No matter how models are driven, forwards and backwards compatibility that can fully capture uncertainty should be a goal of such assessments. Emission-driven

scenarios can then be part of the core experimental protocols allowing results to be presented across a range of different model types, ESMs operating at different resolutions, intermediate complexity models and simple climate models.

As a single model, MAGICC cannot explicitly consider structural uncertainty (Knutti *et al.*, 2008). Furthermore, constraining its parameters using historical data does not necessarily form a reliable basis for future projections as one or more of those parameters may be state dependent. This is probably the case for climate sensitivity (Armour *et al.*, 2013) and the future behavior of the carbon cycle, particularly as the temperature feedback effects are only poorly constrained by historical data. As a simplified ESM, MAGICC also lacks certain processes that could increase uncertainty (such as water and nutrient cycles, the release of carbon from permafrost and albedo changes due to ice cover and vegetation, as well as changes in ocean ventilation and stratification and changes on the ocean's biological carbon cycle). Nevertheless, it is a valuable tool for evaluating and comparing emission scenarios.

Acknowledgements

We gratefully thank Malte Meinshausen and Jens Kattge for supplying the MAGICC and MCMH code, respectively. We also thank the anonymous reviewers for their helpful comments. We acknowledge the World Climate Research Programmes Working Group on Coupled Modeling, which is responsible for CMIP and thanks to Scott Power and Greg Kociuba, Australian Bureau of Meteorology, for extracting and summarizing the RCP ΔGMST change projections from the CMIP5 archive. RB is funded by a Victoria University Postdoctoral Research Fellowship. PR is in receipt of an Australian Professorial Fellowship (DP1096309).

References

Armour KC, Bitz CM, Roe GH. 2013. Time–varying climate sensitivity from regional feedbacks. *Journal of Climate* **26**: 4518–4534, doi: 10.1175/JCLI-D-12-00544.1.

Bodman RW, Karoly DJ, Rayner PJ. 2013. Uncertainty in temperature projections reduced using carbon cycle and climate observations. *Nature Climate Change* **3**: 725–729, doi: 10.1038/NCLIMATE1903.

Booth BBB, Bernie D, McNeall D, Hawkins E, Caesar J, Boulton C, Friedlingstein P, Sexton DMH. 2013. Scenario and modelling uncertainty in global mean temperature change derived from emission-driven global climate models. *Earth System Dynamics* **4**: 95–108, doi: 10.5194/esd-4-95-2013.

Collins M, Knutti R, Arblaster J, Dufresne J-L, Fichefet T, Friedlingstein P, Gao X, Gutowski WJ, Johns T, Krinner G, Shongwe M, Tebaldi C, Weaver AJ, Wehner M. (2013). Long–term climate change: projections, commitments and irreversibility. Stocker TF, Qin D, Plattner G-K, Tignor M, Allen SK, Boschungand J, Nauels A, Xia Y, Bex V, Midgley PM. Climate Change 2013: The Physical Science Basis. Contribution of Working Group I to the Fifth Assessment Report of the Intergovernmental Panel on Climate Change. Cambridge, London and New York, NY: Cambridge University Press.

Friedlingstein P, Cox P, Betts R, Bopp L, von Bloh W, Brovkin V, Cadule P, Doney S, Eby M, Fung I, Bala G, John J, Jones C, Joos F, Kato T, Kawamiya M, Knorr W, Lindsay K, Matthews HD, Raddatz T, Rayner P, Reick C, Roeckner E, Schnitzler K-G, Schnur R, Strassmann K, Weaver AJ, Yoshikawa C, Zeng G. 2006. Climate–carbon cycle feedback analysis: results from the C4MIP model intercomparison. *Journal of Climate* **19**: 3337–3353.

Friedlingstein P, Meinshausen M, Arora VK, Jones CD, Anav A, Liddicoat SK, Knutti R. 2014. Uncertainties in CMIP5 climate projections due to carbon cycle feedbacks. *Journal of Climate* **27**: 511–526, doi: 10.1175/JCLI-D-12-00579.1.

Hibbard KA, Meehl GA, Cox PM, Friedlingstein P. 2007. A strategy for climate change stabilization experiments. *Eos* **88**: 217–219.

Huntingford C, Lowe JA, Booth BBB, Jones CD, Harris GR, Gohar LK, Meir P. 2009. Contributions of carbon cycle uncertainty to future climate projection spread. *Tellus* **61B**: 355–360, doi: 10.1111/j.1600-0889.2009.00414.x.

IPCC. 2007. Climate Change 2007: The Physical Science Basis. Contribution of Working Group I of the Fourth Assessment Report, Cambridge, London: Cambridge University Press.

IPCC. 2013a. Climate Change 2013: The Physical Science Basis. Working Group I Contribution to the Fifth Assessment Report, Geneva, Switzerland: IPCC Secretariat.

IPCC. 2013b. Summary for policymakers. Stocker TF, Qin D, Plattner G-K, Tignor M, Allen SK, Boschung J, Nauels A, Xia Y, Bex V, Midgley PM. Climate Change 2013: The Physical Science Basis. Working Group I Contribution to the Fifth Assessment Report of the Intergovernmental Panel on Climate Change. Cambridge, London and New York, NY: Cambridge University Press.

IPCC. 2014. Climate Change 2014: Synthesis Report. Contribution of Working Groups I, II and III to the Fifth Assessment Report of the Intergovernmental Panel on Climate Change, Geneva, Switzerland: IPCC.

Knutti R, Allen MR, Friedlingstein P, Gregory JM, Hegerl GC, Meehl G, Meinshausen M, Murphy JM, Plattner G-K, Raper SCB, Stocker TF, Stott PA, Teng H, Wigley TML. 2008. A review of uncertainties in global temperature projections over the twenty-first century. *Journal of Climate* **21**: 2651–2662, doi: 10.1175/2007JCLI2119.1.

Meehl GA, Stocker TF, Collins W, Friedlingstein P, Gaye A, Gregory JM, Kitoh A, Knutti R, Murphy J, Noda A, Raper SCB, Watterson I, Weaver A, Zhao Z-C. 2007. Global climate projections. Solomon S, Qin D, Manning M, Chen Z, Marquis M, Averyt KB, Tignor M, Miller HL. Climate Change 2007: The Physical Science Basis. Contribution of Working Group I to the Fourth Assessment Report of the Intergovernmental Panel on Climate Change. Cambridge University Press: Cambridge, London and New York, NY.

Meinshausen M, Meinshausen N, Hare W, Raper SCB, Frieler K, Knutti R, Frame DJ, Allen MR. 2009. Greenhouse-gas emission targets for limiting global warming to 2°C. *Nature* **458**: 1158–1162, doi: 10.1038/nature08017.

Meinshausen M, Raper SCB, Wigley TML. 2011a. Emulating coupled atmosphere–ocean and carbon cycle models with a simpler model, MAGICC6 – part 1: model description and calibration. *Atmospheric Chemistry and Physics* **11**: 1417–1456, doi: 10.5194/acp-11-1417-2011.

Meinshausen M, Smith SJ, Calvin K, Daniel JS, Kainuma M. Lamarque J-F, Matsumoto K, Montzka S, Raper SCB, Riahi K, Thomson A, Velders GJM, van Vuuren D. 2011b. The RCP greenhouse gas concentrations and their extensions from 1765 to 2300. *Climatic Change* **109**: 213–241, doi: 10.1007/s10584-011-0156-z.

Meinshausen M, Wigley TML, Raper SCB. 2011c. Emulating coupled atmosphere-ocean and carbon cycle models with a simpler model, MAGICC6 - part 2: applications. *Atmospheric Chemistry and Physics* **11**: 1457–1471, doi: 10.5194/acp-11-1457-2011.

Moss RH, Edmonds JA, Hibbard K, Manning M, Rose SK, van Vuuren DP, Carter TR, Emori S, Kainuma M, Kram T, Meehl GA, Mitchell JFB, Nakicenovic N, Riahi K, Smith SJ, Stouffer RJ, Thomson AM, Weyant JP, Wilbanks TJ. 2010. The next generation of scenarios for climate change research and assessment. *Nature* **463**: 747–756, doi: 10.1038/nature08823.

Nakicenovic N, Davidson O, Davis G, Grübler A, Kram T, La Rovere EL, Metz B, Morita T, Pepper W, Pitcher H, Sankovski A, Shukla P, Swart R, Watson R, Dadi Z. (2000). IPCC Special Report on Emissions Scenarios, Cambridge, London: Cambridge University Press.

Rogelj J, Hare W, Lowe J, van Vuuren DP, Riahi K, Matthews B, Hanaoka T, Jiang K, Meinshausen M. 2011. Emission pathways consistent with a 2 °C global temperature limit. *Nature Climate Change* **1**: 413–417, doi: 10.1038/NCLIMATE1258.

Rogelj J, Meinshausen M, Knutti R. 2012. Global warming under old and new scenarios using IPCC climate sensitivity range estimates. *Nature Climate Change* **2**: 248–253, doi: 10.1038/NCLIMATE1385.

Shao P, Zeng X, Zeng X. 2014. Differences in carbon cycle and temperature projections from emission- and concentration-driven earth system model simulations. *Earth System Dynamics Discussions* **5**: 991–1012, doi: 10.5194/esdd-5-991-2014.

Stocker TF, Qin D, Plattner G-K, Alexander LV, Allen SK, Bindoff NL, Bréon F-M, Church JA, Cubasch U, Emori S, Forster P, Friedlingstein P, Gillett N, Gregory JM, Hartmann DL, Jansen E, Kirtman B, Knutti R, Kumar KK, Lemke P, Marotzke J, Masson-Delmotte V, Meehl GA, Mokhov II, Piao S, Ramaswamy V, Randall D, Rhein M, Rojas M, Sabine C, Shindell D, Talley LD, Vaughan DG, Xie S-P. 2013. Technical summary. Stocker TF, Qin D, Plattner G-K, Tignor M, Allen SK, Boschung J, Nauels A, Xia Y, Bex V, Midgley PM. Climate Change 2013: The Physical Science Basis. Contribution of Working Group I to the Fifth Assessment Report of the Intergovernmental Panel on Climate Change. Cambridge, London and New York, NY: Cambridge University Press.

Taylor KE, Stouffer RJ, Meehl GA. 2012. An Overview of CMIP5 and the Experiment Design. *Bulletin of the American Meteorological Society* **93**: 485–498, doi: 10.1175/BAMS-D-11-00094.1.

Tebaldi C, Knutti R. 2007. The use of the multi-model ensemble in probabilistic climate projections. *Philosophical Transactions of the Royal Society* **65**: 2053–2075, doi: 10.1098/rsta.2007.2076.

Wigley TML, Raper SCB. 1987. Thermal expansion of sea level associated with global warming. *Nature* **330**: 127–131.

Wigley TML, Raper SCB. 2001. Interpretation of high projections for global–mean warming. *Science* **293**: 451–454, doi: 10.1126/science.1061604.

3

Using feedback from summer subtropical highs to evaluate climate models

Carmen Sánchez de Cos,[1] Jose M. Sánchez-Laulhé,[1] Carlos Jiménez-Alonso[1] and Ernesto Rodríguez-Camino[2]*

[1]Centro Meteorológico de Málaga, AEMET, Málaga, Spain
[2]Servicios Centrales, AEMET, Madrid, Spain

*Correspondence to:
E. Rodríguez-Camino, AEMET,
Servicios Centrales, Leonardo
Prieto Castro 8, 28040 Madrid,
Spain.
E-mail: erodriguezc@aemet.es

Abstract

This letter aims to broaden the spectrum of methods for model evaluation by providing new physically based metrics that focus on accurate couplings between subsystems of the climate system. A simplified version of the feedback scheme that describes the dynamics of subtropical high-pressure systems is applied to evaluate how well CMIP5 climate models can simulate atmosphere–ocean–land interactions and resulting feedbacks in the Azores high-pressure system during summer, which affects climate throughout the Atlantic near southern Europe and North Africa.

Keywords: climate model evaluation; Azores high; feedback

1. Introduction

Climate models are numerical representations of the climate system that are based on the physical, chemical and biological properties of its components; they include the interactions and feedback processes between these components and account for some of the known properties of the climate system. The complex internal feedbacks of the climate system that determine its highly nonlinear behavior can either amplify (via 'positive feedbacks') or dampen (via 'negative feedbacks') the effects of a perturbation of one climate variable. Climate models, to the extent that they are faithful representations of the climate system, should be able to simulate the major feedbacks in the system (Flato *et al.*, 2013).

In this study, we base the evaluation of climate models on the importance of subtropical anticyclones for general circulation because these features influence climate over extensive regions. Subtropical anticyclones are characterized by strong atmospheric descent on their eastern flanks. Below and associated with this descent (Klein and Hartmann, 1993; Klein *et al.*, 1995) are marine stratus clouds; owing to their persistence, low altitude and high reflectivity, these clouds are important for the global radiation budget, providing unique and critical pathways for radiative cooling of the tropics (they have also been called 'radiator fins'; Pierrehumbert, 1995). During winter, the descending branch of the Hadley cell gives rise to a belt of high pressure in the subtropics of both hemispheres. However, different processes are required to explain the existence and intensification of these subtropical highs in summer, when the Hadley circulation is weakened (Liu *et al.*, 2001), although they are always linked to the land–sea thermal contrast. A pronounced land–sea pressure gradient develops during summer as the continents warm, leading to the formation of equatorward winds along the western continental coasts. These winds enhance evaporation off the coast and trigger the upwelling of cool waters from below. The cool surface waters help to stabilize the lower atmosphere and favor the development of reflective stratiform clouds, which in turn cool the lower atmosphere, effectively forcing the development of near-surface highs. This positive feedback loop, triggered by land-mass warming in late spring or early summer, mainly derives from the much smaller heat capacity of the land than of the oceanic mixed layer (Miyasaka and Nakamura, 2005).

High-pressure systems over the subtropical oceans largely shape the low- and mid-latitude climate. In particular, the Azores high-pressure system influences summertime climate in south-western Europe and north-western Africa. In this study, we seek to validate the performance of CMIP5 climate models in simulating atmosphere–ocean–land interactions, evaluating whether they can create the amplifying feedback loop that explains summertime Azores high-pressure system dynamics. The relationships between pairs of physical parameters involved in the feedback loop are described by 2D scatter-plots (Betts, 2004) and quantified with a metric based on the Hellinger coefficient (Sanchez de Cos *et al.*, 2013). This metric allows us to quantitatively estimate the resemblance between the same empirical relationship obtained from a climate model simulation and from reanalyses. Contrary to most metrics for evaluating climate models, which frequently focus on outcome variables (usually precipitation and temperature), this approach goes directly to the representation of physical processes, in particular the coupling between subsystems and climate system feedbacks.

2. Data

We use the ERA-Interim (EI) atmospheric reanaly-sis (Dee *et al.*, 2011) as a reference to establish the observed relationships between pairs of variables that describe the feedback mechanism underlying sum-mertime Azores high dynamics. The climate models to be evaluated were selected from CMIP5 histori-cal runs (Taylor *et al.*, 2012). Table S1, Supporting Information provides information on the 28 models (selected based on the availability of the variables listed below) considered in this study (the data were obtained from the Earth System Grid Federation web server: http://pcmdi9.llnl.gov/esgf-web-fe/). We note that this study emphasizes methodology and is not intended to comprehensively cover all CMIP5 models.

Although some objections could be raised concerning the use of reanalysis energy budget fluxes as an accu-rate representation of the observations, as well as other magnitudes that we did not directly analyse, we suggest that our approach is practical, circumventing the lack of spatial coverage of *in situ* data and the inaccuracy of satellite data for certain surface variables. Further infor-mation is provided by Sanchez de Cos *et al.* (2013), who discussed and validated EI land surface data against *in situ* and satellite observations over a domain close to our region of interest.

The following variables – from both EI and CMIP5 – were used in this study: monthly means of the daily averaged (at 0000, 0600, 1200, 1800 UTC) 10 m meridional wind component (V10m), 2 m air tem-perature (T2m) and mean sea level pressure (MSLP); monthly means of daily 12 h forecast accumulations (at 0000 and 1200 UTC) of cloudy-sky downward surface solar radiation (SW_{down}) as well as clear-sky ($SW_{down(clear)}$), top of atmosphere (TOA) net solar radiation (SW_{net}), TOA net thermal radiation (LW_{net}), clear-sky TOA net solar radiation ($SW_{net(clear)}$) and clear-sky TOA net thermal radiation ($LW_{net(clear)}$). As the EI data (covering 1979–2011) and the CMIP5 data (covering 1961–2000) do not completely overlap, we compare their common period (1979–2000). The EI data and the 28 CMIP5 models were interpolated to a common grid (1.0° latitude × 1.0° longitude) over the domain (34.0°N, 2.0°W, 20.0°N, 26.0°W) (see Figure 1).

3. Methodology

The complexity of the feedback loops explaining summertime subtropical high-pressure system dynam-ics has been confirmed by many studies (Hoskins, 1996; Rodwell and Hoskins, 2001; Seager *et al.*, 2003; Liu *et al.*, 2004) and was synthesized by Miyasaka and Nakamura (2005) using the simplified diagram in Figure 2. Our study focuses on the loop marked with red lines, which defines an amplifying feedback, and on the corresponding relationships between pairs of variables. This loop encompasses: (1) increasing

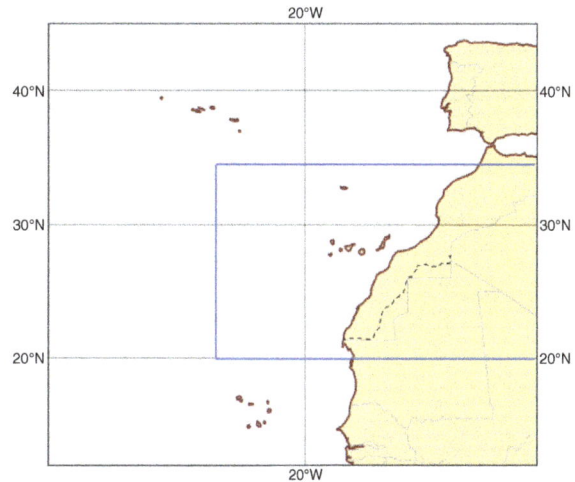

Figure 1. Map showing the area selected for this study (blue rectangle).

of land–sea thermal contrast; (2) intensification of along-shore wind; (3) development of reflective strat-iform clouds favored by a stable lower atmosphere due to evaporation and upwelling of cool waters from below; and (4) increasing of radiative cooling (see Figure 2). The choice of this single loop was mainly determined by the availability of data (both from EI and CMIP5 datasets) that describe the different processes summarized in Figure 2. Because our main interest in this paper is to study the summertime intensification of the Azores high-pressure system (Li *et al.*, 2012), we restrict our analysis to the months of May, June and July and to the domain defined by Figure 1. We use only ocean grid points, except for the thermal contrast, which is calculated as the difference between land and ocean grid points. Figure 3 shows EI data for the period of 1979–2011, emphasizing the reinforcing of all magnitudes that intervene in the positive feedback loop from May to July.

Monthly means of the thermal contrast (the difference between sea and land grid point averages of 2 m air tem-peratures) are computed for the specified EI and CMIP5 periods. Monthly along-shore winds are computed as the meridional component of the 10 m wind averaged for the sea points. Marine stratus extension is estimated using effective cloud albedo (α_{cloud}) (Betts and Viterbo, 2005, Betts, 2007, Fasullo and Trenberth, 2012) by

$$\alpha_{cloud} = \left[SW_{down(clear)} - SW_{down} \right] / SW_{down(clear)} \quad (1)$$

where SW_{down} and $SW_{down(clear)}$ stand for, respectively, cloudy-sky and clear-sky downward surface solar radi-ation.

Radiative cooling is considered to be well represented by cloud radiative forcing (CRF) at the top of the atmosphere (TOA) (Charlock and Ramanathan, 1985, Ramanathan *et al.*, 1989, Wang and Su, 2013). It is computed by

$$CRF = \left(SW_{net} - SW_{net(clear)} \right) + \left(LW_{net} - LW_{net(clear)} \right) \quad (2)$$

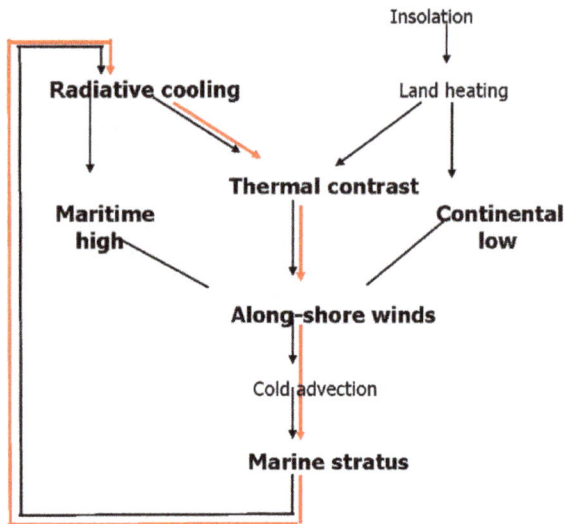

Figure 2. Schematic diagram of the feedbacks associated with the Azores high-pressure system (based on Miyasaka and Nakamura, 2005).

where the first term corresponds to the shortwave forcing and the second to the longwave forcing. If CRF > 0 (<0), clouds are a warming (cooling) mechanism. Although stratus clouds should ideally cause radiative cooling throughout the entire boundary layer, we use 2 m air temperature instead of the mean layer temperature. Allan (2011) compared CRF obtained from EI data with CRF retrieved from satellite data and showed remarkable agreement.

We note that when we refer to coupling between two variables or magnitudes, we mean that one variable controls the other (following Seneviratne *et al.* (2010)) in a way determined by the underlying processes. To quantify the similarities or differences in the empirical relationships defining the coupling between two variables (both for EI data and for each of the CMIP5 models), we use the Hellinger coefficient (Hellinger, 1909), following Sanchez de Cos *et al.* (2013). The Hellinger coefficient, originally designed to estimate the proximity of probability density functions (pdfs), can be thought of as a measure of the 'overlap' between two distributions. It gives information about the differences or similarities in the relative position, shape and orientation of the pdfs, returning values between 0 (fully disjoint distributions) and 1 (identical distributions). The Hellinger coefficient is defined as follows:

$$\mathrm{dHell}^{(s)}(\mathrm{P.Q}) = \int_R q(x)^s \, p(x)^{(1-s)} \, \mathrm{d}x \quad (3)$$

where $q(x)$ and $p(x)$ are the pdfs to be compared, and s is a parameter ($0 < s < 1$). We calculate the Hellinger coefficient with $s = 1/2$, which yields a symmetric measure with values between zero and one. R stands for the phase space where the pdfs are defined.

4. Results

Scatterplots of EI data for the months of May and July (Figure 4) of the four relationships defining the

feedback loop show that some pairs of variables are highly correlated, indicating a strong coupling [e.g. the amount of marine stratus clouds and radiative cooling in July (Figure 4(c))]. However, others are only weakly correlated (e.g. the land–sea thermal contrast and radiative cooling in July (Figure 4(d))). We note that in the latter case, a land–sea contrast value that falls below a certain threshold, as usually occurs in May, may not be able to represent the feedback loop. In fact, the positive correlation between the land–sea thermal contrast and radiative cooling in May evolves to a slightly negatively correlation in July, which properly simulates the amplifying feedback loop.

After examining the corresponding four relationships in the 28 CMIP5 models, we find that only 11 of them can properly simulate the amplifying feedback loop. Therefore, the original 28 models can be initially classified into two main categories depending on their ability to simulate the intensification of the Azores high-pressure system at the beginning of boreal summer. Figures S1–S4 show scatterplots of the four relationships between pairs of magnitudes for EI data and for the 11 models able to simulate the amplifying feedback loop during the month of July. The 'wind–thermal contrast' relationship (see Figure S1) is properly simulated by all 28 CMIP5 models here considered (not shown). The 'wind–marine stratus' relationship only shows a positive correlation in 14 out of 28 models. In the remaining 14 models, the amount of marine stratus clouds decreases when the wind strengthens, breaking the amplifying feedback loop. This behavior contrasts with the results of previous studies (e.g. Klein and Hartmann, 1993, Seager *et al.*, 2003, Miyasaka and Nakamura, 2005, Nakamura, 2012) and with EI reanalysis data. The coupling between wind and cloud albedo is not very strong in most models, and the range of the variables also differs between the various models (see Figure S1). The scatterplot for radiative cooling and the amount of marine stratus clouds [represented by the equivalent relationship between CRF TOA and cloud albedo (see Figure S3)] shows that the two variables are strongly coupled in both the EI data and the CMIP5 models; the maximum value of CRF TOA in the EI data is lower than that in any of the CMIP5 models. The relationship between thermal contrast and radiative cooling (which is represented by CRF TOA) does not show a strong coupling in either the CMIP5 models or the EI data (see Figure S4). We note that there are large differences in the upper limit of the CRF TOA values, especially between the EI data and the CMIP5 models. Although most of the CMIP5 models use a solar constant of approximately 1365 Wm^{-2} (higher than the EI value of 1377 Wm^{-2}) we estimate that the difference of approximately 1% can hardly have effect in the results. These differences largely derive from the discrepancies in shortwave forcing between the EI data and the CMIP5 models. Nevertheless, some models, including BNU-ESM, MIROC-ESM-CHEM and NorESM1-M,

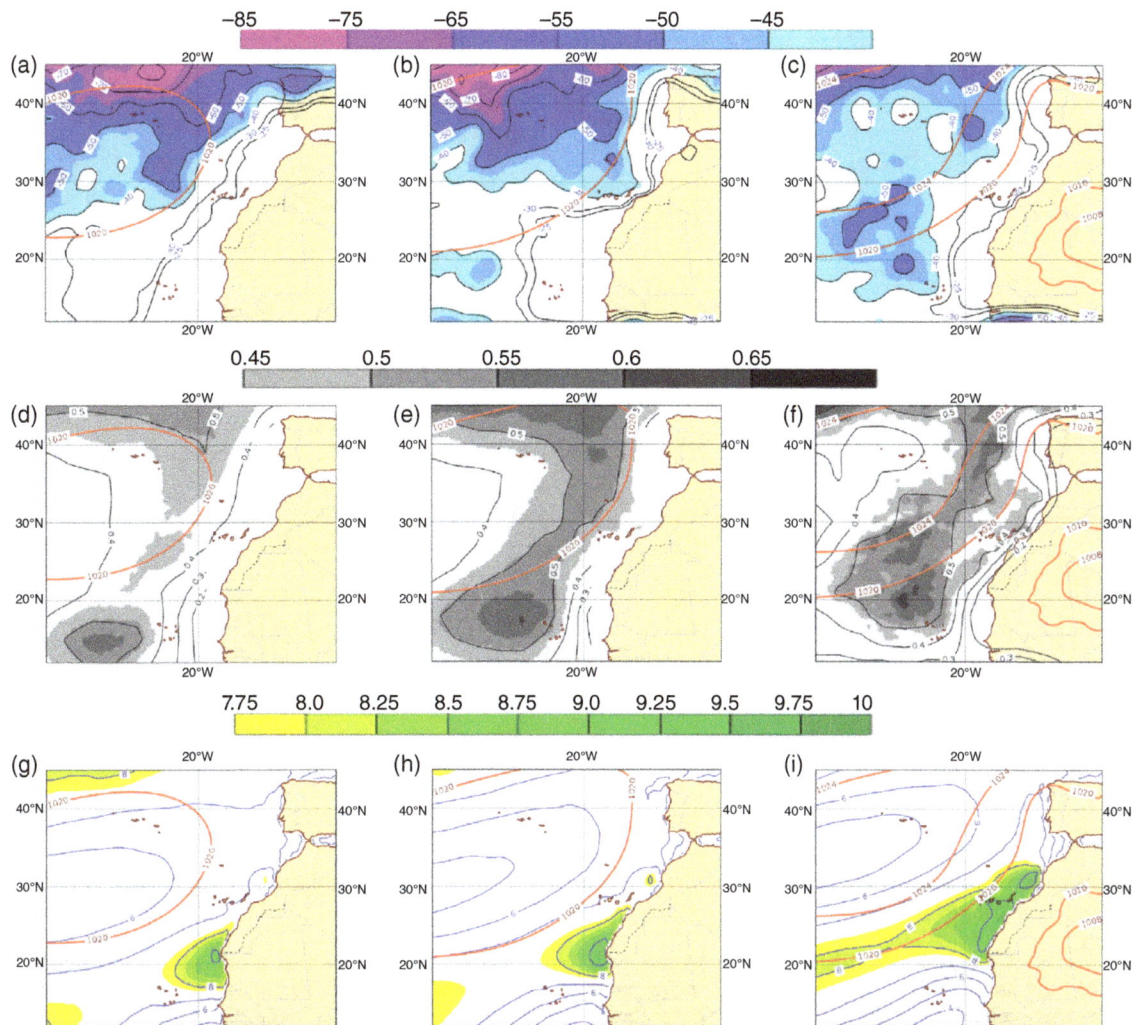

Figure 3. Radiative forcing by clouds (black lines and magenta–clear blue areas in Wm^{-2}) for May (a), June (b) and July (c). Cloudiness (black lines and grey areas in albedo units ranging from 0 to 1) for May (d), June (e) and July (f). Wind speed (blue lines and green–yellow areas in ms^{-2}) for May (g), June (h) and July (i). Note that sea level pressure (red lines in hPa) is included in all maps for the sake of completeness.

show a relatively strong coupling between both variables.

For each of the four relationships and each of the 11 CMIP5 models simulating the positive feedback loop, we compute Hellinger distances with respect to the EI data (see Table 1). We can use this metric to rank the 11 CMIP5 models that can simulate the positive feedback loop without considering any additional combination of the four coefficients, and we note that the metric was computed separately for the four relationships. Some models rank well based on a majority of the four distances, such as IPSL-CM5A-MR, ACCESS1-3 and BNU-ESM, whereas others have difficulty reproducing the processes described by the reanalysis data, such as MRI-CGCM3 and MIROC-ESM.

Although it is not easy to explain why so many models (17 out of 28) fail to simulate the positive feedback loop that describes the dynamics of subtropical high-pressure systems during summer, a preliminary analysis of deficiencies in these models suggests that they are unable to simulate the marine stratus layer, the correct thermal contrast between land and sea or both. A more detailed analysis of each of the simulations is beyond the scope of this letter.

5. Conclusions

The evaluation of climate models conducted in this letter is based on the relevance of subtropical anticyclones to both global and regional climate. In particular, the intensification of the Azores high plays an important role in the climate of south-western Europe and north-western Africa during summer. We found that only 11 out of the 28 models examined were able to represent the positive feedback loop similarly to the reanalysis data. The relatively small number of models (11 out of 28) able to simulate the amplifying feedback loop indicates that climate models have difficulty representing some of the relevant physical processes that occur in the land–ocean–atmosphere subsystem, in particular those related to marine stratus clouds, which are underestimated by most of the CMIP5 models.

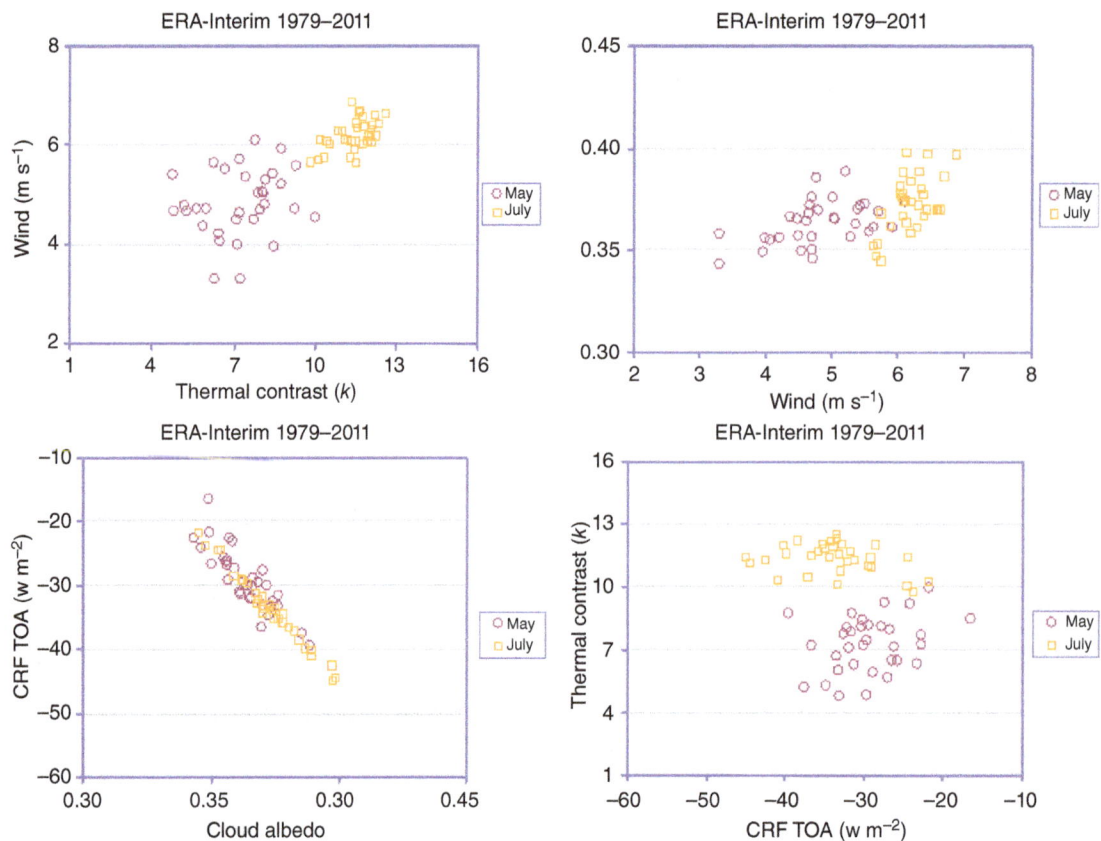

Figure 4. Relationships among pairs of variables over the area shown in Figure 1 (based on ERA-Interim data from 1979 to 2011) for the months of May and July.

Table 1. Values of the Hellinger coefficient for the following relationships: thermal contrast-wind, wind-cloud albedo, cloud albedo-CRF TOA and CRF TOA-thermal contrast for the month of July (from 1979 to 2000). Variables are expressed as differences from their mean value. The model acquiring the highest (bold) and the lowest value (underlined) of the Hellinger coefficient for each relationship is indicated, and ranking positions are noted within brackets.

Models	Thermal contrast-V10m	V10m-cloud albedo	Cloud albedo-CRF_TOA	CRF_TOA-thermal contrast
ACCESS1-0	0.96 [2]	0.94 [2]	0.90 [8]	0.92 [9]
ACCESS1-3	0.91 [5]	0.92 [3]	**0.99 [1]**	0.97 [2]
BNU-ESM	0.88 [7]	0.89 [6]	0.98 [3]	**0.98 [1]**
HadGEM2-ES	0.94 [3]	0.87 [7]	0.88 [9]	0.94 [5]
IPSL-CM5A-LR	0.88 [8]	0.91 [5]	0.96 [4]	0.92 [7]
IPSL-CM5A-MR	**0.97 [1]**	**0.99 [1]**	0.98 [2]	0.96 [3]
MIROC-ESM	0.86 [10]	0.77 [10]	0.80 [10]	0.89 [10]
MIROC-ESM-CHEM	0.84 [11]	0.82 [9]	0.91 [5]	0.94 [6]
MRI-CGCM3	0.87 [9]	0.67 [11]	0.60 [11]	0.75 [11]
NorESM1-M	0.90 [6]	0.85 [8]	0.90 [7]	0.96 [4]
NorESM1-ME	0.93 [4]	0.92 [4]	0.91 [6]	0.92 [8]

Our proposed method of evaluation aims to explore and quantify how well various models simulate the coupling between atmosphere, ocean and land surface. Climate models are based on sound and well-established physical laws, and their success in simulating the climate system depends on an accurate representation of relevant processes such as the dynamics of subtropical highs. We emphasize the importance of evaluation studies that focus on physical processes, particularly the features at the interface between subsystems. Analysis of the simulated coupling between subsystems, as in this study of the dynamics of the Azores anticyclone, could help to diagnose modeling deficiencies in representing climate variables that underlie poor performance.

Acknowledgements

We thank the WCRP's Working Group on Coupled Modeling, which is responsible for CMIP, and all groups that have developed and made available model outputs. We also thank ECMWF for providing the ERA-Interim reanalysis dataset. Finally, we appreciate many useful comments on this manuscript from our colleagues Beatriz Navascués and María J. Casado. The authors are also grateful to two referees for comments and corrections that have helped to improve the text.

Supporting information

The following supporting information is available:

Table S1. The 28 CMIP5 models considered in this study.

Figure S1. Scatterplots of the relationship between V10m and thermal contrast for ERA-Interim data and for the 11 CMIP5 models able to simulate the positive feedback loop in July.

Figure S2. The same as Figure S1, but for cloud albedo and V10m.

Figure S3. The same as Figure S1, but for CRF TOA and cloud albedo.

Figure S4. The same as Figure S1, but for thermal contrast and CRF TOA.

References

Allan RP. 2011. Combining satellite data and models to estimate cloud radiative effect at the surface and in the atmosphere. *Meteorological Applications* **18**: 324–333, doi: 10.1002/met.285.

Betts AK. 2004. Understanding hydrometeorology using global models. *Bulletin of the American Meteorological Society* **85**: 1673–1688, doi: 10.1175BAMS-85-11-1673.

Betts AK. 2007. Coupling of water vapor convergence, clouds, precipitation, and land-surface processes. *Journal of Geophysical Research* **112**: D10108, doi: 10.1029/2006JD008191.

Betts AK, Viterbo P. 2005. Land-surface boundary layer and cloud-field coupling over the southwestern Amazon in ERA-40. *Journal of Geophysical Research* **110**: D14108, doi: 10.1029/2004JD005702.

Charlock TP, Ramanathan V. 1985. The albedo field and cloud radiative forcing produced by a general circulation model with internally generated cloud optics. *Journal of the Atmospheric Sciences* **42**: 1408–1429, doi: 10.1175/1520-0469(1985)042<1408:TAFACR>2.0.CO;2.

Dee DP, Uppala SM, Simmons AJ, Berrisford P, Poli P, Kobayashi S, Andrae U, Balmaseda MA, Balsamo G, Bauer P, Bechtold P, Beljaars ACM, van de Berg L, Bidlot J, Bormann N, Delsol C, Dragani R, Fuentes M, Geer AJ, Haimberger L, Healy SB, Hersbach H, Hólm EV, Isaksen L, Kållberg P, Köhler M, Matricardi M, McNally AP, Monge-Sanz BM, Morcrette JJ, Park BK, Peubey C, de Rosnay P, Tavolato C, Thépaut JN, Vitart F. 2011. The ERA-Interim reanalysis: configuration and performance of the data assimilation system. *Quarterly Journal of the Royal Meteorological Society* **137**: 553–597.

Fasullo JT, Trenberth KE. 2012. A less cloudy future: the role of subtropical subsidence in climate sensitivity. *Science* **338**: 792–794, doi: 10.1126/science.1227465.

Flato GJ, Marotzke J, Abiodun B, Braconnot P, Chou SC, Collins W, Cox P, Driouech F, Emori S, Eyring V, Forest C, Gleckler P, Guilyardi E, Jakob C, Kattsov V, Reason C, Rummukainen M. 2013. Evaluation of climate models. In *Climate Change 2013: The Physical Science Basis. Contribution of Working Group I to the Fifth Assessment Report of the IPCC*, Stocker TF, Qin D, Plattner GK, Tignor M, Allen SK, Boschung J, Nauels A, Xia Y, Bex V, Midgley PM (eds). Cambridge University Press: Cambridge, UK and New York, NY.

Hellinger E. 1909. Neue Begründung der Theorie quadratischer Formen von unendlich vielen Veränderlichen. *Journal für die reine und angewandte Mathematik* **136**: 210–271.

Hoskins BJ. 1996. On the existence and strength of the summer subtropical anticyclones. *Bulletin of the American Meteorological Society* **77**: 1287–1292.

Klein SA, Hartmann DL. 1993. The seasonal cycle of low stratiform clouds. *Journal of Climate* **6**: 1587–1606, doi: 10.1175/1520-0442(1993)006<1587:TSCOLS>2.0.CO;2.

Klein SA, Hartmann DL, Norris JR. 1995. On the relationships among low-cloud structure, sea surface temperature, and atmospheric circulation in the summertime northeast Pacific. *Journal of Climate* **8**: 1140–1155.

Li W, Li L, Ting M, Liu Y. 2012. Intensification of Northern Hemisphere subtropical highs in a warming climate. *Nature Geoscience* **5**: 830–834, doi: 10.1038/ngeo1590.

Liu Y, Wu G, Liu H, Liu P. 2001. Condensation heating of the Asian summer monsoon and the subtropical anticyclone in the eastern hemisphere. *Climate Dynamics* **17**(4): 327–338.

Liu Y, Wu G, Ren R. 2004. Relationship between the subtropical anticyclone and diabatic heating. *Journal of Climate* **17**: 682–698, doi: 10.1175/1520-0442(2004)017<0682:RBTSAA>2.0.CO;2.

Miyasaka T, Nakamura H. 2005. Structure and formation mechanisms of the Northern Hemisphere summertime subtropical highs. *Journal of Climate* **18**: 5046–5065, doi: 10.1175/JCLI3599.1.

Nakamura H. 2012. Atmospheric science: future oceans under pressure. *Nature Geoscience* **5**: 768–769, doi: 10.1038/ngeo1623.

Pierrehumbert RT. 1995. Thermostats, radiator fins, and the local runaway greenhouse. *Journal of the Atmospheric Sciences* **52**: 1784–1806.

Ramanathan V, Cess RD, Harrison EF, Minnis P, Barkstrom BR, Ahmad E, Hartmann D. 1989. Cloud-Radiative Forcing and Climate: Results from the Earth Radiation Budget Experiment. *Science* **243**: 57–63.

Rodwell MJ, Hoskins BJ. 2001. Subtropical anticyclones and summer monsoons. *Journal of Climate* **14**: 3192–3211, doi: 10.1175/1520-0442(2001)014<3192:SAASM>2.0.CO;2.

Sanchez de Cos C, Sanchez-Laulhe JM, Jimenez-Alonso C, Sancho-Avila JM, Rodriguez-Camino E. 2013. Physically based evaluation of climate models over the Iberian Peninsula. *Climate Dynamics* **40**: 1969–1984, doi: 10.1007/s00382-012-1619-2.

Seager R, Murtugudde R, Naik N, Clement A, Gordon N, Miller J. 2003. Air–sea interaction and the seasonal cycle of the subtropical anticyclones. *Journal of Climate* **16**: 1948–1966, doi: 10.1175/1520-0442(2003)016<1948:AIATSC>2.0.CO;2.

Seneviratne SI, Corti T, Davin E, Hirschi M, Jaeger EB, Lehner I, Orlowsky B, Teuling AJ. 2010. Investigating soil moisture climate interactions in a changing climate: a review. *Earth Science Reviews* **99**(3–4): 125–161, doi: 10.1016/j.earscirev.2010.02.004.

Taylor KE, Stouffer RJ, Meehl GA. 2012. An overview of CMIP5 and the experiment design. *Bulletin of the American Meteorological Society* **93**: 485–498, doi: 10.1175/BAMS-D-11-00094.1.

Wang H, Su W. 2013. Evaluating and understanding top of the atmosphere cloud radiative effects in intergovernmental panel on climate change (IPCC) fifth assessment report (AR5) coupled model intercomparison project phase 5 (cmip5) models using satellite observations. *Journal of Geophysical Research – Atmospheres* **118**: 683–699, doi: 10.1029/2012JD018619.

4

Impacts of land use and land cover changes on the climate over Northeast Brazil

Ana Paula M. A. Cunha,[1]* Regina C. S. Alvalá,[1] Paulo Y. Kubota[2] and Rita M. S. P. Vieira[2]

[1] Brazilian Centre for Monitoring and Warning of Natural Disasters, São José dos Campos, SP, Brazil
[2] National Institute for Space Research, Cachoeira Paulista, SP, Brazil

*Correspondence to:
A. P. M. A. Cunha, Brazilian Centre for Monitoring and Warning of Natural Disasters, Av dos Astronautas 1758, Jd. Granja, CEP 12227-010, São José dos Campos, SP, Brazil.
E-mail: apaulama2011@gmail.com

Abstract

Two numerical experiments using the Atmospheric Global Circulation Model by the Center for Weather Prediction and Climate Studies and Integrated Biosphere Simulator were performed to investigate the impacts of land use and land cover changes on the climate in the semiarid area of Northeast Brazil due to replacing natural vegetation with pasture and degraded areas. Such a disturbance led to a decrease in the mean rainfall during the dry season at the study area. A meridional dipole pattern with a near surface temperature increase (reduction) in the northern (southern) areas of the semiarid region was found. The results also highlight that land use/cover change led to changes in the surface energy components and carbon balance.

Keywords: land use change; semiarid; regional climate change

1. Introduction

Changes in the use and occupation of land, caused by human actions, have been shown to have an important effect on climate (Pielke *et al.*, 2002, 2011). Numerous modelling studies conducted in different ecosystems have demonstrated that converting natural vegetation to crop or pasture lands affects the exchanges of energy, water and carbon between the atmosphere and surface (Chase *et al.*, 1996, 2000; Bonan, 1997; Betts, 2012; Hoffmann *et al.*, 2000; Claussen *et al.*, 2001; Matthews *et al.*, 2003; Twine *et al.*, 2004; Pongratz *et al.*, 2006; Cotton and Pielke, 2007; Findell *et al.*, 2007, 2009; Diffenbaugh, 2009; Bathiany *et al.*, 2010; Lawrence *et al.*, 2012; Mahmood *et al.*, 2013). Semiarid and arid regions are the most vulnerable regions to these changes due to frequent years of below average rainfall and severe drought. A number of recent studies have examined the influence of land use/cover change (LULCC) on climate in semiarid regions around the globe (Taylor *et al.*, 2002; Gao *et al.*, 2003; Liu *et al.*, 2008; Deo *et al.*, 2009; Paeth *et al.*, 2009; Otieno and Anyah, 2012). Although many modelling studies have considered LULCC in different semiarid regions, only limited attention has been given to the climate impact caused by changes in the use and occupation of land in the semiarid regions of South America, especially in Brazil.

In recent decades, most of Brazil's territory has shown indications of land use/cover spatial transformations. According to the land use/cover map developed by the National Institute for Space Research (INPE), the typical vegetation of the semiarid area of Northeast Brazil (SANEB), known as *caatinga* (closed xeric shrubland), has been replaced by agricultural activities and pasture

area in only a few years. Approximately 40% of the *caatinga* has been converted to these uses, and the remaining area is being transformed at a rate of 0.3% per year. Currently, more than 10% of the semiarid area has already undergone a high degree of environmental degradation. The driest areas, which are those with an average annual precipitation of <500 mm, are the most susceptible to desertification (Oyama and Nobre, 2004).

The objective of this paper is to quantify the possible biogeophysical (Betts, 1999) and biogeochemical (Defries *et al.*, 1999) effects of LULCC for land use planning and climate change adaptation in the semiarid region of Northeast Brazil. Although some studies on the climate effects of LULCC show that the biogeophysical and biogeochemical mechanisms are of the same order of magnitude (Brovkin *et al.*, 2004, 2006), it is necessary to consider both the mechanisms to appropriately quantify the impacts of LULCC (Bathiany *et al.*, 2010).

In this paper, experiments of natural vegetation conversion were performed using the surface model IBIS ('Integrated Biosphere Simulator', Foley *et al.*, 1996) coupled to the Atmospheric Global Circulation Model by the Center for Weather Prediction and Climate Studies (AGCM/CPTEC). The differences in various simulated climatic variables among three experiments with different land surface conditions were analysed.

2. Model and experiment design

In this study, the numerical experiments were performed with AGCM/CPTEC. Its main features can be found in Bonatti (1996) and the modification of the original AGCM version by the Center for Ocean/Land

and Atmosphere Studies (COLA) can be found in Kubota (2012). Recently, Kubota (2012) performed the surface scheme IBIS coupled with the AGCM/CPTEC using a boundary layer scheme that allowed the coupling of different surface schemes. IBIS is part of a new generation of global biosphere models, termed Dynamic Global Vegetation Models, and it represents a wide range of processes, including land surface physics, canopy physiology, plant phenology, vegetation dynamics and competition, and carbon and nutrient cycling. The AGCM/CPTEC/IBIS version of the model has been validated for the short (7 days) and long term (50 years). A more detailed description regarding the coupling and validation of AGCM/CPTEC/IBIS is provided in Kubota (2012).

The simulations were performed using AGCM/CPTEC/IBIS model, with a horizontal resolution of approximately 100 km and 28 vertical levels (T126L28), in global domain. Reanalysis data from the 'European Reanalysis Atmospheric' (ERA-40) at a 1.250° resolution and climatological values of the sea surface temperature data from the 'National Oceanic and Atmosphere Administration' at a 1° resolution were used in the initialization of the numerical simulations. In all simulations, atmospheric CO_2 concentrations are set to 370 ppmv. All simulations were performed for a period of 10 years using the same initial conditions. The first 5 years of each integration were neglected due to soil moisture spin-up. The numerical integrations began in January 2003 and extended until the last day of December 2012. For each run, three-member ensemble averages were taken to filter out the AGCM intermember variability. The statistical significance of the anomalies was evaluated using Student's t-test. Statistical significance was determined using the annual and seasonal (wet season: February, March–May; dry season: July–October) averages for semiarid region of Northeast Brazil.

The climate impacts of LULCC were assessed using climate simulations considering three scenarios of vegetation distribution:

(a) *Potential vegetation (control)*: 'Global Potential Vegetation Dataset' was described in Ramankutty and Foley (1999), originally with a resolution of 5 min.

(b) *Actual vegetation over SANEB*: Data were provided by the PROVEG-INPE project (Vieira *et al.*, 2013), which included the natural vegetation (natural *caatinga* and deciduous and evergreen forests) and pasture lands. This map was accomplished by the visual interpretation of a digital mosaic of ETM + Landsat 7 and TM Landsat 5 images from 1999 to 2000. According to the new vegetation map (2000), a large *caatinga* natural area was noticeably devastated, indicating that approximately 22% of the natural vegetation has been degraded within a 7-year period.

(c) *Future scenario of the vegetation distribution over SANEB*: In a semiarid region, water availability is the main obstacle to agricultural production, which can be compounded with future climate conditions. Thus, as a possible future land use/cover scenario, this work

considered that the expansion of a pasture may be limited by desert and semi-desert areas. These degraded areas were obtained from the future projection of the biome distribution in South America developed by Salazar Velásquez (2009) using CPTEC PVMReg and emission scenario A2 from the Intergovernmental Panel on Climate Change. It is worth emphasizing that the procedure adopted for the creation of a future scenario is only an abstract attempt to extrapolate the future distribution of vegetation in the semiarid of Brazil.

The actual vegetation and future scenario simulations were compared with the potential vegetation (control) simulation. The main difference between the actual vegetation/future scenario and control experiment was the presence of pasture lands (Figure 1). Among the 76 grid cells within the semiarid boundary that were changed in the actual vegetation scenario (compared with the control), approximately 70% was due to the exchange of *caatinga* to pasture. In the future scenario, 81 grid cells were modified, of which 40% was due to the conversion of the *caatinga* to pasture and approximately 30% was due to the conversion to open *caatinga* (or degraded *caatinga*).

3. Results

3.1. Biogeophysical effects of LULCC

The main changes in the roughness length (z_0, parameter related to vegetation structure) were due to the conversion of evergreen and deciduous forest to pasture (z_0 was reduced by approximately 1.5 m). In the IBIS model, z_0 was dependent on the canopy heights and vegetation density, which depended on the leaf area index (LAI). As for z_0, converting evergreen and deciduous forest to pasture produced the largest changes in LAI (reduced by approximately 4 $m^2 m^{-2}$).

The conversion of *caatinga* and evergreen forest to pasture (the difference between the actual vegetation and control experiments) resulted in increases in rainfall in the southern semiarid region (+0.3 mm day^{-1}) and the central region (+0.6 mm day^{-1}; Figure 2(a)). Regarding the conversion of *caatinga* to pasture and degraded *caatinga* (the difference between the future and control experiments), there was a reduction in rainfall in the central semiarid (-0.3 mm day^{-1}) and part of the eastern region (-0.6 mm day^{-1}; Figure 2(b)). During the wet season, increased precipitation over the central region (1.5 mm day^{-1}, Figure 2(c)) was consistent with anomalous vertically integrated moisture flux convergence and increased ascent in the midtroposphere (not shown).

A dipole pattern with a reduction (increase) in precipitation in the northern (southern) semiarid areas was found (Figure 2(d)). In the dry season, the impacts on precipitation were smaller for most semiarid areas; however, the precipitation was significantly reduced in the eastern region (1.5 mm day^{-1}) (Figure 2(e) *and* (f)). It is worth noting that in all scenarios the precipitation differences tended to spread to other regions, with a statistically significant reduction in rainfall in the northern

Change in vegetation

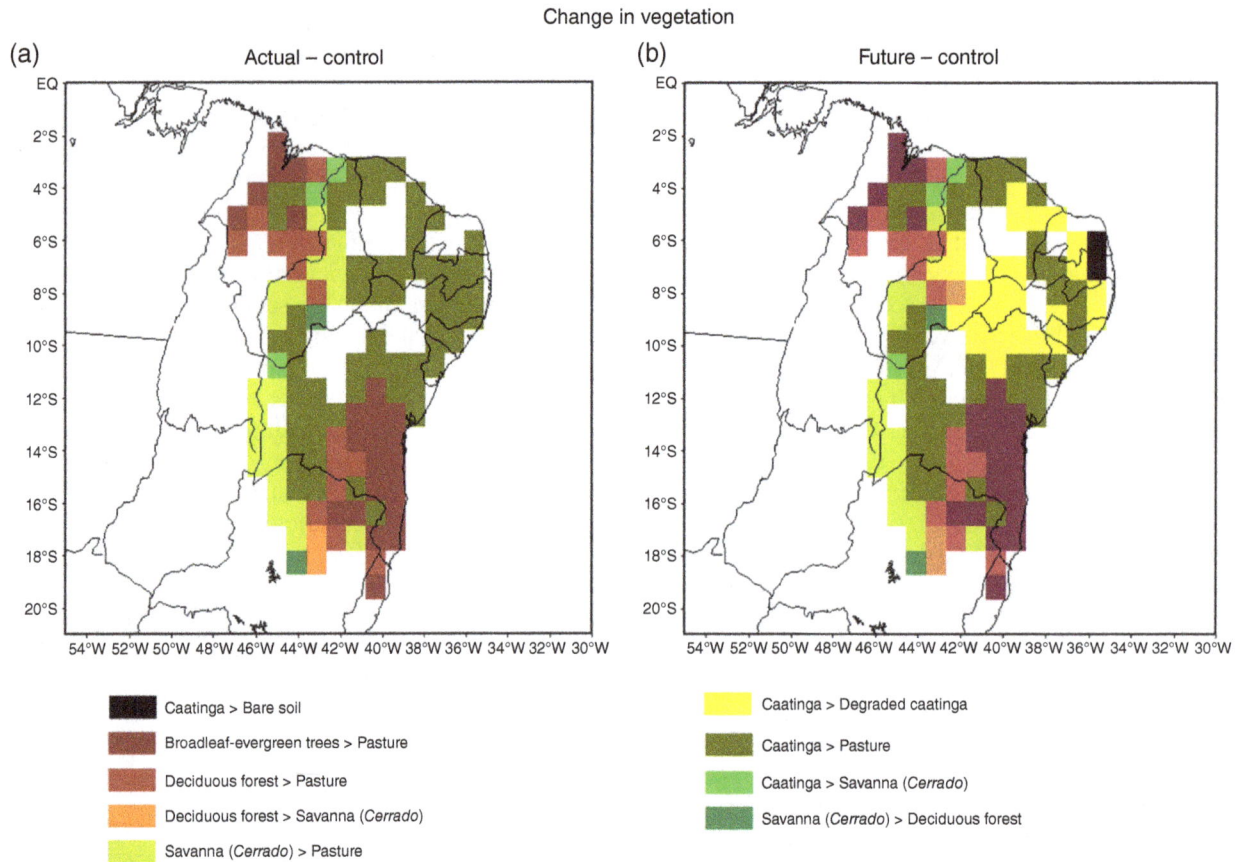

Figure 1. Differences in the vegetation distribution for Brazilian semiarid between the actual future and control experiments.

region and an increase in the Midwest, South and Southeast of Brazil (Figure 2).

The conversion of forests to pasture also led to changes in wind speed intensity and direction and moisture content (Figure 3). In both scenarios and seasons, an intensification in the moisture transported from northwest to southeast of South America was observed, with an increase in moisture content between Bolivia, Paraguay and southern Brazil. These atmospheric conditions tend to create a greater availability of moisture at low levels, consistent with increased rainfall, particularly in the La Plata basin. Moreover, the moisture content reduction also explains the reduced precipitation, particularly for the North of Brazil. During the wet season, the conversion of forests to pasture (z_0 reduction) allows an intensification in moisture transported from the Atlantic Ocean to the continent (approximately between 22° and 12°S) and an amplification of the moisture transported from the northwest towards the southeast of Brazil.

LULCC in the semiarid areas also led to statistically significant changes in the near surface temperature. Converting forests to pasture causes an annual average surface cooling (-1.8 to -0.3 °C) in the southern semiarid region, according to the potential vegetation scenario (Figure 4(a) and (b)). Furthermore, in areas where *caatinga* was converted to degraded *caatinga* the surface temperature increased (0.3 to 0.9 °C) (Figure 4(b)). A meridional dipole pattern was observed between the degraded areas and adjacent southern areas. Souza and

Oyama (2011) found a similar pattern for the semiarid regions of Brazil resulting from partial degradation. This dipole pattern can partially counterbalance the climatic impacts.

The impacts of LULCC on energy balance components were significantly different between the experiments. The conversion of forests to pasture and degraded *caatinga* resulted in a reduction of available surface energy. In the first case, it was mainly due to increased reflected solar radiation (albedo), while in the second case it was due to an increase in emitted long-wave radiation (Table 1). The reduction of surface net radiation (R_n) represents a decrease in available surface energy and in the sum of the sensible (H) and latent (LE) turbulent energy fluxes. The lower available energy at the surface associated with the lower roughness length and changes in rooting depth also caused the decrease in canopy transpiration (Table 1). The increase in the evaporation from the surface (1.4 mm day^{-1}) did not counterbalance the decrease in transpiration (2.4 mm day^{-1}), which indicates that the evapotranspiration was decreased, a common result of all tropical deforestation studies.

Finally, surface sensible fluxes are main forcing to drive the PBL development. In both scenarios (Figure 4(c) and (d)), the reduced turbulent heat flux leads to shallower PBL.

Regarding the neighbouring regions, a surface warming in the northern region of the Brazil (Amazon) was found in both the experiments. Surface warming is

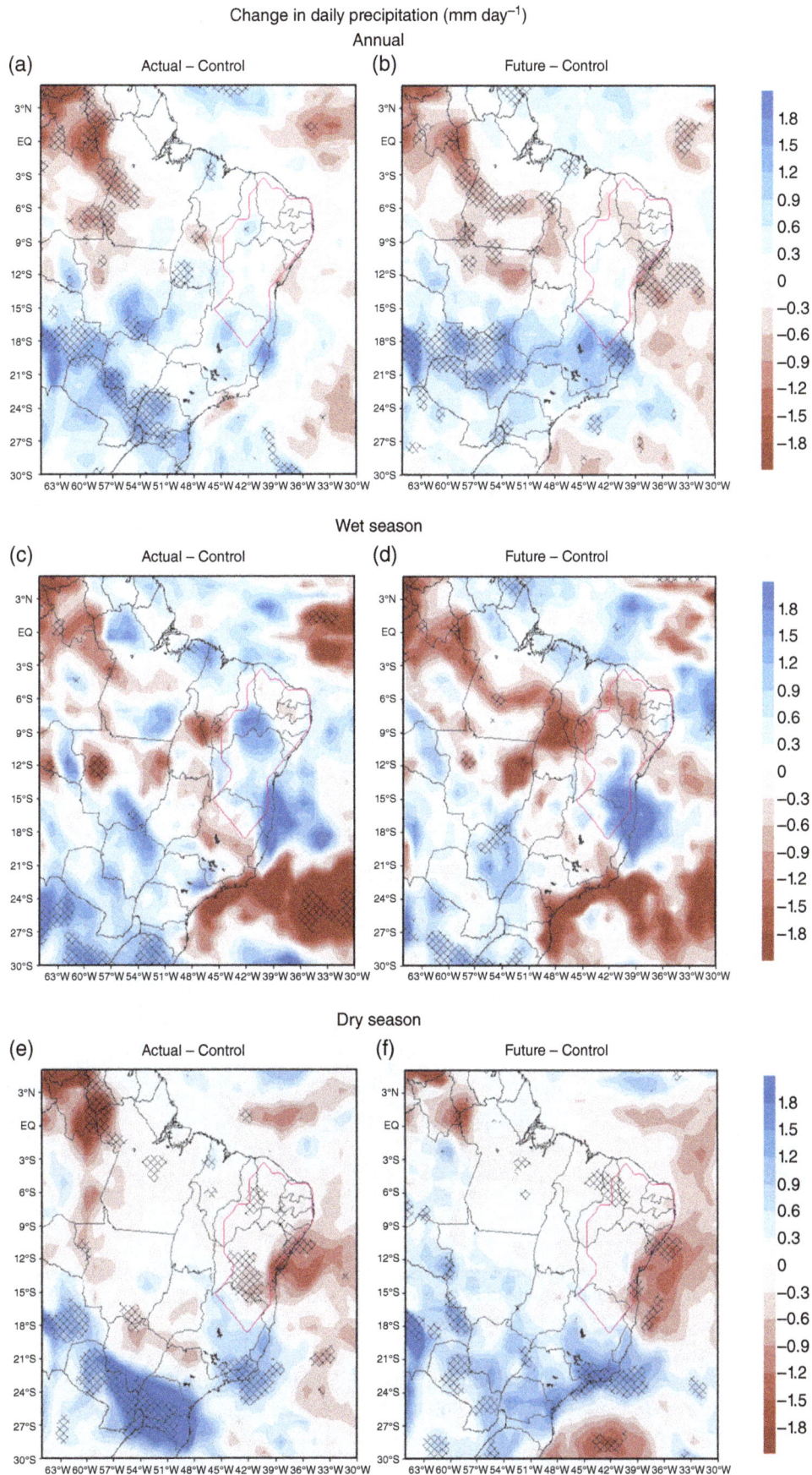

Figure 2. Differences in daily precipitation for (actual minus control experiments and future scenario minus control): (a) and (b) annual average; (c) and (d) wet season (February to May); (e) and (f) dry season (August to October). Cross-hatching shows areas where changes are statistically significant (95% confidence level in Student's *t*-test).

Change in 850-mb atmospheric circulation (ms^{-1}) and moisture content ($\times 10^{-4}$ kg kg^{-1})
(Difference in wind vectors and mixing ratio at 850 mb)

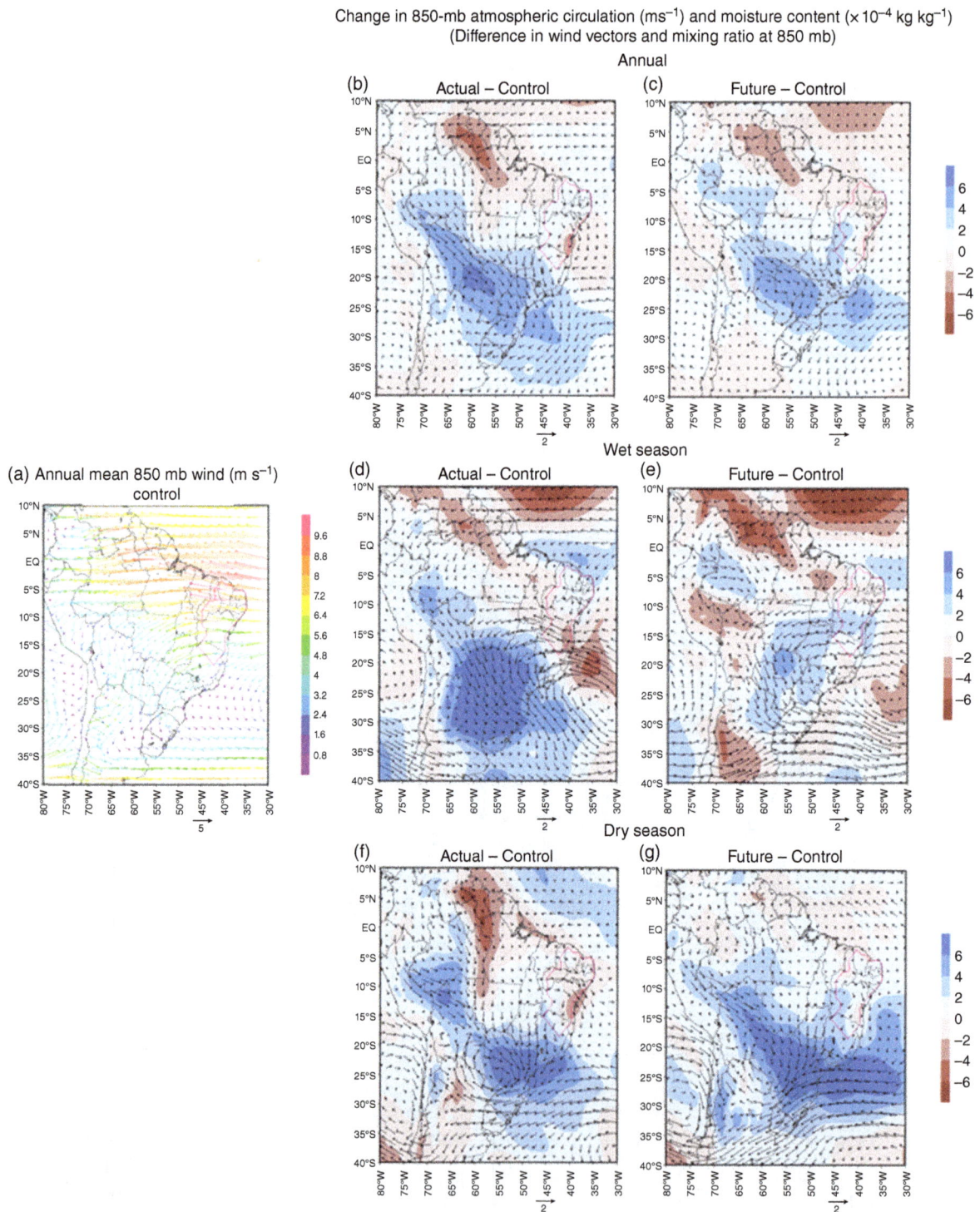

Figure 3. Annual mean of the wind vectors at 850 mb in control experiment (a); differences in atmospheric circulation and moisture content (actual scenario minus control scenario and future scenario minus control scenario): (b) and (c) annual average; (d) and (e) wet season (February to May); (f) and (g) dry season (August to October).

generally associated with decreased moisture content (Figure 3) because the lower atmosphere becomes hotter and drier over these areas.

3.2. Biogeochemical effects of LULCC

Evaluating the sources and sinks of carbon from LULCC is important in the global carbon budget

(Houghthon *et al.*, 2012). With respect to the LULCC effects on the vegetation–atmosphere CO_2 exchanges, it was found that the conversion from forest to pasture led to decreases of approximately 90% (9 kg C m^{-2}) for the total canopy biomass; therefore, there is significant carbon release to the atmosphere when forest areas are converted to pasture lands. A biomass reduction also occurred in the *caatinga* degradation scenario (60%,

Change in near-surface temperature (°C)
Annual

(a) Actual – Control

(b) Future – Control

Change in planetary boundary layer (m)
Annual

(c) Actual – Control

(d) Future – Control

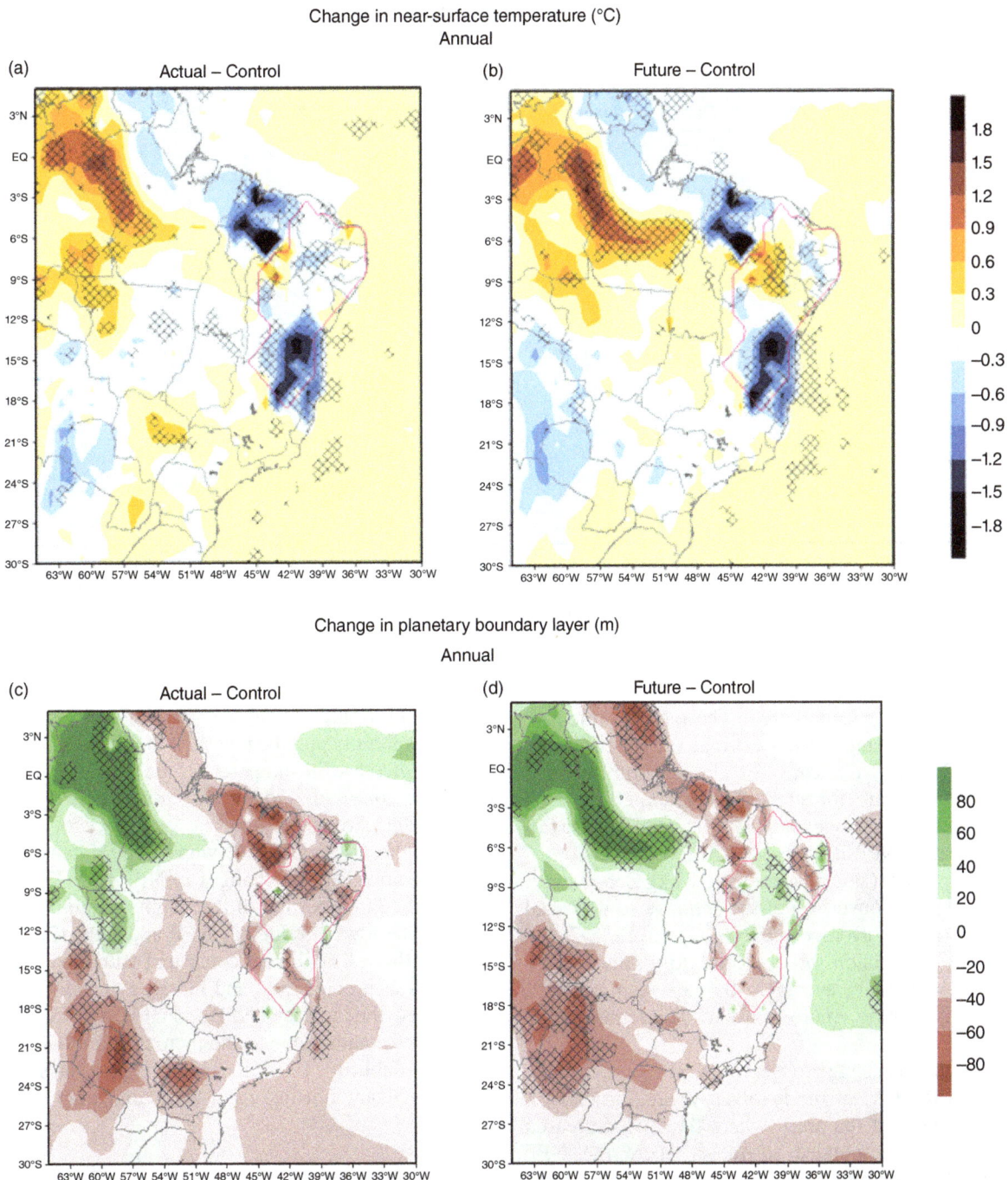

Figure 4. Annual differences in near surface temperature and planetary boundary layer height (m) (a) and (c) Actual scenario minus control scenario. (b) and (d) Future scenario minus control scenario. Cross-hatching shows areas where changes are statistically significant (95% confidence level in Student's t-test).

$0.3 \, kg \, C \, m^{-2}$). These results are similar to those found by Houghton (1991). When forests are cleared for pasture lands, a large proportion of the aboveground biomass may be burned, rapidly releasing most of its carbon into the atmosphere. This knowledge of biomass is important for calculating the sources (and sinks) of carbon that resulted from LULCC.

The gross primary production (GPP) is the total amount of carbon dioxide 'fixed' by land plants through the processes of photosynthesis. The numerical experiments showed statistically significant changes in GPP

due the LULCC. The conversion of forest to pasture and *caatinga* to degraded *caatinga* caused decrease in GPP by $-3 \times 10^6 \, mol \, CO_2 \, m^{-2} \, s^{-1}$ (38%) and $-1.72 \times 10^6 \, mol \, CO_2 \, m^{-2} \, s^{-1}$ (68%), respectively. The opposite was found for the conversion of *caatinga* to pasture, i.e. an increase of $2 \times 10^6 \, mol \, CO_2 \, m^{-2} \, s^{-1}$ (106%).

Figure 5 shows the impact of LULCC on net primary productivity (NPP), which represents the plant growth (difference between GPP and autotrophic respiration). In the control experiment, the annual average, NPP for natural *caatinga* was $0.5 \, kg \, C \, m^{-2} \, year^{-1}$, whereas for

Table 1. Local impacts of LULCC on the components of the radiation, energy water and carbon balances (differences represents averages over the mask of vegetation types within Northeast Brazil region).

Variable	Forest → pasture			Caatinga → pasture			Caatinga → degrad. caatinga		
	Annual	Wet	Dry	Annual	Wet	Dry	Annual	Wet	Dry
T2m	−1.2	−1.4	−1.0	−0.2	−0.3	−0.2	+0.2	+0.2	+0.3
Cloud	+0.2	+0.5	−4.6	+1.7	+0.4	−1.4	−1.3	−1.1	−2.8
S↓	+1.9	−2.5	+4.4	−2.1	−5.2	+0.4	+1.7	+2.0	+0.8
S↑	+9.8	+10.3	+9.8	−3.7	−1.5	−4.1	+3.2	−0.19	+1.0
ΔS	−7.9	−12.8	−5.4	+1.8	−3.7	4.5	−1.5	2.19	−0.2
L↓	+1.4	+1.8	+0.8	+0.3	+0.5	−0.4	+1.0	+0.5	+1.8
L↑	+6.7	+4.4	+6.6	−3.4	−4.0	−3.8	+7.7	+8	+6.2
ΔL	−5.3	−2.6	−5.8	+3.7	4.5	3.4	−6.7	−7.8	−4.4
R_n	−13.2	−15.4	−11.2	+5.5	+0.8	+7.9	−8.2	−5.6	−4.6
H	−1.6	+0.3	− −1.0	−2.6	−1.8	−2.3	−4.8	−5.5	−2.4
LE	−14.7	−17.5	−16.3	+12.4	+10.2	+12.2	−15.6	−12.0	−18.6
β	+0.17	+0.17	+0.34	−0.49	−0.21	−0.97	+0.69	+0.16	+1.83
Prec	+0.16	+1.02	−0.44	+0.17	+0.33	−0.23	−0.19	−0.22	−0.12
E	+1.39	+0.87	+0.37	−1.03	−0.24	−0.20	+0.02	+0.31	+0.10
Tran	−2.37	−3.03	−1.89	+0.97	+0.91	+0.94	−1.04	−0.98	−1.08
GPP	−2.88			+2.01			−1.72		
NPP	−1.71			+1.32			−1.02		

T2m, near surface temperature (°C); cloud, cloud cover (%); S↓, incident solar radiation (W m^{-2}); S↑, reflected solar radiation (W m^{-2}); ΔS, net surface shortwave radiation (W m^{-2}); L↓, incident longwave radiation (W m^{-2}); L↑, reflected longwave radiation (W m^{-2}); ΔL, net surface longwave radiation (W m^{-2}); R_n, net surface radiation (W m^{-2}); H, sensible heat flux (W m^{-2}); LE, latent heat flux (W m^{-2}); β, Bowen ratio; Prec, precipitation (mm day^{-1}); E, soil evaporation (mm day^{-1}); Tran, canopy transpiration (mm day^{-1}); GPP, gross primary productivity (×10^6 mol CO_2 m^{-2} s^{-1}); NPP, net primary productivity (×10^6 mol CO_2 m^{-2} s^{-1}).

forest was 2.16 kg C m^{-2} year^{-1}. These NPP values are similar to the average values estimated by the MODIS images: 0.4 kg C m^{-2} year^{-1} for closed shrubland and 1.15 kg C m^{-2} year^{-1} for forest (Zhao *et al.*, 2005).

The pattern of changes in NPP was similar to that of GPP, because NPP decreased with the conversion of forests to pasture (annual average: -0.82 kg C m^{-2} year^{-1}, 38%) and with conversion of the *caatinga* to degraded *caatinga* (annual average: -0.36 kg C m^{-2} year^{-1}, 70%). With the conversion of *caatinga* to pasture, NPP increased (annual average: 0.51 kg C m^{-2} year^{-1}, 118%), indicating more carbon sequestration by vegetation.

In general, the changes observed in NPP were directly related to the LAI of the involved vegetation classes. It was found that the areas where the LAI decreased (increased) corresponded to those where the NPP decreased (increased). Thus, the increase or decrease in canopy productivity can be a result of the increase or decrease of LAI and photosynthetic activity of the leaves. Moreover, one can infer that the GPP and NPP changes are also associated with canopy transpiration (Table 1) and conditions of low humidity. In conditions of low humidity, the stomata close to prevent excessive water loss through transpiration, implying the reduction of CO_2 assimilation. The regions where the NPP and GPP decreased match those where the canopy transpirations decreased (not shown).

Another variable that can be associated with GPP and NPP modifications is the R_n, because the areas where R_n decreases (increases) correspond to those where GPP and NPP decrease (increase) (Table 1). This pattern was indeed found by Cao and Woodward (1998). For the degraded scenario of *caatinga*, the

changes in rainfall pattern (which affects the plant water availability and the length of the growing season) and the increase in temperature are also factors that may have contributed to the decrease in NPP and GPP in this water-stressed region.

The impacts of LULCC were more intense over SANEB; however, the changes in NPP and GPP were also verified in a portion of the Southern and Northern Brazil. The lowest values of NPP and GPP simulated over Northern Brazil were consistent with the decreasing trend of precipitation and increase of temperature. In contrast, the increased NPP and GPP in the Southern Brazil were mainly due to the increased rainfall (more water availability), which means the vegetation showed more efficiency in using water to obtain carbon.

4. Conclusion

The main objective of this study was to investigate the impacts of LULCC on surface processes considering atmospheric feedbacks. In general, the results showed that LULCC, through alterations in surface processes, caused climate changes in both local and neighbouring regions. Considering that the initial conditions of the experiments are the same, the changes in the water, energy and carbon balance as well as in atmospheric circulation suggest that the land cover change can influence the climate processes on a large scale. In general, most of the LULCC responses were more intense during the wet season and showed opposite results for the experiments of taller vegetation (forest conversion to grassland) and short vegetation (conversion of *caatinga* to grassland and to degraded *caatinga*).

Change in GPP ($\times 10^6$ MolCO$_2$ m^{-2} S^{-1})

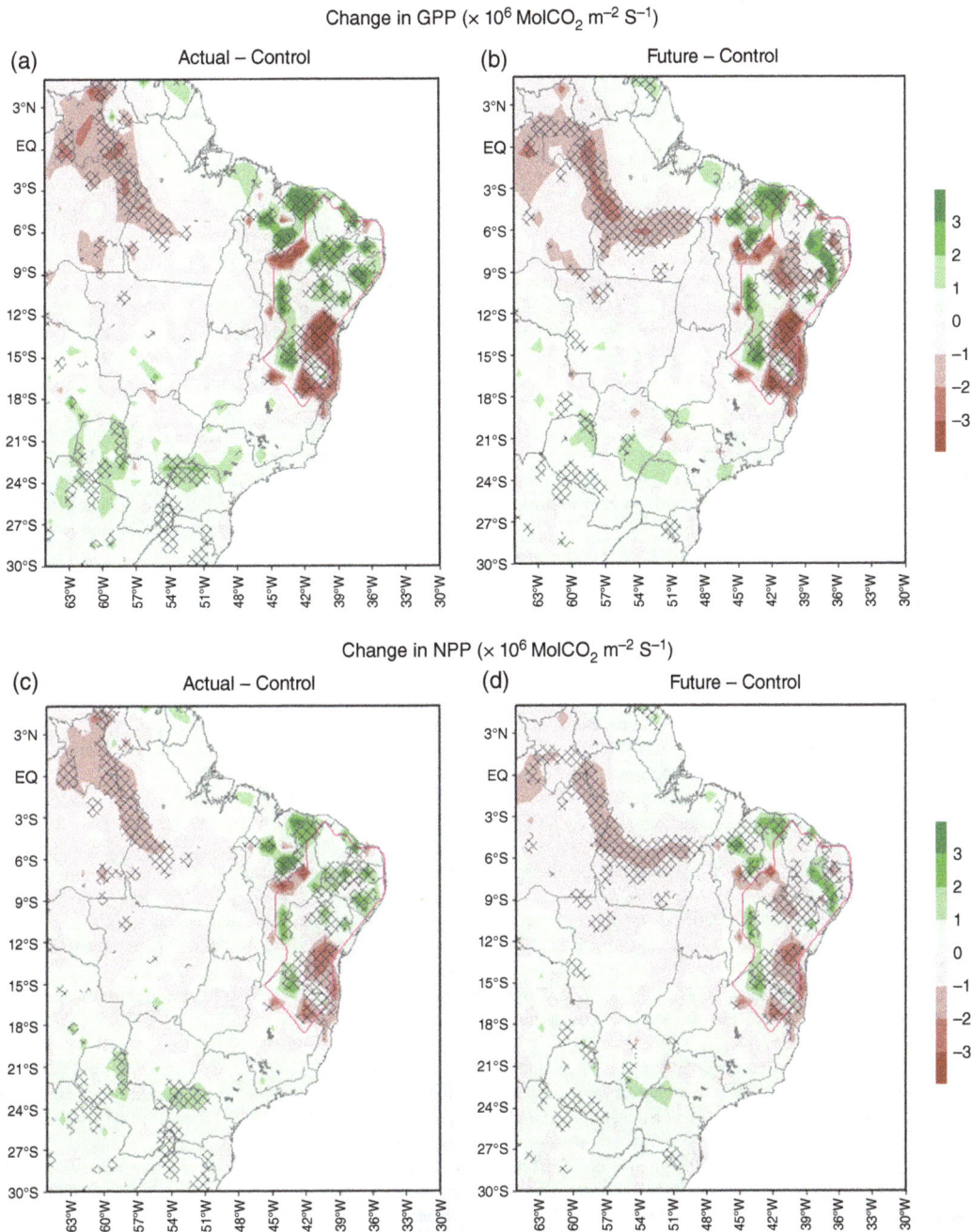

Figure 5. Annual differences in GPP and NPP for: (a) and (c) actual scenario minus control scenario and (b) and (d) future scenario minus control scenario. Cross-hatching shows areas where changes are statistically significant (95% confidence level in Student's *t*-test).

In this study, we assessed the impact of vegetation changes in SANEB. We considered the land use/cover with the more realistic approach of an actual vegetation scenario based on the interpretation of high-resolution satellite images, from 2000. For the future scenario, pasture expansion in the semiarid region was considered to be limited by degradation processes (Marengo, 2008; Salazar Velásquez, 2009). However, LULCC between scenarios were smaller compared with previous studies (Oyama and Nobre, 2004; Souza, 2006). These results suggest that LULCC in Brazil semiarid region, even on a small scale, can cause climate impacts, on the local and regional scale. Finally, we highlight that the

diagnosis of the evolution of LULCC and its climatic implications are essential to guide policy makers with regard to resource applications and policy development to achieve better management and planning for this important region of the country.

References

Bathiany S, Claussen M, Brokvin V, Raddatz T, Gayler V. 2010. Combined biogeophysical and biogeochemical effects of large-scale forest cover changes in the MPI earth system model. *Biogeosciences* **7**: 1383–1399.

Betts RA. 1999. The impact of land use on the climate of present-day. In *Research Activities in Atmospheric and Oceanic Modelling*, Richie

H (ed). World Meteorological Organization: Geneva, Switzerland; 7.11–7.12 CAS/JSC WGNE Report No. 28.

Betts RA. 2000. Offset of the potential carbon sink from boreal forestation by decreases in surface albedo. *Nature* **408**: 187–190

Bonan GB. 1997. Effects of land use on the climate of the United States. *Climate Change* **37**: 449–486.

Bonatti JP. 1996. Modelo de Circulação Geral Atmosférico do CPTEC. *Climanálise Especial*, edição comemorativa de 10 anos.

Brovkin VS, Sitch W, Bloh VM, Claussen EB, Cramer W. 2004. Role of land cover changes for atmospheric CO_2 increase and climate change during the last 150 years. *Global Change Biology* **10**: 1253–1266.

Brovkin V, Claussen M, Driesschaert E, Fichefet T, Kicklighter T, Loutre MF, Matthews HD, Ramankutty N, Schaeffer M, Sokolov A. 2006. Biogeophysical effects of historical land cover changes simulated by six Earth system models of intermediate complexity. *Climate Dynamics* **26**: 587–600.

Chase TN, Pielke RA, Kittel TGF, Nemani R, Running SW. 1996. Sensitivity of a general circulation model to global changes in leaf area index. *Journal of Geophysical Research* **101**: 7393–7408.

Chase TN, Pielke RA, Kittel TGF, Nemani R, Running SW. 2000. Simulated impacts of historical land cover changes on global climate in northern winter. *Climate Dynamics* **16**: 93–105.

Claussen M, Brovkin V, Ganopolski A. 2001. Biogeophysical versus biogeochemical feedbacks of large-scale land cover. *Geophysical Research Letters* **28**(6): 1011–1014.

Cotton WR, Pielke RA. 2007. *Human Impacts on Weather and Climate*, 2nd ed. Cambridge University Press: New York, NY.

Defries RS, Field CB, Fung J, Collatz GJ, Bounoua L. 1999. Combining satellite data and biogeochemical to estimate global effects of human-induced land cover change on carbon emissions and primary productivity. *Global Biogeochemical Cycles* **13**: 803–815.

Deo RC, Syktus JI, McAlpine CA, Lawrence PJ, McGowan HA, Phinn SR. 2009. Impact of historical land cover change on daily indices of climate extremes including droughts in eastern Australia. *Geophysical Research Letters* **36**: L08705.

Diffenbaugh NS. 2009. Influence of modern land cover on the climate of the United States. *Climate Dynamics* **33**(7-8): 945–958.

Findell KL, Elena S, Stouffer RJ. 2007. Modeled impact of anthropogenic land cover change on climate. *Journal of Climate* **20**(14): 3621–3634.

Findell KL, Pitman AJ, England MH, Pegion PJ. 2009. Regional and global impacts of land cover change and sea surface temperature anomalies. *Journal of Climate* **22**(12): 3621–3634.

Foley JA, Prentice IC, Ramankutty N, Levis S, Pollard D, Sitch S, Haxeltine A. 1996. An integrated biosphere model of land surface processes, terrestrial carbon balance and vegetation dynamics. *Global Biogeochemical Cycles* **10**(4): 603–628.

Gao XJ, Luo Y, Lin WT, Zhao ZC, Giorgi F. 2003. Simulation of effects of land use change on climate in China by a regional climate model. *Advances in Atmospheric Sciences* **20**(4): 583–592.

Hoffmann WA, Jackson RB. 2000. Vegetation-climate feedbacks in the conversion of tropical savanna to grassland. *Journal of Climate* **13**: 1593–1602.

Houghton RA. 1991. Tropical deforestation and atmospheric carbon cycle. *Climate Change* **19**: 99–118.

Houghton RA, House JI, Pongratz J, Werf GR, DeFries RS, Hansen MC, Quere LE, Ramankutty N. 2012. Carbon emissions from land use and land-cover change. *Biogeosciences* **9**(12): 5125 - 5142.

Kubota PY. 2012. Variabilidade da energia armazenada na superfície e o seu impacto na definição do padrão de precipitação na América do Sul (sid.inpe.br/mtc-m19/2012/08.02.02.42-TDI). Tese (Doutorado em Meteorologia), INPE, São José dos Campos, Brazil, 309 pp.

Lawrence PJ, Feddema JJ, Bonan GB, Meehl GA, O'Neill BC, Oleson KW, Levis S, David ML, Kluzek E, Lindsay K, Thornton PE. 2012. Simulating the biogeochemical and biogeophysical impacts of transient land cover change and wood harvest in the Community

Climate System Model (CCSM4) from 1850 to 2100. *Journal of Climate* **25**: 3071–3095.

Liu MH, Tian G, Chen W, Ren C, Liu J. 2008. Effects of land use and land cover change on evapotranspiration and water yield in China during 1900–2000. *Journal of the American Water Resources Association* **44**(5): 1193-1207.

Mahmood R, Pielke RA, Hubbard KG, Niyogi D, Dirmeyer P, McAlpine C, Carleton A, Hale R, Gameda S, Beltran-Przekurat A, Baker B, McNider R, Legates DR, Shepherd M, Du J, Blanken P, Frauenfeld OW, Nair US, Fall S. 2013. Land cover changes and their biogeophysical effects on climate. *International Journal of Climatology* **34**: 929–953.

Marengo J. 2008. Vulnerabilidade, impactos e adaptação à mudança do clima no semi-árido do Brasil. Parcerias Estratégicas, Brasília, DF.

Matthews HD, Weaver AJ, Eby MK, Meissner KJ. 2003. Radiative forcing of climate by historical land cover change. *Geophysical Research Letters* **30**: 1055.

Otieno VO, Anyah RO. 2012. Observed and simulated influence of land use changes on the Greater Horn of Africa climate. Case study over Kenya. *Climate Research* **52**: 77–95.

Oyama MD, Nobre CA. 2004. Climatic consequences of a large-scale desertification in Northeast Brazil: a GCM simulation study. *Journal of Climate* **17**(16): 3203–3213.

Paeth H, Born K, Girmes R, Podzun R, Jacob D. 2009. Regional Climate change in tropical and northern Africa due to greenhouse forcing and land use changes. *Journal of Climate* **22**(1): 114.

Pielke RA, Marland G, Betts RA, Chase TN, Eastman JL, Niles JO, Niyogi D, Running S. 2002. The influence of land-use change and landscape dynamics on the climate system – relevance to climate change policy beyond the radiative effect of greenhouse gases. *Philosophical Transactions of the Royal Society of London A* **360**: 1705–1719.

Pielke RA, Pitman A, Niyogi D, Mahmood R, McAlpine C, Hossain F, Klein K, Nair U, Betts R, Fall S, Reichstein M, Kabat P, Noblet N. 2011. Land use/land cover changes and climate: modeling analysis and observational evidence. *WIREs Climate Change* **2**: 828–850.

Pongratz J, Bounoua L, Defries RS, Morton DC, Anderson LO, Mauser W, Klink CA. 2006. The impact of land cover change on surface energy and water balance in Mato Grosso, Brazil. *Earth Interactions* **10**: 1–17.

Ramankutty N, Foley JA. 1999. Estimating historical changes in global land cover: croplands from 1700 to 1992. *Global Biogeochemical Cycles* **13**(4): 997–1027.

Salazar Velásquez LF. 2009. Consequências das mudanças climáticas na distribuição dos biomas na América do Sul, com ênfase na Amazônia e nordeste (INPE-16573-TDI/1566). Tese (Doutorado em Meteorologia), São José dos Campos, SP, Brazil, 277 pp.

Souza SS. 2006. Impactos climáticos regionais da mudança de vegetação no semi-árido do nordeste brasileiro (INPE-14432-TDI/1131). Tese (Doutorado em Meteorologia), São José dos Campos, SP, Brazil, 209 pp.

Souza DC, Oyama MD. 2011. Climatic consequences of gradual desertification in the semi-arid area of Northeast Brazil. *Theoretical and Applied Climatology* **103**: 345–357.

Taylor CM, Lambin EF, Stephenne N, Harding RJ, Essery RLH. 2002. The influence of land use change on climate in the Sahel. *Journal of Climate* **15**: 3615–3629.

Twine TE. 2004. Effects of land cover change on the energy and water balance of the Mississipi River Basin. *Journal of Hydrometeorology* **5**: 640–655.

Vieira RMSP, Cunha APMA, Alvalá RCS, Carvalho VC, Ferraz Neto S, Sestini MF. 2013. Land use and land cover map of a semiarid region of Brazil for meteorological and climatic models. *Revista Brasileira de Meteorologia* **28**(2): 129–138.

Zhao M, Heinsch FA, Nemani R, Running SW. 2005. Improvements of the MODIS terrestrial gross and net primary production global data set. *Remote Sensing of Environment* **95**:164–176.

5

Surface air temperature variability in global climate models

Richard Davy* and Igor Esau

Nansen Environmental and Remote Sensing Centre, Centre for Climate Dynamics (SKD), 5006, Bergen, Norway

*Correspondence to:
Richard Davy, Nansen
Environmental and Remote
Sensing Centre, Centre for
Climate Dynamics (SKD),
Thormøhlensgt. 47, 5006
Bergen, Norway.
E-mail: richard.davy@nersc.no

Abstract

New results from the Coupled Model Intercomparison Project phase 5 (CMIP5) and multiple global reanalysis datasets are used to investigate the relationship between the mean and standard deviation (SD) in the surface air temperature (SAT) at intra- and inter-annual timescales. A combination of a land–sea mask and orographic filter was used to investigate the geographic region with the strongest correlation and in all cases this was found to be for low-lying over-land locations. This result is consistent with the expectation that differences in the effective heat capacity of the atmosphere are an important factor in determining the SAT response to forcing.

Keywords: climate; climate change; planetary boundary layer; temperature; CMIP5

1. Introduction

The surface air temperature (SAT) and its various properties (diurnal and seasonal range and extremes, trends and variability) are some of the most widely used metrics of global climate and climate change (IPCC AR4, 2007). This is to be expected given that the SAT is such a readily measurable, and anthropically important, variable. The variability of the SAT at a given timescale is determined by three factors: the magnitude of the forcing, any feedback processes and the effective heat capacity of the system (Hasselmann, 1976; Frankignoul and Hasselmann, 1977; Hansen *et al.*, 2011). Through consideration of the near-surface energy balance it has been demonstrated that there is an inverse dependency of the magnitude of changes in SAT on the effective heat capacity of the atmosphere (Esau *et al.*, 2012, hereafter EDO12) and soil (Seneviratne *et al.*, 2010). The effective heat capacity of the atmosphere has been shown to be directly proportional to the depth of the planetary boundary layer (PBL) (Esau and Zilitinkevich, 2010) whereas the soil heat capacity depends on the soil moisture available (deVries, 1963). In both cases an increased heat capacity acts to dampen near-surface temperature trends. It may be expected that an increase in forcing will affect the PBL depth as well as temperature. For example, on the diurnal timescale much of the heating in the morning transition goes into increasing the PBL depth from the relatively shallow nocturnal boundary layer to a much deeper convective boundary layer. However, in our analysis such changes are already accounted for in the base-line climatology; and it has been shown that increases in the boundary-layer depth due to enhanced forcing is a relatively small effect on climatological timescales (Figure S1, Supporting Information).

This mechanism – PBL response – suggests that some relations between metrics of the SAT are innate to the thermodynamics of the climate system. For example, the strong positive correlation between SAT trends and variability that exists across the globe, as has been shown in observations, reanalysis and CMIP3 datasets (EDO12). This may be expected given that SAT trends and variability are both manifestations of the magnitude of SAT response to forcing. EDO12 noted that such features of the PBL response, e.g. the linear relationship between mean and standard deviation (SD) of SAT, were not to be found in high-altitude locations, such as Greenland. At such locations the surface air is exposed to the free atmosphere, and as such one would not expect to see the same PBL response. Here, we make use of this characteristic of PBL-response theory to provide a test of this theory in comparison with alternate explanations of this relationship between SAT mean and SD.

The strongest signal of PBL response is expected to be seen in locations with shallow boundary layers, and thus a low effective heat capacity. The shallowest boundary layers are found over land in locations with consistent negative surface sensible heat flux, which leads to a strong stable thermal stratification. Therefore, we expect to see the strongest signal of PBL response at over-land locations. As this mechanism is driven by the turbulent mixing that characterizes the PBL, we expect there to be a weaker signal in high-altitude locations where the PBL is exposed to the free atmosphere. This prediction opens the possibility to distinguish the effects of the PBL response from the response of soil moisture (Quesada *et al.*, 2012) as they are not expected to change with altitude. There has been some discussion of the vertical dependency of the temperature variability and the causes for this; for example Wang *et al.* (2005) demonstrated that the North Atlantic Oscillation signal was a largely tropospheric phenomenon, whereas the Arctic Oscillation signal could penetrate into the stratosphere. This relates to discussions on the vertical profile of recent global warming and as to what role local processes play in determining the profile of atmospheric warming. There has been some indication from the work of Weber *et al.* (1994) that one such pattern predicted from PBL-response theory – the asymmetric response of diurnal minimum and maximum temperatures – is constrained to the PBL, but the reason for this asymmetry was not identified in their work. We further investigate the cause for such asymmetries in temperature trends by extending the analysis of SAT response in mountainous and low-lying locations from station data, as in Weber *et al.* (1994), to a global analysis using reanalyses and climate models. Such asymmetries may appear in observations, but not in global climate models (GCMs) where the vertical exchange is not explicitly resolved and is instead represented through parameterization schemes. Such schemes have fundamental problems in representing turbulence as they are based on eddy diffusivity schemes (K-schemes) which means the turbulence is modelled as a diffusion operator, not as an advection operator. The prediction that these relationships should be weaker over oceans and in locations exposed to the free atmosphere allows us to falsify the PBL-response mechanism by comparing the strength of such relationships with and without the use of land and orographic filters.

One of the biggest challenges in assessing the role of the PBL in Earth's climate system is the limitations on assessing the climatology of the PBL (Seidel *et al.*, 2010). Seidel *et al.* (2012) compared the climatology of the PBL over the continental United States and Europe between climate models, reanalysis and radiosonde observations. They found diurnal, seasonal and geographical structures of the PBL climatology in models and observations. However, there were significant biases in the model PBL depth for shallow PBLs as the models have trouble simulating conditions of stable stratification and over estimate the PBL depth. Thus, the diurnal and seasonal cycles are only reproduced by models because the PBL is so much deeper during the day than at night and during the summer than in winter. There are still large uncertainties in the depth and variability of shallow PBLs in models, and they have trouble capturing the morning and evening transition periods (Denning *et al.*, 2008). PBL depth data were available in the ERA-Interim reanalysis, and we have used this for verification of the proposed relationship between the depth of the mixed layer and SAT trends and variability. However, it should be noted that the PBL data are generally not available for GCM results, therefore it is necessary to use a proxy for the PBL mixing, such as the mean SAT (EDO12). It has been established that PBL response predicts an inverse relationship between the mean and SD of

SAT (EDO12). We have used a simple least squares linear regression analysis to investigate this relationship on intra- and inter-annual timescales, using data from Coupled Model Intercomparison Project phase 5 (CMIP5) results and various reanalysis datasets.

2. Data sets

The data used for this study were taken from the CMIP5 results and various global reanalysis projects, the data for which is hosted by the National Center for Atmospheric Research (NCAR). The temperature data were taken from the last 30 years of the 'historical' runs of the CMIP5, and from the same years in the reanalysis datasets with the exception of the Japanese 25-year reanalysis project (JRA25).

The historical simulations of the CMIP5 are run using changing conditions consistent with observations including: atmospheric composition, including anthropogenic and volcanic influences; solar forcing; land use changes and emissions or concentrations of short-lived species and natural and anthropogenic aerosols (Taylor *et al.*, 2009). These simulations are performed by coupled ocean–atmosphere GCMs. We have used results from eight of the groups that contributed to CMIP5, which offered a reasonable cross-section of GCM results: CSIRO Australia's CSIRO Mk3.5 (Gordon *et al.*, 2010); the Japan agency for Marine-Earth Science and Technology and National Institute for Environmental Studies MIROC-ESM-CHEM (Watanabe *et al.*, 2011) and MIROC5 (Watanabe *et al.*, 2010); the French National Centre for Meteorological Research's CNRM-CM5 (Voldoire *et al.*, 2010); the UK Met Office Hadley Centre's HadCM3 and HadGEM2-AO (Collins *et al.*, 2011); the Institute for Numerical Mathematics' INMCM4 (Volodin, *et al.*, 2010) and the Norwegian Climate Centre's NorESM (Iversen *et al.*, 2012).

The reanalysis datasets used were: the JRA25 (Onogi *et al.*, 2007); the National Centers for Environmental Prediction (NCEP) Climate Forecast System Reanalysis (CFSR) (Saha *et al.*, 2010); the European Center for Medium-range Weather Forecast (ECMWF) Interim reanalysis (ERA-Interim) (Dee *et al.*, 2011); the National Oceanic and Atmospheric Administration (NOAA) 20th Century Reanalysis v2 (Compo *et al.*, 2011) and the NCEP/NCAR reanalysis I (Kalnay *et al.*, 1996).

3. Methods

Each of the GCMs and reanalysis has associated orographic data and a land/sea mask. An orographic filter, 'Orog', was created that removed all data corresponding to locations where the surface elevation is greater than 1 km. This height threshold was chosen as a characteristic height for mountainous terrain, and the results are insensitive to a 20% variation in this threshold. Locations with surface elevation greater than 1 km ('highland') are assumed to be exposed to the free atmosphere. It should be noted that this filtering technique will also remove high-altitude, but relatively flat, areas which would not be expected to have the same exposure to the free atmosphere. However, since there was not a ready method to differentiate between mountainous and high-altitude, plain locations, we chose to use a simplified filtering technique which would ensure the removal of mountainous locations. The land–sea mask of each dataset was used as a 'land' filter, to select only those grid points which lay over land. The combination of the orographic and land filters was used to select only low-lying, over-land ('lowland') locations, where we anticipate the strongest signal of the PBL response.

Monthly means of the diurnal-mean SAT were obtained for each dataset, covering the last 30 years of the CMIP5 'historical' simulations (1976–2005). For each dataset, a time series of temperature anomalies was calculated by removing the 30-year monthly means from the monthly temperature time series. So for each month the climatological mean was removed to obtain the anomaly:

$$T'_{\text{month}} = T_{\text{month}} - \frac{1}{n} \sum_{n \text{ years}} T_{\text{month}}$$

The temperature statistics (SD) were calculated from the monthly anomalies after the removal of a fourth order polynomial trend from the time series (Braganza et al., 2003).

The inter-annual correlations were calculated by taking the climatological (30 years) mean and the SD of the SAT at each location, and correlating these across all locations, using an area-weighted correlation. This is given by:

$$C_{\text{mean}}^{\text{SD}} = \frac{\text{cov}\,(T_{\text{mean}}, \sigma_T; A)}{\sqrt{\text{cov}\,(T_{\text{mean}}, T_{\text{mean}}; A)\,\text{cov}\,(\sigma_T, \sigma_T; A)}}$$

where T_{mean} is the climatological mean temperature (K), σ_T is the SD (K), A is the surface area of each grid box (m^2) and $\text{cov}(x, y; w)$ is the covariance of x and y with weighting function, w, and is given by:

$$\text{cov}(x, y; w) = \frac{\sum_i w_i \left(x_i - \frac{\sum_j x_j w_j}{\sum_j w_j}\right)\left(y_i - \frac{\sum_j y_j w_j}{\sum_j w_j}\right)}{\sum_j w_j}$$

The intra-annual correlations were calculated at each location by correlating the mean and SD of each month. These correlations were filtered for significance ($p < 0.05$) prior to using an area-weighted mean to obtain a single, global-mean correlation.

For the ERA-Interim reanalysis dataset, the climatological mean (average over indicated years) of the reciprocal of the 3-hourly boundary-layer depth

was calculated. This averaging of the reciprocal boundary-layer depth reflects the observation that the temperature response to a given forcing is reciprocally dependant on the boundary-layer depth on all timescales. This averaging puts a strong emphasis on locations and periods with shallow boundary layers which are generally poorly defined by model parameterization schemes. The PBL depth in the ERA-Interim reanalysis is defined from a threshold Flux-Richardson number, integrated between model levels. Analysis of different methods for determining the PBL depth have shown that definitions based on the Richardson number are the most reliable (Seibert et al., 2000; Zilitinkevich et al., 2007). The Global Energy and Water EXchanges project's (GEWEX) Atmospheric Boundary Layer Study (GABLS) projects showed that model schemes give an uncertainty in the PBL depth between 50 and 100% throughout the diurnal cycle (Svensson et al., 2011) and that for a given uncertainty in the PBL depth, averaging over reciprocal boundary-layer depth gives a lower absolute uncertainty in the mean PBL depth.

4. Results

It is the shallower boundary layers, found over land and at high latitudes (Figure 1(a)), where we expect to see the strongest temperature response. We also see a strong seasonal variation in the depth of the PBL, with the shallower boundary layers found over land during the winter months (Figure 1(b)). Note that the averaging of the reciprocal of boundary-layer depth gives heavy weight to the periods with shallow boundary layers. As such, the characteristic boundary-layer depth over land is much shallower (100 m) than that found over ocean (1 km), as it is only over land where you find the very shallow boundary layers associated with strongly stably stratified conditions. In the winter boundary layers over land can be of the order 10 m, and it is these very shallow boundary layers which dominate the climatological mean of reciprocal boundary-layer depth. For the ERA-Interim reanalysis, we find a strong correlation between the reciprocal of the PBL depth and both the SAT trends ($r = 0.84$, $p < 0.05$) and SD ($r = 0.97$, $p < 0.05$) intra-annually (Figure 1(a)). We also find strong relationships inter-annually by comparing the climatology of the reciprocal PBL depth and SAT variability at different locations around the globe (Figure 1(b)).

Figure 2(a) shows the correlation of the mean and inter-annual SD of the SAT from various GCMs. A negative correlation means that the locations with the colder climatological mean temperature have greater variability than the warmer locations. Application of the orographic filter always improves the correlation and the strongest correlations are found for lowland locations, with the exception of the results from NorESM where the over-land correlation is especially

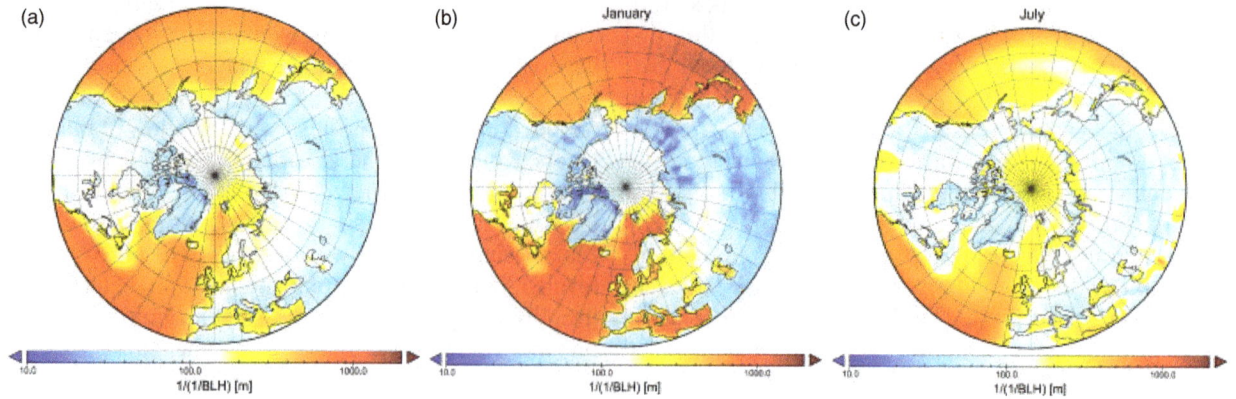

Figure 1. The reciprocal of the climatological mean of the reciprocal boundary-layer depth for (a) annual and (b) January and July data for the Northern Hemisphere, from ERA-Interim.

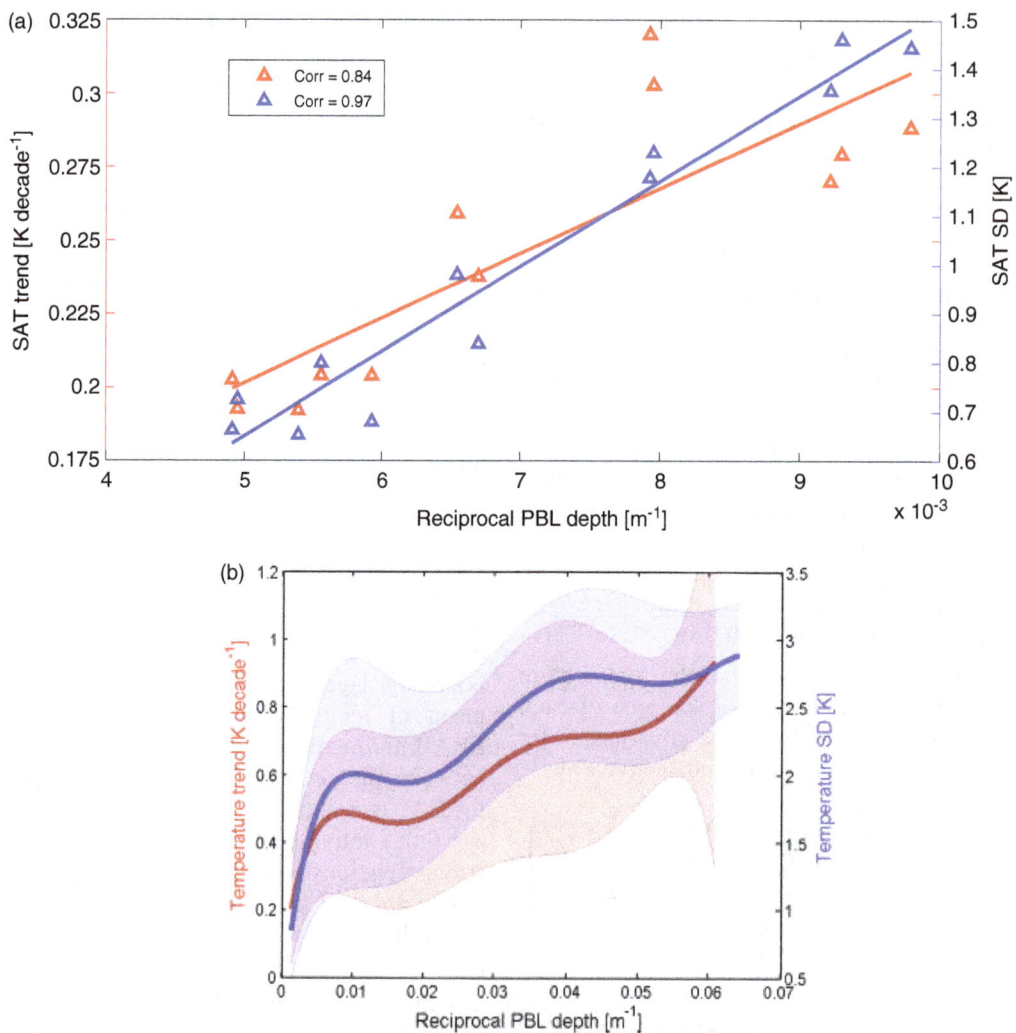

Figure 2. (a) The area-weighted SD and linear trend in SAT as a function of the monthly mean of the reciprocal boundary-layer depth for the Northern Hemisphere, from ERA-Interim reanalysis. (b) The inter-annual standard deviation and trend in SAT as a function of the mean of the reciprocal boundary-layer depth from ERA-Interim. The thick line represents the mean of the binned value, and the shaded area is one standard deviation within each bin. Inter-annually there is an overall correlation of $r = 0.56$, $p < 0.05$ between the reciprocal PBL depth and SAT trends, and a correlation of $r = 0.60$, $p < 0.05$ between reciprocal PBL depth and SD in SAT.

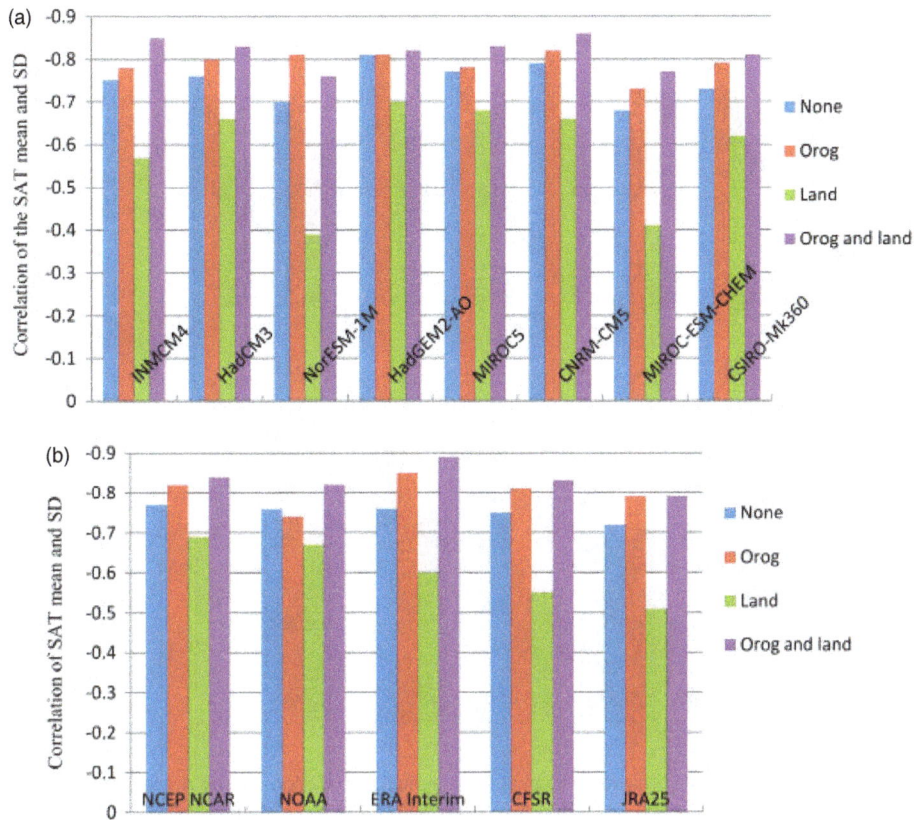

Figure 3. The inter-annual correlation of the mean and SD in SAT from (a) CMIP5 models and (b) reanalysis calculated after applying a land–sea mask (Land) and an orographic filter (Orog).

poor. We find the same result in the reanalysis datasets where the strongest correlations are always found for lowland locations (Figure 2(b)).

In all cases, the biggest difference in the correlation comes with the addition of the orographic filter to the land-filtered data. The significance of these differences in the correlations can be seen from the difference in the number of data points used in each case. The highland regions represent 13% of the total number of grid points so the addition of the orographic filter to the global data does not significantly improve the correlation. However, the highland locations represent 41% of the over-land grid points; hence the largest improvement in the correlations is seen in the application of the orographic filter to the over-land data. This is also why we see a drop in the correlation after the application of the land filter to the global data: the fraction of grid points which correspond to highland locations is increased.

We find a similar pattern in the area-weighted average intra-annual correlations (Figure 3). A positive correlation indicates that the warmer months have the greater variability and a negative correlation indicates that the colder months have the higher variability – the result anticipated from PBL-response theory. Therefore in the regions where the PBL response plays a more dominant role in determining SAT variability we would expect a stronger negative correlation. Such strong negative correlations are found for overland

locations and, with the exception of the INMCM4 and NOAA 20th Century Reanalysis results, the strongest correlations are found for lowland locations. This pattern of intra-annual correlations can be readily understood from the associated geographical distribution.

There is a very similar geographical pattern to the intra-annual correlations in all CMIP5 models and reanalysis with strong negative correlations across the middle- and high-latitude continental regions and over the Arctic Ocean (Figure 4). There are two bands of strong positive correlation over the Atlantic and Pacific oceans. Hence the biggest difference in the geographical mean correlations (Figure 3) comes from the application of the land filter. These bands of positive correlation – suggesting the warmer months are more variable than colder months – lie over the storm tracks and reflect the high inter-annual variations in the summer air temperatures over these regions. This is to be expected as SATs over open ocean are governed by the sea-surface temperature (SST). In winter there is a relatively deep ocean mixed layer and relatively small inter-annual variations in the radiative forcing, and so the SST is relatively consistent from year to year. However, in the summer time the water is warmer, the mixed layer is shallower (and so the effective heat capacity is lower), and so the strong variations in the radiative forcing from inter-annual changes in the cloud cover can strongly affect the SST, and thus the SAT (Figure 5).

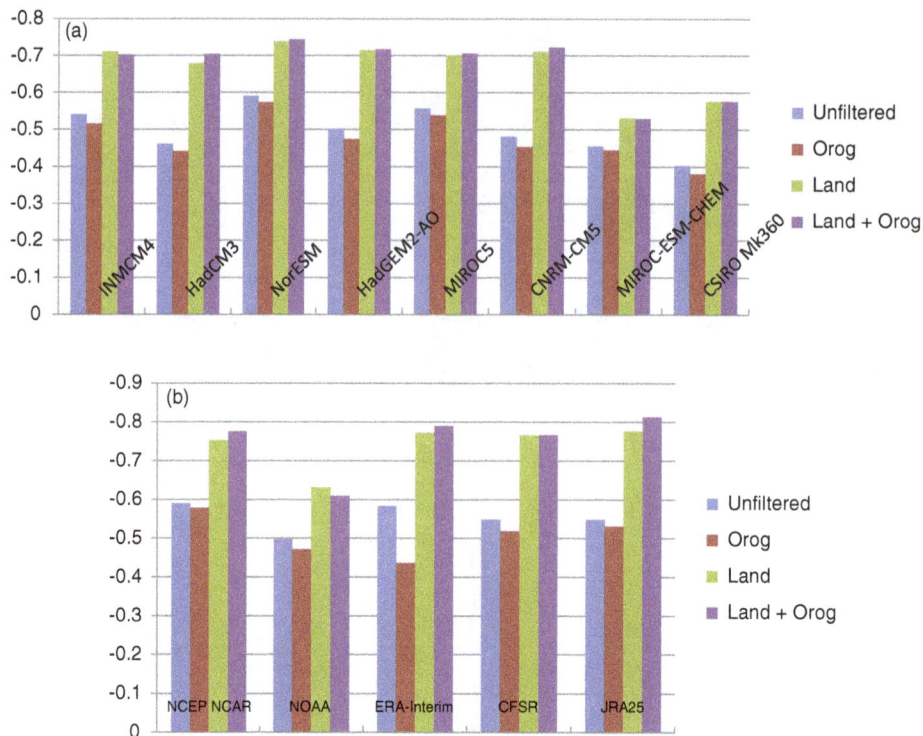

Figure 4. The Northern Hemisphere area-weighted mean of intra-annual correlations of the mean and SD in SAT from (a) CMIP5 models and (b) reanalysis calculated after applying a land–sea mask (Land) and an orographic filter (Orog).

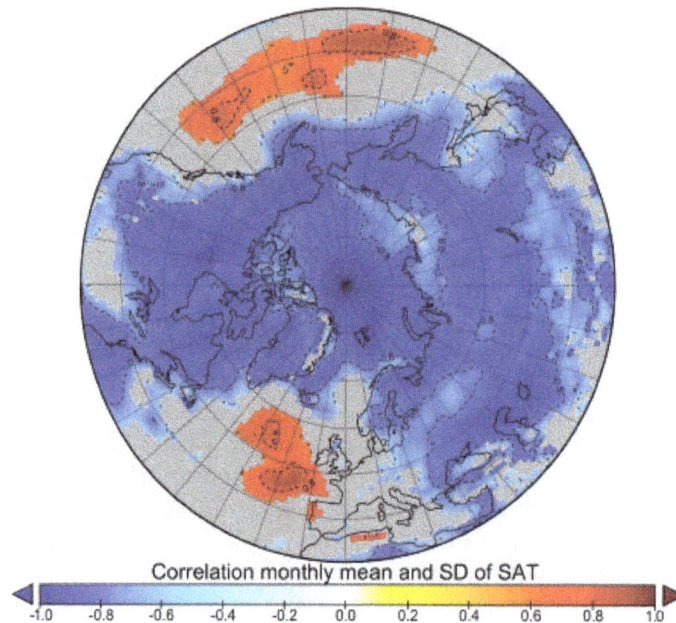

Figure 5. The monthly intra-annual correlation of mean and SD of SAT from ERA-Interim.

5. Conclusions

In this study, we used the results from the latest phase of the CMIP to identify one of the climatological features of SAT variability: the strong negative correlation between SAT mean and variability. This result was confirmed in multiple reanalysis datasets on inter-annual and intra-annual timescales. This relation is consistent with the PBL-response hypothesis that the differences in the SAT variability are, in part, due to variations in the effective heat capacity of the atmosphere such that, the lower the effective heat capacity, the greater the SAT response to a given forcing (EDO12). This was confirmed using the PBL depth data available in the ERA-Interim reanalysis dataset, where we found a strong link between the climatology of the PBL and the SAT response to forcing on intra- and inter-annual timescales. Indeed, these results are difficult to explain on the basis of soil moisture and cloud cover changes alone. When we include the highland locations where the PBL is exposed to the free atmosphere – and as such is not expected to produce the same response – we find a weaker correlation between the SAT mean and SD. This is in agreement with our expectation that this relationship is a consequence of the PBL response.

Acknowledgements

This work has been funded by the Norwegian Research Council FRINAT project *PBL-feedback* 191516/V30.

References

Braganza K, Karoly DJ, Hirst AC, Mann ME, Stott P, Stouffer RJ, Tett SFB. 2003. Simple indices of global climate variability and change: Part I, variability and correlation structure. *Climate Dynamics* **20**: 491–502.

Collins WJ, Bellouin N, Doutriaux-Boucher M, Gedney N, Halloran P, Hinton T, Hughes J, Jones CD, Joshi M, Liddicoat S, Martin G, O'Connor F, Rae J, Senior C, Sitch S, Totterdell I, Wiltshire A, Woodward S. 2011. Development and evaluation of an Earth-system model – HadGEM2. *Geoscientific Model Development* **4**: 1051–1075, DOI: 10.5194/gmd-4-1051-2011.

Compo GP, Whitaker JS, Sardeshmukh PD, Matsui N, Allan RJ, Yin X, Gleason BE, Vose RS, Rutledge G, Bessemoulin P, Bronnimann S, Brunet M, Crouthamel RI, Grant AN, Groisman PY, Jones PD, Kruk MC, Kruger AC, Marshall GJ, Maugeri M, Mok HY, Nordli Ø, Ross TF, Trigo RM, Wang XL, Woodruff SD, Worley SJ. 2011. The twentieth century reanalysis project. *Quarterly Journal of the Royal Meteorological Society* **137**: 1–28, DOI: 10.1002/qj.766.

Dee DP, Uppala SM, Simmons AJ, Berrisford P, Poli P, Kobayashi S, Andrae U, Balmaseda MA, Balsamo G, Bauer P, Bechtold P, Beljaars ACM, van de Berg L, Bidlot J, Bormann N, Delsol C, Dragani R, Fuentes M, Geer AJ, Haimberger L, Healy SB, Hersbach H, Holm EV, Isaksen L, Kållberg P, Kohler M, Matricardi M, McNally AP, Monge-Sanz BM, Morcrette J-J, Park B-K, Peubey C, de Rosnay P, Tavolato C, Thepaut J-N, Vitart F. 2011. The ERA-Interim reanalysis: configuration and performance of the data assimilation system. *Quarterly Journal of the Royal Meteorological Society* **137**: 553–597, DOI: 10.1002/qj.828.

Denning AS, Zhang N, Yi C, Branson M, Davis K, Kleist J, Bakwin P. 2008. Evaluation of modelled atmospheric boundary layer depth at the WLEF tower. *Agricultural and Forest Meteorology* **148**: 206–215.

DeVries DA. 1963. Thermal properties of soils. In *Physics of Plant Environment*, van Wijk WR (ed). North-Holland Publishing Company: Amsterdam.

Esau I, Zilitinkevich S. 2010. On the role of the planetary boundary layer depth in the climate system. *Advances in Science and Research* **4**: 63–69, DOI: 10.5194/asr-4-63-2010.

Esau I, Davy R, Outten S. 2012. Complementary explanation of temperature response in the lower atmosphere. *Environmental Research Letters* **7**: 044026, DOI: 10.1088/1748-9326/7/4/044026.

Frankignoul C, Hasselmann K. 1977. Stochastic climate models, part II application to sea-surface temperature anomalies and thermocline variability. *Tellus* **29**: 4.

Gordon HB, O'Farrell S, Collier M, Dix M, Rotstayn L, Kowalczyk E, Hirst T, Watterson I. 2010. The CSIRO Mk3.5 Climate Model, technical report No. 21. The Centre for Australian Weather and Climate Research, Aspendale, Victoria, Australia.

Hansen J, Sato M, Kharecha P, von Schuckmann K. 2011. Earth's energy imbalance and implications. *Atmospheric Chemistry and Physics* **11**: 13421–13449, DOI: 10.5194/acp-11-13421-2011.

Hasselmann K. 1976. Stochastic climate models. *Tellus* **28**: 6.

Iversen T, Bentsen M, Bethke I, Debernard JB, Kirkevåg A, Seland Ø, Drange H, Kristjansson JE, Medhaug I, Sand M, Seierstad IA. 2012. The Norwegian Earth System Model, NorESM1-M – Part 2: climate response and scenario projections. *Geoscientific Model Development Discussions* **5**: 2933–2998.

Kalnay E, Kanamitsu M, Kistler R, Collins W, Deaven D, Gandin L, Iredell M, Saha S, White G, Woollen J, Zhu Y, Chelliah M, Ebisuzaki W, Higgins W, Janowiak J, Mo KC, Ropelewski C, Wang J, Leetmaa A, Reynolds R, Jenne R, Joseph D. 1996. The NCEP/NCAR 40-year reanalysis project. *Bulletin of the American Meteorological Society* **77**: 437–470.

Onogi K, Tsutsui J, Koide H, Sakamoto M, Kobayashi S, Hatsushika H, Matsumoto T, Yamazaki N, Kamahori H, Takahashi K, Kadokura S, Wada K, Kato K, Oyama R, Ose T, Mannoji N, Taira R. 2007. The JRA-25 reanalysis. *Journal of the Meteorological Society of Japan* **85**(3): 369–432.

Quesada B, Vautard R, Yiou P, Hirschi M, Seneviratne SI. 2012. Asymmetric European summer heat predictability from wet and dry southern winters and springs. *Nature Climate Change* **2**: 736–741, DOI: 10.1038/nclimate1536.

Saha S, Moorthi S, Pan H-L, Wu X, Wang J, Nadiga S, Tripp P, Kistler R, Woollen J, Behringer D, Liu H, Stokes D, Grumbine R, Gayno G, Wang J, Hou Y-T, Chuang H-Y, Juang H-MH, Sela J, Iredell M, Treadon R, Kleist D, Van Delst P, Keyser D, Derber J, Ek M, Meng J, Wei H, Yang R, Lord S, Van Den Dool H, Kumar A, Wang W, Long C, Chelliah M, Xue Y, Huang B, Schemm J-K, Ebisuzaki W, Lin R, Xie P, Chen M, Zhou S, Higgins W, Zou C-Z, Liu Q, Chen Y, Han Y, Cucurull L, Reynolds RW, Rutledge G, Goldberg M. 2010. The NCEP climate forecast system reanalysis. *Bulletin of the American Meteorological Society* **91**: 1015–1057, DOI: 10.1175/2010BAMS3001.1.

Seibert P, Beyrich F, Gryning S-E, Joffre S, Rasmussen A, Tercier P. 2000. Review and intercomparison of operational methods for the determination of the mixing height. *Atmospheric Environment* **34**(7): 1001–1027.

Seidel DJ, Ao CO, Li K. 2010. Estimating climatological planetary boundary layer heights from radiosonde observations: comparison of methods and uncertainty analysis. *Journal of Geophysical Research* **115**: D16113, DOI: 10.1029/2009JD013680.

Seidel DJ, Zhang Y, Beljaars ACM, Golaz J-C, Jacobson AR, Medeiros B. 2012. Climatology of the planetary boundary layer over the continental United States and Europe. *Journal of Geophysical Research*, DOI: 10.1029/2012JD018143.

Seneviratne SI, Corti T, Davin EL, Hirschi M, Jaeger EB, Lehner I, Orlowsky B, Teuling AJ. 2010. Investigating soil moisture-climate interactions in a changing climate: a review. *Earth Science Reviews* **99**: 125–161, DOI: 10.1016/j.earscirev.2010.02.004.

IPCC. 2007. Climate change 2007. In *The Physical Science Basis. Contribution of Working Group I to the Fourth Assessment Report of the Intergovernmental Panel on Climate Change*, Solomon S, Qin D, Manning M, Chen Z, Marquis M, Averyt KB, Tignor M, Miller HL (eds). Cambridge University Press: Cambridge, United Kingdom/New York, NY, USA.

Svensson G, Holtslag AAM, Kumar V, Mauritsen T, Steeneveld GJ, Angevine WM, Bazile E, Beljaars A, de Bruijn EIF, Cheng A, Conangla L, Cuxart J, Ek M, Falk MJ, Freedman F, Kitagawa H, Larson VE, Lock A, Mailhot J, Masson V, Park S, Pleim J, Soderberg S, Weng W, Zampieri M. 2011. Evaluation of the diurnal cycle in the atmospheric boundary layer over land as represented by a variety of single-column models: the second GABLS experiment. *Boundary-Layer Meteorology* **140**: 177–206.

Taylor KE, Stouffer RJ, Meehl GA. 2009. An overview of CMIP5 and the experiment design. *Bulletin of the American Meteorological Society* **93**: 485–498.

Voldoire A, Sanchez-Gomez E, Salas Y Melia D, Decharme B, Cassou C, Senesi S, Valcke S, Beau I, Alias A, Chevallier M, Deque M, Deshayes J, Douville H, Ferneandez E, Madec G, Maisonnave E, Moine M-P, Planton S, Saint-Martin D, Szopa S, Tyteca S, Alkama R, Belamari S, Braun A, Coquart L, Chauvin F. 2012. The CNRM-CM5.1 global climate model: description and basic evaluation. *Climate Dynamics*, DOI: 10.1007/s00382-011-1259-y.

Volodin EM, Dianskii NA, Gusev AV. 2010. Simulating present-day climate with the INMCM4.0 coupled model of the atmospheric and oceanic general circulations. *Izvestiya Atmospheric and Oceanic physics* **46**(4): 448–466.

Vose RS, Easterling DR, Gleason B. 2005. Maximum and minimum temperature trends for the globe: an update through 2004. *Geophysical Research Letters* **32**: L23822.

Wang D, Wang C, Yang X, Lu J. 2005. Winter Northern Hemisphere surface air temperature variability associated with the Arctic Oscillation and North Atlantic Oscillation. *Geophysical Research Letters* **32**: L16706, DOI: 10.1029/2005GL022952.

Watanabe M, Suzuki T, O'ishi R, Komuro Y, Watanabe S, Emori S, Takemura T, Chikira M, Ogura T, Sekiguchi M, Takata K, Yamazaki D, Yokohata T, Nozawa T, Hasumi H, Tatebe H, Kimoto M. 2010. Improved climate simulation by MIROC5: mean states, variability and climate sensitivity. *Journal of Climate* **23**: 6312–6335, DOI: 10.1175/2010JCLI3679.1.

Watanabe S, Hajima T, Sudo K, Nagashima T, Takemura T, Okajima H, Nozawa T, Kawase H, Abe M, Yokohata T, Ise T, Sato H, Kato E, Takata K, Emori S, Kawamiya M. 2011. MIROC-ESM: model description and basic results of CMIP5-20c3m experiments. *Geoscientific Model Development* **4**: 845–872, DOI: 10.5194/gmd-4-845-2011.

Weber RO, Talkner P, Stefanicki G. 1994. Asymmetric diurnal temperature change in the Alpine region. *Geophysical Research Letters* **21**: 673–676.

Zilitinkevich S, Esau I, Baklanov A. 2007. Further comments on the equilibrium height of neutral and stably stratified atmospheric boundary layers. *Quarterly Journal of the Royal Meteorological Society* **133**: 265–271.

Zilitinkevich SS, Elperin T, Kleeorin N, Rogachevskii I, Esau I, Mauritsen T, Miles MW. 2008. Turbulence energetics in stably stratified geophysical flows: strong and weak mixing regimes. *Quarterly Journal of the Royal Meteorological Society* **134**: 793–799, DOI: 10.1002/qj.264.

6

Long-term changes in Australian tropical cyclone numbers

Andrew J. Dowdy*

The Centre for Australian Weather and Climate Research, Bureau of Meteorology, Docklands, 3007, Australia

*Correspondence to:
A. J. Dowdy, The Centre for Australian Weather and Climate Research, Bureau of Meteorology, 700 Collins Street, Docklands 3007, Australia.
E-mail: a.dowdy@bom.gov.au*

Abstract

Tropical cyclone (TC) observations are used to examine changes in the TC climatology of the Australian region. The ability to investigate long-term changes in TC numbers improves when the El Niño-Southern Oscillation (ENSO) is considered. Removing variability in TC numbers associated with ENSO shows a significant decreasing trend in TC numbers at the 93–98% confidence level. Additionally, there is some indication of a temporal change in the relationship between ENSO and TC numbers, with ENSO accounting for about 35–50% of the variance in TC numbers during the first half of the study period, but only 10% during the second half.

Keywords: hurricane; tropical; cyclone; typhoon; climate; change

1. Introduction

The influence of a changing climate on tropical cyclone (TC) activity is currently a highly active area of research. Many previous studies have examined whether or not long-term changes can be detected in TC activity based on the observed records (e.g. Emanuel, 2005; Webster *et al.*, 2005; Chan, 2006; Kossin *et al.*, 2007; Kuleshov *et al.*, 2010), with Knutson *et al.* (2010) concluding from studies such as these that it is uncertain whether or not past changes in TC activity have exceeded variability due to natural causes. Recently, Holland and Bruyère (2013) reported no anthropogenic signal in annual global TC or hurricane frequencies, but that a strong signal occurs in proportions of both weaker and stronger hurricanes, with weaker systems becoming less frequent and stronger systems becoming more frequent.

Future projections based on theory and modeling tend to indicate fewer but more intense TCs could be expected due to increasing atmospheric greenhouse gas concentrations (e.g. Walsh and Ryan, 2000; Walsh *et al.*, 2004; Sugi and Yoshimura, 2013). However, a recent study of TC activity (Emanuel, 2013) downscaled from the Coupled Model Intercomparison Project 5 (CMIP5) suite of GCMs (Taylor *et al.*, 2012), indicates an increase in both the frequency and intensity of TCs in most locations, with the exception of the southwestern Pacific region.

There are a number of reasons for the considerable difficulty in identifying climatological changes in observed TC datasets such as inhomogeneities in data relating to changes in observation and analysis techniques (Landsea *et al.*, 2006; Landsea and Franklin, 2013). For example, TC observations prior to the satellite-era generally have a higher uncertainty than observations from recent years, effectively limiting the period of high-quality data available for climatological examinations (Dowdy and Kuleshov, 2012). Another reason for the difficultly in identifying significant changes in observed TC activity is the considerable degree of temporal variability in TC occurrence (Chan, 2006).

Numerous studies have documented the relationship between the temporal variability in TC numbers in the region near Australia and the El Niño-Southern Oscillation (ENSO) (e.g. Nicholls, 1979, 1984, 1985; Evans and Allan, 1992; Basher and Zheng, 1995; McBride, 1995; Camargo *et al.*, 2007; Dowdy *et al.*, 2012). This well-established relationship suggests the possibility of an improved ability to examine potential changes in TC numbers by using ENSO to account for some of this variability. This study examines this possibility, using a newly updated dataset of observed TCs in the Australian region from the 1981/1982 to 2012/2013 Southern Hemisphere TC seasons. Additionally, the possibility of a long-term change in the relationship between ENSO and TC numbers is also examined. Results are also discussed in relation to their implications for the seasonal prediction of TCs.

2. Data and methodology

The Southern Hemisphere TC season is typically considered as running from November in one year, to April in the following year, while noting that TCs can also sometimes occur outside of this range. TC seasons are referred to in this study based on the year for January (e.g. the 2012/2013 Southern Hemisphere TC season is referred to as the 2013 TC season, as January corresponds to 2013 in this case). A TC archive for the Southern Hemisphere has been compiled at the National Climate Centre (NCC) of the Australian Bureau of

Meteorology with records of central pressure magnitude and location back to the 1970 TC season (Kuleshov *et al.*, 2008). Dowdy *et al.* (2012) recently updated this dataset to include TC seasons up to 2010 and presented maps of TC activity near Australia and the South Pacific region. This dataset is further updated in this study for the Australian region to include TC numbers (hereafter, 'TC #s') for the three most recent TC seasons (from 2011 to 2013). In an effort to minimize the impact of satellite-induced inhomogeneities, data are not used prior to the 1982 TC season (following Kuleshov *et al.* (2010)). Consequently, the study period used here consists of the 32-year period of data from the 1982 to 2013 TC seasons. This study examines TC #s, given that this measure of TC activity is commonly used for purposes such as seasonal forecasting in the Australian region (by the NCC), while noting that there is a range of other measures of TC activity including those that account for TC duration and intensity (e.g. the accumulated cyclone energy (ACE) index (Bell *et al.*, 2000)).

TC observations are considered here for the Australian region from 90°E to 160°E in the Southern Hemisphere. This longitude range represents the region defined by the World Meteorological Organization as being Australia's area of responsibility for monitoring TC occurrence, as well as being the region used operationally by the NCC to produce seasonal predictions of TC occurrence. A minimum pressure threshold of 995 hPa is applied to avoid very weak systems. Additionally, data are only used equatorward of 25°S following the results of studies such as Sinclair (2002) who indicate that TCs generally become more characteristic of baroclinic midlatitude storms than TCs when they move poleward of this latitude. This study is based on TC occurrence, rather than TC genesis. For example, if a TC has its genesis location outside of the Australian region, but then moves into the Australian region, it is counted as occurring in the Australian region if it meets all of the criteria described above at some point in time while in the Australian region.

The influence of ENSO is examined based on anomalies of the NINO3.4 index (hereafter 'NIN') as an oceanic measure of ENSO, with data obtained from the Climate Prediction Centre (CPC) of the National Oceanic and Atmospheric Administration. The Southern Oscillation Index [SOI, Troup (1965)], calculated by the NCC, is used as an atmospheric measure of ENSO. Three-month averages for July, August and September are used in this study. This 3-month period shows strong correlations with TC occurrence in the subsequent TC season (November to April) in the Australian region (Nicholls 1984) and is used operationally by the NCC for producing seasonal TC occurrence predictions. It is for these reasons that the two indices (NIN and SOI) were selected for use here. It is also noted that there are a variety of different indices that could be used to represent ENSO and the influence of other types of El Niño such as Modoki (Trenberth and Stepaniak, 2001; Larkin and Harrison, 2005; Ashok *et al.*, 2007). In the later parts of this study, the El Niño Modoki Index

(EMI) is used as described by Ashok *et al.* (2007), with data obtained from the Japan Agency for Marine-Earth Science and Technology (JAMSTEC).

Linear fits are performed based on minimizing the Chi-squared error statistic. The use of alternative methods, such as Poisson regression, is not warranted as the average number of TCs per season in this region is small (10.7 TCs on average per season) relative to the sample size. The Pearson product–moment correlation coefficient, r, is used to examine the dependence between datasets. The significance of results is determined using a nonparametric bootstrap method based on 200 000 random permutations of the data, with two-sided confidence intervals based on percentiles. Setting the degree of confidence threshold is a somewhat arbitrary decision, thus the confidence interval is given for each individual result.

3. Results

3.1. Climatological features

To provide points of reference for the subsequent analyses, some features of the TC climatology based on the entire study period are given. The frequency distribution of the number of TCs per season in the Australian region (Figure 1) ranges from 5 up to 18 during the study period. Although there is considerable variation related to the limited period of available data (32 years), there is some indication that the distribution is close to normal in shape (based on a 3-point running mean to smooth the data).

Consistent with previous studies, there is a clear relationship between ENSO and the interannual variability of TC #s (Figure 2). There does not appear to be a strong non-linearity in this relationship based on the available data, such that the linear fits shown in Figure 2 are used throughout the study to represent this relationship. The correlation coefficients are $r = -0.50$ between NIN and TC #s, and $r = 0.41$ between SOI and TC #s. The square of these correlation coefficients indicates that 25% (17%) of the interannual variability in TC #s is associated with NIN (SOI).

3.2. Trend analyses

The relationship between ENSO and TC #s is used here to help examine the significance of the long-term trend in TC #s in the Australian region. The lines of best fit shown in Figure 2 provide an indication of the most likely number of TCs that could be expected to occur based on the influence of ENSO. The difference, D, between the expected number of TCs (defined by the lines of best fit in Figure 2) and the average number of TCs per season, A, is described by Equations 1 and 2, for NIN and SOI, respectively.

$$D_{\mathrm{NIN}} = -2.15 \times \mathrm{NIN} + 10.7 - A \qquad (1)$$

$$D_{\mathrm{SOI}} = 0.13 \times \mathrm{SOI} + 10.9 - A \qquad (2)$$

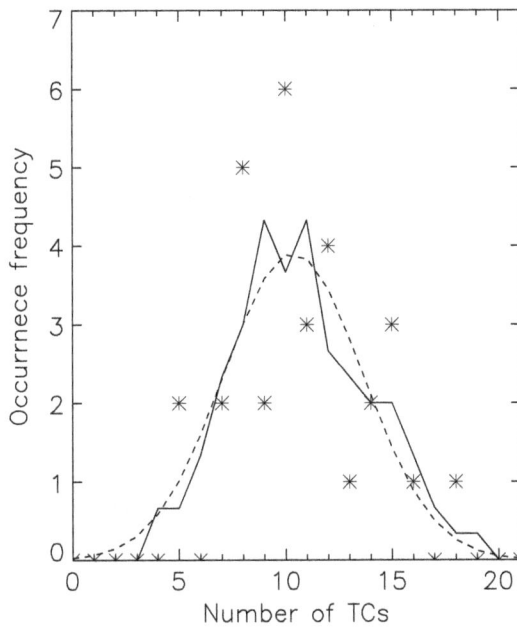

Figure 1. Frequency distribution of the number of TCs per season (star symbols), with a smoothing applied (solid line) and a Gaussian fit (dashed line).

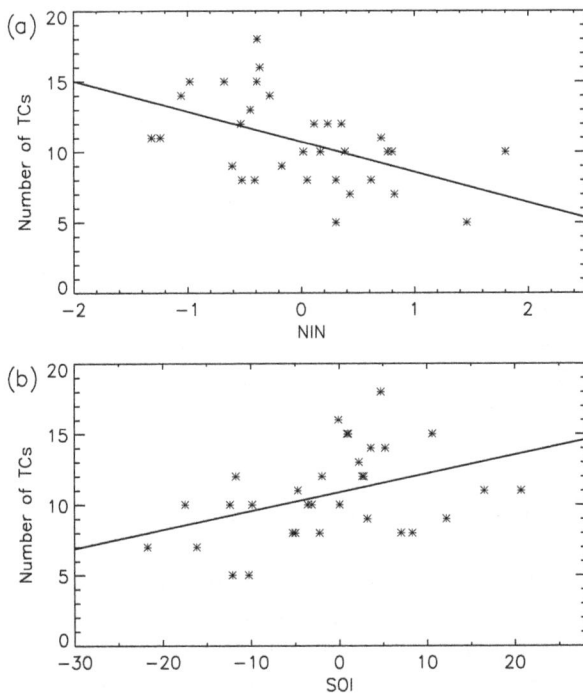

Figure 2. Scatter plots of the number of TCs per season versus NIN (a) and SOI (b). Linear fits to these data are shown.

The variability in TC #s relating to ENSO [given by D, from Equations 1 and 2] is subtracted from the actual number of TCs that occurred in each season, so as to focus on the variability in TC #s not associated with ENSO. This has the effect of reducing the standard deviation in TC #s from 3.2 down to 2.7 based on NIN and to 2.9 based on SOI. As expected, squaring the standard deviations shows that the variance in TC #s is reduced by an amount consistent with the degree of

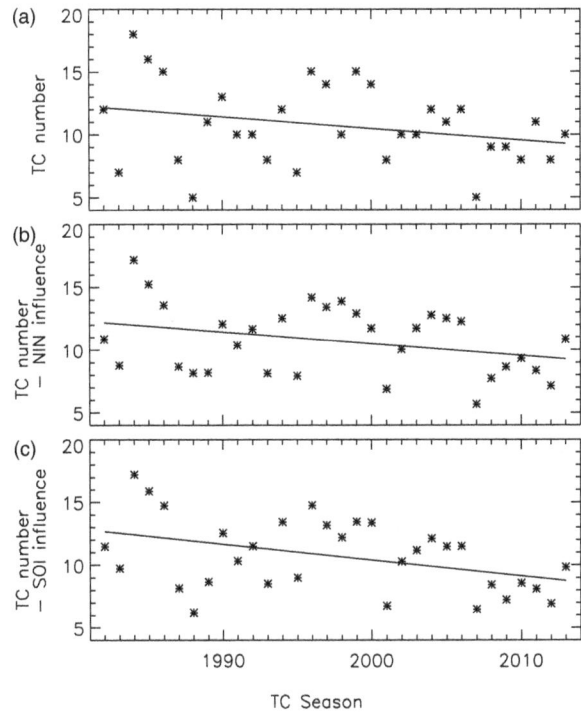

Figure 3. Time series of the number of TCs in the Australian region (a) with the influence of ENSO removed based on NIN (b) and SOI (c). Linear fits to these data are shown.

correlation (noted previously) between the TC #s and the ENSO indices.

A time series of TC #s from the 1982 to 2013 TC seasons is shown in Figure 3(a). The slope of the line of best fit to this time series is equal to $-0.09\,\mathrm{year}^{-1}$, with the downward trend being significant at the 87% confidence level. When the influence of ENSO is removed [Figure 3(b) and (c)], the confidence level is increased to 93% based on NIN and 98% based on SOI.

The slope of the line of best fit shows an increase in magnitude (from -0.09 to $-0.13\,\mathrm{year}^{-1}$) when TC variance is accounted for based on SOI. This is due to a small positive trend in SOI over the study period of $0.25\,\mathrm{year}^{-1}$ (not significant at the 90% confidence level) that amplifies the magnitude of the downward trend. In contrast to SOI, NIN has virtually no linear change over the study period ($-0.0007\,\mathrm{year}^{-1}$). Consequently, there is virtually no change in the best fit line that accounts for ENSO variability based on NIN, as the increased confidence (i.e. a 6% increase from 87 to 93%) is almost entirely attributable to the decreased variance of the TC #s. For SOI, the increase in confidence is 11%, while noting that this increase is partly due to the increased magnitude of the trend, as well as due to the reduced variance of the time series (as shown quantitatively in Section 3.4).

3.3. Variability of the relationship between ENSO and TC #s

To investigate potential changes in the relationship between ENSO and TC #s, scatter plots of TC number versus the ENSO indices are presented separately

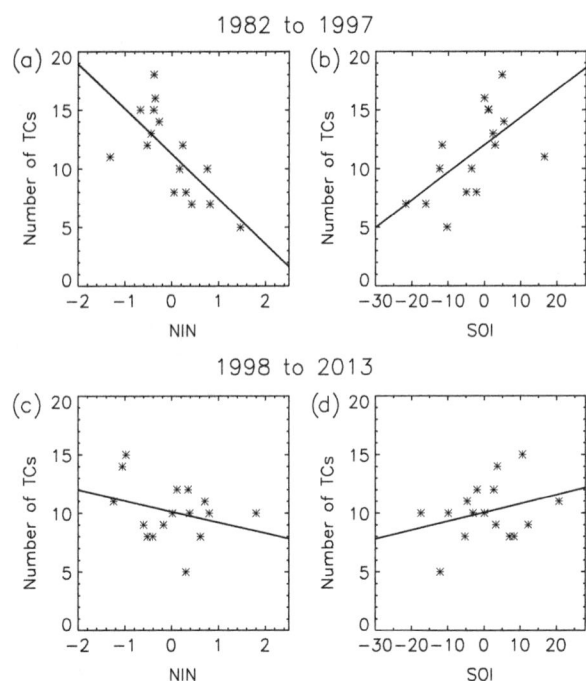

Figure 4. Scatter plots of the number of TCs per season versus NIN and SOI shown separately for the first (a, b) and second (c, d) halves of the study period. Linear fits to these data are shown.

Table I. Correlation coefficients between TC #s and ENSO indices (NIN and SOI) for the first and second halves of the study period.

Time period	NIN	SOI
1982–1997	−0.70	0.60
1998–2013	−0.30	0.30

for the first and second halves of the study period (Figure 4). The scatter plots indicate that the relationship between ENSO and TC #s is stronger during the first half of the study period (from the 1982 to 1997 TC seasons) than during the second half (from the 1998 to 2013 TC seasons).

Correlation coefficients for the relationship between TC #s and the ENSO indices are listed in Table I for the first and second halves of the study period. During the first half of the study period, the correlation coefficients are −0.70 between NIN and TC #s and 0.60 between SOI and TC #s. During the second half of the study period, the correlation coefficients are reduced in magnitude to −0.30 for NIN and 0.30 for SOI. Squaring the correlation coefficients shows that NIN (SOI) accounts for 49% (36%) of the interannual variability in TC #s during the first half of the study period, but only 9% (9%) during the second half.

3.4. Robustness of the methodology

The interannual and spatial variability in TC #s, as well as variability in the relationship between ENSO and TC #s (as presented in Section 3.3), suggests the possibility

that variations in the study region or time period could have some influence on the results. Consequently, this section examines the robustness of the study method (i.e. a trend analysis that accounts for ENSO) to variations in the input parameters, including variations to the dataset of TC #s and variations to the ENSO measures. Table II shows the effect of small changes in the study time period or geographic region, while Table III shows the effect of changes to the ENSO indices including de-trending the indices or using a different measure of ENSO (the EMI).

The effects of reducing the size of the study region (by 5 degrees and 10 degrees) or shifting the study period backwards in time (by 1 and 2 years) are detailed in Table II. Some variation occurs in the magnitude of the trend in TC #s (ranging from −0.07 to −0.10 year^{-1}), as well as some variation in the significance of trend (ranging from 81 to 90%), but in each individual case the significance of the trend is increased after removing the variability relating to ENSO. The increases in significance range from 5 to 7% for NIN and from 9 to 15% for SOI.

The larger increase in significance for SOI than NIN relates to the small (i.e. non-significant) trend in SOI during the study period (based on average values for the months July, August and September), with virtually no trend in the NIN data. If a linear fit is removed from the SOI (to de-trend its time series), Table III shows that accounting for ENSO variability based on the de-trended SOI does not change the magnitude of the trend in TC #s, but the significance of the trend in TC #s is increased to 92% (i.e. a 5% increase, similar to the case for the de-trended NIN time series). This indicates that about half of the increase in significance when ENSO is accounted for based on the SOI (i.e. from 87 to 98%) is related to the reduction in variability of the TC #s, and the other half is related to the increase in magnitude of the trend in TC #s due to the small trend in SOI values.

As another means of examining the robustness of the method, the EMI is examined here as an alternative to the SOI and NIN. Modoki events are characterized by warming in the central Pacific, with less warming in the eastern Pacific than occurs for canonical El Niño events. The results shown in Table III based on the EMI are similar to the results examined previously for the other ENSO indices (NIN and SOI) in that there is an increase in significance of the trend in TC #s when ENSO is accounted for. The increase in significance of the trend is 6%, from 87 to 93%, for the EMI. There is virtually no trend in the EMI values over the study period such that the magnitude of the trend in TC #s does not change when ENSO variability is accounted for based on a de-trended version of the EMI time series (similar to the case for the NIN).

Table II. The robustness of the study method to small changes in study region or time period. The trend in TC #s (magnitude and significance) is shown prior to the removal of variability relating to ENSO. The correlation between TC #s and ENSO indices (NIN and SOI) is shown. The trend in TC #s (magnitude and significance) is also shown after the removal of ENSO variability (based on each of the two ENSO indices).

Region	Time period	TC Trend (year^{-1})	Significance (%)	Correlation		TC trend (year^{-1})		Significance	
				NIN	SOI	NIN	SOI	NIN (%)	SOI (%)
90°E–160°E	1982–2013	−0.09	87	−0.50	0.41	−0.09	−0.13	93	98
92.5°E–157.5°E	1982–2013	−0.09	90	−0.56	0.45	−0.09	−0.13	96	99
95°E–155°E	1982–2013	−0.07	81	−0.51	0.41	−0.07	−0.10	88	96
90°E–160°E	1981–2012	−0.10	89	−0.50	0.41	−0.10	−0.13	95	98
90°E–160°E	1980–2011	−0.09	87	−0.52	0.44	−0.09	−0.12	92	97

Table III. Correlation coefficients for the entire study region (90°E–160°E) and time period (1982–2013) between TC #s and various different ENSO indices: de-trended NIN and de-trended SOI, as well as the EMI. The trend in TC #s (magnitude and significance) is shown after removing the variability in TC #s relating to ENSO (for each one of the ENSO indices).

Index	Correlation	TC trend (year^{-1})	Significance (%)
De-trended NIN	−0.50	−0.09	92
De-trended SOI	0.49	−0.09	92
EMI	−0.49	−0.09	93

4. Discussion and conclusions

The possibility that climatological changes can be identified in observed TC data was examined using an updated dataset and a new methodology. The trend in TC #s is more strongly confirmed by accounting for ENSO in TC variability. A decreasing trend is suggested by the results presented here for the Australian region. It is noted that this is not representative of other regions or the global average.

Accounting for variability in TC #s relating to ENSO increased the probability that the downward trend is significant from 87 to 93% based on NIN and to 98% based on SOI. This indicates that the observed trends have higher confidence than is the case when the ENSO influence is not considered. This result (i.e. an increase in trend significance) was shown to be robust for small changes to study region and time period, as well as to changes in the indices used to represent ENSO. This method of TC trend analysis accounting for ENSO could potentially be applicable to other measures of TC activity (e.g. the ACE index), as well as to other regions of the world, noting that the applicability of this method is dependent on a significant relationship occurring between ENSO and TC activity in a given region.

To ensure high data quality, TC #s were examined only from the 1982 TC season to the present, due to potential satellite-induced data inhomogeneities prior to this period. The average number of TCs per season during the period of data that was excluded (i.e. from the 1970 to 1981 TC seasons) is about two TCs higher than for the study period used here. Assuming that poorer satellite reconnaissance in the pre-1982 period would result in fewer TCs, the two TCs higher value indicates that the decreasing trend found in this study is not an artifact of excluding the data prior to the 1982 TC season. The use of the study region from 90°E to 160°E is beneficial for ensuring a high quality of data, given that this region is the sole responsibility of a single organization (i.e. the Australian Bureau of Meteorology), rather than consisting of the combination of observations from multiple different organizations that could potentially have some variation in their observation and analysis procedures.

Numerous data quality checks were used in this study, as noted above, and there are no known systematic errors in the observations or analyses that would be likely to produce the decreasing trend in TC #s examined here. This suggests the possibility that the downward trend may be associated with changes in environmental conditions. Studies such as Emanuel (2005) indicate that the changes in various different environmental conditions that are occurring due to anthropogenic influences are having an influence on TC activity in some regions of the world. In the Australian region, one possible physical explanation for a decreasing trend in TC #s could be a potential long-term shift in environmental conditions towards a more El Niño-like state, given that fewer TCs tend to occur in the Australian region during El Niño than La Niña conditions. Although the NIN and SOI data used here did not show a significant long-term trend, some studies suggest the possibility of a shift towards more El Niño-like conditions, relating to a possible weakening of the Walker circulation due to increasing atmospheric greenhouse gas concentrations (Vecchi et al., 2006; Power and Smith, 2007; Huang et al., 2013; DiNezio et al., 2013). However, it is also noted that considerable uncertainty exists in observed changes to the Walker circulation based on reanalyses (Compo et al., 2011).

The correlation between TC #s and the ENSO indices is weaker during the second half of the study period (from 1998 to 2013) than during the first half (from 1982 to 1997). ENSO was found to account for about 35–50% of the variance in TC #s during the first half of the study period (36% for SOI and 49% for NIN), but only about 10% during the second half (9% for both NIN and SOI). One potential explanation for a variation in the relationship between TC #s and the ENSO indices between the first and second halves of the study period, could be that the NIN and SOI indices are not representing all of the different types of ENSO variability that influence TC activity, including multiple different types of the El Niño Modoki or neutral ENSO phases (Chand *et al.*, 2013; Wang and Wang, 2013). Some studies have hypothesized that increasing greenhouse gas concentrations could potentially influence the occurrence of Modoki events, with Ashok *et al.* (2012) suggesting that further increases in global warming in the tropical Pacific may result in more basin-wide warm events in place of canonical El Niño events, along with the possible occurrence of more intense El Niño Modoki events.

To examine the potential influence of Modoki events on the decreasing trend in Australian TC #s, the EMI was used as an alternative to NIN. The application of the study method resulted in an increase in trend significance when ENSO was accounted for using the EMI, similar to the case for the other ENSO indices (NIN and SOI), indicating that the results and conclusions presented here are robust for a range of different measures of ENSO variability. There was virtually no trend in the EMI data used here, such that a temporal change in the occurrence frequency of Modoki events is not likely to be the cause of the downward trend in TC #s (Figure 3) or the variation between the two halves of the study period (Figure 4). Variation between the two halves of the study period could potentially be caused by either a change in environmental conditions not associated with ENSO (i.e. environmental conditions not related to the ENSO indices examined here: NIN, SOI or EMI), or a change in the relationship between ENSO and TC #s in this region.

The decreasing trend in TC #s shown here based on high quality observations is broadly consistent with results of previous studies examining TC occurrence in the region around Australia. Nicholls *et al.* (1998) presented a decline in TC #s near Australia using observations from the 1970 to 1996 TC seasons, although it is noted that satellite data inhomogeneities during the early part of this time period could potentially have a significant influence on trend analyses (Kuleshov *et al.* (2010)). Callaghan and Power (2010) also reported a decreasing trend in TCs making landfall over eastern Australia, significant at the 90% level, based on reports from numerous historical sources (including peer-reviewed publications, newspapers, sea-faring observations and other media reports dating back to the late 19th century).

The results of this study have implications for seasonal prediction of TCs. Long-term changes in the TC climatology present significant challenges for prediction methodologies, such as the downward trend in TC #s indicating a considerable degree of uncertainty in the average number of TCs that could be expected to occur in the current climate. Additionally, statistical methods of producing TC seasonal predictions could produce considerably different results depending on the period of data used. For example, a method based on the relationship between TC #s and ENSO (e.g. as represented by NIN or SOI) would have a considerably lower sensitivity to ENSO if based on the latter half of the study period than the earlier half. Results such as these highlight the difficulties associated with using statistical models in a changing climate, with dynamical models presumably more adaptable to long-term changes in the underlying climatology.

Acknowledgements

The author would like to thank John McBride, Savin Chand and Surendra Rauniyar from the Australian Bureau of Meteorology for providing comments on earlier drafts of this manuscript. The comments of the reviewers Greg Holland from NCAR and Mike Fiorino from NOAA are gratefully acknowledged for helping to improve the quality of this manuscript.

References

Ashok K, Behera SK, Rao SA, Weng H, Yamagata T. 2007. El Niño Modoki and its possible teleconnection. *Journal of Geophysical Research* **112**: C11007, DOI: 10.1029/2006JC003798.

Ashok K, Sabin TP, Swapna P, Murtugudde RG. 2012. Is a global warming signature emerging in the tropical Pacific? *Geophysical Research Letters* **39**: L02701, DOI: 10.1029/2011GL050232.

Basher RE, Zheng X. 1995. Tropical cyclones in the Southwest Pacific: spatial patterns and relationships to Southern Oscillation and sea surface temperature. *Journal of Climate* **8**: 1249–1260, DOI: 10.1175/1520-0442(1995)008<1249:TCITSP>2.0.CO;2.

Bell GD, Halpert MS, Schnell RC, Higgins RW, Lawrimore J, Kousky VE, Tinker R, Thiaw W, Chelliah M, Artusa A. 2000. Climate assessment for 1999. *Bulletin of the American Meteorological Society* **81**: 1328, DOI: 10.1175/1520-0477(2000)081<1328:CAF>2.3.CO;2.

Callaghan J, Power S. 2010. Variability and decline in the number of severe tropical cyclones making land-fall over eastern Australia since the late nineteenth century. *Climate Dynamics* **37**: 647–662, DOI: 10.1007/s00382-010-0883-2.

Camargo S, Emanuel K, Sobel A. 2007. Use of a genesis potential index to diagnose ENSO effects on TC genesis. *Journal of Climate* **20**: 4819–4834, DOI: 10.1175/JCLI4282.1.

Chan JCL. 2006. Comment on "Changes in tropical cyclone number, duration, and intensity in a warming environment". *Science* **311**: 1713, DOI: 10.1126/science.1121522.

Chand S, Tory K, McBride J, Wheeler M, Dare R, Walsh K. 2013. The different impact of positive-neutral and negative-neutral ENSO regimes on Australian tropical cyclones. *Journal of Climate* **26**: 8008–8016, DOI: 10.1175/JCLI-D-12-00769.1.

Compo GP, Whitaker JS, Sardeshmukh PD, Matsui N, Allan RJ, Yin X, Gleason Jr BE, Vose RS, Rutledge G, Bessemoulin P, Bronnimann S, Brunet M, Crouthamel RI, Grant AN, Groisman PY, Jones PD, Kruk MC, Kruger AC, Marshall GJ, Maugeri M, Mok HY, Nordli Ø, Ross TF, Trigo RM, Wang RL, Woodruff SD, Worley SJ. 2011. The twentieth century reanalysis project. *Quarterly Journal of the Royal Meteorological Society* **137**: 1–18, DOI: 10.1002/qj.776.

DiNezio PN, Vecchi GA, Clement AC. 2013. Detectability of changes in the Walker circulation in response to global warming. *Journal of Climate* **26**: 4038–4048, DOI: 10.1175/JCLI-D-12-00531.1.

Dowdy AJ, Kuleshov Y. 2012. An analysis of TC occurrence in the Southern Hemisphere derived from a new satellite-era dataset. *International Journal of Remote Sensing* **23**: 7382–7397, DOI: 10.1080/01431161.2012.685986.

Dowdy AJ, Qi L, Jones D, Ramsay H, Fawcett R, Kuleshov Y. 2012. TC climatology of the South Pacific Ocean and its relationship to the El Niño-Southern Oscillation. *Journal of Climate* **25**: 6108–6122, DOI: 10.1175/JCLI-D-11-00647.1.

Emanuel KA. 2005. Increasing destructiveness of tropical cyclones over the past 30 years. *Nature* **436**: 686–688, DOI: 10.1038/nature03906, doi:10.1038/nature03906.

Emanuel KA. 2013. Downscaling CMIP5 climate models shows increased tropical cyclone activity over the 21st century. *Proceedings of the National Academy of Sciences of the United States of America* **110**: 12219–12224, DOI: 10.1073/pnas.1301293110.

Evans JL, Allan RJ. 1992. El Niño/Southern Oscillation modification to the structure of the monsoon and tropical cyclone activity in the Australian region. *International Journal of Climatology* **12**: 611–623, DOI: 10.1002/joc.3370120607.

Holland GJ, Bruyère C. 2013. Recent intense hurricane response to global climate change. *Climate Dynamics* **42**: 617–627, DOI: 10.1007/s00382-013-1713-0.

Huang X, Chuang H-W, Dessler A, Chen X, Minschwaner K, Ming Y, Ramaswamy V. 2013. A radiative–convective equilibrium perspective of weakening of the tropical Walker circulation in response to global warming. *Journal of Climate* **26**: 1643–1653, DOI: 10.1175/JCLI-D-12-00288.1.

Knutson TR, McBride JL, Chan J, Emanuel K, Holland G, Landsea C, Held I, Kossin JP, Srivastava AK, Sugi M. 2010. Tropical cyclones and climate change. *Nature Geoscience* **3**: 157–163, DOI: 10.1038/ngeo779.

Kossin JP, Knapp KR, Vimont DJ, Murnane RJ, Harper BA. 2007. A globally consistent reanalysis of hurricane variability and trends. *Geophysical Research Letters* **34**: L04815, DOI: 10.1029/2006GL028836.

Kuleshov Y, Qi L, Fawcett R, Jones D. 2008. On the El Niño-Southern Oscillation and TC activity and trends in the Southern Hemisphere. *Geophysical Research Letters* **35**: L14S08, DOI: 10.1029/2007GL032983.

Kuleshov Y, Fawcett R, Qi L, Trewin B, Jones D, McBride J, Ramsay H. 2010. Trends in TCs in the South Indian Ocean and the South Pacific Ocean. *Journal of Geophysical Research* **115**: D01101, DOI: 10.1029/2009JD012372.

Landsea C, Franklin J. 2013. Atlantic Hurricane Database Uncertainty and Presentation of a New Database Format. *Monthly Weather Review* **141**: 3576–3592, DOI: 10.1175/MWR-D-12-00254.1.

Landsea CW, Harper BA, Hoarau K, Knaff JA. 2006. Can we detect trends in extreme tropical cyclones? *Science* **313**: 452–454, DOI: 10.1126/science.1128448.

Larkin NK, Harrison DE. 2005. Global seasonal temperature and precipitation anomalies during El Niño autumn and winter. *Geophysical Research Letters* **32**: L16705, DOI: 10.1029/2005GL022860.

McBride J. 1995. Tropical cyclone formation. Global view of tropical cyclones. WMO: Geneva, Switzerland, Tech. Rep. TCP-38.

Nicholls N. 1979. A possible method for predicting seasonal TC activity in the Australian region. *Monthly Weather Review* **107**: 1221–1224, DOI: 10.1175/1520-0493(1979)107<1221:APMFPS>2.0.CO;2.

Nicholls N. 1984. The Southern Oscillation, Sea surface temperature, and interannual fluctuations in Australian tropical cyclone activity. *Journal of Climatology* **4**: 661–670, DOI: 10.1002/joc.3370040609.

Nicholls N. 1985. Predictability of interannual variations of Australian seasonal TC activity. *Monthly Weather Review* **113**: 1144–1149, DOI: 10.1175/1520-0493(1985)113<1144:POIVOA>2.0.CO;2.

Nicholls N, Landsea C, Gill J. 1998. Recent trends in Australian region tropical cyclone activity. *Meteorology and Atmospheric Physics* **65**: 197–205.

Power SB, Smith IN. 2007. Weakening of the Walker circulation and apparent dominance of El Niño both reach record levels, but has ENSO really changed? *Geophysical Research Letters* **34**: L18702, DOI: 10.1029/2007GL030854.

Sinclair MR. 2002. Extratropical transition of Southwest Pacific tropical cyclones. Part 1: climatology and mean structure changes. *Monthly Weather Review* **130**: 590–609, DOI: 10.1175/1520-0493(2002)130<0590:ETOSPT>2.0.CO;2.

Sugi M, Yoshimura J. 2013. Decreasing trend of tropical cyclone frequency in 228-year high-resolution AGCM simulations. *Geophysical Research Letters* **39**: L19805, DOI: 10.1029/2012GL053360.

Taylor KE, Stouffer RJ, Meehl GA. 2012. An Overview of CMIP5 and the experiment design. *Bulletin of the American Meteorological Society* **93**: 485–498, DOI: 10.1175/BAMS-D-11-00094.1.

Trenberth KE, Stepaniak DP. 2001. Indices of El Niño evolution. *Journal of Climate* **14**: 1697–1701, DOI: 10.1175/1520-0442(2001)014<1697:LIOENO>2.0.CO;2.

Troup AJ. 1965. The 'southern oscillation'. *Quarterly Journal of the Royal Meteorological Society* **91**: 490–506, DOI: 10.1002/qj.49709139009.

Vecchi GA, Soden BJ, Wittenberg AT, Held IM, Leetmaa A, Harrison MJ. 2006. Weakening of tropical Pacific atmospheric circulation due to anthropogenic forcing. *Nature* **441**: 73–76, DOI: 10.1038/nature04744.

Walsh KJE, Ryan BF. 2000. Tropical cyclone intensity increase near Australia as a result of climate change. *Journal of Climate* **13**: 3029–3036, DOI: 10.1175/1520-0442(2000)013<3029:TCIINA>2.0.CO;2.

Walsh KJE, Nguyen KC, McGregor JL. 2004. Finer-resolution regional climate model simulations of the impact of climate change on tropical cyclones near Australia. *Climate Dynamics* **22**: 47–56, DOI: 10.1007/s00382-003-0362-0.

Wang C, Wang X. 2013. Classifying El Niño Modoki I and II by different impacts on rainfall in Southern China and typhoon tracks. *Journal of Climate* **26**: 1322–1338, DOI: http://dx.doi.org/10.1175/JCLI-D-12-00107.1.

Webster PJ, Holland GJ, Curry JA, Chang H-R. 2005. Changes in tropical cyclone number, duration, and intensity in a warming environment. *Science* **309**: 1844–1846, DOI: 10.1126/science.1116448.

7

A new perspective on Australian snow

Sonya Louise Fiddes,* Alexandre Bernardes Pezza and Vaughan Barras

School of Earth Sciences, University of Melbourne, Parkville, VIC, 3010, Australia

*Correspondence to:
S. Fiddes, School of Earth
Sciences, University of
Melbourne, Parkville, VIC 3010,
Australia.
E-mail:
sonya.fiddes@unimelb.edu.au

Abstract

The Australian Alps have a unique climatological, ecological and hydrological environment and play a key role in water supply for southeastern Australia. Using resort observations we compile a new and robust snow accumulation data set. Both maximum snow depth and total snow accumulation have declined over the last 25 years. A significant decreasing trend was observed for the total number of light snow days, whereas the total number of heavy snow occurrences has remained constant. Maximum temperatures are highly related to all snow variables. It is suggested that global warming is already impacting light snowfall events, while heavy events are less affected.

Keywords: snowfall; Australia; climate; trends

1. Introduction

The Australian Alps, running down the east coast of Australia, are an important source of freshwater and home to a multi-million dollar ski industry. Snowfall in the region, whilst already in a fragile state, is highly susceptible to precipitation and temperature regimes, and is currently experiencing declines due to changes in these parameters.

Duss (1992), using the Snowy Hydro Limited (SHL) data set for Spencers Creek, southwest of Canberra, created an integrated snow profile (by integrating snow depth over time) in which he found no declining trend from 1910 to 1991. It was shown by Whetton *et al.* (1996) that the variation of snow cover is related to the prevailing precipitation and temperature anomalies. For many years the most basic snow models were based on a monthly data set of these two parameters alone (Galloway, 1988; Duss, 1992; Whetton *et al.*, 1996; Hennessy *et al.*, 2003).

In a 2003 report, Hennessy *et al.* (2008) discussed findings that temperatures were increasing at a greater rate in the higher altitudes of southeastern Australia than that of lower altitudes, whilst rainfall in the Victorian Alps had experienced a small decrease. A weak decline in the maximum snow depth was observed across several sites over the 1957 to 2002 period. Hennessy *et al.* (2008) also noted a decline in late season (August and September) maximum snow depths.

Nicholls (2005), in a comprehensive analysis of the snow depth records at Spencers Creek, showed that whilst there was a slight decline in the maximum snow depth since 1962 (10%), the weak decline in wintertime precipitation was not strong enough to explain this trend, whereas the increase of wintertime temperature experienced was too strong. Analysis of the springtime snow depths showed a much stronger decline of up to 40% over the same period (1962–2002), which the author attributed largely to the increase of temperature.

In 2009, Nicholls updated this work across the period 1954 to 2008, to confirm a continuing, albeit insignificant, decline of maximum snow depths and a strong, significant decline in spring time snow depths.

Bhend *et al.*, 2012, using snow depth data from the Rocky Valley Damn site, found that maximum snow depths have declined over the 1954–2011 period, with the majority of this occurring in the recent past. Furthermore, they indicated that snow seasons are ending earlier.

Davis (2013) found at Spencers Creek, snow depths have decreased by about 15% in the last decade. By using a statistical linear regression of temperature and precipitation data, it was estimated that snow depths in the beginning of the 20th century were 5–14% greater than the 1961–1990 average.

When considering the future, the simulation of snowfall has received considerably less attention than the simulation of rainfall in southeastern Australia or snowfall elsewhere in the world. The snow cover model first developed by Galloway (1988), and updated by Whetton *et al.* (1996), has been the most predominantly used for the region. This model uses monthly precipitation and temperature, the daily temperature standard deviation and an empirically derived relationship specific to the Australian Alps to estimate snowfall and ablation (melt). By using a high-resolution temperature and precipitation grid, Whetton *et al.* (1996) (and hence Hennessy *et al.*, 2003, 2008; Bhend *et al.*, 2012) were able to model snow cover and duration. The most recent work by Bhend *et al.* (2012), focusing on the 2020 and 2050 periods, indicates that the current decline in maximum snow depth is likely to continue. Shorter seasons are projected, mainly owing to earlier spring melt, with later season onset also contributing, while the area of snow cover across the Victorian Alps is likely to decline.

In this study, a new daily snowfall accumulation data set is presented for the first time, which is more closely related to rainfall and temperature variations than the

Australian Alpine weather stations

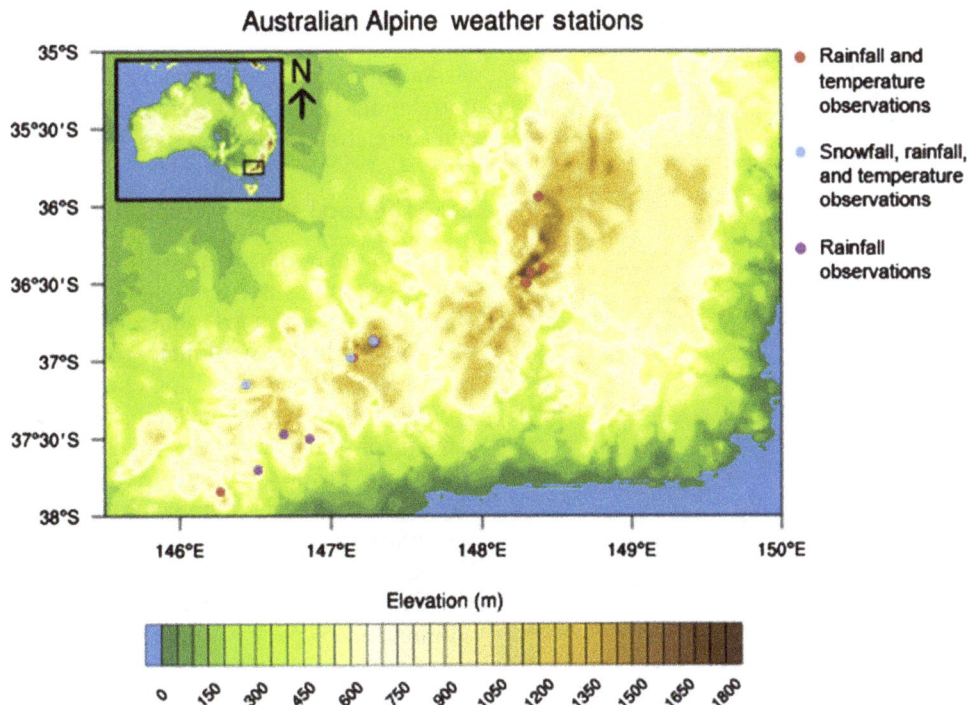

Figure 1. Topographical map of the Australian Alps with the locations of weather stations marked in red for rainfall and temperature observations, blue for rainfall, temperature and snowfall observations and purple for rainfall observations only.

maximum snow depth traditionally used. This data is explored in terms of trends and variability and its relationships to climate drivers known to influence rainfall in the southeast Australian region. Also for the first time we are able to explore the frequency of certain snowfall events. Finally, using projections from the Climate Model Intercomparison Project 3 (CMIP3), the future of snowfall events greater than 20 cm in southeastern Australia is explored.

2. Data and methods

2.1. Data selection

Data used in this study consists of temperature and precipitation received from the Bureau of Meteorology (BOM). Weather stations must be above 1400 m (with Thredbo Village the exception) in altitude and within the box 35–38°S and 145.5–150°E (Figure 1 and Table 1). The time period of interest is from 1988 to 2013, chosen due to the limitation of the snowfall data, during the June, July, August, September (JJAS) season. New South Wales weather station data is included in this analysis as Victorian data alone does not cover the entire time period satisfactorily. All weather station data underwent quality control (performed by the authors), which is detailed below.

Snow depth observations were obtained from the Mount Hotham Resort Management, Mount Buller and Mount Stirling Resort Management/Mount Buller Ski Lift PTY LTD, Falls Creek Resort Management Board/Falls Creek Ski Lift Company, each located in

the Victoria Alps, and the Alpine Resorts Coordinating Council (ARCC). The data received begins in 1988 for Mount Buller and Mount Hotham and 1993 for Falls Creek. It is an average of three sites across the mountain, in an area unaffected by snowmaking, wind barriers and shade. The locations and elevations of the nine sites are provided in Table 1.

The European Centre for Medium-range Weather Forecasting's (ECMWF) ERA-Interim data is used as an estimate of the average daily 850 hPa temperature (T850) during heavy snowfall events. By averaging the T850 over the box of interest defined above, such an estimate can be made. The ERA-Interim data has a resolution of $0.75° \times 0.75°$ and was obtained from the ECMWF Data Server (Dee *et al.*, 2011).

Climate drivers analysed in this study include the Southern Annular Mode (SAM), defined as the leading empirical orthogonal function of monthly mean 700 hPa height anomalies north of 20°S (Gong and Wang, 1999), and obtained from the National Oceanic and Atmospheric Administration's Climate Prediction Centre. The Southern Oscillation Index (SOI) was used as a proxy of the El Niño Southern Oscillation (ENSO) and was obtained from the BOM. SOI is defined as the standardized pressure difference between Tahiti and Darwin (Troup, 1965). The Indian Ocean Dipole (IOD) was explored via the Dipole Mode Index (DMI), defined as the difference between sea surface temperatures (SSTs) over the west Indian Ocean (10°S–10°N, 50–70°N) and the southeastern Indian Ocean (10°S–0°, 90–110°E) (Saji *et al.*, 1999). This data set was obtained from the Japanese Agency for Marine-Earth Science and Technology database. Lastly,

Table 1. Latitude, longitude, elevation and time span of snow plots and weather stations used for this study.

Name	Latitude	Longitude	Elevation	Time period
Snow depth plots				
Falls Creek 1	36.86	147.26	1639 m	1993–2013
Falls Creek 2	−36.86	147.26	1638 m	1993–2013
Falls Creek 3	−36.87	147.27	1675 m	1993–2013
Mount Hotham 1	−36.99	147.145	1760	1988–2013
Mount Hotham 2	−36.99	147.15	1755	1988–2013
Mount Hotham 3	−36.98	147.14	1675	1988–2013
Mount Buller 1	−37.14	146.44	1708	1988–2013
Mount Buller 2	−37.15	146.43	1731	1988–2013
Mount Buller 3	−37.15	146.43	1689	1988–2013
Weather stations				
Charlotte Pass[a]	−36.43	148.33	1755	1930–2013
Thredbo AWS[a]	−36.49	148.29	1957	1966–2013
Thredbo Village[a]	−36.50	148.30	1380	1969–2013
Perisher Valley Ski Centre[a]	−36.40	148.41	1735	1976–2010
Cabramurra SMHEA	−35.94	148.38	1475	1955–1999
Cabramurra SHMEA AWS[a]	−35.94	148.38	1482	1996–2013
Mt Buller[a]	−37.15	146.44	1707	1948–2013
Falls Creek SEC	−36.86	147.28	1510	1947–1990
Falls Creek Rocky Valley	−36.88	147.29	1661	1951–2013
Mount Hotham[a]	−36.98	147.15	1750	1977–1990
Falls Creek[a]	−36.87	147.28	1765	1990–2013
Mount Hotham[a]	−36.98	147.13	1849	1990–2013
Mount Useful	−37.70	146.52	1440	2002–2013
Mount Tamboritha	−37.47	146.69	1446	1989–2013
Mount Baw Baw[a]	−37.84	146.27	1561	1991–2013
Mount Wellington	−37.50	146.86	1559	1997–2013

[a]Weather stations that provide rainfall and temperature data, and whether stations with no indicators denote just rainfall data. Note that the time span does not necessarily indicated data was available or suitable for this research during the entire period.

the subtropical ridge (STR) position and intensity, as defined by the Drosdowsky (2005), index is explored. This data was obtained from the BOM. This index measures the local maxima of monthly mean surface pressure for the area of 10–44°S and 145–150°E. The mean surface pressure is derived from station data that has been interpolated over a 1° grid.

2.2. Quality control

Only one station (Cabramurra SMHEA for temperature) was available in the BOM's high quality data set that fitted the above criteria. For this reason the raw station data was used and put through the following quality tests. Having removed null values, quality control of the observational data was undertaken following the method of Chubb *et al.* (2011) in order to determine how well the stations related to one another. This was performed by creating a daily average of the rainfall, temperature and snowfall data sets, then correlating each individual station back to this daily average. If the correlation was below 0.65 for rainfall or temperature stations that had undergone quality control by the BOM, then the station was not used. For temperature, snowfall and rainfall stations that had not undergone quality control by the BOM the correlation was required to be greater than 0.75. Furthermore, the BOM data had to

span 10 years with at least 8 years having more than 80% of daily data available. This quality control method was also used in the Fiddes *et al.*'s (2014) work. All correlations in this work were performed using the Pearson correlation method (Wilks, 1995).

2.3. The total snow accumulation data set

Unlike the majority of snow data used in previous literature, we now have daily snow depth measurements allowing us, for the first time, to estimate the amount of snow that had either fallen or melted each day. This is done simply by subtracting one day from the next in order to find the change in depth. Subsequently, by summing up only days when this change is positive (indicating snowfall) for the entire JJAS season and ignoring any negative changes (snowmelt), we developed a data set similar to that of the total JJAS rainfall data set. We have named this new data set the snow accumulation data set. This data set is less susceptible to changes within the snowpack (compression).

2.4. Future climates

Future climate projections were selected using the CSIRO's Representative Future Climate framework; a simplified model selection process using the CMIP3 array, whereby models that represent a range of future climatic scenarios based on parameters of interest are analysed as opposed to analysing the full complement of CMIP3 models (Whetton *et al.*, 2012). In this work, changes in precipitation and temperature were considered important; significant warming and a reduction in precipitation are considered to be detrimental to the occurrence of snowfall, whereas little change in these parameters is considered to be a 'best' case scenario.

The models that most closely represent these scenarios were the CSIRO Mk3.5 model, representing a *hotter, drier climate* and the Japanese Model for Interdisciplinary Research on Climate (MIROC) Medres model, representing a *warmer climate, with little change in precipitation*. These selection processes were conducted by the authors and the CSIRO together.

The MIROC Medres model has 43 vertical levels and a horizontal grid of 1.4° longitude by 0.5°–1.4° latitude. The CSIRO Mk3.5 model has 18 vertical levels and a horizontal grid of 1.875° x 1.875° (CSIRO, 2013). The emission scenarios A1B, A1FI and B1 were chosen; for information on these see the work done by Nakicenovic *et al.* (2000). The years 2030, 2050 and 2070 are focused on in this study.

3. Results

3.1. Trends and variability

Table 2 displays the detrended correlations of snowfall variables with rainfall and temperature in the region, and key climate drivers. It can be observed that the

Table 2. Correlations of snowfall parameters total JJAS snow accumulation, maximum snow depth and number of days of snowfall with total rainfall, number of days of rain, the maximum observed temperature and the climate drivers: SOI, SAM, DMI, STR position and STR intensity.

	Total snow accumulation	Maximum snow depth	Number of snowfall days >1 cm	Number of snowfall days 1–10 cm	Number of snowfall days 11–20 cm	Number of snowfall days >20 cm
			Rainfall parameters			
Total JJAS rainfall	**0.59**	**0.54**	0.29	0.12	0.37	**0.60**
Number of rainfall days	**0.73**	**0.66**	**0.62**	**0.45**	**0.62**	**0.65**
			Temperature parameters			
Observed minimum temperature (°C)	−0.27	−0.34	**−0.54**	−0.59	−0.20	−0.10
Observed maximum temperature (°C)	**−0.64**	**−0.65**	**−0.85**	**−0.82**	**−0.49**	**−0.42**
			Climate drivers			
SOI	0.07	0.13	−0.03	−0.06	0.04	0.09
SAM	−0.11	−0.12	−0.17	−0.16	−0.20	0.07
DMI	−0.29	−0.22	−0.31	−0.28	−0.21	−0.20
STR position	−0.17	−0.26	−0.15	−0.08	−0.34	−0.02
STR intensity	−0.14	−0.12	−0.15	−0.07	−0.35	−0.04

Bold indicates significance to the 95th percentile.

total snow accumulation is more highly correlated with the rainfall variables than the maximum snow depth, indicating that it is a powerful measure of snowfall. Both snow accumulation and maximum snow depth show a strong relationship with maximum temperature, as do all the snow frequency variables. No significant correlations were apparent with snowfall variables and relevant climate drivers using the detrended data.

Figure 2 plots the average total snow accumulation and maximum snow depth from 1988 to 2013, as a percentage of their respective averages. The average maximum snow depth is 151 cm per season whilst the average total snow accumulation for this period is 234 cm. Declining trends of 23 cm per decade or 15% per decade with respect to the average are apparent for the maximum snow depth, in strong agreement with the work by Davis (2013). For snow accumulation, a 21 cm per decade (9%) decline is shown. Both of these trends reflect large changes in snowfall in the recent decades.

Figure 3 plots the total number of snowfall days (greater than 0 cm), the number of days between 1 and 10 cm, 11 and 20 cm and those above 20 cm experienced per JJAS season. The average number of snow days experienced per season for each of these parameters is 31, 24, 5 and 2 respectively. Interestingly, a strong decline in the frequency of snowfalls greater than 1 cm and of between 1 and 10 cm per JJAS season of approximately 5 days per 10 years (significant to the 95th percentile) exists for both. No trends are apparent in the number of snowfall events of 11–20 cm and greater than 20 cm.

The strong declines in the 1–10 cm group suggest that the decline in snow accumulation and depth is largely attributed to the loss of small scale snowfall events. In effect, these more marginal events are already being impacted to a greater extent by global warming than the larger snow events which occur during strong cutoff lows or embedded lows (troughs) (Fiddes et al., 2014), which also tend to bring cold temperatures to the region.

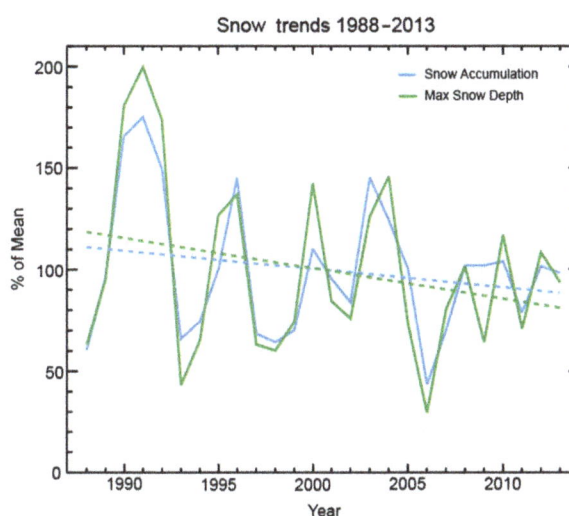

Figure 2. Time series (solid lines) of the snow accumulation (blue) and maximum snow depth (green) as a percentage of their respective averages over 1988–2013. The trends (in dotted lines), while not significant to the 95th percentile, show decreases of 9 and 15% respectively.

Correlations of events greater than 0 cm and events greater than 20 cm to snow accumulation yields $R = 0.80$ and $R = 0.84$ respectively. Analysis of the percentage of snow accumulation that each group makes up per season indicates that the 1–10 cm, 11–20 cm and greater than 20 cm groups make up 40.2, 34.6 and 23.5 percent on average (not shown). These percentages experience high interannual variability and have insignificant trends (of differing signs).

3.2. Future snowfall

Currently, global climate models are showing a poor ability to simulate regional precipitation, especially in southeastern Australia (Irving et al., 2012). Using the MIROC Medres model as the best-case scenario and the

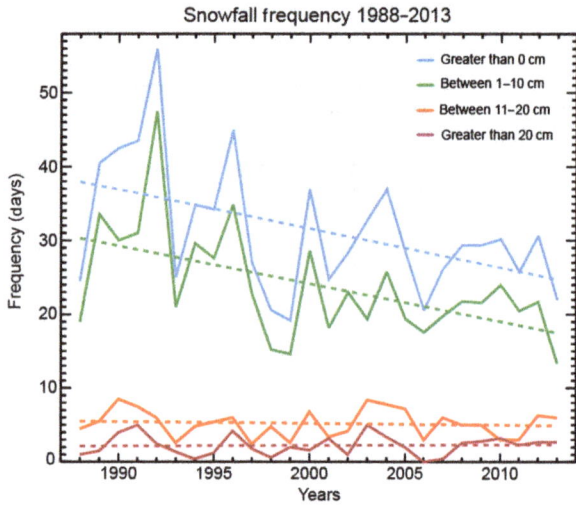

Figure 3. Time series of the total number of snowfall days (solid lines) for all events (greater than 0 cm) (blue), events between 1 and 10 cm (green), events between 11 and 20 cm (orange) and events greater than 20 cm (maroon) experienced per JJAS season. Trends are shown in dotted lines; both the number of events greater than 0 cm and between 1 and 10 cm are significant to the 95th percentile.

Table 3. The current average T850 estimated from ERA-Interim reanalysis over the box 145.5–150°E and 35-38°S, and the average T850 projected by the CMIP3 climate models for the years 2030, 2050, 2070 and emission scenarios AIB, AIFI and BI over the same box.

Current			
Reanalysis T850			2.88
	CSIRO Mk3.5 projections		
	AIB	AIFI	BI
2030	3.45	3.41	3.24
2050	4.10	4.41	3.62
2070	4.82	5.67	4.05
	MIROC Medres projections		
	AIB	AIFI	BI
2030	2.50	2.47	2.31
2050	2.92	3.17	2.54
2070	3.56	4.27	2.93

Temperatures are in degree Celsius.

CSIRO Mk3.5 model as the worst, we compare the projected temperatures to those of today in order to determine how changes in temperature alone are likely to affect heavy snowfalls in the Australian Alps. Projected average changes in 850 hPa temperature (T850) over the region bounded by 145.5–150°E and 35–38°S are provided in Table 3, along with the average reanalysis T850 over the same region.

Figure 4 plots the histogram of reanalysis daily average T850 for heavy snowfall days. It is important to note that already, the distribution of T850 of heavy snowfall lies over the 0°C boundary, with the mean at 0.96°C and a standard deviation at 1.59°C. This implies that just a small warming can have a dramatic effect

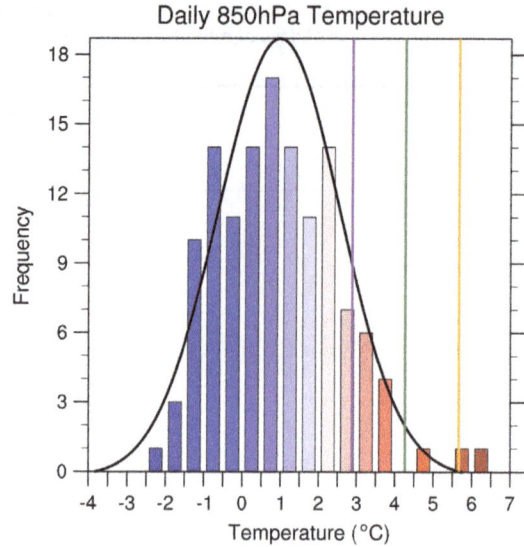

Figure 4. Histograms of the average T850 over the Alpine area estimated by the reanalysis data experienced on days when greater than 20 cm of snow fell at Mt Hotham, Mt Buller or Falls Creek. The left axis showing probability corresponds to the normal distribution in black, and the right axis shows the frequency of events corresponding to the coloured histogram. Vertical lines indicate the JJAS season averages for the reanalysis at 2.88°C (purple), the A1FI 2070 CSIRO projected average at 5.67°C (orange) and the A1B 2070 MIROC projected average at 4.27°C (green).

on the already fragile snowfall conditions in Alpine Australia.

Focusing now on the orange and green vertical lines, which represent the A1FI (worst case emission scenario) 2070 projected average T850 by the CSRIO Mk 3.5 and the MIROC Medres models respectively, it can be seen that currently at these temperatures, whilst heavy snowfall has still been known to occur, such events are rare. This suggests that heavy snowfall events will become limited to occurring with intense synoptic events bringing heavy precipitation and cold temperatures to the region. Values for all the other projected temperatures can be seen in Table 3.

4. Discussion and conclusion

In this study a new snowfall data set has been presented for the first time. A new snow parameter (snow accumulation) is shown to be a highly effective measure of snowfall in Alpine Australia. Declining trends have been found in both the maximum snow depth, in support of recent work done by Nicholls (2009) and Davis (2013), and snow accumulation. Both are highly related to rainfall variables and maximum temperature. Interestingly, no significant correlations were found between snowfall variables and key climate drivers. This is surprising as work done by Fiddes and Pezza (2014) indicates that alpine rainfall has some relationship with the SOI and DMI indices.

With the use of daily snow depth data, a measure of the frequency of snowfall has been explored for the first time. The total number days of snowfall greater than 0 cm and those between 1 and 10 cm have been shown to be decreasing significantly, whilst days of between 11 and 20 cm and greater than 20 cm show no declining trend with time. The percentage of snow accumulation the four of these groups make up individually is not changing significantly. A more detailed analysis, using smaller bin sizes, is needed to fully understand how Australian snowfall is changing.

Future projections show that the average temperature is expected to increase by up to 3 °C by 2070 over the Alpine area. If these temperatures are to become the new climatological average, then the occurrence of snowfall events is likely to be significantly reduced and would only be observed during intense cold outbreaks, such as those described by Ashcroft *et al.* (2009) or the most intense systems found to bring snow in the study by Fiddes *et al.*, 2014. This shift in snowfall regime is found to already be occurring, with significant declines in marginal snow events being observed.

These results say nothing of the changes expected in precipitation in the area, especially extreme precipitation. Previous studies have shown that there has been an increase in the amount of seasonal rainfall attributed to extreme rainfall events over eastern Australia (Haylock and Nicholls, 2000). Furthermore, Alexander and Arblaster (2009) show that although very noisy, some proxies of extreme rainfall are projected to increase significantly with global warming. With an increase in the number of extreme precipitation events, often accompanied by cool weather in southeast Australia, the chance for heavy snowfall events to occur may also increase. Such suggestions are consistent with the finding in this study of no change to heavy snowfall events.

Work by Timbal and Drosdowsky (2013) indicates that with a strengthening of the STR (and to a lesser degree a southwards shift), the ability for weaker cold fronts to penetrate northwards over the Australian continent will decline. While only a weak link with the STR was found in this work, such a shift could have further serious impacts on light snowfalls in the Australian Alps and again suggests that snowfall may only occur as a result of the most intense cold fronts.

Acknowledgements

S. Fiddes and A. Pezza would like to acknowledge and thank Blair Trewin (BOM), Leanne Webb (CSIRO), Falls Creek Resort Management Board/Falls Creek Ski Lift Company, Mount Hotham Resort Management, Mount Buller and Mount Stirling Resort Management/Mount Buller Ski Lift PTY LTD and the Alpine Resorts Coordinating Council for access to the rainfall, climate projection and snowfall data used in this study. A. Pezza would also like to thank the Australian Antarctic Division for funding parts of this work. The authors have no conflict of interest to declare.

References

Alexander LV, Arblaster JM. 2009. Assessing trends in observed and modelled climate extremes over Australia in relation to future projections. *International Journal of Climatology* **29**: 417–435, doi: 10.1002/joc.1730.

Ashcroft LC, Pezza AB, Simmonds I. 2009. Cold events over Southern Australia: synoptic climatology and hemispheric structure. *Journal of Climate* **22**: 6679–6698, doi: 10.1175/2009JCLI2997.1.

Bhend J, Bathols J, Hennessy K. 2012. Climate change impacts on snow in Victoria. CAWCR Report. http://www.climatechange.vic.gov.au/__data/assets/pdf_file/0005/200795/cawcr_report_on_climate_change_and_Victorian_snow_final-dec12_web.pdf (accessed 8 September 2014).

Chubb TH, Siems ST, Manton MJ. 2011. On the decline of wintertime precipitation in the snowy mountains of Southeastern Australia. *Journal of Hydrometeorology* **12**: 1483–1497, doi: 10.1175/JHM-D-10-05021.1.

CSIRO. 2013. Science. OzClim. https://wiki.csiro.au/confluence/display/ozclim/Science#Science-ClimateModels (accessed 17 March 2014).

Davis C. 2013. Towards the development of long term winter records for the Snowy Mountains. *Australian Meteorological and Oceanographic Journal* **63**: 303–313.

Dee DP, Uppala SM, Simmons AJ, Berrisford P, Poli P, Kobayashi S, Andrae U, Balmaseda MA, Balsamo G, Bauer P, Bechtold P, Beljaars ACM, van de Berg L, Bidlot J, Bormann N, Delsol C, Dragani R, Fuentes M, Geer AJ, Haimberger L, Healy SB, Hersbach H, Holm EV, Isaksen L, Kallberg P, Kohler M, Matricardi M, McNally AP, Monge-Sanz BM, Morcrette J-J, Park B-K, Peubey C, de Rosnay P, Tavolato C, Thepaut J-N, Vitart F. 2011. The ERA-Interim reanalysis: configuration and performance of the data assimilation system. *Quarterly Journal of the Royal Meteorological Society* **137**: 553–597.

Drosdowsky W. 2005. The latitude of the subtropical ridge over eastern Australia: the L index revisited. *Journal of Climatology* **25**: 1291–1299.

Duss AL. 1992. Estimation and analysis of snowcover in the Snowy Mountains between 1910 and 1991. *Australian Meteorology Magazine* **40**: 195–204.

Fiddes SL, Pezza AB. 2014. Current and future climate variability associated with wintertime preciptiatoin in alpine Australia. *Climate Dynamics*, doi: 10.1007/s00382-014-2177-6.

Fiddes SL, Pezza AB, Barras V. 2014. Synoptic climatology of extreme precipitation in Alpine Australia. *International Journal of Climatology*, doi: 10.1002/joc.3970.

Galloway RW. 1988. The potential impact of climate changes on Australian ski fields. In *Greenhouse: Planning for Climate Change*, Pearman GI (ed). CSIRO/Brill: Melbourne; 428–437.

Gong D, Wang S. 1999. Definition of Antarctic oscillation index. *Geophysical Research Letters* **26**: 459–462.

Haylock M, Nicholls N. 2000. Trends in extreme rainfall indices for an updated high quality data set for Australia, 1910-1998. *International Journal of Climatology* **20**: 1533–1541.

Hennessy K, Whetton P, Smith I, Bathols J, Hutchinson M, Sharples J. 2003. The impact of climate change on snow conditions in mainland Australia. A Report for the Victorian Department of Sustainability and Environment, Victorian Greenhouse Office, Parks Victoria, New South Wales National Parks and Wildlife Service, New South Wales Department of Infrastructure, Planning and Natural Resources, Australian Greenhouse Office and Australian Ski Areas Association. CSIRO Atmospheric Research, Aspendale; 50 pp. http://www.climatechange.vic.gov.au/__data/assets/pdf_file/0015/73212/TheImpactofClimateChangeonSnowConditions2003.pdf (accessed 18 November 2013).

Hennessy KJ, Whetton PH, Walsh K, Smith IN, Bathols JM, Hutchinson M, Sharples J. 2008. Climate change effects on snow conditions in mainland Australia and adaptation at ski resorts through snowmaking. *Climate Research* **35**: 255–270, doi: 10.3354/cr00706.

Irving DB, Whetton P, Moise AF. 2012. Climate projections for Australia: a first glance at CMIP5. *Australian Meteorological and Oceanographic Journal* **62**: 211–225.

Nakicenovic N, Davidson O, Davis G, Grübler A, Kram T, La Rovere EL, Metz B, Morita T, Pepper W, Pitcher H, Sankovski A, Shukla P, Swart R, Watson R, Dadi Z. 2000. Special Report on Emissions Scenarios : a special report of Working Group III of the Intergovernmental Panel on Climate Change. https://www.ipcc.ch/pdf/special-reports/spm/sres-en.pdf (accessed 28 August 2013).

Nicholls N. 2005. Climate variability, climate change and the Australian snow season. *Australian Meteorology Magazine* **54**: 177–185.

Nicholls N. 2009. The continuing decline of Australian snow depths – an update to 2008. *Bulletin of the Australian Meteorological and Oceanographic Society* 22, 3 pp. http://www.amos.org.au/documents/item/165 (accessed 8 September 2014).

Saji NH, Goswami BN, Vinayachandran PN, Yamagata T. 1999. A dipole mode in the tropical Indian Ocean. *Letters to Nature* **401**: 360–363.

Timbal B, Drosdowsky W. 2013. The relationship between the decline of Southeastern Australian rainfall and the strengthening of the subtropical ridge. *International Journal of Climatology* **33**: 1021–1034, doi: 10.1002/joc.3492.

Troup AJ. 1965. Southern Oscillation. *Quarterly Journal of the Royal Meteorological Society* **91**: 490–506.

Whetton PH, Haylock MR, Galloway R. 1996. Climate change and snow-cover duration in the Australian Alps. *Climatic Change* **32**: 447–479.

Whetton P, Hennessy K, Clarke J, McInnes K, Kent D. 2012. Use of representative climate futures in impact and adaptation assessment. *Climatic Change* **115**: 433–442, doi: 10.1007/s10584-012-0471-z.

Wilks DS. 1995. *Statistical Methods in the Atmospheric Sciences*. International Geophysical Series, Vol. 59. Academic Press: San Diego, London.

8

Observed and projected precipitation variability in Athens over a 2.5 century period

D. Founda,[1] C. Giannakopoulos,[1]* F. Pierros,[1] A. Kalimeris[2] and M. Petrakis[1]

[1] National Observatory of Athens, Institute for Environmental Research and Sustainable Development, Athens, Greece
[2] Department of Environmental Technology and Ecology, Technological and Educational Institute of the Ionian Islands, Zakynthos, Ionian Islands, Greece

*Correspondence to:
C. Giannakopoulos, National
Observatory of Athens, Institute
for Environmental Research and
Sustainable Development,
Athens, Greece.
E-mail: cgiannak@meteo.noa.gr

Abstract

The observed and projected precipitation variability in Athens is assessed for a 2.5 century period. The observed variability is examined using the 150-year historical record of the National Observatory of Athens, while future variability using an ensemble of regional climate models. A good agreement between observations and the model ensemble was found for annual and seasonal precipitation. The model ensemble overestimated the number of rainy days and precipitation amount corresponding to light precipitation and underestimated them corresponding to moderate–heavy precipitation. A 35% decrease in total annual precipitation coupled with an increase in extreme precipitation was projected by the model ensemble up to 2100.

Keywords: precipitation; variability; Athens; trends; climate models

1. Introduction

The Mediterranean climate is controlled by interactions between mid-latitude and tropical processes, and thus sensitive to even minor modifications of the general circulation (Giorgi and Lionello, 2008). The Mediterranean area has been identified as one of the 'hotspots' in future climate change projections as regards temperature rise and precipitation decrease (Giorgi, 2006). Furthermore, reduced soil moisture inherent to water deficits is expected to increase temperature variability through land surface–atmosphere coupling and thus result to more extreme hot weather (Seneviratne et al., 2006). Alcamo et al. (2007) report on increased drought risk in the Mediterranean, mainly due to a decrease in winter precipitation and increase in summer temperature. Apart from changes in mean precipitation, changes in daily variability and precipitation extremes are of particular interest because of their direct impact on society and water resources. Kendon et al. (2010) discuss the mechanisms responsible for projected changes in daily precipitation statistics over Europe. In winter, the 'warming' mechanism – reinforced in southern Europe and Mediterranean by soil moisture changes – makes the major contribution to decreases in rainy day frequency and increases in the extreme/rainy day intensity ratio over eastern Europe and Mediterranean. In summer, 'warming' leads to significant increases in mean precipitation and rainy day and extreme intensities over northern Europe and decreases over the Mediterranean.

The Greater Athens area enjoys a Mediterranean climate with wet mild winters and hot dry summers. The area, however, belongs to one of the zones of the Mediterranean basin, which are most vulnerable to global warming (Alcamo et al., 2007). In this work, the evolution of the precipitation regime in Athens has been studied, using the historical daily precipitation record of the National Observatory of Athens (NOA). The record extends back to the mid-19th century and constitutes a valuable source of historical meteorological information for south-eastern Europe and eastern Mediterranean, where long-term (pre-1900) data are missing. Although precipitation measurements in Athens have existed since 1839, systematic observations started after 1858 (Eginitis, 1907). The analysis of the daily precipitation record sheds light to the characteristics of the precipitation regime in the area during the past and more recent period not only in terms of total precipitation amounts but also precipitation intensity, number of rainy days and frequency of extreme precipitation events.

Different periods of the record have been analysed by several authors in studies concerning precipitation characteristics in Athens or the Greek and Mediterranean area, for instance: Katsoulis and Kambezidis (1989), period:1858–1985; Pnevmatikos and Katsoulis (2006), period: 1900–1999; Nastos and Zerefos (2007), period 1891–2004; Feidas et al. (2007), period 1955–2001, etc.

In this study, the entire historical record of NOA from 1860 to present is analysed, while future climate projections extending to 2100 are performed using output from several regional climate models (RCMs), undertaken in the context of the EU project ENSEMBLES (www.ensembles-eu.org). Observed and predicted daily precipitation data, span a period of approximately 2.5 centuries. The findings from observations and future projections are important for the evaluation of future water availability and drought risk in the area.

2. Analysis of NOA's precipitation time series

2.1. Homogeneity tests

Several inhomogeneities are inherent in long-term observational records reflecting abrupt changes (e.g. relocation of the station, instrument change, changes in observational practices etc.) or gradual changes of the surrounding environment. Such inhomogeneities can lead to climate series biases and to erroneous conclusions regarding climate change and trends. The station of NOA was relocated several times (at distances within 2 km within the city) until 1890, when it was permanently established at its present location (Eginitis, 1907). Changes in the surrounding environment are mostly related to the urbanization of the area. NOA's precipitation series was found to be homogeneous from 1890 to 1985 (Katsoulis and Kambezidis, 1989), from 1900 to 1999 (Pnevmatikos and Katsoulis, 2006) and from 1955 to 2001 (Feidas et al., 2007), using several statistical methods and homogeneity tests.

In the following, homogeneity of the entire time series of the annual precipitation at NOA (1860–2008) is examined. Four well-established statistical tests were selected to detect departures from homogeneity, according to the widely used procedure of Wijngaard et al. (2003). These tests are: the standard normal homogeneity test (SNHT; Alexandersson, 1986), the Buishand range test (Buishand, 1982), the Pettitt test (Pettitt, 1979) and the Von Neumann ratio test (Von Neumann, 1941). The tests suppose under the null hypothesis that the annual values of precipitation are independent and identically distributed, while under the alternative hypothesis the tests assume that a break is likely. Only the three first tests provide information on the year of the break. According to Winjgaard et al. (2003), a time series is labelled as 'useful' if one or zero tests reject the null hypothesis at the 1% level, 'doubtful' if two tests reject the null hypothesis at 1% level and 'suspect' if more than two tests reject the null hypothesis at 1% level.

Although it is recommended to use relative homogeneity tests, considering a 'reference' time series – built from data of neighbouring stations which are supposed to be homogeneous – this was not possible for NOA, due to the lack of neighbouring historical stations. Thus, homogeneity analysis was restricted to absolute tests. Figure 1(a)–(c) displays the results of SNHT, Buishand range and Pettitt tests, along with the corresponding 1 and 5% critical values. The Von Neumann ratio, N, was found to be 1.98 which is very

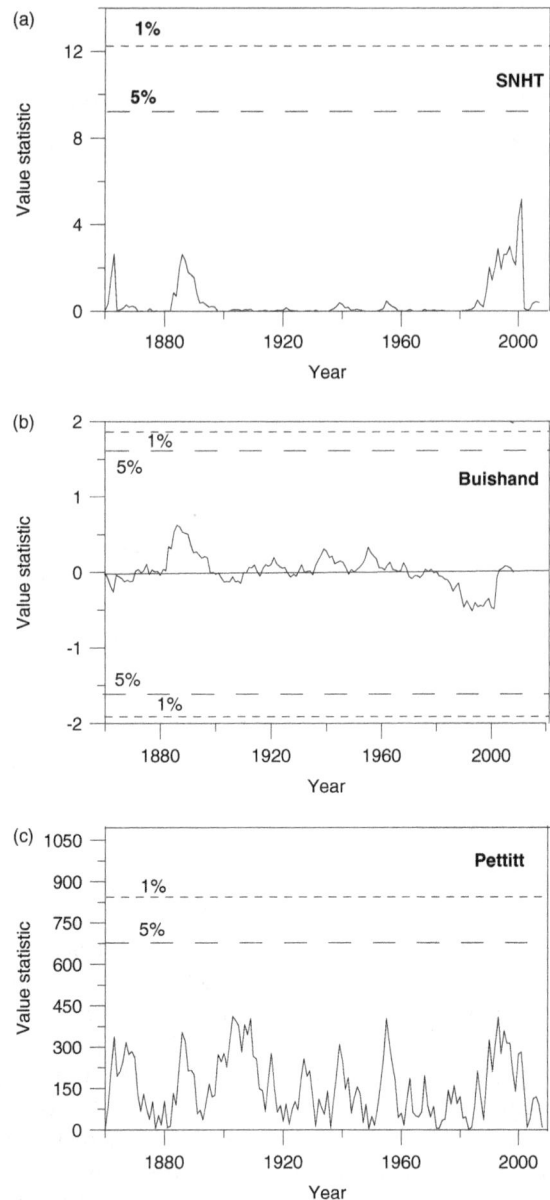

Figure 1. Results of the SNHT (a), Buishand (b) and Pettitt (c) tests applied to NOA annual precipitation time series (1860–2008). Dashed lines represent the 5 and 1% critical levels for the tests.

close to the expected value of $N = 2$ for homogeneous records.

Given the successive relocations of the station of NOA before 1890, the time series is not expected to be homogeneous. However, as it comes out from Figure 1(a)–(c) all three statistical tests, when applied to the annual precipitation time series, did not reject the null hypothesis, supposing that values are independent and identically distributed. It is concluded that any inhomogeneities, present in the series, may be sufficiently small with respect to the inter-annual standard deviation of the series that largely escape detection. Thus, the time series can be labelled as 'useful' for trend analysis and variability analysis, according to the classification developed by Wijngaard et al. (2003).

2.2. Trend analysis

The long-term (1860–2008) climatic mean of the annual precipitation in Athens was estimated equal to 399.4 mm (standard deviation = 115 mm). It is remarkable that this value is very close to the corresponding value of the World Meteorological Organization (WMO) reference period 1961–1990 (397 mm) (Annual Climatological Bulletins, NOA) and also very close to the mean value of the recent 30-year period 1979–2008 (396 mm), indicating no long-term precipitation water deficit in the area. Nevertheless, there are intermediate drier and wetter periods. Figure 2(a) shows the variation of the annual precipitation in Athens from 1860 to 2008 along with the 11-year running means and the trend line. The annual amount of precipitation varies greatly from year to year, with some extremely rainy years before 1900, for instance, 1864 (720 mm), 1883 (846 mm) and 1885 (717 mm). The year 2002 stands out as the absolute record year of the time series with 987 mm total precipitation amount. The exceptional precipitation amount in 2002 is the result of a combination of several reasons, such as the occurrence of three flash flooding episodes during 2002 summer (Lasda et al., 2010), the frequent and intense precipitation events in September due to atmospheric instability and also the extreme precipitation events during November and December 2002. The driest year of the record is 1898 (115 mm), while the period 1989–1990 was the driest period of the whole record (accumulated precipitation, 350 mm).

A least square regression analysis and the non-parametric Mann-Kendall test (Sneyers, 1990) were applied to the annual precipitation series for trend detection. The analysis did not detect any statistically significant trend in the annual precipitation time series (slope of the regression line: −0.0121 mm/year, $p > 0.1$). An increasing trend (not statistically significant) is observed in the time series after 1989 (Figure 2(a)).

The same analysis was applied to the seasonal precipitation time series (Figure 2(b)–(e)). The negative long-term trends, which were observed in winter, autumn and summer precipitation and the positive trend observed in spring precipitation, were not found to be statistically significant. However, a statistically significant increase ($p < 0.001$) in the inter-annual variability of spring precipitation is obvious during the last 30 years (Figure 2(d)). It was estimated that the standard deviation of spring precipitation during the last 30 years is 50% larger than its long-term climatic value. Increased variability is also observed during the first 30 years of the time series. A decrease of year-to-year variability is observed in summer precipitation after the 1950s (Figure 2(e)). Figure 2(f) depicts the anomalies of the number of rainy days/decade from their long-term climatic value which was estimated equal to 644 days. A rainy day is defined here as a day with total precipitation >0.1 mm. Consecutive negative anomalies are observed after the mid-1970s which are more pronounced during the last two decades.

In Figure 2(g), the percentage of rainy days/decade with heavy (>30 mm day^{-1}) or extreme (50 mm day^{-1}) precipitation is shown. A pronounced increase in the number of extreme precipitation episodes is evident after the 1990s. This is in agreement with other extreme rainfall observations along the Mediterranean basin (Alpert et al., 2002; Millan et al., 2005). Alpert et al. (2002) comment on the increase of Mediterranean extreme daily precipitation in spite of the decrease in annual totals.

3. Future climate simulations

Daily output from seven RCM simulations, undertaken in the context of the EU project ENSEMBLES (www.ensembles-eu.org) and employing the A1B scenario, was used for future climate projections in the area. More specifically, the RCMs used in transient simulations from 1950 to 2100 or 2050 are (http://ensemblesrt3.dmi.dk/) as follows:

RCA3, developed at **C4I** Institute in Ireland and employing the ECHAM5 parent GCM,

RM5, developed at **CNRM** in France and employing the ARPEGE parent Global Climate Model (GCM),

CLM developed at **ETHZ** in Switzerland and employing the HadCM3Q0 parent GCM,

RACMO2, developed at **KNMI** in the Netherlands and employing the ECHAM5-r3 parent GCM,

HadRM3Q0, developed at the **METO** in the UK and employing the HadCM3Q0 parent GCM,

HIRHAM, developed at **METNO** in Norway and employing the HadCM3Q0 parent GCM and

REMO developed at **MPI** in Germany and employing the ECHAM5-r3 parent GCM.

In this study, the institute name acronyms shown above in bold have been used to identify the RCMs throughout the text and in the figures. The selection of these RCMs ensures a good multi-model ensemble of climate projections with simulations performed using different combinations of GCMs and RCMs. All models have a horizontal resolution of 25 km × 25 km. By taking output from these seven models, an ensemble mean pattern (ENSEMBLE-7) of the precipitation regime for the city of Athens for a control (1951–2000) and a near future period (2001–2050) was produced. Simulations for a more distant future period 2051–2100 were also performed using three models, KNMI, METO and MPI (ENSEMBLE-3).

Validation of the ensemble model was performed by comparing output of RCM simulations at the grid-point closest to NOA station with observations at NOA for a long reference period, 1951–2000. The choice of a longer averaging period makes the comparison with RCMs climatology less prone to the choice of the period. Kostopoulou et al. (2009) show that it is feasible to evaluate a regional model using the closest

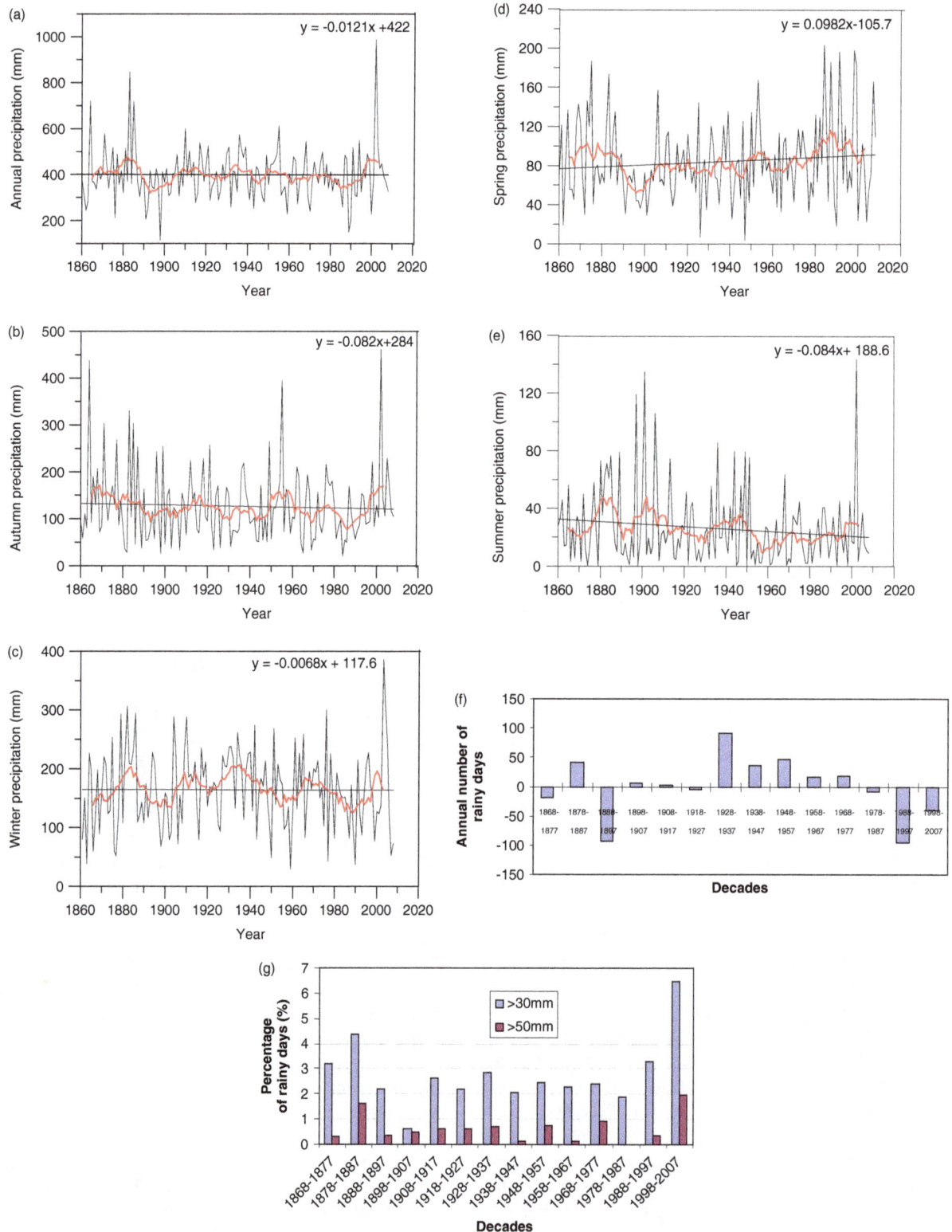

Figure 2. Variation of the annual (a) and seasonal (b)−(e) precipitation at NOA from 1860 to 2008 along with linear trend and 11-points running average curve. Anomalies of the number of rainy days/decade from its long-term climatic value (f). Percentage of rainy days/decade with daily precipitation >30 mm and >50 mm (g).

grid point to the observing meteorological station. The authors have indicated that if an average of more grid boxes is used in an attempt to get a better representation of precipitation, the results worsen significantly. Moreover, a statistical downscaling from 25 km down to say 2 km does not necessarily yield more encouraging results in the distribution of precipitation (Kostopoulou *et al.*, 2010). There is also the approach of using several neighbouring gridboxes and determining the best performing one by a selection procedure as in Hadjinicolaou *et al*. (2011), instead of taking the average. We tested the procedure followed in Hadjinicolaou *et al*. (2011) but for each RCM a different grid cell would have to be selected. For consistency, we kept the closest model grid cell to the observation and avoided averaging between gridpoints as this would also affect precipitation results, especially extremes.

Hence, it was decided to compare the closest 25 km × 25 km RCM grid box with the observing station data but given the episodic nature of precipitation the results and the future projections should always be viewed with a certain degree of caution and a smaller degree of confidence compared with other variables (e.g. temperature). Assessments show that models are more capable of reproducing temperature extremes than precipitation extremes (IPCC, 2007; Alexander *et al*. 2009).

For the definition of a rainy day, a fixed threshold for precipitation of 1 mm both in RCMs data and in observations has been used. This ensures a reasonable and measurable amount of precipitation leaving out the occasional drizzle, often erroneous in RCMs, as these tend to generate too many light precipitation events. This was shown by Sun *et al*. (2006) who investigated the intensity of daily precipitation simulated by 18 models and found that most of the models produce too much light precipitation (below 10 mm day^{-1}), too few heavy precipitation events and too little precipitation in heavy events (over 10 mm day^{-1}).

Figure 3 displays the 11-year running means of the observed annual precipitation at NOA from 1860 to 2008 along with the 11-year running means of the simulated annual precipitation from the seven RCMs and the ensemble mean (ENSEMBLE-7) for the period 1951–2050 and ENSEMBLE-3 for the period 2051–2100. Large discrepancies are apparent among the models' simulations, with some of them providing an overestimation (e.g. ETHZ, METNO) and others an underestimation (e.g. MPI, C4I) of the annual precipitation at NOA. A good performance is provided not only by KNMI and CNRM regarding the mean annual precipitation but also by the ensemble mean ENSEMBLE-7/ENSEMBLE-3. A decreasing trend of the simulated annual precipitation at NOA is evident in Figure 3, which is more pronounced during the second half of the 21st century.

Figure 4(a) presents the annual and seasonal precipitation at NOA from observations and simulations for the control period as well as projections for the two future periods. Observed and simulated (ENSEMBLE-7) mean annual precipitations for the reference period 1951–2000 are in good agreement. The simulated annual precipitation from the seven models for the control period ranged from 504 mm (ETHZ) to 279 mm (MPI). The average value of the annual precipitation from all models (ENSEBLE-7) for the control period was found to be 400 mm ± 85 mm, while the observed value at NOA for the same period equals 386 mm. The comparison of the

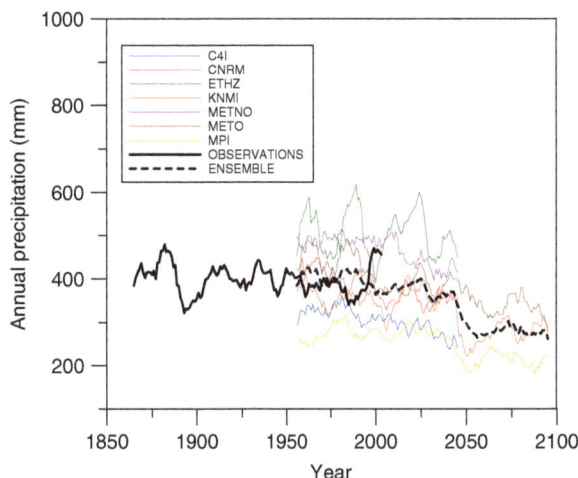

Figure 3. Eleven-year running means of the annual precipitation at NOA from observations (black solid line), output of seven RCMs (coloured lines) and a mean pattern of ENSEMBLE-7 and ENSEMBLE-3 models (black dashed line).

two means (observed and simulated) resulted in a *p* value <0.15. On a seasonal basis, good agreement was also found between simulations and observations in autumn and winter. ENSEMBLE-7/ENSEMBLE-3 simulations result in a decrease of the mean annual precipitation by 10%/35% for the periods 2001–2050/2051–2100 respectively, with respect to the reference period.

For a better interpretation of the results derived from future projections, a categorization of daily precipitation amount (mm day^{-1}) in five categories was conducted. It is important to know whether a future change in precipitation regime is due to an increase or a decrease in light, moderate or heavier precipitation. Five categories were employed as follows: light (1–5 mm), light–moderate (5–15 mm), moderate–heavy (15–30 mm), heavy (30–50 mm) and extreme (>50 mm). Although various threshold values exist in the literature, the categorization suggested here is similar to what most researchers use for the Mediterranean area (e.g. Alpert *et al*., 2002; Bocheva *et al*., 2009). It was estimated that precipitation <15 mm day^{-1} accounts for 60%/50% of total precipitation from ENSEMBLE-7/NOA's observations. ENSEMBLE-7 overestimates the percentage of precipitation amount in the category of light precipitation and underestimates it in the category of moderate–heavy or heavy and extreme precipitation (Figure 4(b)). Similar findings are observed in the number of rainy days per rain category (Figure 4(c)).

Figure 4(d) presents the percentage of precipitation amount per rain category for the control and the two future periods, as well as from NOA's observations. For comparison purposes, all calculations were performed using the ENSEMBLE-3 results. As with ENSEMBLE-7, ENSEMBLE-3 overestimates the precipitation percentage in the category of light precipitation and underestimates it in the category of

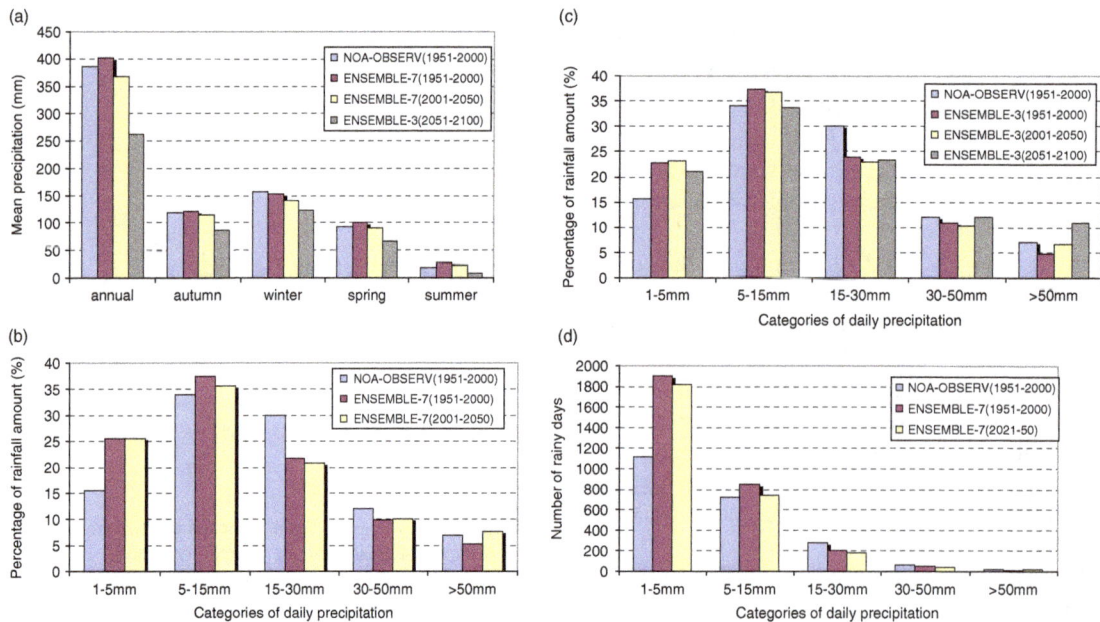

Figure 4. Comparison of observations and simulations for the control period as well as projections for the two future periods for (a) annual and seasonal precipitation, and (b) percentage of precipitation amount per rain category using ENSEMBLE-7. Part (c) is as (b) but using ENSEMBLE-3. (d) Number of rainy days per rain category using ENSEMBLE-7.

moderate–heavy or heavy precipitation when compared with observations of the reference period. The model is in better agreement with observations as regards the category of light–moderate precipitation. As it comes out from Figure 4(d), ENSEMBLE-3 simulations indicate a decrease in the percentage of precipitation amount corresponding to the category of light and moderate precipitation for the period 2051–2100, but an increase in the percentage of heavy and extreme precipitation.

More specifically, despite the simulated reduction in the mean annual precipitation by almost 35% for the future period 2051–2100, the percentage of precipitationl amount due to extreme precipitation for the same period is double its reference period value.

4. Conclusions

The historical daily precipitation time series of the NOA from 1860 to 2008 was used for trend and variability analysis. Four absolute homogeneity tests were applied to the annual precipitation series. The tests did not reject the null hypothesis at 1 and 5% significance levels, concluding that the series can be used for trend analysis.

According to the Mann-Kendall test results, the annual and seasonal precipitation at NOA does not exhibit any statistically significant long-term trend. However, the number of rainy days/decade exhibits negative anomalies during the last three decades with respect to the long-term climatic value. A pronounced increase in the percentage of precipitation amount due to heavy ($>30\,mm\,day^{-1}$) and extreme ($>50\,mm\,day^{-1}$) precipitation is also observed during the last decade.

Future projections were assessed using an ensemble of several RCMs. The validation of the ensemble models through observations, suggested an excellent simulation of precipitation on an annual basis, but with pronounced deviations from the observed data in the category of light and moderate–heavy precipitation. A decrease of the mean annual precipitation by almost 35% was simulated for the future period 2051–2100, combined with increased percentage of precipitation amount due to heavy and extreme precipitation. This combination is expected to lead to precipitation water deficit and increased drought risk in the area by the end of the century.

In this work it has also been shown that an ensemble average of several regional climate models succeeds in representing the total annual amount. Yet some discrepancies remain with respect to distribution of precipitation intensity, but the use of an ensemble average certainly improves the picture of precipitation portrayed by the climate models.

References

Alcamo J, Moreno JM, Nováky B, Bindi M, Corobov R, Devoy RJN, Giannakopoulos C, Martin E, Olesen JE, Shvidenko A. 2007. Europe. Climate Change 2007: Impacts, Adaptation and Vulnerability. Contribution of Working Group II to the Fourth Assessment Report of the Intergovernmental Panel on Climate Change, Parry ML, Canziani OF, Palutikof JP, van der Linden PJ, Hanson CE (eds). Cambridge University Press, Cambridge; 541–580.

Alexander LV, Tapper N, Zhang X, Fowler HJ, Tebaldi C, Lynch A. 2009. Climate extremes: progress and future directions. *International Journal of Climatology* **29**: 317–319.

Alexandersson H. 1986. A homogeneity test applied to precipitation data. *Journal of Climatology* **6**: 661–675.

Alpert PT, Ben-gai T, Baharad A, Benjamini Y, Yekutieli D, Colacino M, Diodato L, Ramis C, Homar V, Romero R, Michaelidis S. 2002. The paradoxical increase of Mediterranean extreme daily rainfall in spite of decrease in total values. *Geophysical Research Letters* **29**(11): X1–X4, DOI: 10.1029/2001GLO13554.

Bocheva L, Marinova T, Simeonov P, Gospodinov I. 2009. Variability and trends of extreme precipitation events over Bulgaria (1961–2005). *Atmospheric Research* **93**: 490–497.

Buishand TA. 1982. Some methods for testing the homogeneity of rainfall records. *Journal of Hydrology* **58**: 11–27.

Eginitis D. 1907. *The climate of Greece, Part A*. P. Sakkelariou Editions: Athens.

Feidas H, Noulopoulou N, Makrogiannis T, Bora-Senta E. 2007. Trend analysis of precipitation time series in Greece and their relationship with circulation using surface and satellite data: 1955–2001. *Theoretical and Applied Climatology* **87**: 155–177.

Giorgi F. 2006. Climate change Hot – spots. *Geophysical Research Letters* **33**: L08707.

Giorgi F, Lionello P. 2008. Climate change projections for the Mediterranean region. *Global and Planetary Change* **63**: 90–104.

Hadjinicolaou P, Giannakopoulos C, Zerefos C, Lange M, Pashiardis S, Lelieveld J. 2011. Mid-21st century climate and weather extremes in Cyprus as projected by six regional climate models. *Regional Environmental Change* **11**(3): 441–457, http://dx.doi.org/10.1007/s10113-010-0153-1.

IPCC. 2007. Climate models and their evaluation. Randall *et al.* (Chapter 9). In *Climate Change 2007: The Physical Scientific Basis*. Contribution of Working Group I to the Fourth Assessment Report of the Intergovernmental Panel on Climate Change. Solomon S, Qin D, Manning M, Chen Z, Marquis M, Averyt KB, Tignor M, Miller HL (eds). Cambridge University Press: Cambridge, UK.

Katsoulis BD, Kambezidis HD. 1989. Analysis of the long term precipitation series in Athens, Greece. *Climatic Change* **14**: 263–290.

Kendon E, Rowell D, Jones RG. 2010. Mechanisms and reliability of future projected changes in daily precipitation. *Climate Dynamics* **35**(2): 489–509, http://dx.doi.org/10.100 /s00382-009-0639-z.

Kostopoulou E, Giannakopoulos C, Holt T, Le Sager P. 2010. Assessment of interpolated ERA-40 reanalysis temperature and precipitation against observations of the Balkan Peninsula. *Theoretical and Applied Climatology*, DOI: 10.1007/s00704-009-0249-z.

Kostopoulou E, Tolika K, Tegoulias I, Giannakopoulos C, Somot S, Anagnostopoulou C, Maheras P. 2009. Evaluation of a regional climate model using in situ temperature observations over the Balkan Peninsula. *Tellus A* **61**(3): 357–370.

Lasda O, Dikou A, Papapanagiotou E. 2010. Flash flooding in Attika, Greece: climatic change or urbanization? *Ambio* **39**: 608–611, DOI: 10.1007/s/3280-010-00503.

Millan MM, Estrela JJ, Miro J. 2005. Rainfall components: variability and spatial distribution in a Mediterranean Area (Valencia Region). *Journal of Climate* **18**: 2682–2705.

Nastos PT, Zerefos CS. 2007. On extreme daily precipitation totals in Athens, Greece. *Advances in Geoscience* **10**: 59–66.

Pettitt AN. 1979. A non-parametric approach to the change-point detection. *Applied Statistics* **28**: 126–135.

Pnevmatikos JD, Katsoulis BD. 2006. The changing rainfall regime in Greece and its impact on climatological means. *Meteorological Applications* **13**: 331–345.

Seneviratne SI, Lüthi D, Litschi M, Schär C. 2006. Land-atmosphere coupling and climate change in Europe. *Nature* **443**: 205–209.

Sneyers R. 1990. On the statistical analysis of series of observations. Technical Note 143, WMO No. 415, Geneva, 192.

Sun Y, Solomon S, Dai A, Portmann R. 2006. How often does it rain? *Journal of Climate* **19**: 916–934.

Von Neumann J. 1941. Distribution of the ratio of the mean square successive difference to the variance. *Annals of Mathematical Statistics* **13**: 367–395.

Wijngaard JB, Klein Tank AMG, Konnen GP. 2003. Homogeneity of 20th century European daily temperature and precipitation series. *International Journal of Climatology* **23**: 679–692.

9

Clausius–Clapeyron-like relationship in multidecadal changes of extreme short-term precipitation and temperature in Japan

Fumiaki Fujibe*

Meteorological Research Institute, JMA, Tsukuba, Japan

*Correspondence to:
F. Fujibe, Meteorological
Research Institute, 1-1
Nagamine, Tsukuba 305-0052,
Japan.
E-mail: ffujibe@mri-jma.go.jp

Abstract

Long-term changes in extreme precipitation were studied using data of daily maximum 10-min and hourly precipitation, as well as daily total precipitation for 60 years (1951–2010) in Japan. It was found that extreme 10-min and hourly precipitations, defined from 5-year averages of annual maximum and 95th percentile values, have a high correlation with temperature with a rate of $9.7 \pm 4.1\%$ and $8.8 \pm 8.3\%\,K^{-1}$, respectively. These values are close to the Clausius–Clapeyron (CC) rate of change in saturation vapor pressure (about 6% per degree), indicating that the CC relation roughly holds for multidecadal changes in extreme short-time precipitation in Japan.

Keywords: extreme precipitation; Clausius–Clapeyron relation; climate change; Japan

1. Introduction

It is believed that heavy precipitation is increasing in many regions of the world, as a result of global warming (IPCC, 2007; Min *et al.*, 2011). A number of statistical studies have examined the relationship between precipitation extremes and temperature with respect to the Clausius–Clapeyron (CC) relation, which states that saturated vapor pressure increases with temperature by approximately $7\%\,K^{-1}$ at 5 °C and $6\%\,K^{-1}$ at 25 °C. Some of the studies have found the applicability of the CC relation in low-to-moderate temperature ranges, especially for short-term precipitation (Hardwick Jones *et al.*, 2010; Utsumi *et al.*, 2011). Some others have found a higher increase of precipitation extremes with temperature than is expected from the CC relation, possibly due to increase of convective precipitation with temperature, or dynamical feedbacks owing to excess latent heat release in extreme showers (Lenderink and van Meijgaard, 2008; Haerter and Berg, 2009; Berg and Haerter, 2013; Berg *et al.*, 2013). On the other hand, a sub-CC relation or even negative dependence on temperature has been found for precipitation extremes in a high temperature range, and/or for summer and tropics, presumably as a result of decreasing moisture availability with temperature (Berg *et al.*, 2009; Hardwick Jones *et al.*, 2010; Utsumi *et al.*, 2011). Haerter *et al.* (2010) stated that 'the CC relation may not provide an accurate estimate of the temperature relationship of precipitation at any temporal resolution', in the sense that the relationship between temperature and precipitation is not so simple as to follow the CC relation. Yu and Li (2012) have indicated the diversity in precipitation–temperature relationships in China according to the precipitation intensity and regions.

As for long-term changes, Lenderink *et al.* (2011) have shown a good agreement between the interdecadal changes of dew-point temperature and hourly precipitation extremes (average over 95th, 99th, and 99.9th percentiles based on a 15-year window) for De Bilt in the Netherlands and for the intermediate season (October and February to April) in Hong Kong, at a rate twice larger than the CC relation (above 10% per degree), although they found no significant relationship for the wet season (May to September) in Hong Kong. Kao and Ganguly (2011) have shown the similarity of the increasing trends in saturated vapor density and 24-h extreme precipitation (30-year return value) from reanalysis data over the world. Studies based on climate model simulation have also shown a future increase of precipitation extremes with temperature at a rate comparable to or higher than the CC rate (Kharin *et al.*, 2007; Pall *et al.*, 2007; Allan and Soden, 2008; Sugiyama *et al.*, 2010; Kao and Ganguly, 2011; Muller *et al.*, 2011). However, observational studies on the relationship between long-term changes in extreme precipitation and temperature have been relatively few so far, probably owing to limitation in data availability.

Increasing trends in extreme precipitation have also been found for Japan. Fujibe *et al.* (2005a, 2006) showed the increase of extreme daily precipitation for 1901–2004, as well as extreme four-hourly and hourly precipitation for 1898–2003, using data at about 50 stations. Data from the Automated Meteorological Data Acquisition System (AMeDAS), which has a network of more than a 1000 rain gauges that have operated since the late 1970s, also indicate an increase in the frequency of $\geq 50\,\mathrm{mm\,h^{-1}}$ precipitation events (Japan Meteorological Agency, 2011). However, information of changes in short-time rainfalls have been relatively few, including those with duration of less than an hour, in spite of a widespread public view that intense precipitation has increased in recent decades. In this study, changes of extreme precipitation with time scales from 10 min to a day were examined with focus on their relation to temperature changes, using data for 60 years from 1951 to 2010.

2. Data and procedure of analysis

2.1. Data

Data of daily maximum 10 min precipitation (P_10m), daily maximum hourly precipitation (P_hour), and daily precipitation (P_day) were obtained from the database of the Japan Meteorological Agency for the warm season (May to October) from 1951 to 2010. A reason for limiting the analysis to the warm season is that wintertime P_10m and P_hour data were unavailable in the 1950s, due to lack of observation under snowfall in northern Japan. In fact, the lack of cold season data is not a serious problem because most of extreme precipitation events occur in summer in Japan (Miyajima and Fujibe, 2011). The number of missing records at each station was required to be less than 3% during the warm season of each year (namely, less than 6 days for May to October every year) for all the three quantities of P_10m, P_hour, and P_day. On this requirement, 92 stations were selected for the analysis. Figure 1 shows the location of stations.

P_10m is defined by the maximum value of precipitation in a 10-min interval that ends between 0000 and 2400 JST (Japan Standard Time, UTC +9 h) in a day. P_hour is defined similarly. P_day is defined by the total precipitation from 0000 to 2400 JST. It is to be noted that P_10m, P_hour, and P_day were defined for 0900–0900 JST of the next day for the period from 1953 to 1963 at many stations, although this is not likely to affect the analysis seriously. It is more important to note that observation of P_10m and P_hour was made using the siphon recording rain gauge, and that of P_day was made by the cylinder-type rain gauge with precision of 0.1 mm until the middle of the 1960s, but these rain gauges were replaced by the automated tipping-bucket rain gauge having precision of 0.5 mm around 1968.

The temperature time series was obtained by the analysis of Fujibe (2012), who combined the data at nonurban stations (with population density less than 100 people per square kilometer) of AMeDAS and a local observation network that operated before the deployment of AMeDAS.

2.2. Indices of extreme precipitation

As the essence of the CC effect is intensification of precipitation due to increase of atmospheric moisture

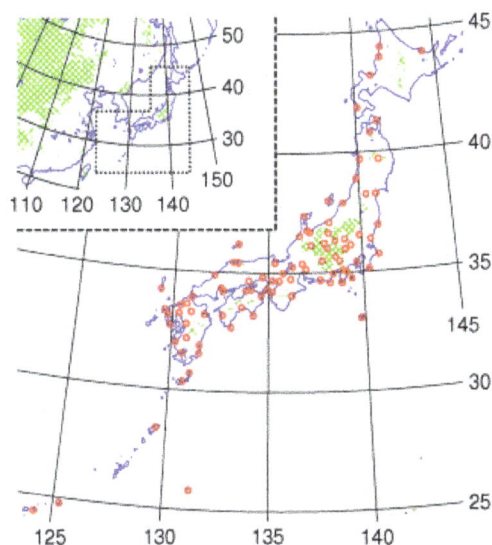

Figure 1. Stations used for analysis. Hatching in light green indicates elevation above 1000 m.

content with temperature, it will manifest itself in the change of absolute precipitation intensity, although many of previous studies have been based on relative increase of the precipitation intensity of a given percentile. In this study, only daily maximum values of 10-min and hourly precipitation are available, which makes it impossible to obtain percentiles among 10-min and hourly values. Alternatively, the analysis was based on the following two indices. One is absolute intensity defined by annual maximum values of P_10m, P_hour, and P_day at each station. Analysis of annual tenth values was also performed. The other index is 95th and 90th percentiles among daily maximum values of P_10m and P_hour, and daily values of P_day.

The percentiles were defined from annual number of days on which each of P_10m, P_hour, and P_day had a value of 0.5 mm or more. In order to account for the change of observation precision due to the change in rain gauges around 1968, the 0.1 mm precision data before 1968 were reduced to the resolution of 0.5 mm by considering the probability at which they would have given a value of 0.5 mm if a tipping-bucket rain gauge had been used (Fujibe et al., 2006). As the tipping bucket rain gauge incrementally measures precipitation accumulation, precipitation of 0.2 mm, for example, would give a value of 0.5 mm with a chance of 40% (=0.2/0.5) if a tipping-bucket rain gauge was used. Thus, a 0.2 mm value was given a weight of 0.4 when calculating the annual number of $\geq 0.5\,\mathrm{mm}$ days.

3. Results

3.1. Recent changes in precipitation extremes

Figure 2 shows the time series of extreme indices for 1951–2010 averaged over the 92 stations, as well as

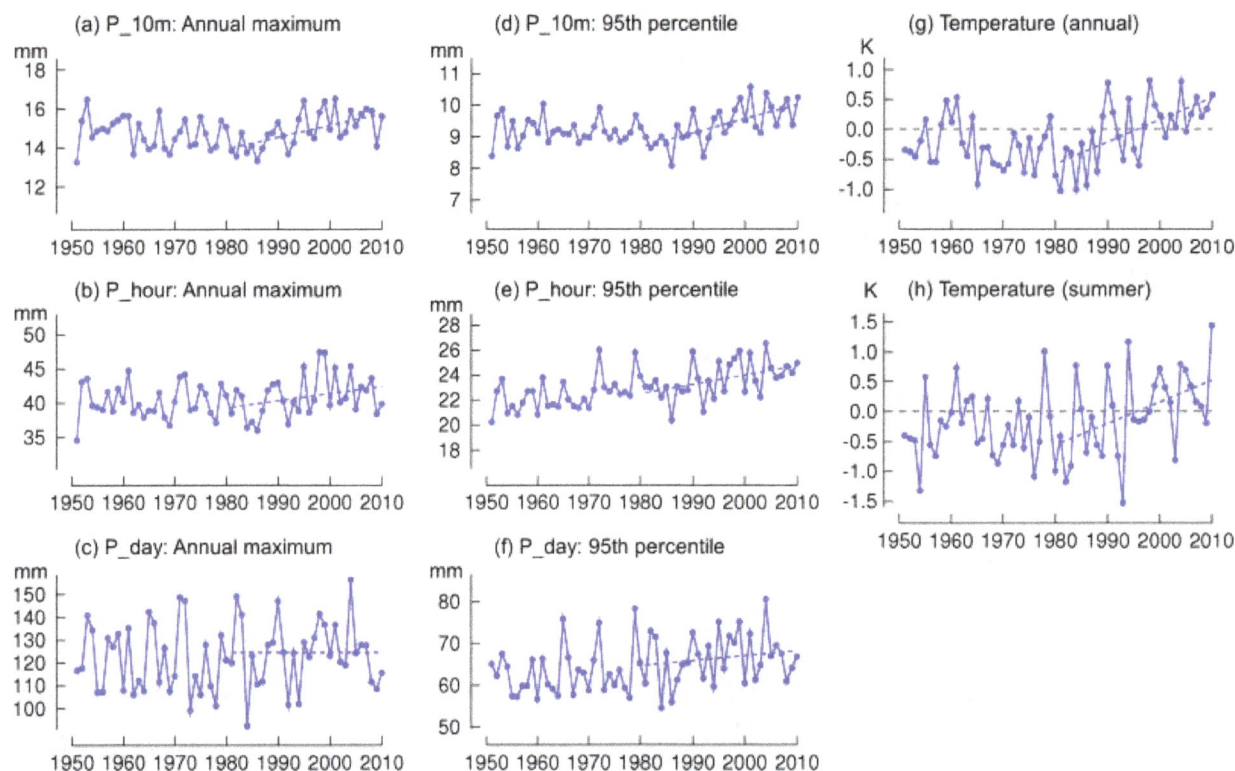

Figure 2. Time series of precipitation indices (a–f) and temperature (g, h). Temperature is shown in deviation from the 1981–2010 average. Dashed lines indicate linear regression for 1981–2010.

those of annual and summertime (June to August) mean temperature over Japan. For the latter three decades (1981–2010), both annual maximum and 95th percentile precipitations of P_10m have positive trends that are significant at the 1% level. The increase is less evident for P_hour, but is significant at the 10% level (an analysis using data at 144 stations for which data are available for 1981–2010 yields an increase of P_hour extremes with 5% significance). On the other hand, no significant trends are found for indices about P_day. Table I shows the value of the trend for each index for 1981–2010. The trends of annual tenth and 90th percentile values do not deviate largely from those of annual maximum and the 95th percentile values, although statistical significance is lower. On the other hand, trends are insignificant for all the indices for P_day.

There are also increasing trends in temperature after 1980, with larger year-to-year variations in summertime temperature than in annual temperature. The increasing rate of annual temperature for 1981–2010 is 0.37 K per decade with a 95% confidence range of 0.18 K per decade (hereafter denoted by 0.37 ± 0.18 K per decade), and that of summer temperature is 0.36 ± 0.27 K per decade. These values are close to the global temperature trend for the period (0.29–0.34 K per decade for land over the Northern Hemisphere for 1979–2004, according to IPCC (2007, Chap. 3)). The ratios of the trend of annual maximum precipitation and that of annual temperature are 12.0% K^{-1} for P_10m and 7.3% K^{-1} for P_hour.

Before 1980, both annual and summer temperatures show a decreasing trend, and so are extreme indices in P_10m and P_hour, although these trends are not statistically significant. Figure 3(a) and (b) shows scatter diagrams for 5-year averages of annual mean temperature and annual 95th percentile values of P_10m and P_hour for 1951–2010. Both P_10m and P_hour have a high correlation to temperature. The correlation coefficients are 0.86 and 0.60, respectively, which are significant at the 1% level. The slope of the linear regression (hereafter $\Delta P/\Delta T$) is 9.7 ± 4.1% K^{-1} for P_10m and 8.8 ± 8.3% K^{-1} for P_hour. These are slightly higher than the increment given by the CC rate (6 % K^{-1} at 25 °C), but the difference from the CC rate is not significant. Table II shows the values of $\Delta P/\Delta T$ for other indices, as well as the results based on summer temperature. It can be seen that the value of $\Delta P/\Delta T$ does not vary largely with extremity, although $\Delta P/\Delta T$ for the 90th percentile of P_10m is significantly larger than 6% K^{-1}. It is also to be noted that 95th and 90th percentile values of P_day have a significant correlation with summer temperature, although insignificant for annual temperature.

For values of each year, the correlations between annual mean temperature and annual 95th percentiles of P_10m and P_hour are 0.61 and 0.48, and the slopes of the linear regression are 7.1 ± 2.4% K^{-1} and 6.6 ± 3.2% K^{-1}, respectively. The smaller slope in comparison to that obtained from 5-year averages can be explained by the large year-to-year scatter, which reduces least-squares coefficients. On the other

Table I. Average and trend of each index for 1981–2010.

	P_10m		P_hour		P_day	
	Average (mm)	Trend (% per decade)	Average (mm)	Trend (% per decade)	Average (mm)	Trend (% per decade)
Annual maximum	14.0	4.5 ± 2.3*	39.3	2.7 ± 3.2***	124.5	0.0 ± 5.1
Annual tenth	5.0	5.0 ± 3.8**	12.5	3.1 ± 4.1	33.8	1.0 ± 5.8
95th percentile	8.7	5.4 ± 2.3*	22.5	3.4 ± 2.5*	64.7	1.9 ± 4.0
90th percentile	6.1	6.1 ± 2.3*	15.5	3.6 ± 2.6**	43.0	1.8 ± 4.0

*1% significant.
**5% significant.
***10% significant.

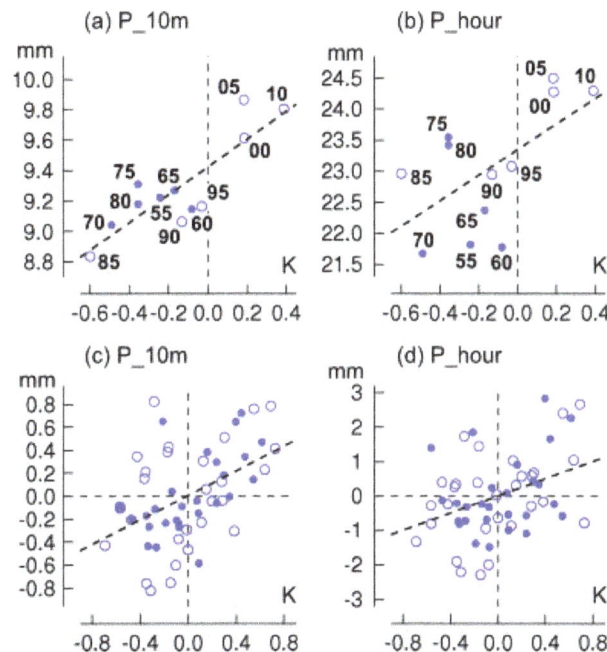

Figure 3. Relation between 5-year averages (a, b) and departures from the 5-year running average (c, d) in annual temperature (deviation from the 1981–2010 average) and annual 95th percentile precipitation (P_10m and P_hour). Open and closed circles indicate the values until and after 1980, respectively. Numbers in panels (a) and (b) indicate the last two digits of the end of 5-year period. The dashed line indicates the linear regression.

hand, Figure 3 (c) and (d) shows scatter diagrams of deviations from 5-year running averages in annual mean temperature and annual extremes of P_10m and P_hour, in order to examine their relationship with respect to year-to-year variation. Both P_10m and P_hour extremes have a positive correlation to temperature with a coefficient of 0.47 and 0.38, and linear regression corresponding to $5.8 \pm 3.3\%\,K^{-1}$ and $5.4 \pm 4.1\%\,K^{-1}$, respectively. This fact implies that, rather unexpectedly, extremes in short-time precipitation have a CC-like dependence on temperature even in year-to-year variations.

4. Concluding remarks

This study has revealed an increase of extreme short-time precipitation over Japan after 1980 in accordance with the rapid temperature increase. The increase is more conspicuous for P_10m than P_hour. Analysis for 1951–2010 has shown a close relationship between

temperature and extremes in P_10m and P_hour, with a rate comparable to or slightly higher than the CC relation of about 6 K per degree. In this respect, our analysis has confirmed a CC-like dependence of short-time precipitation extremes on temperature in the course of multidecadal climate change.

In comparison to the result of Lenderink *et al.* (2011), who showed that hourly extreme precipitation had a dependence of over $10\%\,K^{-1}$ on temperature, this study has yielded somewhat smaller values of $\Delta P/\Delta T$. A possible explanation to the difference is the use of temperature in our analysis, instead of dew-point temperature used by Lenderink *et al.* (2011). As relative humidity is found to have decreased in the latter 20th century over Japan (Fujibe *et al.*, 2005b), the long-term trend of dew-point temperature will be smaller than that of temperature, so that a larger value of $\Delta P/\Delta T$ may have been obtained if dew-point temperature is used. Moreover, the percentiles of 10-min and hourly precipitations used in

Table II. Correlation coefficient and $\Delta P/\Delta T$ between temperature and each precipitation index.

Precipitation extremes		Temperature	P_10m Correlation	P_10m $\Delta P/\Delta T$ (%/K)	P_hour Correlation	P_hour $\Delta P/\Delta T$ (%/K)	P_day Correlation	P_day $\Delta P/\Delta T$ (%/K)
Five-year averages	Annual maximum	Annual	0.86*	8.6 ± 3.6*	0.71*	6.7 ± 4.6*	0.21	2.8 ± 9.1
		Summer	0.63**	6.4 ± 5.6**	0.67*	6.3 ± 5.0**	0.32	4.3 ± 8.9
	Annual tenth	Annual	0.81*	11.1 ± 5.6*	0.75*	8.6 ± 5.3*	0.26	3.1 ± 7.9
		Summer	0.75*	10.3 ± 6.4*	0.69**	7.9 ± 5.9**	0.06	0.7 ± 8.3
	95th percentile	Annual	0.86*	9.7 ± 4.1*	0.60**	8.8 ± 8.3**	0.48	6.4 ± 8.2
		Summer	0.87*	9.9 ± 3.9*	0.76*	11.1 ± 6.8*	0.66**	8.8 ± 7.1**
	90th percentile	Annual	0.84*	11.1 ± 5.0*	0.62**	9.9 ± 8.8**	0.49	5.7 ± 7.3
		Summer	0.90*	12.0 ± 4.0*	0.75*	12.1 ± 7.4*	0.60**	7.2 ± 6.7**
Annual values	95th percentile	Annual	0.61*	7.1 ± 2.4*	0.48*	6.6 ± 3.2*	0.31**	6.0 ± 4.9**
		Summer	0.60*	5.2 ± 1.8*	0.33**	3.4 ± 2.6**	−0.04	−0.5 ± 3.9
Deviation from 5-year running mean	95th percentile	Annual	0.47*	5.8 ± 3.3*	0.38*	5.4 ± 4.1**	0.27**	6.6 ± 7.3***
		Summer	0.56*	4.2 ± 2.0*	0.19	1.7 ± 2.7	−0.24***	−3.6 ± 4.6

*1% significant.
**5% significant.
***10% significant.

the present analysis are those based on daily maximum values, and not obtained from continuous records as done in the study by Lenderink *et al.* (2011). This may be another reason for the different values of $\Delta P/\Delta T$ in this study and that of Lenderink *et al.* (2011), because this study may have underestimated the trends of percentile values if there are fewer precipitation events per day in recent years. A study using continuous records, for which digitization is underway by the Japan Meteorological Agency for hourly precipitation, will be needed to clarify the situation.

Unlike P_10m and P_hour, extremes in P_day show no significant dependence on temperature, except for the correlation of percentiles with summer temperature. This may reflect the high interannual variability of P_day, which is more strongly affected by the sporadicity of synoptic disturbances than short-time precipitation. However, the analysis of Fujibe *et al.* (2006) based on data for over 100 years (1901–2004) showed a statistical significant increase of annual maximum values of P_day at a rate of $0.89 \pm 0.72\%$ per decade. Because the warming rate of the last century is of the order of 0.1 K per decade, the annual maximum of P_day can be regarded to have increased at a rate of the order of $10\% \, K^{-1}$, namely, at a rate of the same order as the CC relation. This fact implies a CC-like increase of extreme daily precipitation corresponding to long-term warming, although it may not be detectable from data for a few decades.

Acknowledgements

The author is grateful to an anonymous reviewer for valuable comments. This study is part of 'Tokyo Metropolitan Area Convection Study for Extreme Weather Resilient Cities (TOMACS)' funded by 'Strategic Funds for the Promotion of Science and Technology' of JST/MEXT of Japan.

References

Allan RP, Soden BJ. 2008. Atmospheric warming and the amplification of precipitation extremes. *Science* **321**: 1481–1484, DOI: 10.1126/science.1160787.

Berg P, Haerter JO. 2013. Unexpected increase in precipitation intensity with temperature. A result of mixing of precipitation types? *Atmospheric Research* **119**: 56–61, DOI: 10.1016/j.atmosres.2011.05.012.

Berg P, Haerter JO, Thejll P, Piani C, Hagemann S, Christensen JH. 2009. Seasonal characteristics of the relationship between daily precipitation intensity and surface temperature. *Journal of Geophysical Research* **114**(D18102), 9pp, DOI: 10.1029/2009JD012008.

Berg P, Moseley C, Haerter JO. 2013. Strong increase in convective precipitation in response to higher temperatures. *Nature Geoscience* **6**: 181–185, DOI: 10.1038/ngeo1731.

Fujibe F. 2012. Evaluation of background and urban warming trends based on centennial temperature data in Japan. *Papers in Meteorology and Geophysics* **63**: 43–56, DOI: 10.2467/mripapers.63.43.

Fujibe F, Yamazaki N, Katsuyama M, Kobayashi K. 2005a. The increasing trend of intense precipitation in Japan based on four-hourly data for a hundred years. *SOLA* **1**: 41–44, DOI: 10.2151/sola.2005-012.

Fujibe F, Yamazaki N, Katsuyama M. 2005b. Long-term trends in the diurnal cycles of precipitation frequency in Japan. *Papers in Meteorology and Geophysics* **55**: 13–19, DOI: 10.2467/mripapers.55.13.

Fujibe F, Yamazaki N, Kobayashi K. 2006. Long-term changes of heavy precipitation and dry weather in Japan (1901–2004). *Journal of the Meteorological Society of Japan* **84**: 1033–1046, DOI: 10.2151/jmsj.84.1033.

Haerter JO, Berg P. 2009. Unexpected rise in extreme precipitation caused by a shift in rain type? *Nature Geoscience* **2**: 372–373, DOI: 10.1038/ngeo523.

Haerter JO, Berg P, Hagemann S. 2010. Heavy rain intensity distributions on varying time scales and at different temperatures. *Journal of Geophysical Research* **115**(D17102), 7pp, DOI: 10.1029/2009JD013384.

Hardwick Jones R, Westra S, Sharma A. 2010. Observed relationships between extreme sub-daily precipitation, surface temperature, and relative humidity. *Geophysical Research Letters* **37**(L22805), 5pp, DOI: 10.1029/2010GL045081.

IPCC. 2007. Contribution of Working Group I to the Fourth Assessment Report of the IPCC. In: *Climate Change 2007: The Physical Science Basis.* Solomon S, Qin D, Manning M, Chen Z, Marquis M, Averyt K, Tignor M, Miller HL (eds). Cambridge University Press: New York, NY; 1056.

Japan Meteorological Agency. 2011: Climate Change Monitoring Report 2010. 99. (available from http://ds.data.jma.go.jp/tcc/tcc/products/gwp/gwp.html)

Kao S-C, Ganguly AR. 2011. Intensity, duration, and frequency of precipitation extremes under 21st-century warming scenarios. *Journal of Geophysical Research* **116**(D16119), 14pp, DOI: 10.1029/2010JD015529.

Kharin VV, Zwiers FW, Zhang X, Hegerl GC. 2007. Changes in temperature and precipitation extremes in the IPCC ensemble of global coupled model simulations. *Journal of Climate* **20**: 1419–1444, DOI: 10.1175/JCLI4066.1.

Lenderink G, van Meijgaard E. 2008. Increase in hourly precipitation extremes beyond expectations from temperature changes. *Nature Geoscience* **1**: 511–514, DOI: 10.1038/ngeo262.

Lenderink G, Mok HY, Lee TC, van Oldenborgh GJ. 2011. Scaling and trends of hourly precipitation extremes in two different climate zones – Hong Kong and the Netherlands. *Hydrology and Earth System Sciences* **15**: 3033–3041, DOI: 10.5194/hess-15-3033-2011.

Min S, Zhang X, Zwiers FW, Hegerl GC. 2011. Human contribution to more-intense precipitation extremes. *Nature* **470**: 378–381, DOI: 10.1038/nature09763.

Miyajima J, Fujibe F. 2011. Climatology of extreme precipitation in Japan for different time scales. *SOLA* **7**: 157–160, DOI: 10.2151/sola.2011-040.

Muller CJ, O'Gorman PA, Back LE. 2011. Intensification of precipitation extremes with warming in a cloud-resolving model. *Journal of Climate* **24**: 2784–2800, DOI: 10.1175/2011JCLI3876.1.

Pall P, Allen MR, Stone DA. 2007. Testing the Clausius-Clapeyron constraint on changes in extreme precipitation under CO_2 warming. *Climate Dynamics* **28**: 351–363, DOI: 10.1007/s00382-006-0180-2.

Sugiyama M, Shiogama H, Emori S. 2010. Precipitation extreme changes exceeding moisture content increases in MIROC and IPCC climate models. *Proceedings of the National Academy of Sciences* **107**: 571–575, DOI: 10.1073/pnas.0903186107.

Utsumi N, Seto S, Kanae S, Maeda EE, Oki T. 2011. Does higher surface temperature intensify extreme precipitation? *Geophysical Research Letters* **38**(L16708), DOI: 10.1029/2011GL048426.

Yu R, Li J. 2012. Hourly rainfall changes in response to surface air temperature over eastern contiguous China. *Journal of Climate* **25**: 6851–6861, DOI: 10.1175/JCLI-D-11-00656.1.

10

Decadal change of the connection between summer western North Pacific Subtropical High and tropical SST in the early 1990s

Chao He[1,2]* and Tianjun Zhou[2,3]

[1]*Guangzhou Institute of Tropical and Marine Meteorology, China Meteorological Administration, Guangzhou, China*
[2]*LASG, Institute of Atmospheric Physics, Chinese Academy of Sciences, Beijing, China*
[3]*CCRC, Institute of Atmospheric Physics, Chinese Academy of Sciences, Beijing, China*

Correspondence to:
*C. He, Guangzhou Institute of
Tropical and Marine Meteorology,
Guangzhou 510080, China.
E-mail: hechao@mail.iap.ac.cn*

Abstract

A decadal change has occurred on the relationship between the summertime western North Pacific subtropical High (WNPSH) and the tropical Sea Surface Temperature anomalies (SSTA) in the early 1990s. After this decadal change, the interannual variability of the WNPSH is more strongly regulated by the SSTA over the equatorial central Pacific and the maritime continent. The earlier decay of El Niño in strong WNPSH years and the earlier development of El Niño in weak WNPSH years both contributed to this change. Such decadal change is found in the pre-industrial control simulation of a coupled model, suggesting it is probably an internal variability of the climate system.

Keywords: western North Pacific Subtropical High; ENSO; decadal change; internal climate variability

1. Introduction

The western North Pacific Subtropical High (WNPSH) in the lower troposphere is a crucial system to the East Asian climate in summer. The rainfall anomalies over the East Asia and the associated floods or droughts are regulated by the interannual variation of the WNPSH (e.g. Chang *et al.*, 2000; Yang and Sun, 2003; Zhang and Tao, 2003). The decadal change of the WNPSH also contributes to the climate regime shift in the late 1970s (Zhou *et al.*, 2009; Huang *et al.*, 2010; Xie *et al.*, 2010).

The interannual variability of the WNPSH is regulated by the SST anomalies (SSTA) within the tropical Indo-Pacific Ocean in summer, including the tropical Indian Ocean (TIO), the maritime continent (MC) and the equatorial central Pacific (CP). As a response to the El Niño in the previous winter, the warm SSTA over the TIO persists into the summer and reinforces the WNPSH via Kelvin wave induced Ekman divergence (Terao and Kubota, 2005; Li *et al.*, 2008; Xie *et al.*, 2009; Wu *et al.*, 2010a; Kosaka *et al.*, 2013). The summertime cold SSTA in the CP associated with developing La Niña drives an anomalous strong WNPSH as a Rossby wave response to its northwest (Wang *et al.*, 2013; Xiang *et al.*, 2013). The warm SSTA over the MC can also induce an anomalous strong WNPSH by enhancing the meridional overturning circulation, which ascends over the MC and descends over the western North Pacific (WNP) (Lu *et al.*, 2006; Sui *et al.*, 2007; Wu and Zhou, 2008; Wu *et al.*, 2009, Chung *et al.*, 2011).

It is recently discovered that the relationship between the anomalous WNPSH and the TIO SSTA forcing has strengthened after the late 1970s (Huang *et al.*, 2010; Xie *et al.*, 2010; Chowdary *et al.*, 2012), possibly due to the delayed decay of El Niño events and prolonged TIO capacitor effect (Huang *et al.*, 2010; Li *et al.*, 2012). However, it remains unknown whether the contributions of the MC and the CP to the WNPSH have experienced any decadal change. This motivates this study focusing on the following questions: is there any decadal change in the relative contributions of the SSTA over the TIO, the MC and the CP to the WNPSH in recent decades? If so, what is the mechanism? These questions will be addressed in this study.

2. Data and methods

This study employs the National Centers for Environmental Prediction-National Center for Atmospheric Research (NCEP-NCAR) reanalysis (hereafter NCEP1) (Kalnay *et al.*, 1996), the NCEP-DOE AMIP-II reanalysis (hereafter NCEP2) (Kanamitsu *et al.*, 2002) and the Extended Reconstruction SST (ERSST) v3 (Smith *et al.*, 2008) data. These data are referred to as 'observations' in the following. Linear trends of these data are removed before analysis. For the atmospheric phenomenon, we focus on the seasonal mean of summer [June, July and August (JJA)]. Running correlation is adopted to reveal the timing for the decadal change. Composite analysis (mean value of strong WNPSH years minus mean value of weak WNPSH

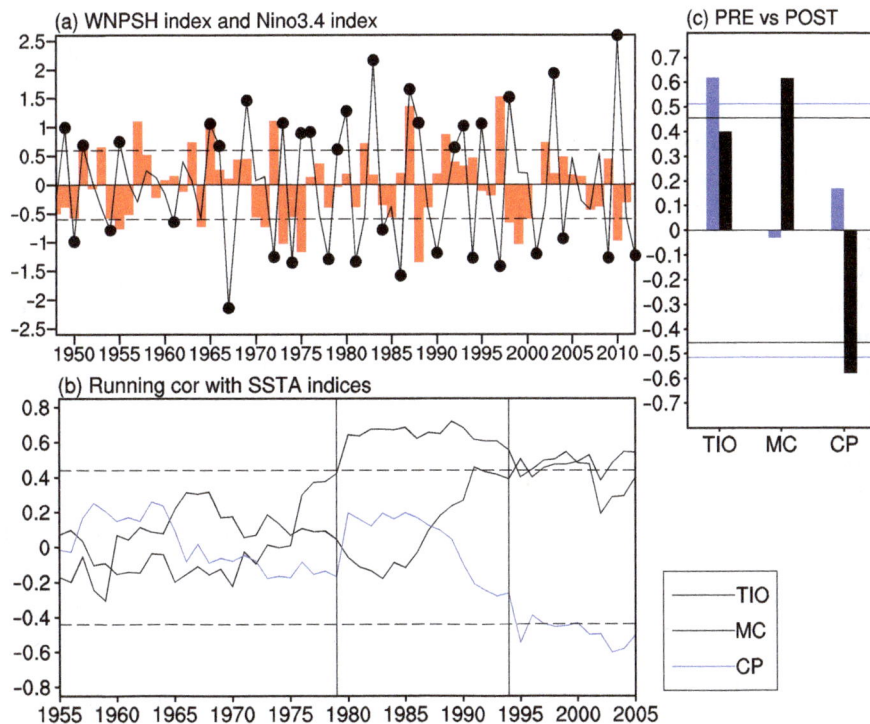

Figure 1. (a) The time series of the standardized WNPSH indices (non-dimensional) defined as the regional mean sea level pressure of 10°−25°N, 120°−150°E (black line). The anomalous strong (weak) WNPSH years with the WNPSH indices above 0.6 (below −0.6) are marked with black dots. The red bars are the time series of the JJA Nino3.4 index (unit: °C) (b) The 15-year running correlation coefficients between the WNPSH index and the three SST indices. The three SST indices include the TIO (10°S−10°N, 40°−100°E, black line), the MC (10°S−10°N, 100°−150°E, red line) and the CP (5°S−5°N, 170°−120°W, blue line). The horizontal dashed lines are the 90% confidence level for the correlation coefficients with 15 samples. The vertical solid lines represent the years 1979 and 1994. (c) The correlation coefficients of the WNPSH index with the three SST indices for 1979−1993 (blue bar) and 1994−2012 (red bar). The blue (red) horizontal lines are the 95% confidence level for correlation coefficients in 1979−1993 (1994−2012).

years) is performed for the periods before and after the decadal change, respectively. The pre-industrial control run (319 years in total) of the Community Earth System Model-Community Atmosphere Model version 5 (CESM1-CAM5) is adopted from the Coupled Model Intercomparison Project Phase 5 (CMIP5, https://pcmdi9.llnl.gov) (Taylor *et al.*, 2012).

3. Results

A WNPSH index is defined as the standardized mean sea level pressure over the WNP (10°−25°N, 120°−150°E), similar as the indices in many previous studies (e.g. Xie *et al.*, 2009; Xie *et al.*, 2010; Chowdary *et al.*, 2012). The JJA WNPSH indices derived from the NCEP1 is shown in Figure 1(a), together with the JJA Nino3.4 index. To identify the time for the decadal change, three SST indices are defined as the regional averaged SSTA in JJA, including the TIO (40°−100°E, 10°S−10°N), the MC (100°−150°E, 10°S−10°N) and the CP (170°−120°W, 5°S−5°N, i.e. Nino3.4 region) indices. The 15-year running correlation of the WNPSH index with these three SST indices are shown in Figure 1(b).

The correlation of the WNPSH with these three SST indices are all below the 90% significance level before

1979, suggesting their weak relationship. The positive correlation for the TIO index has become significant at the 90% confidence level after 1979, which was well studied in many previous studies (Huang *et al.*, 2010; Xie *et al.*, 2010; Chowdary *et al.*, 2012). A more remarkable decadal change has occurred around 1994, when the positive correlations for the MC index and the negative correlation for the CP index both become stronger and significant at the 90% level, while the positive correlation for the TIO index becomes slightly weaker but still important. The decadal change in the early 1990s seems not the opposite to the decadal change in late 1970s, so we only focus on the period 1979−2012. As many previous studies claimed that 1993/1994 is the time for multiple decadal changes over the WNP (e.g. Kwon *et al.*, 2005, 2007; Park *et al.*, 2010; Wu *et al.*, 2010b; Kajikawa and Wang, 2012), we compare the 1979−1993 (PRE) epoch with the 1994−2012 (POST) epoch in this study. The NCEP2 data is used in the following analyses because of its improved quality compared with the NCEP1 data (Kanamitsu *et al.*, 2002).

Figure 1(c) shows the correlation coefficients between WNPSH index and the three SST indices for the two epochs. The WNPSH index is significantly correlated with the TIO SSTA with a correlation coefficient of 0.62, but very weakly correlated with the SSTA over the MC and the CP in the PRE epoch. The correlation of the WNPSH with the TIO SSTA falls to 0.40 in the POST epoch, slightly below the 95% confidence level. Meanwhile, its correlation coefficients with the MC and the CP have risen to 0.61 and −0.58 in the POST epoch, both are significant at the 95% confidence level.

What are the differences in the anomalous SST and circulation patterns associated with the anomalous WNPSH between the two epochs? To address this question, we identified the anomalous strong (weak) WNPSH years whose WNPSH indices are above 0.6 (below −0.6) of the standard deviation (listed in Table 1), and made composite difference between the anomalous strong and weak WNPSH years, for the SSTA, 850 hPa wind, tropospheric temperature anomalies over the Indo-Pacific sector, and meridional overturning circulation anomaly averaged within

120°–150°E. Following previous studies (Xie *et al.*, 2009), the tropospheric temperature is calculated as the difference of the geopotential height between 200 and 850 hPa. The composite analyses are done for the two epochs, separately (Figure 2). Phenomena associated with the anomalous strong WNPSH are described in the following, while they are the opposite for the anomalous weak WNPSH.

In the PRE epoch (Figure 2(a)), the anomalous strong WNPSH is accompanied by significant warm SSTA over the TIO (shading in Figure 2(a)), consistent with Xie *et al.* (2009). There is also weak cold SSTA over the MC and warm SSTA over the CP and the eastern Pacific. The composite tropospheric temperature field (contours in Figure 2(a)) is characterized by a wedge-like warm anomaly over the TIO, which points toward the east and is associated with low-level easterly wind anomaly over

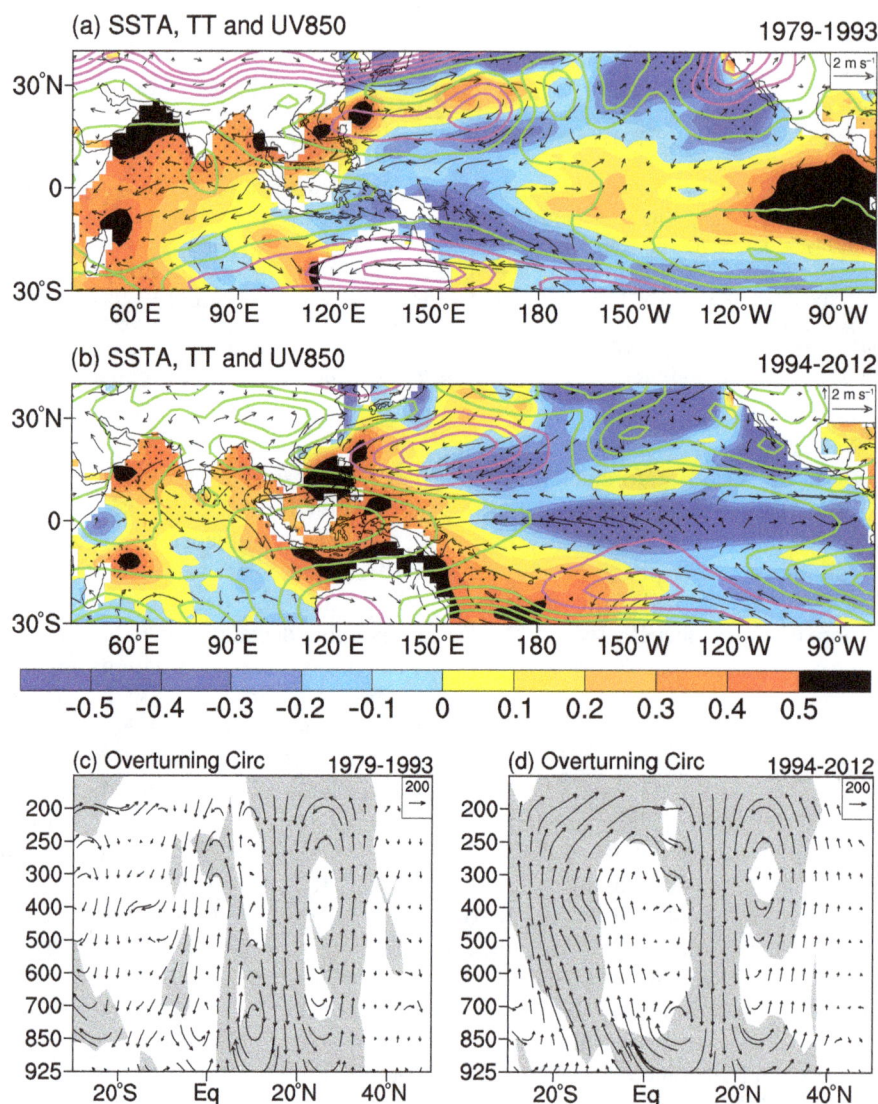

Figure 2. (a) The composite anomalies of SST (shading, unit: °C), tropospheric temperature (contour, unit: °C), and wind at 850 hPa (vector) for the anomalous WNPSH years in the PRE epoch. The contours are ±0.1, ±0.3, ±0.5 and ±0.7 °C with positive/negative contours in green/pink. Composite SSTA significant at the 95% confidence level are dotted. (c) The composite anomalies of the meridional overturning circulation (unit: kg m⁻¹ s⁻¹) averaged within 120°–150°E for the anomalous WNPSH years in the PRE epoch, values significant at the 95% confidence level are in gray shading. (b) and (d) are the same as (a) and (c) but for the POST epoch.

Table 1. Anomalous strong and weak WNPSH years selected for composite analysis.

	Strong WNPSH years	Weak WNPSH years
PRE (1979–1993)	1980, 1983, 1987, 1988 and 1993	1981, 1984, 1986 and 1990
POST (1994–2012)	1995, 1998, 2003 and 2010	1994, 1997, 2001, 2004, 2009, 2011 and 2012

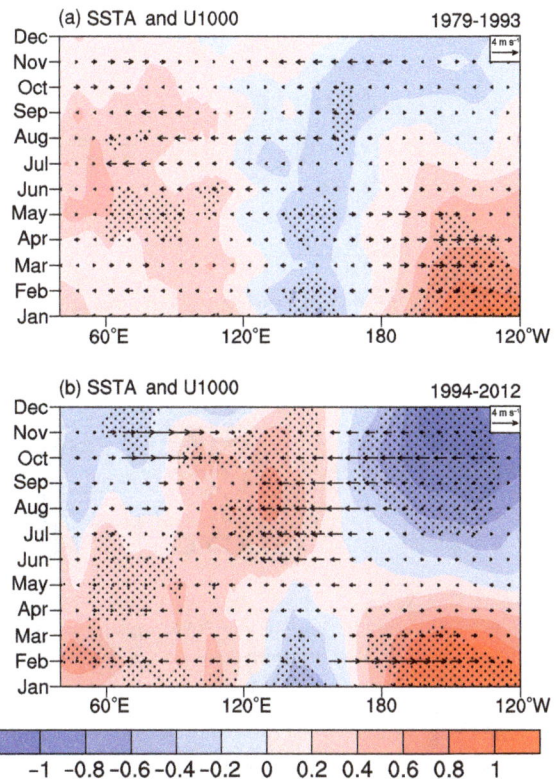

Figure 3. The composite SSTA (unit: °C) and 1000 hPa zonal wind (averaged between 10°S and 10°N) evolutions from January to December in the anomalous WNPSH years. The composite SSTA exceeding the 95% confidence level are marked with black dots. (a) is for the PRE epoch and (b) is for the POST epoch.

TIO and MC (vectors in Figure 2(a)). These features suggest a warm Kelvin wave emanating from the TIO, which is responsible for the anomalous anticyclone over the WNP via Kelvin wave induced Ekman divergence (Terao and Kubota, 2005; Xie et al., 2009). The cold anomaly over the WNP in the PRE epoch is displaced westward for about 10° in longitude than the POST epoch (Figure 2(a) and (b)), suggesting the westward extension of the WNPSH due to TIO warming is more evident in the PRE epoch (Chowdary et al., 2011).

During the POST epoch (Figure 2(b)), the SST and circulation pattern are quite different from the PRE epoch. The strongest warm SSTA is located over the MC instead of the TIO. Correspondingly, the warm anomaly in tropospheric temperature has shifted eastward, centered at the MC. The CP is dominated by significant cold SSTA, which may force the WNPSH via Rossby wave response (Wang et al., 2013). The equatorial western Pacific is dominated by easterly wind anomaly, which may link the cold SSTA over the CP and the warm SSTA over the MC via anomalous Walker circulation (Bjerknes, 1969, Wang et al., 2013). There is a more evident meridional wave-like pattern over the WNP in the tropospheric temperature field in the POST epoch than the PRE epoch, suggesting a stronger Pacific-Japan (PJ) pattern in the POST epoch (Kosaka et al., 2011, Kosaka et al., 2012). As PJ pattern can be invoked by the SSTA over the MC (Wu et al., 2009), it provides evidence for the enhanced impact from the MC in the POST epoch.

As the strongest warm SSTA shifts from the TIO to the MC in the POST epoch, the anomalous meridional overturning circulation connecting the MC and the WNP (Lu et al., 2006; Sui et al., 2007; Wu et al., 2009; Chung et al., 2011) becomes evident. This overturning circulation ascends near the equator and slightly south of the equator, flows northward in the upper troposphere, descends over the WNP at about 20°N, and flows southward toward the equator in the lower troposphere (Figure 2(d)). This overturning circulation is not seen in the PRE epoch (Figure 2(c)), suggesting the impact of MC SSTA on the WNPSH has been enhanced since 1993/1994. In summary, the influences of the MC and the CP on the WNPSH have strengthened after 1993/1994, while the influence of the TIO has weakened slightly.

Why is the WNPSH more strongly influenced by the MC and the CP rather than the TIO in the POST epoch? To answer this question, we show the composite

monthly evolution of the equatorial SSTA and 1000 hPa zonal winds anomalies (averaged between 10°S and 10°N) associated with anomalous WNPSH for the two epochs in Figure 3. The 1000 hPa wind is used instead of 850 hPa to identify possible interaction between SSTA and wind anomaly.

For the PRE epoch (Figure 3(a)), the warm SSTA over the CP associated with anomalous strong WNPSH persists until the early summer, and the warm SSTA over the TIO persists until the early fall. This is in favor of a prolonged influence of the TIO SSTA on the WNPSH (Xie et al., 2009; Xie et al., 2010; Chowdary et al., 2012; Li et al., 2012). Anomalous easterly winds over the MC and the TIO are induced by the warm SSTA of the TIO in summer, as a part of the warm Kelvin wave shown in Figure 2(a).

During the POST epoch (Figure 3(b)), the warm SSTA over the CP dissipates in spring (April), much earlier than the PRE epoch. As a result, the warm SSTA over the TIO decays in early summer (Xie et al., 2009). This explains the slightly weakened influence of the TIO SSTA on the WNPSH after 1993/1994, consistent with previous studies that early decay of El Nino-Southern Oscillation (ENSO) is favorable for the early damping of TIO capacitor (Xie et al., 2010; Chowdary et al., 2012; Li et al., 2012). Furthermore,

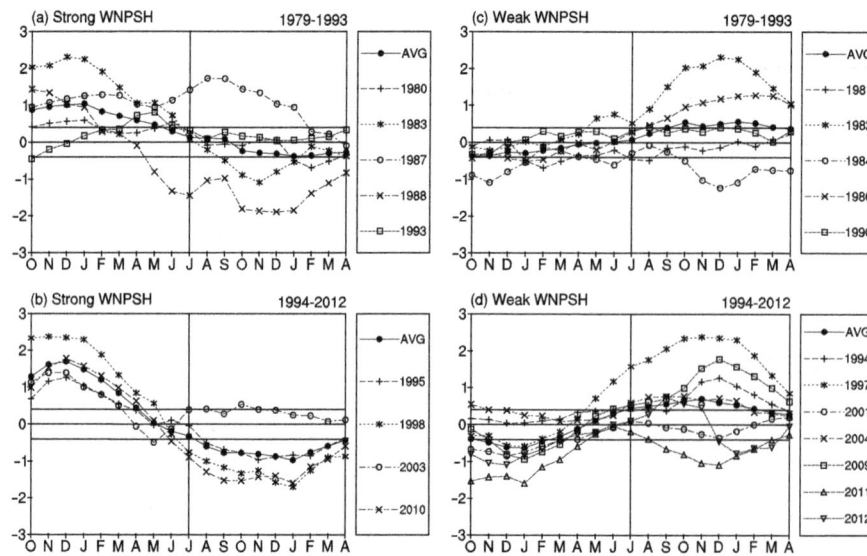

Figure 4. Time series for the monthly evolution of Nino3.4 indices in all of the anomalous WNPSH years. (a) and (b) are for the strong WNPSH years in the PRE and POST epochs, respectively. (c) and (d) are for the weak WNPSH years in the PRE and POST epochs, respectively. Red line with solid dots is the average of the displayed years in each figure.

La Niña events develop in the early summer of anomalous strong WNPSH years, with simultaneous warming of the MC (Figure 3(b)). The warm SSTA over the MC is closely coupled with the developing La Niña in the CP via anomalous Walker circulation, as evidenced by the low-level easterly wind anomalies within 120°–180°E emerging in early summer (Figure 3(b)). This is consistent with previous claims that warm SSTA over the MC is associated with developing La Niña (Wang *et al.*, 2013; Xiang *et al.*, 2013) or phase transition from El Niño to La Niña (Chung *et al.*, 2011).

How about the ENSO conditions in the anomalous strong and weak WNPSH years, respectively? The Nino3.4 indices of each anomalous WNPSH year for the two epochs are shown in Figure 4. According to the definition of ENSO event (Trenberth, 1997), only two (1983 and 1988) of the five anomalous strong WNPSH years in PRE epoch are associated with developing La Niña, but three (1995, 1998 and 2010) of the four anomalous strong WNPSH years in POST epoch are associated with developing La Niña (Figure 4(a) and (b)). On average, the El Niño decays 1 month earlier in the POST epoch than the PRE epoch (red lines in Figure 4(a) and (b)). For the anomalous weak WNPSH years, only two (1982 and 1986) of the five years in PRE epoch are associated with developing El Niño, but five (1994, 1997, 2004, 2009 and 2012) of the seven years in POST epoch are associated with developing El Niño (Figure 4(c) and (d)). On average, the averaged Nino3.4 index exceeds 0.5 °C in September for PRE epoch while in July for POST epoch (red lines in Figure 4(c) and (d)). The earlier development of El Niño is responsible for the strengthened influence of CP on the WNPSH in the POST epoch. Previous studies showed the timing for the ENSO decay is crucial for the tropical SSTA pattern and WNPSH in summer (Xie *et al.*, 2010; Li *et al.*,

2012). Our results show that the timing for the ENSO development is also important.

In summary, our observational results show that the relationship between the WNPSH and the tropical SSTA has undergone a remarkable decadal change in the early 1990s. After this decadal change, the interannual variation of the WNPSH is more strongly regulated by the SSTA over the CP and the MC than the TIO. This decadal change is consistent with the weakened influence of the decaying ENSO in the previous winter and the intensified influence of the developing ENSO.

Is this decadal change related to the external forcing or due to internal climate variability? If such decadal change can be identified in a pre-industrial control run of coupled models, it should probably be the internal variability of the climate system (Taylor *et al.*, 2012). We adopted the 319-year pre-industrial control run of CESM1-CAM5 here. The running correlation of the WNPSH index with the JJA SSTA along the equator is shown in Figure 5(a). The running correlations of the WNPSH index with the Nino3.4 indices in the previous winter (DJF, red line) and the subsequent winter (NDJ, blue line) are shown in Figure 5(b).

The WNPSH indices in summer are closely correlated with the TIO SSTA but weakly correlated with the MC and the CP SSTA in certain epochs, such as the model year 70–110 and 210–245 (Figure 5(a)), reminiscent the PRE epoch in the observation. In these epochs, the positive (negative) correlation of the WNPSH with the ENSO in the previous (subsequent) winter is significant (insignificant) at the 95% confidence level, suggesting the dominant impact of the decaying ENSO (Figure 5(b)). The composite summer SSTA and circulation anomalies associated with anomalous WNPSH in these epochs (figure not shown) are also similar as the PRE epoch in observation (Figure 2(a)). During some other epochs such as the model year 10–50, 110–200

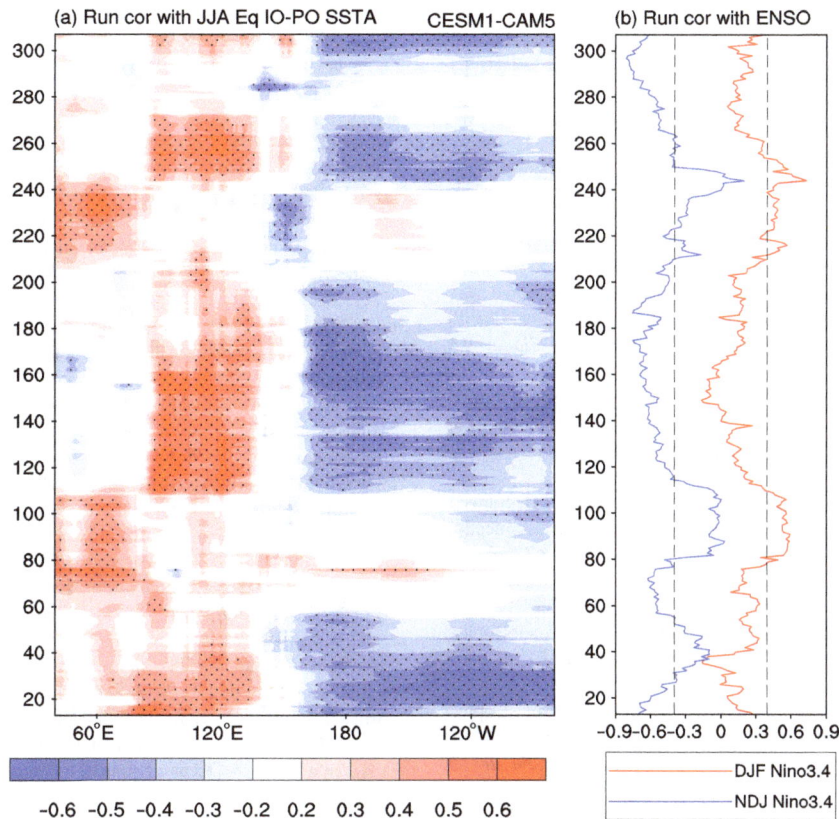

Figure 5. The relationship between the summer WNPSH and the SST in the pre-industrial control run of the CESM1-CAM5. (a) The time-longitude section of the 25-year running correlation between the anomalous WNPSH and the SSTA averaged between 10°S and 10°N in JJA. The abscissa is the longitude and the ordinate is the model-year. The dotted regions indicate the correlation coefficients are significant at the 95% confidence level. (b) The 25-year running correlation of the WNPSH index and the Nino3.4 indices in DJF (red line) and NDJ (blue line). The ordinate is the model-year. The vertical dashed black lines are the 95% confidence level for 25-year correlation.

and 245–270, the WNPSH is more affected by the MC and the CP but less influenced by the TIO (Figure 5(a)), reminiscent the POST epoch in the observation. In these epochs, the WNPSH index is significantly (insignificantly) correlated with the ENSO in the subsequent (previous) winter at the 95% confidence level, suggesting the dominant impact of the developing ENSO (Figure 5(b)). The composite summer SST and circulation anomalies (figure not shown) are similar as the POST epoch in observation (Figure 2(b)). Analogs of the 1993/1994 decadal change can be found around the 110th and 245th years of the pre-industrial control run, suggesting the decadal change in 1993/1994 is an internal variability of the climate system.

4. Summary

The anomalous WNPSH in summer is remotely forced by the SSTA over the tropical Indo-Pacific Ocean, including the TIO, the MC and the CP. Both observational and model results show that the SSTA of the TIO is related to the decaying ENSO of the previous winter, while the SSTA over the MC and the CP are related to the developing ENSO. By investigating the relative contributions of these three regions, it is found that the

relationships of the WNPSH with these three regions have experienced a decadal change in the early 1990s. After this change, the WNPSH is more strongly regulated by the SSTA over the CP and the MC than the TIO. For the strong WNPSH years, the El Niño events in the preceding winter decay earlier after 1993/1994. For the weak WNPSH years, the El Niño events develop earlier in summer after 1993/1994. The changes in the developing and decaying phases of ENSO have both contributed to this decadal change.

By analyzing the pre-industrial control run of the coupled model CESM1-CAM5 without any external forcing changes, it is found that the simulated relationship between the WNPSH and the tropical SST also shows decadal variations, similar as in the observation. Analogs for the 1993/1994 decadal change can be found in the pre-industrial control run of the CESM1-CAM5, suggesting this decadal change probably reflects the internal variability of the climate system.

Acknowledgements

The authors wish to thank the two anonymous reviewers who provided valuable comments and suggestions on the manuscript. Mr Fengfei Song and Miss Lu Dong also helped the authors to improve the quality of the manuscript. This work is jointly supported by National Program on Key Basic Research Project

(2014CB953901), China Meteorological Administration Special Public Welfare Research Fund (GYHY201406001), and National Natural Science Foundation of China (Nos. 41125017, 41205069, 41330423, 41375095).

References

Bjerknes J. 1969. Atmospheric teleconnections from equatorial Pacific. *Monthly Weather Review* **97**: 163–172.

Chang CP, Zhang YS, Li T. 2000. Interannual and interdecadal variations of the East Asian summer monsoon and tropical Pacific SSTs. Part I: Roles of the subtropical ridge. *Journal of Climate* **13**: 4310–4325.

Chowdary JS, Xie SP, Luo JJ, Hafner J, Behera S, Masumoto Y, Yamagata T. 2011. Predictability of Northwest Pacific climate during summer and the role of the tropical Indian Ocean. *Climate Dynamics* **36**: 607–621.

Chowdary JS, Xie SP, Tokinaga H, Okumura YM, Kubota H, Johnson N, Zheng XT. 2012. Interdecadal variations in ENSO teleconnection to the Indo-Western Pacific for 1870-2007. *Journal of Climate* **25**: 1722–1744.

Chung P-H, Sui C-H, Li T. 2011. Interannual relationships between the tropical sea surface temperature and summertime subtropical anticyclone over the western North Pacific. *Journal of Geophysical Research-Atmospheres* **116**: D13111.

Huang G, Hu KM, Xie SP. 2010. Strengthening of tropical Indian Ocean teleconnection to the Northwest Pacific since the mid-1970s: an atmospheric GCM study. *Journal of Climate* **23**: 5294–5304.

Kajikawa Y, Wang B. 2012. Interdecadal change of the South China Sea summer monsoon onset. *Journal of Climate* **25**: 3207–3218.

Kalnay E, Kanamitsu M, Kistler R, Collins W, Deaven D, Gandin L, Iredell M, Saha S, White G, Woollen J, Zhu Y, Chelliah M, Ebisuzaki W, Higgins W, Janowiak J, Mo KC, Ropelewski C, Wang J, Leetmaa A, Reynolds R, Jenne R, Joseph D. 1996. The NCEP/NCAR 40-year reanalysis project. *Bulletin of the American Meteorological Society* **77**: 437–471.

Kanamitsu M, Ebisuzaki W, Woollen J, Yang SK, Hnilo JJ, Fiorino M, Potter GL. 2002. NCEP-DOE AMIP-II reanalysis (R-2). *Bulletin of the American Meteorological Society* **83**: 1631–1643.

Kosaka Y, Xie SP, Nakamura H. 2011. Dynamics of interannual variability in summer precipitation over East Asia. *Journal of Climate* **24**: 5435–5453.

Kosaka Y, Chowdary JS, Xie SP, Min YM, Lee JY. 2012. Limitations of seasonal predictability for summer climate over East Asia and the Northwestern Pacific. *Journal of Climate* **25**: 7574–7589.

Kosaka Y, Xie SP, Lau N-C, Vecchi GA. 2013. Origin of seasonal predictability for summer climate over the Northwestern Pacific. *Proceedings of the National Academy of Sciences* **110**: 7574–7579.

Kwon M, Jhun JG, Wang B, An SI, Kug JS. 2005. Decadal change in relationship between East Asian and WNP summer monsoons. *Geophysical Research Letters* **32**: L16709.

Kwon M, Jhun J-G, Ha K-J. 2007. Decadal change in East Asian summer monsoon circulation in the mid-1990s. *Geophysical Research Letters* **34**: L21706.

Li SL, Lu J, Huang G, Hu KM. 2008. Tropical Indian Ocean Basin warming and East Asian Summer monsoon: a multiple AGCM study. *Journal of Climate* **21**: 6080–6088.

Li Q, Ren RC, Cai M, Wu GX. 2012. Attribution of the summer warming since 1970s in Indian Ocean Basin to the inter-decadal change in the seasonal timing of El Niño decay phase. *Geophysical Research Letters* **39**: L12702.

Lu R, Li Y, Dong BW. 2006. External and internal summer atmospheric variability in the western North Pacific and East Asia. *Journal of the Meteorological Society of Japan Series II* **84**: 447–462.

Park J-Y, Jhun J-G, Yim S-Y, Kim W-M. 2010. Decadal changes in two types of the western North Pacific subtropical high in boreal summer associated with Asian summer monsoon/El Nino-Southern Oscillation connections. *Journal of Geophysical Research: Atmospheres* **115**: D21129.

Smith TM, Reynolds RW, Peterson TC, Lawrimore J. 2008. Improvements to NOAA's historical merged land–ocean surface temperature analysis (1880–2006). *Journal of Climate* **21**: 2283–2296.

Sui CH, Chung PH, Li T. 2007. Interannual and interdecadal variability of the summertime western North Pacific subtropical high. *Geophysical Research Letters* **34**: L11701.

Taylor KE, Stouffer RJ, Meehl GA. 2012. An overview of CMIP5 and the experiment design. *Bulletin of the American Meteorological Society* **93**: 485–498.

Terao T, Kubota T. 2005. East-west SST contrast over the tropical oceans and the post El Niño western North Pacific summer monsoon. *Geophysical Research Letters* **32**: L15706.

Trenberth KE. 1997. The definition of El Niño. *Bulletin of the American Meteorological Society* **78**: 2771–2777.

Wang B, Xiang B, Lee J-Y. 2013. Subtropical High predictability establishes a promising way for monsoon and tropical storm predictions. *Proceedings of the National Academy of Sciences* **110**: 2718–2722.

Wu B, Zhou TJ. 2008. Oceanic origin of the interannual and interdecadal variability of the summertime western Pacific subtropical high. *Geophysical Research Letters* **35**: L13701.

Wu B, Zhou TJ, Li T. 2009. Seasonally evolving dominant interannual variability modes of East Asian climate. *Journal of Climate* **22**: 2992–3005.

Wu B, Li T, Zhou TJ. 2010a. Relative contributions of the Indian Ocean and local SST anomalies to the maintenance of the western North Pacific anomalous anticyclone during the El Niño decaying summer. *Journal of Climate* **23**: 2974–2986.

Wu RG, Wen ZP, Yang S, Li YQ. 2010b. An interdecadal change in Southern China summer rainfall around 1992/93. *Journal of Climate* **23**: 2389–2403.

Xiang B, Wang B, Yu W, Xu S. 2013. How can anomalous western North Pacific subtropical high intensify in late summer? *Geophysical Research Letters* **40**: 2349–2354.

Xie SP, Hu KM, Hafner J, Tokinaga H, Du Y, Huang G, Sampe T. 2009. Indian Ocean capacitor effect on Indo-Western Pacific climate during the summer following El Niño. *Journal of Climate* **22**: 730–747.

Xie SP, Du Y, Huang G, Zheng XT, Tokinaga H, Hu KM, Liu QY. 2010. Decadal shift in El Niño influences on Indo-Western Pacific and East Asian climate in the 1970s. *Journal of Climate* **23**: 3352–3368.

Yang H, Sun SQ. 2003. Longitudinal displacement of the subtropical high in the western Pacific in summer and its influence. *Advances in Atmospheric Sciences* **20**: 921–933.

Zhang QY, Tao SY. 2003. The anomalous subtroical anticyclone in western Pacific and their association with circulation over East Asia during summer. *Chinese Journal of Atmospheric Sciences* **27**: 369–380.

Zhou TJ, Yu RC, Zhang J, Drange H, Cassou C, Deser C, Hodson DLR, Sanchez-Gomez E, Li J, Keenlyside N, Xin XG, Okumura Y. 2009. Why the western Pacific subtropical high has extended westward since the late 1970s. *Journal of Climate* **22**: 2199–2215.

11

Variability of weather regimes in the North Atlantic-European area: past and future

Elke Hertig and Jucundus Jacobeit

Institute of Geography, University of Augsburg, 86159 Augsburg, Germany

*Correspondence to:
E. Hertig, Institute of Geography, University of Augsburg, Universitaetsstrasse 1a, 86159 Augsburg, Germany.
E-mail: elke.hertig@geo.uni-augsburg.de

Abstract

The concept of weather regimes represents a process-oriented method of organizing the varying states of the atmospheric circulation. We define weather regimes as preferred, or recurrent, circulation patterns. We use a suite of reanalysis products and general circulation model (GCM) simulations to assess the reproducibility and variability of the regimes. We find distinct variability of the regimes in observational periods as well as in future projections. Most notable is the high variability of the North Atlantic Oscillation (NAO) regime anomaly patterns in the GCM simulations which is not evident in reanalyses, and the substantial increase of variability regarding the frequency of occurrence of the Atlantic ridge regime and the NAO+ regime.

Keywords: weather regimes; non-stationarities; climate change

1. Introduction

The idea of classifying atmospheric dynamics in the extra-tropics into a defined number of states, so-called weather regimes (Vautard, 1990, Michelangeli *et al.*, 1995), evolves from the principle that to a certain degree the number of possible states of the large-scale circulation is finite. Regimes are defined as recurrent and/or persistent and/or quasi stationary states of the atmosphere (Michelangeli *et al.*, 1995, Stephenson *et al.*, 2004). The concept of weather regimes represents a process-oriented method of organizing the varying states of the atmospheric circulation. Over the North Atlantic-European domain weather regimes are highly correlated to anomalies of local surface temperature and precipitation (Plaut and Simonnet, 2001, Yiou and Nogaj, 2004). There are clear spatial patterns of precipitation and temperature associated with the regimes. The positive North Atlantic Oscillation (NAO) phase during winter, for example, is associated with above-normal precipitation and temperature over northern Europe and Scandinavia and below-normal precipitation over southern parts of central Europe, southern Europe as well as Northwest Africa (Wanner *et al.*, 2001, Hurrell *et al.*, 2003). Opposite patterns of temperature and precipitation anomalies are typically found during strong NAO− phases. During NAO− phases the cold-day frequency increases notably over Scandinavia due to cool or cold air advection from the northwest, whereas it decreases over Iberia due to more cyclonic conditions (Plaut and Simonnet, 2001). Concerning extremes, the NAO+ regime is connected with heavy precipitation over Northern Europe and drought periods over the Mediterranean area (Yiou and Nogaj, 2004). The NAO− regime causes heavy precipitation over Southern Europe. In contrast, the blocking regime controls the drought periods over large parts of Central Europe and Eastern Scandinavia. It also affects maximum temperatures over Scandinavia as well as minimum temperatures over south-eastern Europe (Plaut and Simonnet, 2001, Yiou and Nogaj, 2004). Thus, the practical interest in the classification of the large-scale circulation into a few recurrent patterns lies in the observation that local weather anomalies depend to a large extent on the large-scale atmospheric flow. If weather regimes can be reliably reproduced by general circulation models (GCMs), they provide a tool for inferring regional to local climate change via statistical downscaling which derives statistical relationships between the large scale (e.g. the atmospheric circulation) and regional to local scales (e.g. surface climate variables). The description of the large-scale circulation through weather regimes includes the advantage of being able to provide information on the physical processes of the atmosphere governing regional climate change.

In this study we focus on assessing the degree of correspondence of some GCMs of the most recent generation with reanalyses data. In addition, we address the degree of variability of weather regimes within the individual data sets. Thus, we assess the temporal variability of regimes in the reanalyses, compare GCMs versus reanalyses over the historical period, and evaluate differences between historical, Representative Concentration Pathways (RCP)4.5 and RCP8.5 scenario conditions. Section 2 describes data and methods used in this study. In Section 3 we present and discuss the results of our analysis, which is followed by the conclusions in Section 4.

2. Data and methodology

2.1. Data

National Centers for Environmental Prediction (NCEP) reanalysis (Kalnay *et al.*, 1996, Kistler *et al.*, 2001) for the period 1950–2010 and the European Centre for Medium-Range Weather Forecasts (ECMWF) reanalysis for the period 1979–2012 (ERA-Interim, Dee *et al.*, 2011) were used. Daily geopotential height fields at the 700 hPa level in the North Atlantic-European area (20°N–70°N, 70°W–50°E) were obtained on a 2.5 × 2.5 degree horizontal resolution. Following earlier studies (Vautard, 1990, Michelangeli *et al.*, 1995), we chose the 700 hPa level, which represents a highly relevant level regarding precipitation processes over the European-Mediterranean area (Hertig and Jacobeit, 2013).

700 hPa geopotential height data were taken from a three-member MPI-ESM-LR (Max Planck Institute Earth System Model running on low resolution grid) ensemble with different initial conditions of each run, from HadGEM2-ES (Hadley Global Environment Model 2-Earth System), and from the first member of a five-member CanESM2 (second generation Canadian Earth System Model) ensemble. Model selection is not exhaustive, but serves to exemplarily highlight large-scale circulation variability within and across different GCM data sets. Runs with historical, RCP4.5 scenario, and RCP8.5 scenario (Van Vuuren *et al.*, 2011) conditions performed for the Coupled Model Intercomparison Project round 5 (CMIP5) were downloaded from the CMIP5 archive (http://pcmdi9.llnl.gov/esgf-web-fe/). We used the period 1950–2005 of the historical runs (1980–2005 in case of HadGEM2-ES) and the period 2006–2100 of the scenario runs (2006–2099 for HadGEM2-ES). The horizontal resolution of the model output data was fitted to that of the reanalysis data (2.5 × 2.5 degree) by ordinary kriging.

2.2. Methodology

Weather regimes are obtained by classifying the daily 700 hPa geopotential height fields in the North Atlantic-European area. As a first step we performed a principal component analysis (PCA, Preisendorfer, 1988) of the daily winter data (December to February, DJF), in order to reduce dimensions of the data. As results may be sensitive to the number of principal components (PCs) kept, we considered all solutions from 5 to 15 PCs, capturing between 55 and 90% of the total variance. To the corresponding PCs of each solution we applied the hierarchical clustering method of Ward (Ward, 1963, Cheng and Wallace, 1993) and used the resulting clusters as 'seeds' for the optimising k-means algorithm (Michelangeli *et al.*, 1995). We kept four clusters (regimes), providing a stable partitioning according to earlier publications (Vautard, 1990, Michelangeli *et al.*, 1995, Yiou and Nogaj, 2004). For each regime we calculated the geopotential composite

patterns and the regime anomalies. Composites were derived by calculating the mean 700 hPa geopotential height field from all the days belonging to a specific regime. Anomalies were calculated as the deviations of the standardized composite pattern from the standardized long-term mean geopotential height field of all days in the time period considered. We applied no area weighting within pattern standardization. In addition, we computed the number of days per winter spent in each regime. For each data set considered, the similarity, given by the correlation coefficients, of the particular regime anomalies and the NCEP regime anomalies is used to select the optimum number of PCs.

The whole study period considered for a particular data set was split into 31-year sub-periods, each shifted by 1 year. When the end of the whole time series was reached, years from the beginning of the time series were successively included in order to avoid a more frequent inclusion of years located in the middle of the time series. For each 31-year period, we calculated the weather regimes using the statistical approach presented in the previous paragraph. No analysis was done for the HadGEM2 historical run as data was available only for the years 1980–2005.

For each regime we compare the regime anomalies from the different data sets by using Taylor diagrams (Taylor, 2001). These diagrams can be used to graphically summarize how closely the model anomaly patterns match a reference. NCEP reanalysis regime anomalies are used as 'observational' reference. The similarity between two patterns is quantified in terms of their correlation, their root-mean-square (RMS) difference, and their standard deviations.

3. Results and discussion

Significance testing in previous studies (Michelangeli *et al.*, 1995, Plaut and Simonnet, 2001) yields the result that the optimum solution is four classes (regimes) for the Atlantic domain. It should be noted, however, that there is an ongoing scientific debate on the appropriateness of the description of the large-scale circulation by multiple weather regimes (e.g. Stephenson *et al.*, 2004), and particularly by four classes for the Atlantic domain (e.g. Christiansen, 2007). We classified 5415 NCEP daily winter 700 hPa geopotential height maps in the North Atlantic-European area (20°N–70°N, 70°W–50°E) using seven PCs with 68% of explained variance (EV) and subsequent cluster analysis obtaining the four patterns displayed in Figure 1. The choice of seven PCs is motivated by the highest similarity of the resulting regimes to the patterns obtained in previous studies (Vautard, 1990, Yiou and Nogaj, 2004). We denote them according to Yiou and Nogaj (2004): NAO+ (the positive phase of the North Atlantic Oscillation, zonal pattern of Vautard, 1990), blocking, Atlantic ridge, and NAO− (the negative phase of the North Atlantic Oscillation, Greenland Anticyclone of Vautard, 1990). The anomaly

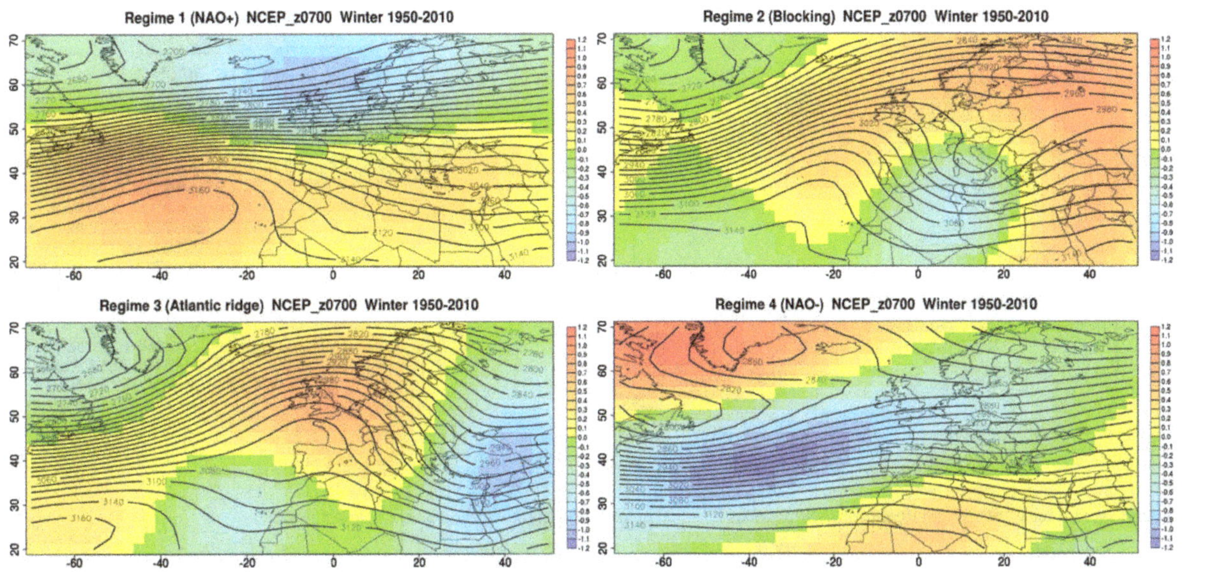

Figure 1. Weather regimes over the North Atlantic-European area. Shown are the regimes estimated from the 700 hPa daily winter (DJF) geopotential heights of the NCEP reanalysis. Colours indicate the regime anomalies and the contour lines show the geopotential composite pattern for each regime.

centre of the Atlantic ridge pattern is somewhat displaced to the east. The amplitude of the positive centre of the blocking pattern is weaker and the centre is more spread out. These differences can be attributed to differences in the clustering technique, height level, spatial domain, and time period considered. The correlation coefficients between the residence times (cumulative number of days spent in a cluster per season) of the two NAO regimes with the NAO index as published by the Climate Prediction Center (http://www.cpc.ncep.noaa.gov/data/teledoc/nao.shtml) are 0.74 (NAO+) and −0.78 (NAO−).

We also computed the weather regimes from the ERA-Interim data using eight PCs (71% EV). In addition, regimes were obtained from various GCM simulations comprising two different emission scenarios (RCP4.5 and RCP8.5), multiple runs for each scenario, and output of three different GCMs (MPI-ESM-LR, HadGEM2-ES, and CanESM2). Figure 2 indicates that the regime anomalies of the two reanalysis data sets are closely related with similar standard deviations and correlation coefficients mostly exceeding 0.9 (0.8 for the Atlantic ridge). The regimes can also be reproduced in the three MPI-ESM-LR historical runs using, depending on the particular run, 7, 12, and 5 PCs with EVs of 68, 85, and 57%, respectively. Figure 2 shows differences in the standard deviations mostly less than 0.1, RMS differences less than 0.3, and correlation coefficients greater than 0.7. Similar characteristics are found for the HadGEM2-ES historical run using 13 PCs (87% EV). In contrast, NAO+ and blocking regime anomalies computed from the CanESM2 historical run using 11 PCs (81% EV) diverge considerably from the NCEP regimes. Correlation coefficients drop below 0.5 and RMS differences reach values of nearly 0.4.

We also assessed the regimes under RCP4.5 and RCP8.5 scenario conditions using 11, 13, and 9 PCs for

the three MPI-ESM-LR RCP4.5 runs (83, 87, and 77% EV), and 15, 10, and 11 PCs in case of the RCP8.5 scenario runs (91, 82, and 85% EV). For HadGEM2-ES 8 PCs (EV 74%) and 14 PCs (EV 90%), for CanESM2 5 PCs (EV 55%) and 9 PCs (EV 78%) are used to obtain the regimes under RCP4.5 and RCP8.5 scenario conditions, respectively. We compare the two scenarios, the different GCMs, and the runs under different initial conditions. The NCEP weather regimes are taken as a reference, again, in this context simply as an operational reference. We find no regularities or systematic behaviour regarding the reproducibility of the regime anomalies. Correlation coefficients, standard deviations, and RMS differences vary in a non-systematic way (Figure 2).

The number of PCs used to obtain the best solution in terms of highest correlation coefficients between the regime anomalies of a specific data set and the NCEP reference regimes, is highly variable across the different data sets. Looking at this characteristic in detail, we find that the methodology is sensitive to the choice of the number of PCs, hence that PCA dimensional reduction influences subsequent cluster results. Some PC solutions within a specific data set do not lead to a selective classification into the four regimes. However, for all data sets considered, at least one PC solution allows for a positive classification. As an alternative approach one could use common centroids (i.e. project anomalies of the GCMs onto the NCEP centroids), as done for example by Cattiaux *et al.* (2013). According to Cattiaux *et al.* (2013) similar patterns are obtained for the majority of the investigated 20 CMIP5 models when computing common centroids compared with centroids for each model individually, and the authors conclude that results are not sensitive to the centroid choice. This gives further confidence in our methodology. When applying the methodology of individual

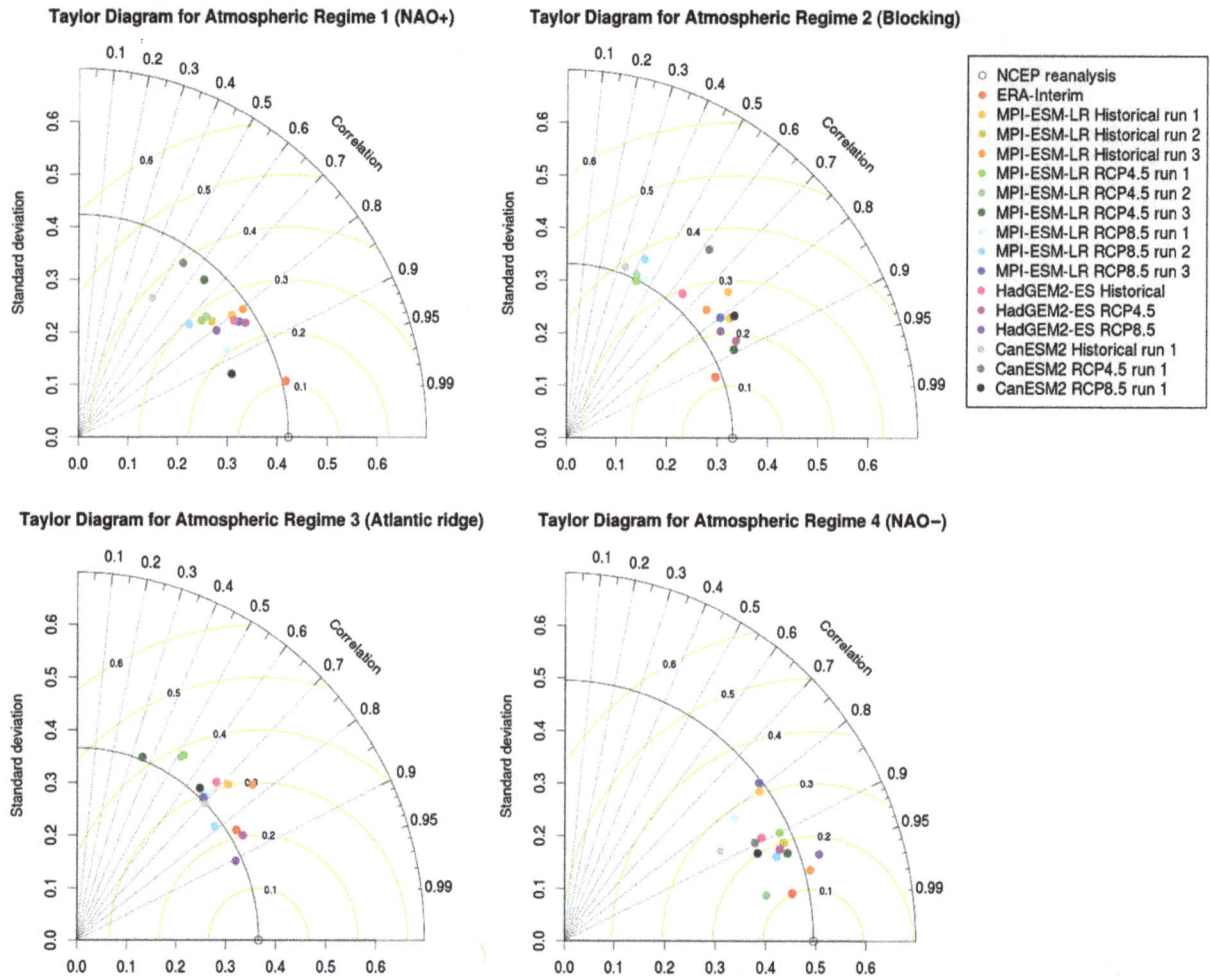

Figure 2. Taylor diagrams displaying the statistical comparison with the NCEP weather regimes of ERAInterim and GCM estimates. GCM data are historical runs, RCP4.5 runs, and RCP8.5 runs from three MPI-ESM-LR members, one HadGEM2-ES run, and the first member of CanESM2. Shown are, for each weather regime, the similarity of the anomaly patterns expressed in terms of the correlation, their root-mean square differences and their standard deviations.

cluster analyses to the 31-year sub-intervals, the choice is also motivated by the potential application of the weather regimes as predictors in a cross-validated statistical downscaling setting. However, the stability of the clustering results may depend on the period considered (Christiansen, 2007). Specific regimes occur more frequently in some sub-intervals due to decadal variability (for the NAO see, e.g. Wanner *et al.*, 2001). In the NCEP reanalysis the range of days belonging to a specific weather regime across the different sub-intervals is 16–28 days per winter (Table I), which yields cluster sizes between 496 and 868 days for a 31-year period. Thus, in all sub-intervals each regime is represented by a minimum of about 500 days. ERA-Interim and the GCM runs show similar results (with the exception of the HadGEM2-ES RCP8.5 run, see Table I).

In terms of overall reproducibility, the consistent stable regime across all data sets is clearly the NAO− regime with similar standard deviations and correlations nearly always greater than 0.8 (Figure 2). The anomalies of the NAO+ regime recur in the GCMs as well, however, with a systematic underestimation of

the variance compared with the reanalyses due to relatively weaker anomalies in the GCMs. The anomalies of the Atlantic ridge regime and the blocking regime show higher variability between the various data sets. This is indicated by a large range of correlations caused by variations in the spatial position and the extent of the anomaly centres. In addition, standard deviations are mostly higher in the GCMs compared with reanalyses, indicating stronger anomalies in the GCMs.

Results from the analysis of running 31-year sub-intervals show further important aspects. A notable feature in Figure 3 and Supporting figures is the considerable intra-dataset spread of the regime anomalies across the 31-year sub-intervals. It points to substantial temporal non-stationarities of the regimes. Within the reanalysis data (NCEP Figure 3, ERA, Figure S4, Supporting Information) the temporal variations of the regimes are lowest for the NAO− regime and highest for the blocking regime. However, for the NAO− regime systematic differences between reanalysis and GCM historical and scenario runs can be seen with overall lower correlation coefficients as well as reduced

Table I. Mean number of days spent in each weather regime in winter (DJF). Indicated are, for each data set, the mean numbers over the whole time period considered and the range resulting from the analysis of 31-year sub-intervals.

Data set	Time period	Regime I (NAO+) Mean number of days	Range in 31 years sub-intervals	Regime 2 (blocking) Mean number of days	Range in 31 years sub-intervals	Regime 3 (Atl. ridge) Mean number of days	Range in 31 years sub-intervals	Regime 4 (NAO−) Mean number of days	Range in 31 year sub-intervals
NCEP reanalysis	1950–2010	24	17–28	23	17–27	23	16–27	20	16–27
ERA-Interim	1979–2012	22	18–29	22	17–28	26	17–29	20	17–27
MPI-ESM-LR historical run 1	1950–2005	27	20–30	19	17–28	22	16–27	22	17–27
MPI-ESM-LR historical run 2	1950–2005	26	21–30	18	18–26	23	14–28	23	16–28
MPI-ESM-LR historical run 3	1950–2005	19	19–30	25	17–28	26	15–25	20	15–27
MPI-ESM-LR RCP4.5 run 1	2006–2100	24	19–30	20	17–26	18	16–26	28	18–27
MPI-ESM-LR RCP4.5 run 2	2006–2100	24	18–30	17	16–27	23	16–28	26	15–27
MPI-ESM-LR RCP4.5 run 3	2006–2100	24	19–30	21	16–29	22	17–27	23	16–28
MPI-ESM-LR RCP8.5 run 1	2006–2100	25	12–32	23	15–30	19	12–29	23	14–26
MPI-ESM-LR RCP8.5 run 2	2006–2100	27	15–31	20	16–27	23	17–27	20	18–30
MPI-ESM-LR RCP8.5 run 3	2006–2100	27	18–29	20	16–29	21	14–30	22	15–29
HadGEM2-ES historical	1950–2005	26	–	21	–	19	–	24	–
HadGEM2-ES RCP4.5	2006–2099	24	18–29	19	17–28	25	17–27	22	16–27
HadGEM2-ES RCP8.5	2006–2099	17	11–32	23	13–29	27	6–30	23	12–28
CanESM2 historical	1950–2005	26	20–37	24	16–25	19	16–30	21	16–27
CanESM2 RCP4.5	2006–2100	25	18–30	24	16–27	19	14–26	22	17–30
CanESM2 RCP8.5	2006–2100	27	19–33	20	15–31	21	10–29	22	14–30

standard deviations in the GCM sub-intervals. This means a considerably reduced temporal consistency of the NAO− regime in the GCM simulations which is not evident in the reanalysis data. The other regimes hold a large but comparable intra-dataset spread in reanalysis and GCM data with regard to the pattern similarities as expressed by the correlation coefficients. However, the variability in the strength of the regime anomalies is higher in the GCM sub-intervals, indicated by a larger overall range of the standard deviations. Additionally, as already outlined in Figure 2, there is a systematic bias towards weaker regime anomalies for the NAO+ regime in the GCM data compared with reanalyses. Higher standard deviations and thus stronger anomalies are present for the blocking regime and the Atlantic ridge regime.

Table I illustrates notable variations in the mean number of days spent in each regime across the data sets as well as in the 31-year sub-intervals. The mean wintertime frequency of the NAO+ regime is 24 and 22 days in NCEP and ERA-Interim, respectively. Across the different GCM simulations it ranges from 17 days in the HadGEM2-ES RCP8.5 run to 27 days mainly in the MPI-ESM-LR and CanESM2 RCP8.5 scenario runs. Looking at the 31-year sub-intervals of the reanalysis data sets, all regimes show mean frequency differences of up to 12 days. A similar value is found for the GCM historical runs with the exception of the NAO+ regime in the CanESM2 sub-intervals with mean frequency differences of up to 17 days. Under scenario conditions, in particular concerning RCP8.5, the total range of the variations increases considerably. The mean number of days spent in the NAO+ regime varies up to 21 days in the different 31-year sub-intervals, up to 24 days for the Atlantic

ridge regime, and up to 16 days for the NAO− regime and the blocking regime. We compare the variability in the frequency of occurrence between historical and scenario conditions by using the variance of the frequencies in the 31-year sub-periods and its 90% confidence limits from 1000 iterations of bootstrapping. Note that HadGEM2-ES RCP8.5, showing highest variability in the frequency of occurrence, is excluded, because data is too short for the historical period. Under RCP8.5 scenario conditions we find that the variances in the frequency of occurrence of the NAO+ regime and of the Atlantic ridge regime lie outside the 90% confidence intervals of the historical runs. We conclude that the variability in the frequency of occurrence increases under stronger greenhouse gas forcing, being significant for the NAO+ regime and for the Atlantic ridge regime. Whether this represents a real climate change signal or is due to model deficiencies has to be resolved yet.

4. Conclusions

Our study shows that the regimes exhibit considerable variability within the data sets as well as across the data sets considered. The NAO− regime specifications are highly varying with time in the GCM historical and scenario runs which is not the case in the reanalysis data. Owing to the consistency of this regime in reanalyses but not in the GCMs, our study suggests that the distinct spatial patterns of temperature and precipitation over Europe associated with this regime are not correctly captured in the GCM simulations. For the other regimes a notable variability can also be identified. However, variability of the regimes within the GCM data sets

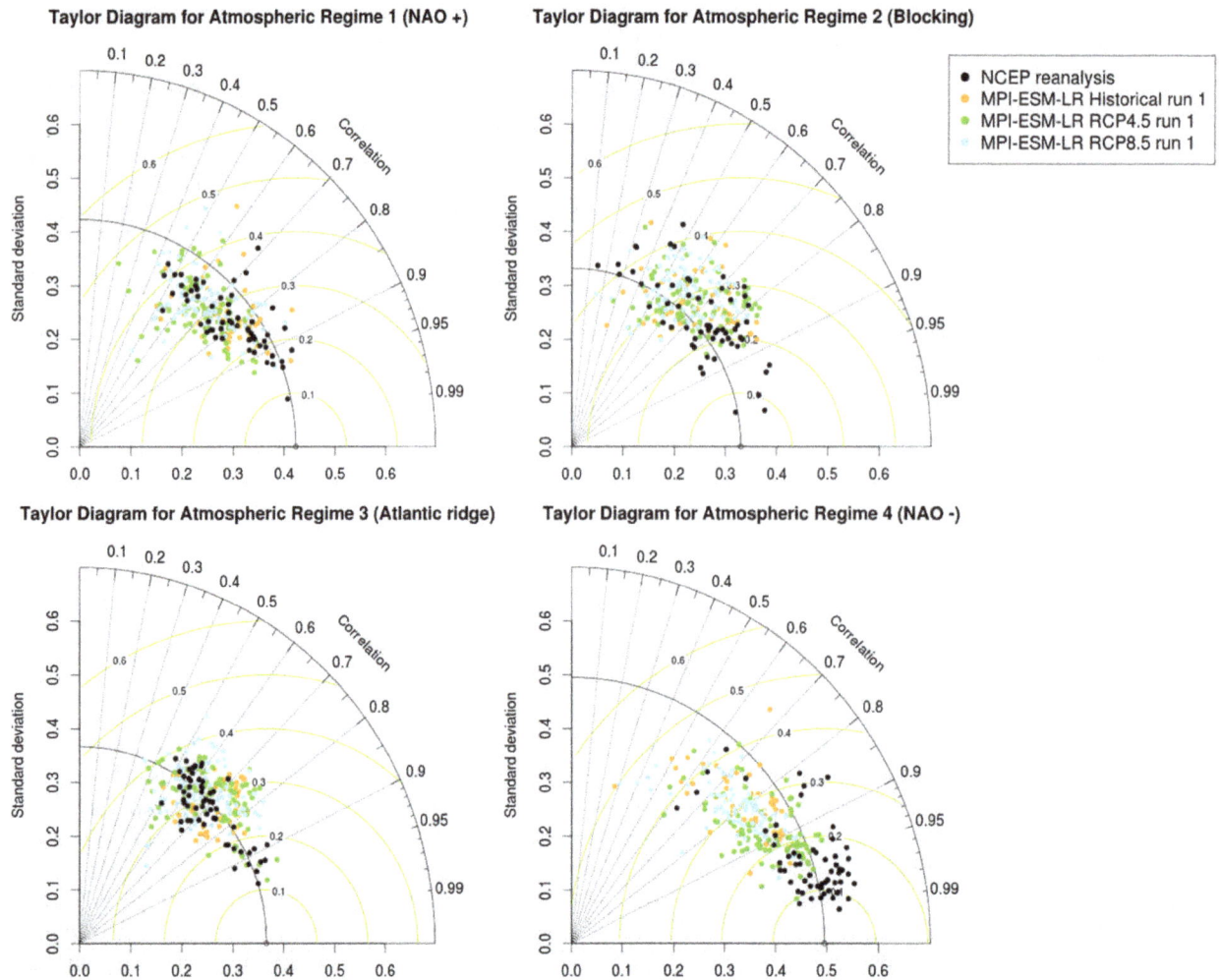

Figure 3. Taylor diagrams showing the similarity of the weather regimes in 31-year sub-intervals. Shown are, for each weather regime, the similarity of the anomaly patterns in 31-year sub-intervals with the NCEP weather regimes of the period 1950–2010. Illustrated are the 31-year sub-intervals of NCEP, and of the historical, RCP4.5, and RCP8.5 simulations of MPI-ESM-LR run 1.

is comparable to the spread seen in the reanalysis sub-intervals. These findings have major implications with respect to downscaling using the weather regimes as large-scale predictors. Despite the substantial variability of the NAO+ regime, of the Atlantic ridge, and particularly of the blocking regime, we expect that the consideration of the full range of observed variability will yield realistic projections of regime-dependent temperature and precipitation patterns under future climate change conditions. In contrast, regime-specific temperature and precipitation assessments will be more problematic for the NAO− regime. The close-confined reanalysis predictor–predictand relationships cannot be transferred one-to-one to the much broader GCM representations to make inferences about future regional climate changes.

We find considerable (real and/or model-induced) variations in the number of days per winter spent in each regime, most pronounced for the Atlantic ridge and the NAO+ regime under RCP8.5 scenario conditions. This leads to substantial modifications of the temperature and precipitation distributions in regions influenced by these regimes. Furthermore, the frequency changes

of the regimes can increase or damp the effects of the above outlined variable regime-climate relationships by changing the proportion to which specific regime characteristics impact on the temperature and precipitation distributions.

Besides the variations of the frequency of occurrence and the modifications of the regime-climate relationships arising from the variability of the regime anomalies, other sources of non-stationarity can be relevant in the scope of future climate change. This concerns, for instance, changes of regime-specific temperature levels or thermodynamic processes. De Vries *et al.* (2013) find for western and central Europe that days in which the daily mean temperature falls below the freezing level, are strongly reduced under SRES A1B scenario conditions. The reduction is a consequence of the shift of the temperature distribution towards higher values. Furthermore, future days below the freezing level occur for more extreme circulation types associated with, on average, drier weather conditions. Hertig and Jacobeit (2013) analysed mean daily precipitation in the Mediterranean area and show that in the observational period non-stationarities occurred within the relationships between precipitation and circulation patterns and their within-type circulation and thermodynamic characteristics. The details of the effects of such 'within-type' (Barry *et al.*, 1981, Beck *et al.*, 2007) changes in the scope of regional climate change assessments using latest generation GCMs remain to be investigated.

Supporting information

The following supporting information is available:

Figure S1. Taylor diagrams showing the similarity of the weather regimes in 31-year sub-intervals of MPI-ESM-LR run 2. Shown are, for each weather regime, the similarity of the anomaly patterns in 31-year sub-intervals with the NCEP weather regimes of the period 1950–2010. Illustrated are the 31-year sub-intervals of the historical, RCP4.5 and RCP8.5 runs.

Figure S2. Taylor diagrams showing the similarity of the weather regimes in 31-year sub-intervals of MPI-ESM-LR run 3. Shown are, for each weather regime, the similarity of the anomaly patterns in 31-year sub-intervals with the NCEP weather regimes of the period 1950–2010. Illustrated are the 31-year sub-intervals of the historical, RCP4.5 and RCP8.5 runs.

Figure S3. Taylor diagrams showing the similarity of the weather regimes in 31-year sub-intervals of CanESM2 run 1. Shown are, for each weather regime, the similarity of the anomaly patterns in 31-year sub-intervals with the NCEP weather regimes of the period 1950–2010. Illustrated are the 31-year sub-intervals of the historical, RCP4.5 and RCP8.5 runs.

Figure S4. Taylor diagrams showing the similarity of the weather regimes in 31-year sub-intervals of HadGEM2-ES and of ERA-Interim. Shown are, for each weather regime, the similarity of the anomaly patterns in 31-year sub-intervals with the NCEP weather regimes of the period 1950–2010. Illustrated are the 31-year sub-intervals of the ERA-reanalysis and of the HadGEM2-ES RCP4.5 and RCP8.5 runs.

Acknowledgements

This project is funded by the German Research Foundation under contract HE 6186/2-1. We acknowledge the climate modelling groups and the World Climate Research Programme's Working Group on Coupled Modelling for making available the CMIP5 data set. Also we acknowledge the free availability of the NCEP reanalysis and the ERA-Interim data.

References

Barry RG, Kiladis G, Bradley RS. 1981. Synoptic climatology of the Western United States in relation to climatic fluctuations during the twentieth century. *Journal of Climatology* **1**: 97–113.

Beck CH, Jacobeit J, Jones PD. 2007. Frequency and within-type variations of large-scale circulation types and their effects on low-frequency climate variability in Central Europe since 1780. *International Journal of Climatology* **27**: 473–491.

Cattiaux J, Douville H, Peings Y. 2013. European temperatures in CMIP5: origins of present-day biases and future uncertainties.

Climate Dynamics **41**: 2889–2907, DOI: 10.1007/s00382-013-1731-y.

Cheng X, Wallace JM. 1993. Cluster analysis of the Northern Hemisphere wintertime 500-hPa height field: spatial patterns. *Journal of the Atmospheric Sciences* **50**: 2674–2696.

Christiansen B. 2007. Atmospheric circulation regimes: can cluster analysis provide the number? *Journal of Climate* **20**: 2229–2250.

De Vries H, Haarsma RJ, Hazeleger W. 2013. On the future reduction of snowfall in western and central Europe. *Climate Dynamics* **41**: 2319–2330.

Dee DP, Uppala SM, Simmons AJ, Berrisford P, Poli P, Kobayashi S, Andrae U, Balmaseda MA, Balsamo G, Bauer P, Bechtold P, Beljaars ACM, van de Berg L, Bidlot J, Bormann N, Delsol C, Dragani R, Fuentes M, Geer AJ, Haimberger L, Healy SB, Hersbach H, Hólm EV, Isaksen L, Kållberg P, Köhler M, Matricardi M, McNally AP, Monge-Sanz BM, Morcrette JJ, Park BK, Peubey C, de Rosnay P, Tavolato C, Thépaut JN, Vitart F. 2011. The ERA-Interim reanalysis: configuration and performance of the data assimilation system. *Quarterly Journal of the Royal Meteorological Society* **137**: 553–597.

Hertig E, Jacobeit J. 2013. A novel approach to statistical downscaling considering nonstationarities: application to daily precipitation in the Mediterranean area. *Journal of Geophysical Research, [Atmospheres]* **118**: 520–533.

Hurrell J, Kushnir Y, Ottersen G, Visbeck M (eds). 2003. *The North Atlantic Oscillation: Climatic Significance and Environmental Impact*. Geophysical Monograph Series, Vol. **134**. AGU: Washington, DC.

Kalnay E, Kanamitsu M, Kistler R, Collins W, Deaven D, Gandin L, Iredell M, Saha S, White G, Woollen J, Zhu Y, Chelliah M, Ebisuzaki W, Higgins W, Janowiak J, Mo KC, Ropelewski C, Wang J, Leetmaa A, Reynolds R, Jenne R, Joseph D. 1996. The NCEP/NCAR 40-year reanalysis project. *Bulletin of the American Meteorological Society* **77**: 437–471.

Kistler R, Kalnay E, Collins W, Saha S, White G, Woollen J, Chelliah M, Ebisuzaki W, Kanamitsu M, Kousky V, van den Dool H, Jenne R, Fiorino M. 2001. The NCEP/NCAR 50-year reanalysis: monthly means CD-ROM and documentation. *Bulletin of the American Meteorological Society* **82**: 247–268.

Michelangeli P-A, Vautard R, Legras B. 1995. Weather regimes: recurrence and quasi stationarity. *Journal of the Atmospheric Sciences* **52**: 1237–1256.

Plaut G, Simonnet E. 2001. Large-scale circulation classification, weather regimes, and local climate over France, the Alps and Western Europe. *Climate Research* **17**: 303–324.

Preisendorfer RW. 1988. Principal component analysis in meteorology and oceanography. In *Developments in Atmospheric Science*, Vol. **17**. Elsevier: Amsterdam.

Stephenson DB, Hannachi A, O'Neill A. 2004. On the existence of multiple climate regimes. *Quarterly Journal of the Royal Meteorological Society* **130**: 583–605.

Taylor KE. 2001. Summarizing multiple aspects of model performance in a single diagram. *Journal of Geophysical Research* **106**: 7183–7192.

Van Vuuren DP, Edmonds J, Kainuma M, Riahi K, Thomson A, Hibbard K, Hurtt GC, Kram T, Krey V, Lamarque J-F, Masui T, Meinshausen M, Nakicenovic N, Smith SJ, Rose SK. 2011. The representative concentration pathways: an overview. *Climatic Change* **109**: 5–31.

Vautard R. 1990. Multiple weather regimes over the North Atlantic: analysis of precursors and successors. *Monthly Weather Review* **18**: 2056–2081.

Wanner H, Brönnimann S, Casty C, Gyalistras D, Luterbacher J, Schmutz C, Stephenson DB, Xoplaki E. 2001. North Atlantic Oscillation – concepts and studies. *Surveys in Geophysics* **22**: 321–382.

Ward J. 1963. Hierarchical grouping to optimize an objective function. *Journal of the American Statistical Association* **58**: 236–244.

Yiou P, Nogaj M. 2004. Extreme climatic events and weather regimes over the North Atlantic: when and where? *Geophysical Research Letters* **31**: L07202.

12

Regional response of annual-mean tropical rainfall to global warming

Ping Huang*

Center for Monsoon System Research, Institute of Atmospheric Physics, Chinese Academy of Sciences, Beijing 100190, China

*Correspondence to:
Dr. P. Huang, Center for Monsoon System Research, Institute of Atmospheric Physics, Chinese Academy of Sciences, Beijing 100190, China
E-mail:
huangping@mail.iap.ac.cn*

Abstract

Regional response of annual-mean tropical rainfall to global warming is investigated based on 18 models from the Coupled Model Intercomparison Project 5. With surface warming, the climatological ascending circulation pumps up increased surface moisture and leads to a rainfall increase over the convergence zone, while the change in ascending flow induces a rainfall increase over the region with sea surface temperature (SST) increase exceeding the tropical mean, with a concomitant modification of background surface moisture and SST. These two effects form a hook-like pattern of rainfall change over the tropical Pacific and an elliptic pattern over the northern Indian Ocean.

Keywords: tropical rainfall; global warming; regional response

1. Introduction

Under global warming, tropical precipitation change plays important role to the changes in atmospheric convection, circulation and energy transportation (Allen and Ingram, 2002; Held and Soden, 2006; Meehl *et al.*, 2007). Projecting rainfall response to global warming is a major challenge (Solomon *et al.*, 2007; Hsu *et al.*, 2012; Kitoh *et al.*, 2013), but the regional patterns of precipitation change have great uncertainties both in observation and simulation (Meehl *et al.*, 2007; Zhang *et al.*, 2007; Ma and Xie, 2013).

Simulated by global climate models (GCMs), the increased precipitation change peaks over the deep tropic (Figure 1(a)) (e.g. Chou *et al.*, 2009). Generally, there are two theories to explain the spatial distribution of precipitation response. One argues that the rainfall increases along with the precipitation climatology (wet-get-wetter) and decreases in the margins of convective regions (upped-ante) (Chou and Neelin, 2004; Held and Soden, 2006; Chou *et al.*, 2009; Chou *et al.*, 2013a); the other suggests increased rainfall where the sea surface temperature (SST) increase exceeds the tropical mean warming (warmer-get-wetter) (Xie *et al.*, 2010). The wet-get-wetter (WeGW) mechanism emphasizes the role of vertical moisture transport change by mean circulation (Chou *et al.*, 2009), while the SST change (ΔSST) pattern overlooked in WeGW is considered to dominate the regional precipitation change by influencing local convective instability in warmer-get-wetter (WaGW) (Xie *et al.*, 2010). As shown in Figure 1(a), however, both climatological rainfall and ΔSST partly overlap the annual-mean rainfall change pattern in GCMs but also have some discrepancies.

More studies notice that the rainfall change can be divided into two components, the thermodynamic and dynamic components, which are related to the specific humidity change and circulation change, respectively (e.g. Chou *et al.*, 2009; Seager *et al.*, 2010). It has obtained a consensus that the thermodynamic component exhibits a WeGW effect, but different views exit regarding the dynamic component (Bony *et al.*, 2013; Chadwick *et al.*, 2013; Huang *et al.*, 2013). Chadwick *et al.* (2013) claim that the dynamic component of rainfall change can be mainly divided into one part related to circulation slowdown (anti-WeGW) and the other related to SST pattern (WaGW), the anti-WeGW part cancels largely out the thermodynamic component (WeGW), and then the rainfall change mainly exhibits a WaGW pattern. To explain the seasonal pattern of zonal-mean precipitation change, Huang *et al.* (2013) suggests that WaGW effect contributes more to the annual mean whereas WeGW effect to the seasonal anomalies because of the almost seasonally independent ΔSST and seasonally varying precipitation climatology.

This study investigates the spatial distribution of precipitation change based on the multi-model ensemble (MME) results of the Coupled Model Inter-comparison Project Phase 5 (CMIP5) model outputs (Taylor *et al.*, 2012). The role of ΔSST pattern is verified using atmospheric model experiments forced by spatial uniform SST increase and spatial patterned SST increase (Bony *et al.*, 2011). A simplified vertical moisture transportation is diagnosed for the contribution of WeGW and WaGW effects as in Huang *et al.* (2013).

2. Models

We use the outputs of 18 models participating CMIP5 including the historical experiment for 1981–2000 and representative concentration pathway 4.5 (RCP 4.5)

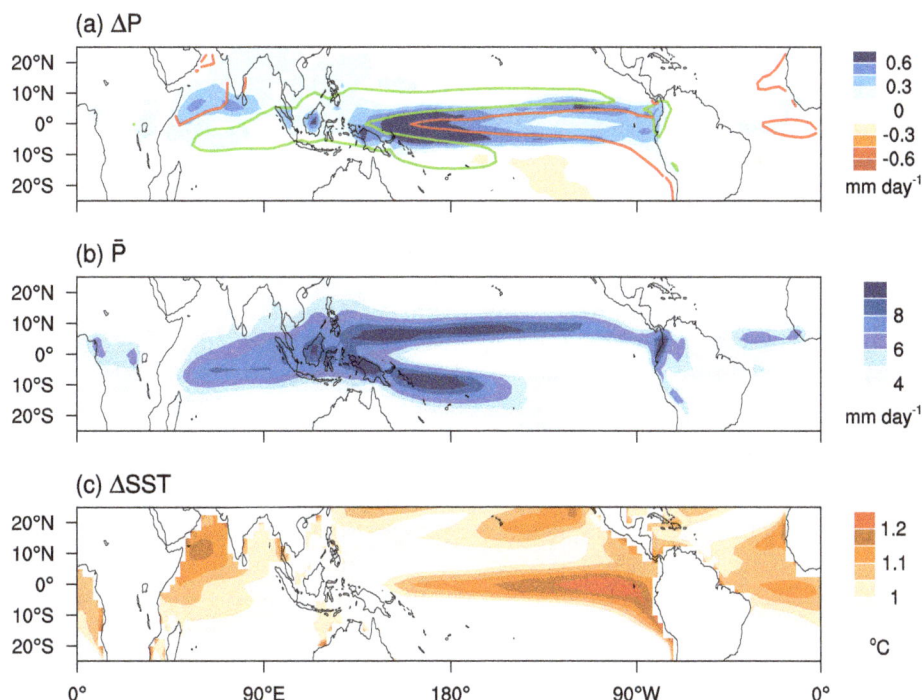

Figure 1. Annual-mean (a) precipitation change, (b) the precipitation climatology and (c) SST change in RCP 4.5 and in MME of 18 CMIP5 models. In (a), the green curves show the 7 mm day^{-1} contour of the precipitation climatology, and the red curves show the 1.1 °C contour of ΔSST.

experiment for 2081–2100 at http://pcmdi9.llnl.gov/ (Taylor *et al.*, 2012). The eighteen models are BCC-CSM1.1, CanESM2, CCSM4, CNRM-CM5, CSIRO-Mk3.6.0, FGOALS-s2, GFDL-CM3, GFDL-ESM2G, GISS-E2-R, HadGEM2-ES, INM-CM4, IPSL-CM5A-LR, IPSL-CM5A-MR, MIROC5, MIROC-ESM, MIROC-ESM-CHEM, MRI-CGCM3 and NorESM1 (see details at http://cmip-pcmdi.llnl.gov/cmip5/ availability.html). The current climatology is defined as the long-term mean for 1981–2000 in historical experiment, whereas the change under global warming is defined as the difference between 2081–2100 in RCP 4.5 and 1981–2000 in historical. The MME is defined as the average results of 18 models.

Six atmospheric models (CanAM4, CNRM-CM5, HadGEM2-A, IPSL-CM5A-LR, MIROC5 and MRI-CGCM3) in CMIP5 that provide the following set of atmospheric experiments are also used (Bony *et al.*, 2011; Taylor *et al.*, 2012): the amip experiment prescribes observed SST (control run), amip4K and amipFuture, respectively, add a spatially uniform SST increase (SUSI) of 4 K and a spatially patterned SST increase (SPSI) from CMIP3 1% per year CO_2 increase to quadrupling experiments as the boundary condition. The ΔSST used in SPSI run is similar to that of coupled GCMs from CMIP5 with peak on the equator. The climate changes in SUSI and SPSI runs are defined as the differences of 20-year monthly climatology for 1981–2000 from the control runs. For comparing the changes among RCP 4.5, SUSI and SPSI runs, all changes are normalized using the tropical (20°S–20°N) mean SST change in corresponding runs.

3. Results

Figure 1 shows the MME and annual-mean precipitation change, mean precipitation ($\overline{P}$, the overbar denotes the mean averaged from 1981 to 2000 in historical run) and ΔSST. The increased precipitation change primitively exhibits a hook-like pattern over the deep equatorial Pacific and an elliptic pattern over the northern Indian Ocean (Figure 1(a)). The hook-like pattern of rainfall change (ΔP) over the tropical Pacific resembles neither the distribution of mean convergence zone (Figure 1(b)) nor the El Niño-like pattern of ΔSST (Figure 1(c)) (e.g. Liu *et al.*, 2005; Xie *et al.*, 2010; Lu and Zhao, 2012), but their combination. The pattern of ΔP over the northern Indian Ocean approximately coincides with the Indian Ocean ΔSST, but the maximum ΔSST is located over the Arabian Sea, east of the central northern Indian Ocean with maximum ΔP (Hsu and Li, 2012). The discrepancies of ΔP pattern with ΔSST and $\overline{P}$ patterns indicate that the spatial distribution of annual-mean ΔP does not simply follow the pattern of climatological rainfall or ΔSST.

To isolate the ΔSST effect, we analyze a group of atmospheric experiments in CMIP5, the SUSI and SPSI runs, forced by uniform and patterned ΔSST, respectively. In SUSI runs, rainfall change generally peaks over the tropical convergence zones, present a quasi-WeGW effect (Figure 2(a)). On the other hand, in SPSI runs, the maximum positive ΔP is located over the equatorial central Pacific with weak negative change over the Maritime Continent, similar to the distribution of ΔP in RCP 4.5 runs

(Figure 2(b)). Some small differences between SPSI and RCP 4.5 could be attributed to the different ΔSST pattern in CMIP3 and CMIP5 (Yeh *et al.*, 2012) or to the lack of radiative forcing in SPSI (Bony *et al.*, 2013).

Vertical velocity change ($\Delta\omega$) plays important role on rainfall change. In SUSI, ascending changes occur over the northwestern Pacific, the cold tongue and southeastern Pacific and descending changes over the Indian Ocean, Maritime Continent, Africa and Amazon (Figure 2(d)). The $\Delta\omega$ pattern in SPSI is very different from that in SUSI over the oceans but similar to the $\Delta\omega$ in RCP 4.5 run (Figures 2(e) and 3(a)). The SST-induced difference of $\Delta\omega$ consists well with that of ΔP, ascending changes corresponding to positive rainfall changes and vice versa (Figure 2(c) and (f)). Therefore, the SST-induced $\Delta\omega$ makes a major contribution to the different ΔP in two runs. As shown in Figure 2(d)–(f), the magnitude of circulation change over oceans in SPSI is much stronger than that in SUSI (the variance of oceanic $\Delta\omega$ in SUSI is only 30% of that in SPSI), and the SST-induced $\Delta\omega$ can explain around 65% variance of oceanic $\Delta\omega$ in SPSI. Therefore, the SST-induced $\Delta\omega$ contributes the majority of $\Delta\omega$ in SPSI. It is consistent with the model result of previous studies, in which the slowdown of the tropical Pacific circulation is attributed to the ΔSST pattern (Tokinaga *et al.*, 2012a, 2012b). The SST pattern can explain little $\Delta\omega$ over the Africa and Amazon, which could be driven by the land–sea warming contrast (Bayr and Dommenget, 2013).

Although the SST-induced $\Delta\omega$ dominates oceanic $\Delta\omega$ in SPSI, the SST-induced rainfall change has significant differences from ΔP in SPSI run and RCP 4.5 run (Figures 1(a) and 2(b) and (c)). The SST-induced rainfall shows an El Niño-like response with negative change over the Maritime Continent and the flanks of tropical Pacific as strong as the positive change over the central Pacific and the northern Indian Ocean (Figure 2(c)), whereas ΔP in SPSI and RCP 4.5 has little negative change over the tropics (DiNezio *et al.*, 2010). Therefore, one other part is needed to ΔP for balancing the negative change induced by ΔSST.

We decompose the tropical rainfall change into the contributions of circulation changes and moisture change as in Huang *et al.* (2013):

$$\Delta P \sim \Delta\omega \cdot \overline{q} + \overline{\omega} \cdot \Delta q \qquad (1)$$

where q denotes surface specific humidity and ω is pressure velocity at 500 hPa. In Equation (1), the dynamic component $\Delta\omega \cdot \overline{q}$ and thermodynamic component $\overline{\omega} \cdot \Delta q$ represent the contributions of circulation change and moisture change, respectively (Chou *et al.*, 2009; Seager *et al.*, 2010). It is a good approximation for the tropics (Held and Soden, 2006), which reproduces the zonal-mean seasonal cycle of ΔP well both in coupled models and atmospheric models (Huang *et al.*, 2013).

The decomposition of Equation (1) well reproduces the spatial distribution of ΔP (Figures 1(a) and 3(c)), including the Pacific hook-like pattern and the elliptic pattern over the northern Indian Ocean. The spatial distributions of both $\Delta\omega \cdot \overline{q}$ and $\overline{\omega} \cdot \Delta q$ are dominated by vertical velocity (Figures 3(a) and 3(b)), because Δq and $\overline{q}$ have flat distribution relative to $\Delta\omega$ and $\overline{\omega}$. The dynamic component $\Delta\omega \cdot \overline{q}$ induces positive rainfall change over the deep-tropical Pacific and the Arabian Sea, and negative change over the warm pool regions, the flanks of equatorial Pacific, Amazon and central Africa (Figure 3(a)). The distribution of dynamic component as well as circulation change generally coincides with the ΔSST pattern as in SUSI run, although the tropical mean convective mass flux must weaken due to the energy constraint (Held and Soden, 2006). It reflects the WaGW effect. On the other hand, the thermodynamic component $\overline{\omega} \cdot \Delta q$ coincides with the climatological rainfall zone, as the mean upward motion pumps up increased moisture under a warming climate (Figure 3(b)). It reflects the WeGW effect. The WeGW effect cancels out the negative change of WaGW effect (dynamic component) over the mean convergence zone, Amazon and central Africa. Except the conspicuous cancellation, the WeGW effect reinforces the positive change over the central Pacific and northern Indian Ocean, and induces the ΔP peaks over the western Pacific and northern central Indian Ocean differing from the dynamic component (Figure 3(c)). It means that the rainfall change with the hook-like pattern over the tropical Pacific and the elliptic pattern over the northern Indian Ocean is a hybrid of WeGW and WaGW effects, like the seasonal pattern of zonal mean (Huang *et al.*, 2013).

Similar decomposition also can be performed on ΔP in SPSI and SUSI (not shown). The WeGW effect dominated by positive change over the tropics also cancels out the negative change induced by SST in SPSI and SUSI. Because of greater $\Delta\omega$ in SPSI, the dynamic component contributes more to total ΔP in SPSI, whereas less contribution of dynamic component to total ΔP in SUSI is due to weaker $\Delta\omega$. Therefore, the ΔP pattern in SUSI generally shows a quasi-WeGW pattern, whereas in SPSI it shows a quasi-WaGW pattern (Xie *et al.*, 2010).

Although the relationship between ΔSST pattern and circulation change has been investigated by previous study (Xie *et al.*, 2010), the circulation change and dynamic component have some discrepancies with ΔSST on close inspection (Figures 4(a) and 1(c)). $\Delta\omega$ shows a hook-like pattern with maximum over the central Pacific, whereas ΔSST has maximum over the eastern Pacific. The effect of ΔSST on circulation change can be understood from the moist instability theory. The moist instability of tropical atmosphere can be evaluated by $I = \delta_p(T + L/C_p \cdot q)$, where C_p is specific heat capacity at constant pressure and L is latent heat of evaporation (Yu *et al.*, 1998; Chou *et al.*, 2013b). Because of the flat temperature change

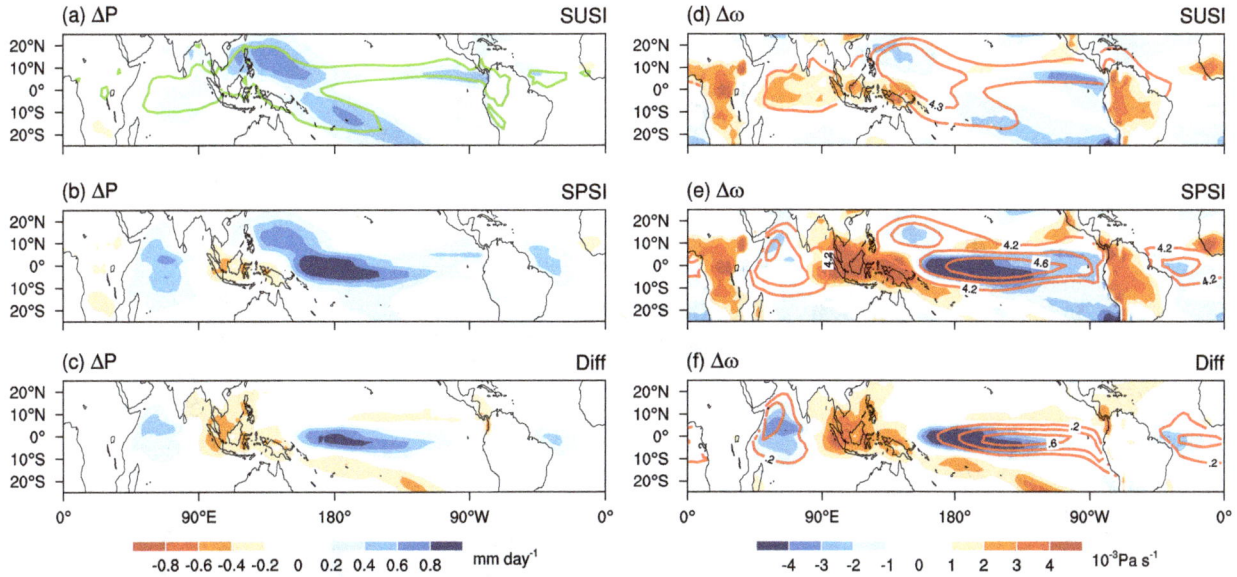

Figure 2. Annual-mean precipitation change (left) and circulation change (right) represented by $\Delta\omega$ at 500 hPa in SUSI and SPSI runs, and their difference. The green curves in (a) show the 6 mm day^{-1} contour of the precipitation climatology in control runs. Red contours in (d)–(f) show the surface component of moist instability change in SUSI and SPSI and their difference (Unit: °C).

in upper troposphere, the distribution of instability change is dominated by its surface component.

$$\Delta I_s = \Delta T_s + L/C_p \cdot \Delta q_s \qquad (2)$$

In Equation (2), Δq also contributes to the moist instability change besides ΔSST (Figure 4(b)). Under the restraint of Clausius–Clapeyron equation, $\Delta q \sim \bar{q} \cdot \Delta SST$ (Xie *et al.*, 2010), the Pacific Δq on

the equator is enhanced by ΔSST pattern, and the maximum Pacific Δq is located in the central Pacific west of ΔSST peak because of $\bar{q}$ with maximum over the warm pool (Figure 4(c)). With the contribution of $\bar{q}$ to Δq, the ΔI_s has almost zonally uniform distribution over the tropical Pacific, which is closer to $\Delta\omega$ compared to ΔSST pattern (Figure 4(a)). The distribution of higher ΔI_s resembles well with ascending circulation change (Figure 4(a) and (b)). Other

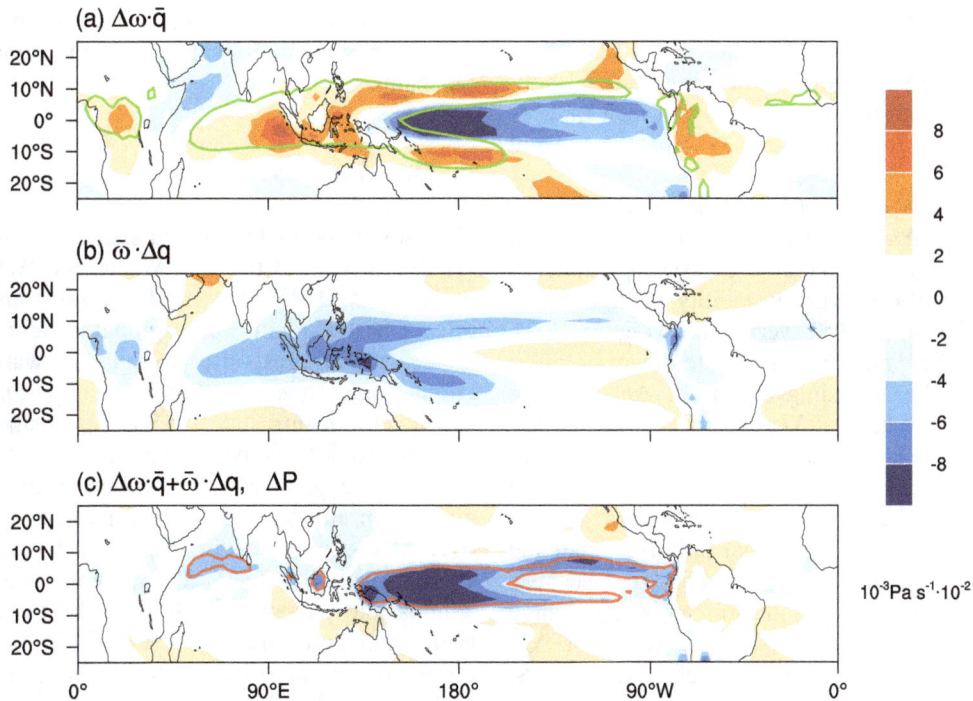

Figure 3. (a) The dynamic ($\Delta\omega \cdot \bar{q}$) and (b) thermodynamic ($\bar{\omega} \cdot \Delta q$) components of ΔP and (c) their sum in RCP 4.5. In (a), the green curves are the -3×10^{-3} Pa s^{-1} 10^{-2} contour of thermodynamic component; in (c), the red curves are the 0.4 mm day^{-1} contour of ΔP.

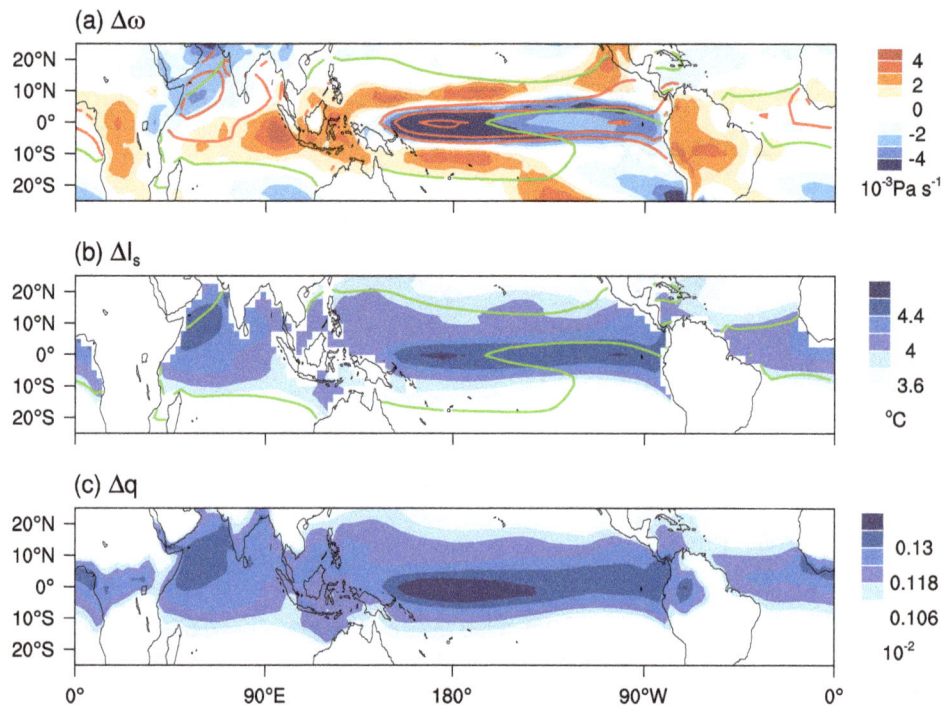

Figure 4. Annual-mean (a) circulation change ($\Delta\omega$ at 500 hPa; shaded), (b) surface component of moist instability change ΔI_s (shaded and red contours in (a)) and (c) specific humidity change Δq in RCP 4.5. Green curves in (a) and (b) are the 26.5 °C contour of annual-mean SST in historical run.

discrepancies between $\Delta\omega$ and ΔI_s are the weak $\Delta\omega$ over the cold tongue with strong instability change. It should be attributed to the background SST (green curves in Figures 4(a) and (b)). Because of the nonlinear relationship between tropical convection and SST and the weaker spatial deviation of ΔSST relative to climatological SST, the relative high ΔI_s cannot lead to large ascending $\Delta\omega$ over the cold tongue (Graham and Barnett, 1987; Johnson and Xie, 2010). Therefore, the forcing of ΔSST pattern (WaGW effect) is modified by the background $\overline{q}$ and SST, and then induces the hook-like pattern of $\Delta\omega$ and ΔP over the tropical Pacific.

The $\Delta\omega$ results in the atmospheric runs verify the explanation of $\Delta\omega$ in RCP 4.5 run by the moist instability theory. The SST-induced $\Delta\omega$ well matches the different surface moist instability change (ΔI_s) between SUSI and SPSI very well (Figure 2(f)). The $\Delta\omega$ in SPSI matches ΔI_s better than in SUSI run. In SUSI, the ascending change of $\Delta\omega$ over the Pacific follows the greater ΔI_s, but the relative greater ΔI_s over the Indian Ocean and Maritime Continent cannot explain the descending change of $\Delta\omega$. It could be attributed to the weak spatial pattern of ΔI_s in SUSI, although ΔI_s is not spatial uniform under the spatial uniform ΔSST forcing (Figure 2(d)) because of the contribution of non-uniform moisture change (Xie *et al.*, 2010). When the spatial gradient of ΔI_s is weak, the relative contribution of other factors, such as the land–sea contrast and mean advection of stratification change, on $\Delta\omega$ will become more significant (Ma *et al.*, 2012; Bayr and Dommenget, 2013).

Although it has been illustrated that the higher ΔI_s associated with higher ΔSST can induce proportional ascending change and the descending change is always located over weaker ΔI_s (Figures 2(d)–(f) and 4(b)), the descending change is not strictly proportional to the weak ΔI_s even if the other factors are removed in Figure 2(f). Therefore, the mechanism of convective suppression under global warming must be more complicated than the simplified ΔI_s mechanism.

4. Conclusions

This study shows the spatial pattern of annual-mean ΔP is a hybrid of the WeGW and the WaGW effects using 18 CMIP5 models. The two complementary effects best explains the regional pattern of annual-mean rainfall response to global warming. The climatological mean circulation pumps up increasing low-level moisture and then induces positive ΔP over current tropical convergence zone (WeGW effect), whereas the greater SST warming drives ascending circulation response and then induces positive ΔP over the central Pacific and northwestern Indian Ocean (WaGW effect). The combined effects form a hook-like pattern of ΔP over the tropical Pacific (with maximum over the western Pacific) and an elliptic pattern of ΔP over the northern Indian Ocean.

Although the dynamic component of ΔP generally exhibits a WaGW pattern, the climatological distribution of surface moisture and SST also contribute to the dynamic component. The background moisture can influence the change in specific humidity that contributes to atmospheric instability change and then to the circulation change. On the other hand, the background SST can modify the influence of instability change on circulation change due to the nonlinear relationship of SST-convection (Graham and Barnett, 1987; Johnson and Xie, 2010). The climatological moisture and SST lead $\Delta\omega$ peaks over the tropical central Pacific, west of the maximum ΔSST over the eastern Pacific. Although the dominant role of ΔSST pattern on dynamic component of ΔP is emphasized in this study, another factors such as weakening circulation induced by mean advection of stratification change, direct role of radiative forcing, the contribution of convection depth change, etc. can also contribute the dynamic component of ΔP (Chou and Chen, 2010; Ma et al., 2012; Bony et al., 2013; Chadwick et al., 2013).

In Huang et al. (2013), the unifying effect of WeGW and WaGW on the seasonal cycle of ΔP was called WeWa effect. Here, it is demonstrated that the unifying WeWa effect including WeGW and modified WaGW also applies to the spatial distribution of annual-mean rainfall change under global warming. Under WeWa mechanism, the WeGW part of ΔP pattern can be projected based on current rainfall, but the WaGW part has great uncertainties due to the uncertain SST response (Deser et al., 2010; Ma and Xie, 2013). However, the modification on WaGW by climatological moisture and SST decreases the uncertainties of WaGW pattern, and thus the uncertainties of ΔP pattern should be decreased relative to around 50% uncertain ΔP estimated by Huang et al. (2013) in which the climatology rainfall and uncertain ΔSST pattern have almost equal contribution to ΔP. This study projects a possible ΔP pattern, a hook-like pattern over the tropical Pacific and an elliptic pattern over the northern Indian Ocean, based on the ΔSST pattern of MME in CMIP5.

Acknowledgements

The work was supported by the National Basic Research Program of China (2012CB955604 and 2014CB953903), and the Natural Science Foundation of China (41105047). I wish to thank Prof. Shang-Ping Xie for helpful discussion. I acknowledge the World Climate Research Programme's Working Group on Coupled Modeling, which is responsible for CMIP5, and the climate modeling groups (listed in Section 2) for producing and making available their model output.

References

Allen MR, Ingram WJ. 2002. Constraints on future changes in climate and the hydrologic cycle. *Nature* **419**: 224–232.

Bayr T, Dommenget D. 2013. The tropospheric land–sea warming contrast as the driver of tropical sea level pressure changes. *Journal of Climate* **26**: 1387–1402.

Bony S, Webb M, Bretherton C, Klein S, Siebesma P, Tselioudis G, Zhang M. 2011. CFMIP: towards a better evaluation and understanding of clouds and cloud feedbacks in CMIP5 models. *CLIVAR Exchanges* **56**: 20–24.

Bony S, Bellon G, Klocke D, Sherwood S, Fermepin S, Denvil S. 2013. Robust direct effect of carbon dioxide on tropical circulation and regional precipitation. *Nature Geoscience* **6**: 447–451.

Chadwick R, Boutle I, Martin G. 2013. Spatial patterns of precipitation change in CMIP5: why the rich don't get richer in the tropics. *Journal of Climate* **26**: 3803–3822.

Chou C, Chen C-A. 2010. Depth of convection and the weakening of tropical circulation in global warming. *Journal of Climate* **23**: 3019–3030.

Chou C, Neelin JD. 2004. Mechanisms of global warming impacts on regional tropical precipitation. *Journal of Climate* **17**: 2688–2701.

Chou C, Neelin JD, Chen C, Tu J. 2009. Evaluating the "rich-get-richer" mechanism in tropical precipitation change under global warming. *Journal of Climate* **22**: 1982–2005.

Chou C, Chiang JCH, Lan C-W, Chung C-H, Liao Y-C, Lee C-J. 2013a. Increase in the range between wet and dry season precipitation. *Nature Geoscience* **6**: 263–267.

Chou C, Wu T-C, Tan P-H. 2013b. Changes in gross moist stability in the tropics under global warming. *Climate Dynamics*, DOI: 10.1007/s00382-013-1703-2.

Deser C, Phillips A, Alexander M. 2010. Twentieth century tropical sea surface temperature trends revisited. *Geophysical Research Letters* **37**: L10701.

DiNezio P, Clement A, Vecchi G. 2010. Reconciling differing views of tropical Pacific climate change. *Eos, Transactions American Geophysical Union* **91**: 141.

Graham N, Barnett TP. 1987. Observations of sea surface temperature and convection over tropical oceans. *Science* **238**: 657–659.

Held IM, Soden BJ. 2006. Robust responses of the hydrological cycle to global warming. *Journal of Climate* **19**: 5686–5699.

Hsu PC, Li T. 2012. Is "rich-get-richer" valid for Indian Ocean and Atlantic ITCZ? *Geophysical Research Letters* **39**: L13705.

Hsu PC, Li T, Luo JJ, Murakami H, Kitoh A, Zhao M. 2012. Increase of global monsoon area and precipitation under global warming: a robust signal? *Geophysical Research Letters* **39**: L06701.

Huang P, Xie S-P, Hu K, Huang G, Huang R. 2013. Patterns of the seasonal response of tropical rainfall to global warming. *Nature Geoscience* **6**: 357–361.

Johnson NC, Xie S-P. 2010. Changes in the sea surface temperature threshold for tropical convection. *Nature Geoscience* **3**: 842–845.

Kitoh A, Endo H, Krishna Kumar K, Cavalcanti IF, Goswami P, Zhou T. 2013. Monsoons in a changing world: a regional perspective in a global context. *Journal of Geophysical Research-Atmospheres* **118**: 3053–3065.

Liu Z, Vavrus S, He F, Wen N, Zhong Y. 2005. Rethinking tropical ocean response to global warming: the enhanced equatorial warming. *Journal of Climate* **18**: 4684–4700.

Lu J, Zhao B. 2012. The role of oceanic feedback in the climate response to doubling CO_2. *Journal of Climate* **25**: 7544–7563.

Ma J, Xie S-P. 2013. Regional patterns of sea surface temperature change: A source of uncertainty in future projections of precipitation and atmospheric circulation. *Journal of Climate* **26**: 2482–2501.

Ma J, Xie S-P, Kosaka Y. 2012. Mechanisms for tropical tropospheric circulation change in response to global warming. *Journal of Climate* **25**: 2979–2994.

Meehl GA, Stocker TF, Collins WD, Friedlingstein P, Gaye AT, Gregory JM, Kitoh A, Knutti R, Murphy JM, Noda A, Raper SCB, Watterson IG, Weaver AJ, Zhao Z-C. 2007. Global climate projections. In *Climate Change 2007: The Physical Science Basis*, Solomon S, Qin D, Manning M, Chen Z, Marquis M, Averyt KB, Tignor M, Miller HL (eds). Cambridge University Press: Cambridge/New York, NY; 747–845.

Seager R, Naik N, Vecchi GA. 2010. Thermodynamic and dynamic mechanisms for large-scale changes in the hydrological cycle in response to global warming. *Journal of Climate* **23**: 4651–4668.

Solomon S, Qin D, Manning M, Chen Z, Marquis M, Averyt KB, Tignor M, Miller HL (eds). 2007. In *Climate Change 2007: The Physical Science Basis*. Cambridge University Press: Cambridge/New York, NY; 996 pp.

Taylor KE, Stouffer RJ, Meehl GA. 2012. An overview of CMIP5 and the experiment design. *Bulletin of the American Meteorological Society* **93**: 485–498.

Tokinaga H, Xie S-P, Deser C, Kosaka Y, Okumura YM. 2012a. Slowdown of the Walker circulation driven by tropical Indo-Pacific warming. *Nature* **491**: 439–443.

Tokinaga H, Xie S-P, Timmermann A, McGregor S, Ogata T, Kubota H, Okumura YM. 2012b. Regional patterns of Tropical Indo-Pacific climate change: evidence of the Walker circulation weakening. *Journal of Climate* **25**: 1689–1710.

Xie S-P, Deser C, Vecchi GA, Ma J, Teng H, Wittenberg AT. 2010. Global warming pattern formation: sea surface temperature and rainfall. *Journal of Climate* **23**: 966–986.

Yeh S-W, Ham Y-G, Lee J-Y. 2012. Changes in the Tropical Pacific SST trend from CMIP3 to CMIP5 and its implication of ENSO. *Journal of Climate* **25**: 7764–7771.

Yu J-Y, Chou C, Neelin JD. 1998. Estimating the gross moist stability of the tropical atmosphere. *Journal of the Atmospheric Sciences* **55**: 1354–1372.

Zhang X, Zwiers FW, Hegerl GC, Lambert FH, Gillett NP, Solomon S, Stott PA, Nozawa T. 2007. Detection of human influence on twentieth-century precipitation trends. *Nature* **448**: 461–465.

13

Multi-GCM by multi-RAM experiments for dynamical downscaling on summertime climate change in Hokkaido

Masaru Inatsu,[1]* Tomonori Sato,[2] Tomohito J. Yamada,[3] Ryusuke Kuno,[4] Shiori Sugimoto,[5] Murad A. Farukh,[3] Yadu N. Pokhrel[6] and Shuichi Kure[7]

[1] Faculty of Science, Hokkaido University, Sapporo, Japan
[2] Faculty of Earth Environmental Sciences, Hokkaido University, Sapporo, Japan
[3] Faculty of Engineering, Hokkaido University, Sapporo, Japan
[4] ARK Information Systems, Yokohama, Japan
[5] Graduate School of Urban Environmental Sciences, Tokyo Metropolitan University, Japan
[6] Department of Civil and Environmental Engineering, Michigan State University, East Lansing, MI, USA
[7] International Research Institute of Disaster Science, Tohoku University, Sendai, Japan

*Correspondence to
 M. Inatsu, Faculty of Science,
Hokkaido University, Rigaku 8th
bldg., N10W8, Sapporo
060-0810, Japan.
E-mail:
inaz@mail.sci.hokudai.ac.jp

Abstract

The experiments with three general circulation models (GCMs) by three regional atmospheric models (RAMs) for the dynamical downscaling (DDS) have been performed to evaluate the uncertainty in the global warming response during summertime in Hokkaido, Japan. The results of a 10-year RAM integration nested into GCM under present or future climate conditions were synthesized after applying bias correction. For the target decades during which the global-mean temperature increases by 2 K in each GCM, the DDS results indicate that surface air temperature and precipitation mostly depend on the GCM imposed as the lateral boundary condition.

Keywords: dynamical downscaling; regional climate change; multi-GCM by multi-RAM experiments

1. Introduction

Researches on climate change adaptation have been widely conducted mostly as governmental or inter-governmental projects, in order to make an optimal policy to adjust society, economy, industries, and life styles to imminent dangers due to the increase in surface air temperatures (SATs), the change of precipitation, snow melting, and sea level rise by global warming. A comprehensive consideration including climate change projection, risk assessment, and decision making is required in such a project. A great hindrance on it is the uncertainty on climate change projection that is inevitable mainly for two reasons: many possible choices in our future society that may control the level of greenhouse gas emissions, and many possible choices in climate model parameterizations that cause the climate sensitivity. Moreover, although the climate change is a universal issue for human being, the adaptation processes ranging from risk assessment to decision making are indeed regional and thus need finer resolution information on the climate change. However, most general circulation models (GCMs) used in climate experiments still use a rather coarse horizontal resolution of hundreds of kilometers.

The dynamical downscaling (DDS) is a modern technique to estimate the climate change in a particular domain by running a regional atmospheric model (RAM) nested in lower-resolution boundary conditions derived from a GCM (Giorgi and Bates, 1989).

Compared with the statistical downscaling, DDS can provide a dataset in which climatic variables are mutually physical consistent, are temporally varying, and are constrained by the synoptic conditions externally given. The DDS result is often sensitive to physical parameterizations implemented in RAMs such as cloud physics, precipitation, radiative transfer, and turbulent mixing (cf. Wang et al., 2004). The GCM is imposed as lateral boundary condition in a RAM experiment, which is equally sensitive to GCM's physical parameterizations. Hence, the multi-model ensemble approach has been advocated in EU ENSEMBLES (van der Linden and Mitchell, 2009), NARCCAP (Wang et al., 2009), and S5-3 (Ishizaki et al., 2012), to evaluate the uncertainty in the DDS estimation. Particularly an effort in the combination of multiple RAMs and multiple GCMs was devoted in EU ENSEMBLES, and the project provided a matrix for assembling their DDS results.

The uncertainty in DDS results can be classified into three aspects with different scales. First, as has been introduced above, there are many possible ways to parameterize physical processes in a RAM. Second, as the RAM is driven by the lateral boundary conditions, the result is apparently influenced by external synoptic states that are more or less uncertain in the climate change projection. Finally the global-mean SAT is highly uncertain due to the socio-economic scenario in the future and the climate sensitivity in the GCM. Without a good organization in the multi-GCM by multi-RAM experiments, however, these aspects in

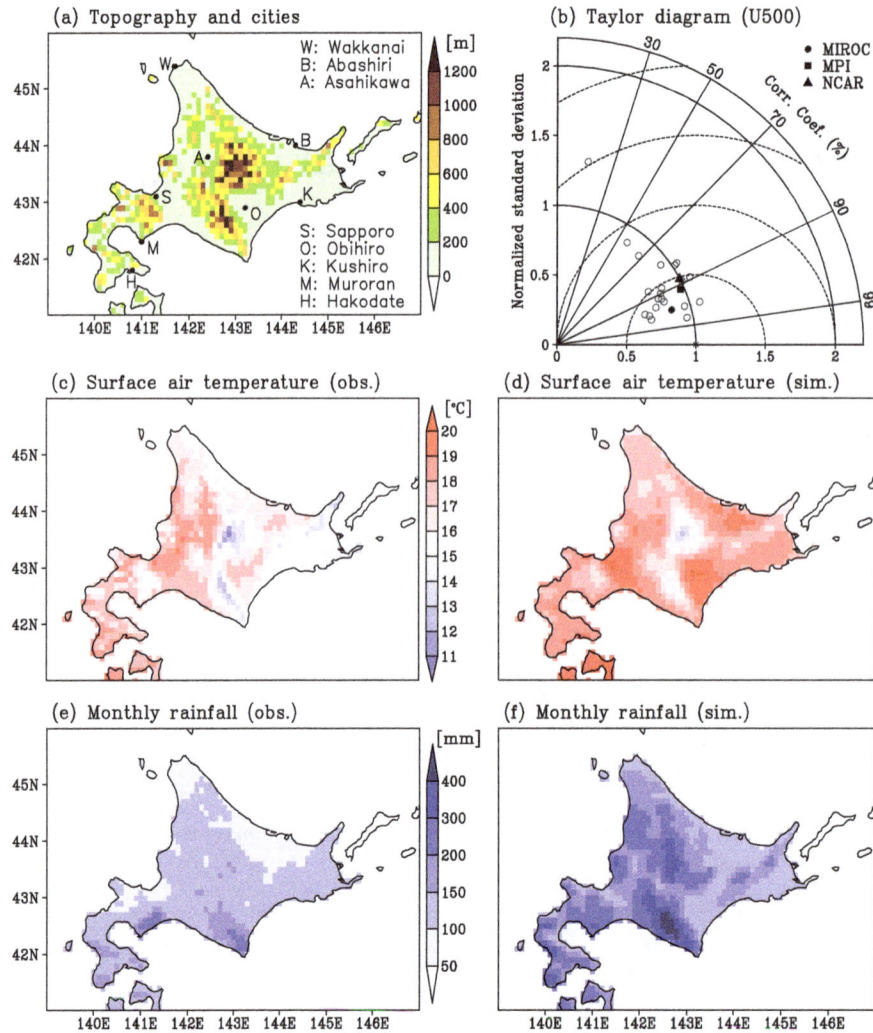

Figure 1. (a) Surface height (m) and major cities in Hokkaido labeled S (Sapporo), A (Asahikawa), H (Hakodate), K (Kushiro), O (Obihiro), M (Muroran), B (Abashiri), and W (Wakkanai). The color shading is as per the reference in the right of the panel. (b) Taylor diagram describing climatological zonal wind at 500 hPa over Northeast Asia to western North Pacific (100°–180°E by 20°–70°N) simulated by 24 models in CMIP3 compared with the observed [cross mark at (1,0)]. MIROC, MPI, and NCAR are denoted circle, square, and triangle, respectively. Standard deviation is normalized by the observed. (c) Climatological surface air temperatures (SATs) (°C) in Hokkaido in June–July–August (JJA) months based on the gridded data interpolated from 30-year Automated Meteorological Data Acquisition System (AMeDAS) observation operated by the Japan Meteorological Agency. (d) The ensemble average of SAT over all DDS experiments. (e) Climatological precipitation (mm month^{-1}) in JJA months based on APHRODITE gridded rainfall data. (f) The ensemble average of precipitation.

uncertainty would get entangled in the DDS results. For example, two DDS experiments by using different GCMs and a single RAM must bring difference in SAT increase due to climate sensitivity of GCMs and some difference in precipitation and wind vector that are probably attributed to the difference of synoptic states in the GCMs. If one added more RAMs in this case, the difference in precipitation could be caused by the difference in RAM parameterizations as well. The multi-GCM by multi-RAM experiments for DDS therefore need an arrangement to clearly separate the aspects in uncertainty, but studies along this line have not been conducted yet.

The purpose of this article is to describe the results of 3-GCM by 3-RAM experiments for DDS, certainly motivated by the EU-ENSEMBLES project. In order that the uncertainty in global-mean SATs due to a

socio-economic scenario in the future and the climate sensitivity in GCMs rules out in advance, we choose a decade during which the global-mean SATs increase by 2 K for a GCM. This article hence focuses on the uncertainty potentially caused by the RAM dynamics and physical parameterizations and the GCM boundary condition. The target domain is Hokkaido (Figure 1(a)) and season is summer, where the present climate is influenced by both dry air settled down in the north and moist air intruding from the south. The Baiu rainband is maintained in the northern fringe of the Bonin high from mid-June to mid-July and is normally dismissed in the end of July (Sampe and Xie, 2010). Because the climate change projection has much uncertainty for the position, duration, and intensity of this front due to model resolution and model parameterizations (Kosaka and Nakamura, 2011), both GCMs and RAMs might potentially bring some uncertainty in the summer rainfall change in Hokkaido. Hence, our DDS strategy may expectedly separate agents that possibly bring uncertainty. We will apply bias correction to all DDS results before analysis for a technical reason, but this is not an essential point in this article.

2. Methods

2.1. Model and experiments

We used three GCMs referred to as MIROC [The high-resolution version of Model for Interdisciplinary Research on Climate 3.2, jointly developed by the University of Tokyo.] (Hasumi and Emori, 2004), MPI [The fifth-generation atmospheric GCM in Max-Planck-Institut für Meteorologie.] (Roeckner et al., 2003), and NCAR [The Community Climate System Model version 3 in the National Center for Atmospheric Research.] (Collins et al., 2006) as initial and boundary conditions. The Taylor diagram for describing climatology of mid-tropospheric jet stream around Hokkaido simulated by CMIP3 models suggested that these three GCMs reproduced the present climate almost identically (Figure 1(b)); however, MIROC did not produce climatological feature of the Okhortsk high; MPI and NCAR provided weaker rainfall over Japan; MPI moreover had a bias with northward-shifted jet stream (Figure 2). The present climate period was set as 1990–1999 in the 20th century experiment (20C3M). It is noted that, as a single decade can be representative of the present climate, 10-year average and 20-year average (1980–1999) are indiscernibly similar (not shown). The future climate period was set as 2050–2059 for MIROC, 2060–2069 for MPI, and 2080–2089 for NCAR, each of which is a decade with an increase by approximately 2 K in global-mean SATs on the Special Report on Emissions Scenarios (SRES) A1b (Meehl et al., 2007). The GCM outputs including sea surface temperatures (SSTs; Figure 2(f)–(h)) were prescribed in the RAM every 6 h; the GCMs under the present climate mostly reproduced the observed SST (Figure 2(e)).

We used three RAMs referred to as NHM [Non-hydrostatic model developed by Japan Meteorological Agency and Meteorological Research Institute.] (Saito et al., 2006), WRF [The Advanced Research Weather Research and Forecasting Model version 3.2.1] (Skamarock et al., 2008) and RSM [The regional spectral model developed by Scripps Experimental Climate Prediction Center.] (Juang and Kanamitsu, 1994) all of which are state-of-the-art models widely used in an operational forecast or a DDS experiment. The physical parameterizations for cloud, precipitation, radiative transfer, and turbulent mixing are implemented in the RAMs with a grid-based framework in NHM and WRF or with a spectral framework in RSM (Table SI in Appendix S1, Supporting Information). The horizontal resolution was set to 10 km

and there were about 30 unequally spacing vertical levels. The domain, which is centered in Hokkaido, covered 135°–150°E and 38°–50°N. The topography was prescribed as in Figure 1(a).

For each combination of three GCMs and three RAMs, a 10-year integration is performed under the present climate condition and another 10-year integration under future climate condition (e.g. Kuno and Inatsu, 2014). Many climatic variables such as SAT and precipitation rate were hourly output, and the daily maximum and minimum temperatures were defined based on the hourly data of SAT. We interpolated the original output from RAMs into an equally spaced grid in longitude and latitude for data handling. This article only describes the results in June–July–August (JJA) months. We will represent this multi-GCM by multi-RAM result in a full 3×3 matrix.

2.2. Bias correction

Figure 1(c) shows the climatology of SAT in JJA months interpolated from the Automated Meteorological Data Acquisition System (AMeDAS) operated by the Japan Meteorological Agency. The mean SAT almost ranges from 16 to 20 °C in average, while it is lower than 15 °C in southeastern Hokkaido because of a frequent fog (Sugimoto et al., 2013). The daily maximum temperature sometimes exceeds 30 °C in Sapporo, Asahikawa, and Obihiro (Table 1). The DDS basically reproduced characteristic of the present-climate temperatures with a systematic bias (Figure 1(d)). The DDS ensemble average shows a 1–2 K warm bias in SAT. We then corrected the bias in SATs for each GCM-RAM combination by offsetting the difference of its present-climate climatology between observation and DDS result. Based on the bias corrected data, we will discuss summer days defined as days with maximum temperature not lower than 25 °C, hot summer days as days with maximum temperature not lower than 30 °C, and tropical night days as days with minimum temperature not lower than 25 °C.

Figure 1(e) shows the climatology of precipitation in JJA months based on APHRODITE rainfall data (Yatagai et al., 2012). The mean precipitation exceeds 200 mm month^{-1} at the mountain near Muroran and at Hidaka mountain range (west of Obihiro) because of the topological uplift for moist air. In contrast, it is about 100 mm month^{-1} along the Okhotsk Sea. The DDS reproduces the observed feature in distribution (Figure 1(f)), but the ensemble average shows a positive bias in precipitation entirely over Hokkaido. We then corrected the bias for each combination of GCM and RAM by mapping the data such that empirical cumulative density function of daily precipitation amount for the observation was matched for each calendar month with that for the DDS result under the present climate condition (Piani et al., 2010). This mapping function was applied to a future-climate DDS result with the same GCM-RAM combination.

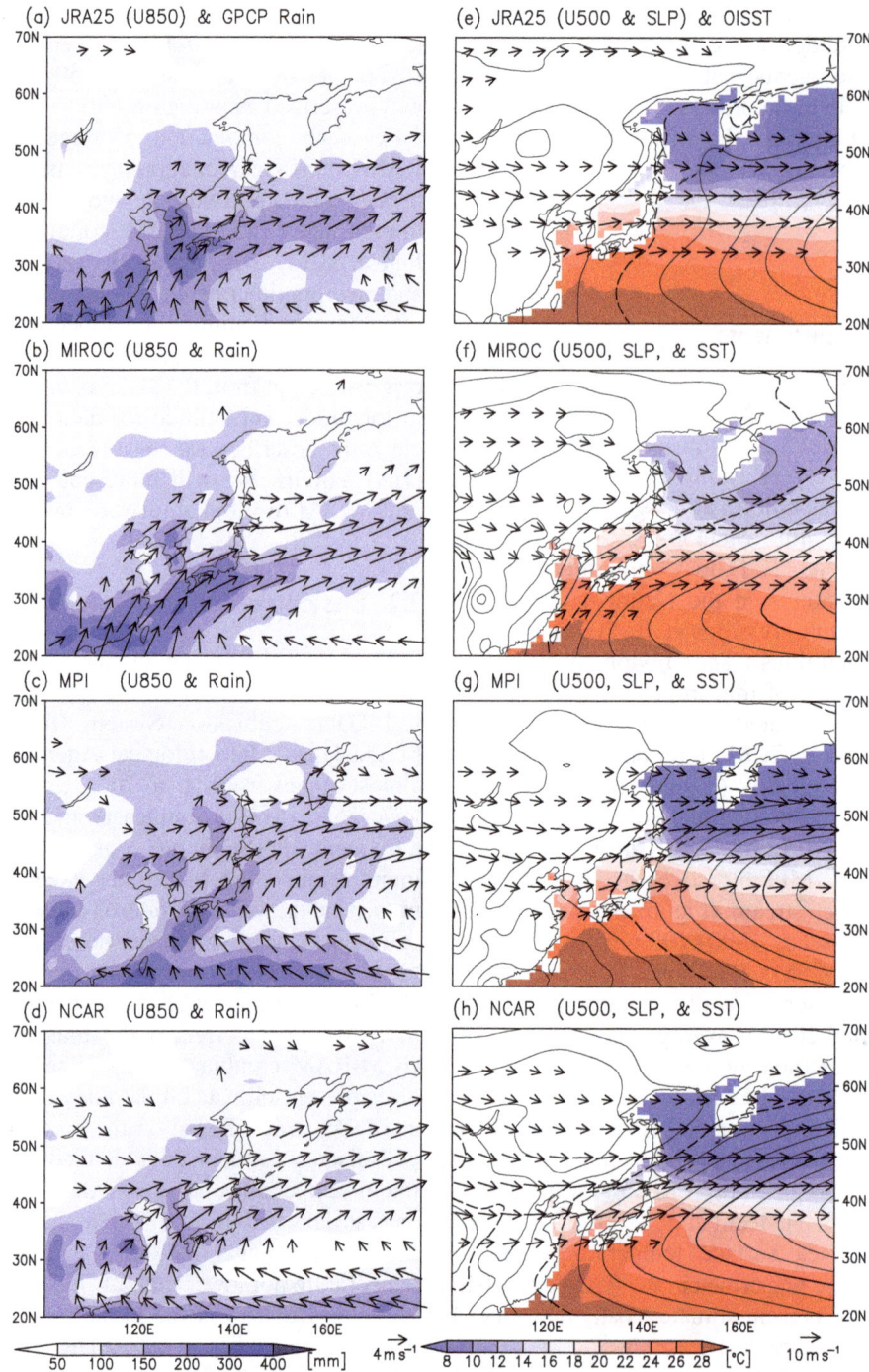

Figure 2. (a)–(d) Ten-year average of (shade) precipitation and (vector) 850-hPa horizontal wind in JJA months. The references for shading and vector are shown in the bottom. (a) The observation in 1990s based on JRA25 reanalysis and Global Precipitation Climatology Project rainfall data; result of 1990s for (b) MIROC, (c) MPI, and (d) NCAR simulations. (e)–(h) Ten-year average of (shade) sea surface temperature (SST), (contour) sea level pressure and (vector) 500-hPa horizontal wind. Contour interval is 2 hPa with a 1010 hPa contour being dotted. (e) The observation in 1990s based on JRA25 reanalysis and National Oceanic and Atmospheric Administration's Optimal Interpolated SST in 1990s; result of 1990s for (f) MIROC, (g) MPI, and (h) NCAR simulations.

3. Results

The 3×3 matrix of the DDS result of the increase in SATs in JJA months is shown in Figure 3(a). During the decade with global-mean SATs increasing by 2 K for each GCM, the temperature over Hokkaido increased by 2.5–3 K for DDS with MIROC boundary, by 2–2.5 K for DDS with MPI boundary, and 1.5–2 K

for DDS with NCAR boundary. The DDS ensemble average is thus higher than 2 K, probably because the lower heat capacity of the Eurasian continent and possibly SST distribution over the North Pacific. Moreover we found the magnitude of increase in RAM variations was relatively small for a given GCM, although the effect of topography was a bit obvious in temperature in some runs. The projection of the result onto the

Table 1. (Left column) Days per year with daily maximum temperature not lower than 25 °C, (middle) days with daily maximum temperatures not lower than 30 °C, and (right) days with daily minimum temperature not lower than 25 °C at S (Sapporo), A (Asahikawa), H (Hakodate), K (Kushiro), O (Obihiro), M (Muroran), B (Abashiri), and W (Wakkanai); cities are shown in Figure 1(a). (Left subcolumn) The days under the current climate are based on the Automated Meteorological Data Acquisition System (AMeDAS) observation. (Right subcolumn) The range of days under the future climate is based on our experiments with a bias correction.

(days year^{-1})	Tmax $\geq$ 25°C		Tmax $\geq$ 30°C		Tmin $\geq$ 25°C	
City	Current	Future	Current	Future	Current	Future
S	45.6	68.6–79.9	7.7	9.8–18.9	0.1	0.2–4.2
A	56.3	75.1–91.4	10.4	8.5–19.8	0.1	0.0–1.0
H	34.3	38.3–55.9	3.1	0.0–3.6	0.0	0.0–0.8
K	4.7	5.7–18.3	0.1	0.0–2.3	0.0	0.0–1.0
O	40.5	52.0–68.3	9.3	5.3–12.0	0.0	0.0–1.9
M	15.8	9.7–46.5	0.3	0.0–3.9	0.0	0.0–3.9
B	22.1	9.8–45.9	3.5	0.2–4.1	0.0	0.0–0.3
W	7.3	3.5–23.5	0.1	0.0–0.1	0.0	0.0–0.1

phase space spanned by two leading empirical orthogonal function (EOF) modes for inter-DDS variations [The EOF analysis performed for the nine DDS variations (temporal variations in an ordinary use of EOF) and we extracted main spatial patterns in DDS departure from the ensemble average.] suggested that the results with NCAR boundary were far from the others (Figure 3(b)). Among NCAR-boundary experiments, the RSM run provided a slightly different result perhaps because its dynamical core is based on a spectral method. The result above means that the GCM boundary condition mostly controls how much SAT increases in the RAM domain. Especially, a smaller SAT increase was found in the DDS with the NCAR boundary where the SST increased by smaller amounts in the subtropics (not shown).

The 3×3 matrix of the DDS result of the increase in precipitation in JJA months is shown in Figure 4(a). The rate of increase in precipitation was similar in RAMs for a given GCM. Only the DDS with MIROC provided many areas with a statistically significant signal

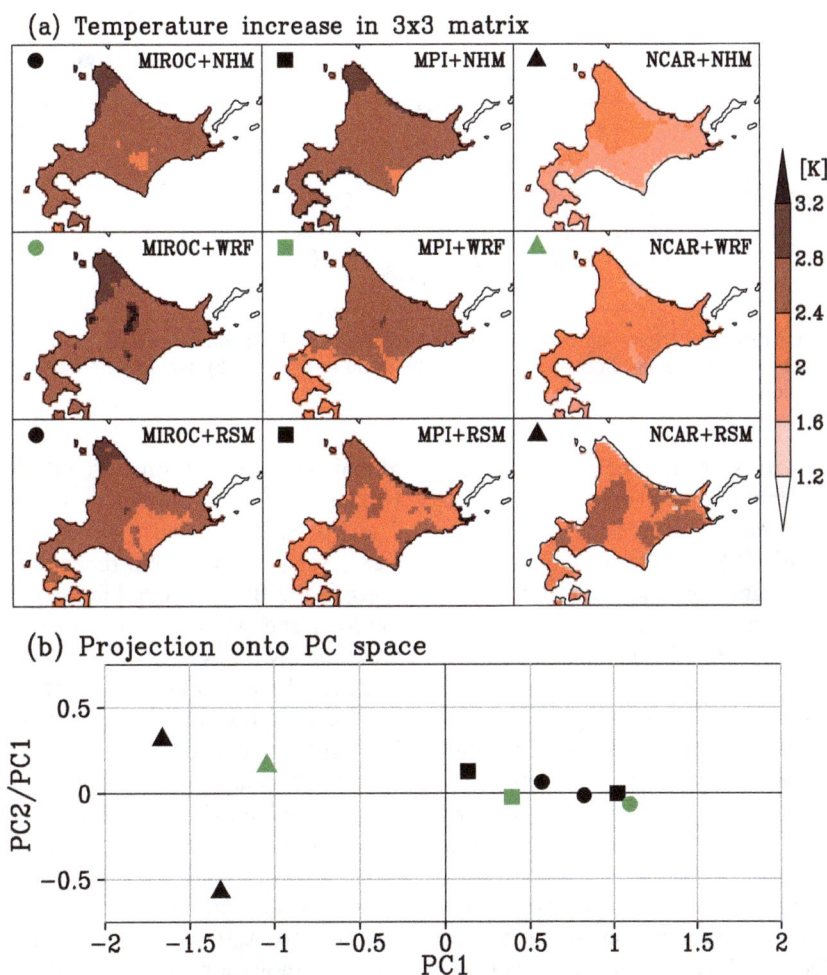

(a) Temperature increase in 3x3 matrix

(b) Projection onto PC space

Figure 3. (a) The increase in surface air temperatures (SATs) by global warming (K) with three general circulation models (GCMs) by three regional atmospheric models (RAMs) for the dynamical downscaling (DDS). The mark type and color denote the GCM-RAM combination: the mark types of circle, square, and triangle respectively mean the GCMs of MIROC, MPI, and NCAR as the lateral boundary condition; red, blue, and green, respectively, denote RAMs of NHM, WRF, and RSM. (b) The horizontal distribution for each GCM-RAM combination projected onto the phase space spanned by two leading principal components (PCs). The second PC is normalized by the ratio of the variance of PC1 to that of PC2.

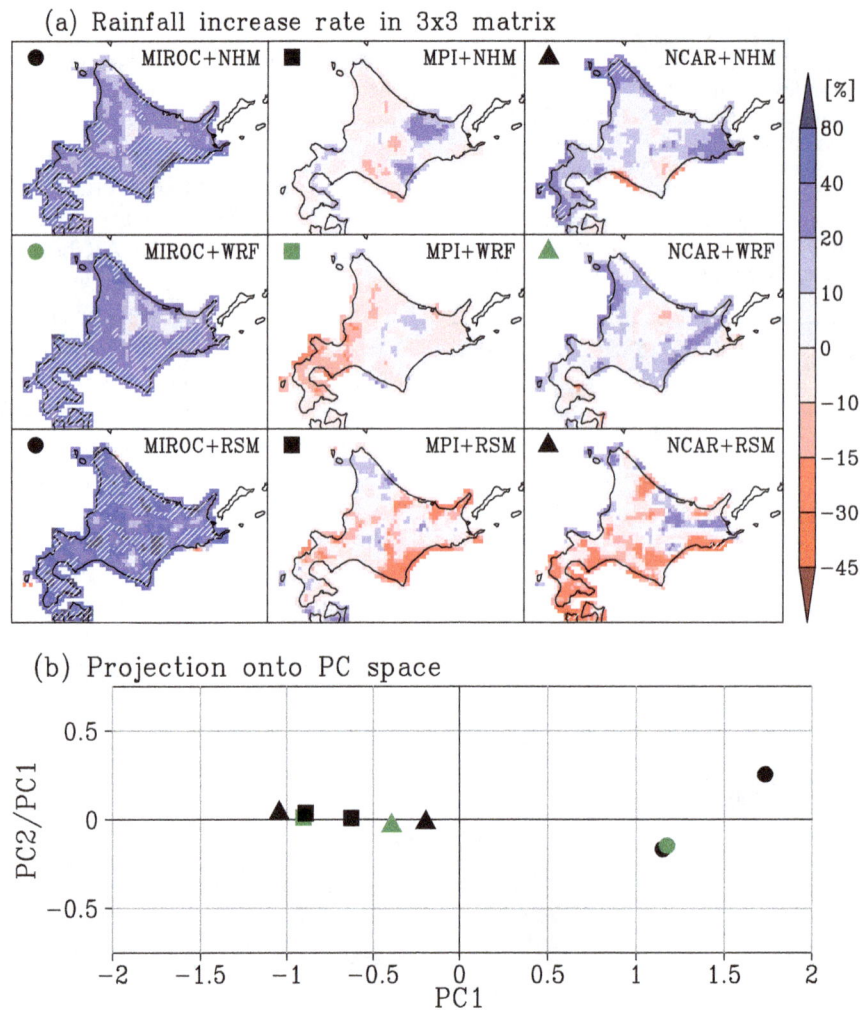

Figure 4. (a) The increase rate in precipitation by global warming (%) for the DDS with three-GCM by three-RAM experiments. The hatches denote areas exceeding a significant level at 10%. (b) Same as Figure 3(b), but for the horizontal distribution of the increase rate in precipitation.

while others did not (not shown). The MIROC-NHM and MIROC-WRF combinations showed a conspicuous signal with ~ 50% rainfall increasing in southern Hokkaido. The MIROC-RSM also showed many areas with a statistical significance but they were a bit more scattered over Hokkaido perhaps because of a different dynamical configuration in RSM. The DDS with MPI boundary provided an insignificant response in precipitation but the tendency is the decrease by ~ 10%. The DDS with NCAR boundary also provided an insignificant signal but the NHM and WRF runs showed a small increase and the RSM run showed a decrease in southern Hokkaido. The GCM boundary condition hence controls how much precipitation increases in the RAM domain. The projection of the results onto the phase space for inter-DDS variations supported the notion that the DDS with MIROC boundary were far from others (Figure 4(b)).

The DDS results of SAT (Figure 3) and precipitation (Figure 4) shown above suggested their strong dependency on the GCM imposed as the lateral boundary condition. In spite of GCM biases (Figure 2) and GCM parameterizations that are different from RAMs, the

global warming response in GCM precipitation over Hokkaido is quite similar to DDS results (Figure 5). The MIROC provided a heavier rainfall from Northeast Asia to western North Pacific along 40°N, a bit north to the Baiu rainband in the present climate, while the MPI and NCAR did not show any significant signals there. Moreover NCAR provided a heavier rainfall just over the Baiu rainband at the present location. The precipitation response is basically related to convective instability attributed to low-level moisture flux and vertical wind shear. MIROC transported more water vapor toward Hokkaido beyond the present-climate Baiu rainband (Figure 5(a); Kimoto, 2005). The upward motion might have been induced by the tendency of upper-level jet stream confluence in the north (not shown; Horinouchi, 2014). In MPI, the anomalous wind appeared to block moist air from the south but instead encouraged the intrusion of dry air from the continent and also by northeasterly from the Okhortsk high (Figure 5(b)), associated with a bit southward shift of the upper-level jet. In NCAR, anomalous easterly was prevailed over Hokkaido and did not contribute to what the moist air straddled over the rainband in the present climate.

(a) MIROC Rainfall & VQ850 2050s/1990s

(b) MPI   Rainfall & VQ850 2060s/1990s

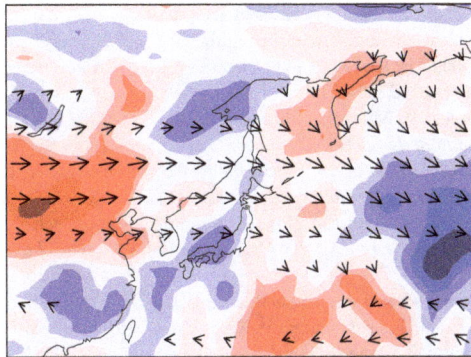

(c) NCAR   Rainfall & VQ850 2080s/1990s

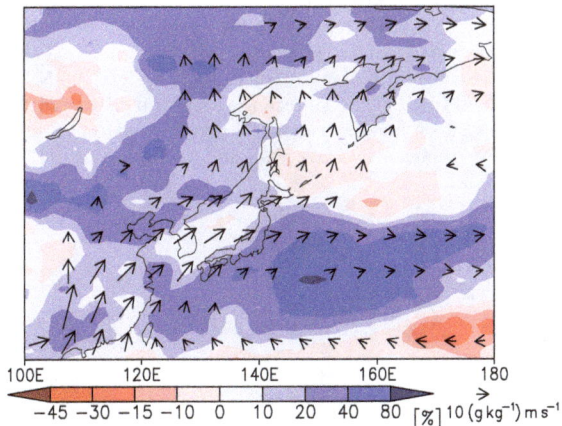

Figure 5. (Shade) The increase rate in precipitation and (vector) the anomalous moisture flux at 850 hPa by global warming for (a) MIROC, (b) MPI, and (c) NCAR simulations.

The moisture flux convergence was however strengthened over south of Japan. As discussed above, the DDS results (Figures 3 and 4) are basically a consequence of the synoptic-scale response to the global warming in GCM (Figure 5).

4. Conclusion and discussion

We have performed the multi-GCM by multi-RAM experiments for DDS over Hokkaido. Focusing on the summertime climate there, the DDS results of SAT and precipitation commonly suggested that the downscaling results were mostly controlled by the GCM boundary condition. All combinations of GCM and RAM increased SATs by 1.5 to 3 K over Hokkaido during the decade in which the global-mean temperature increased by 2 K. Moreover, the DDS results of precipitation were quite uncertain. The experiments with MIROC boundary condition evaluated the increase in precipitation by ~50% over Hokkaido, while other experiments provided an insignificant signal over Hokkaido. This is probably because the global warming changed low-level moisture transport toward the north beyond the present position of the Baiu rainband only in the MIROC GCM.

The increase in summertime temperatures may affect many aspects in human activity. As the present climate in Hokkaido is cool in summer with at most ten hot summer days (Table 1), air conditioners are not mostly installed in houses, buildings, and public space. Hence, a 1.5 to 3 K temperature increase brings solely a slightly hotter climate, but children, elderly people, and health impaired would be exposed to a potential risk of the heat disorder (Ohashi *et al.*, 2014). Together with an urban effect, for example, summer days and hot summer days almost doubled and possibility of tropical night days increased in Sapporo (Table 1), which would indeed deteriorate the quality of life. Temperature increase might also impact agriculture, a major industry in Hokkaido. Cultivation of wheat, potatoes, and sugar beet, and pastoral farming tends to take damages for high temperatures over eastern Hokkaido even now. Such a risk might become higher under the future climate (e.g. Hirota *et al.*, 2011; Nishio *et al.*, 2013).

The possible increase in summer precipitation mainly affects the water resource management. Because of a less summer precipitation in Hokkaido than that in mainland, there is high flood vulnerability to the increase in precipitation even for three major rivers in Hokkaido (Yamada *et al.*, 2012). As the precipitation increases more likely in southern Hokkaido (Figure 4(a)), we can evaluate that Tokachi River has the first priority to reduce vulnerability on possible flood among three river basins.

Acknowledgements

The authors thank M. Kimoto, Y. M. Okumura, N. S. Keenlyside, and R. W. Arritt to help MI to obtain GCM datasets, anonymous reviewers and T. Hirota to give us insightful comments, and Japan Weather Associate to provide gridded climatology data. This study is supported by the Research Program on Climate Change Adaptation and the Program for Risk Information on Climate Change of Ministry of Education, Culture, Sports, Science and Technology, and the Global Environmental Research Fund S-8 of the Ministry of the Environment.

References

Collins WD, Bitz CM, Blackmon ML, Bonan GB, Bretherton CS, Carton JA, Chang P, Doney SC, Hack JJ, Henderson TB, Kiehl JT, Large WG, McKenna DS, Santer BD, Smith RD. 2006. The community climate system model version 3 (CCSM3). *Journal of Climate* **19**: 2122–2143.

Giorgi F, Bates GT. 1989. The climatological skill of a regional model over complex terrain. *Monthly Weather Review* **117**: 2325–2347.

Hasumi H, Emori S. 2004. K-1 coupled GCM (MIROC) description. K-1 Technical Report 1, CCSR, University of Tokyo: Tokyo.

Hirota T, Usuki K, Hayashi M, Nemoto M, Iwata Y, Yanai Y, Yazaki T, Inoue S. 2011. Soil frost control: agricultural adaptation to climate variability in a cold region of Japan. *Mitigation and Adaptation Strategies for Global Change* **16**: 791–802.

Horinouchi T. 2014. Influence of upper tropospheric disturbances on the synoptic variability of precipitation and moisture transport over summertime East Asia and the northwestern Pacific. *Journal of the Meteorological Society of Japan*, doi: 10.2151/jmsj.2014-602.

Ishizaki NN, Takayabu I, Ohízumi M, Sasaki H, Dairaku K, Iizuka S, Kimura F, Kusaka H, Adachi SA, Kurihara K, Murazaki K, Tanaka K. 2012. Improved performance of simulated Japanese climate with a multi-model ensemble. *Journal of the Meteorological Society of Japan* **90**: 235–254.

Juang HM, Kanamitsu M. 1994. The NMC nested regional spectral model. *Monthly Weather Review* **122**: 3–26.

Kimoto M. 2005. Simulated change of the east Asian circulation under global warming scenario. *Geophysical Research Letters* **32**: L16701.

Kosaka Y, Nakamura H. 2011. Dominant mode of climate variability, intermodel diversity and projected future changes over the summertime western North Pacific simulated in the CMIP3 models. *Journal of Climate* **24**: 3935–3955.

Kuno R, Inatsu M. 2014. Development of sampling downscaling: a case for wintertime precipitation in Hokkaido. *Climate Dynamics* **43**: 375–387.

Meehl GA, Stocker TF, Collins WD, Friedlingstein P, Gaye AT, Gregory JM, Kitoh A, Knutti R, Murphy JM, Noda A, Raper SCB, Watterson IG, Weaver AJ, Zhao Z-C. 2007. Global climate projections. In *Climate Change 2007: The Physical Science Basis*, Solomon S et al. (eds). Cambridge University Press: Cambridge, UK and New York, NY; 747–845.

Nishio Z, Ito M, Tabiki T, Nagasawa K, Yamauchi H, Hirota T. 2013. Influence of higher growing-season temperatures on the yield components on winter wheat (*Triticum aestivum* L.). *Crop Science* **53**: 621–628.

Ohashi Y, Kikegawa Y, Ihara T, Sugiyama N. 2014. Numerical simulations of outdoor heat stress index and heat disorder risk in the 23 wards of Tokyo. *Journal of Applied Meteorology and Climatology* **53**: 583–597.

Piani C, Haerter JO, Coppola E. 2010. Statistical bias correction for daily precipitation in regional climate models over Europe. *Theoretical and Applied Climatology* **99**: 187–192.

Roeckner E, Bäuml G, Bonaventura L, Brokopf R, Esch M, Giorgetta M, Hagemann S, Kirchner I, Kornblueh L, Manzini E, Rhodin A, Schlese U, Schulzweida U, Tompkins A. 2003. The atmospheric general circulation model ECHAM5. Part I: model description. MPI for Meteorol Report 349, MPI for Meteorol, Hamburg, Germany.

Saito K, Fujita T, Yamada Y, Ishida J, Kumagai Y, Aranami K, Ohmori S, Nagasawa R, Kumagai S, Muroi C, Kato T, Eito H, Yamazaki Y. 2006. The operational JMA nonhydrostatic mesoscale model. *Monthly Weather Review* **134**: 1266–1298.

Sampe T, Xie SP. 2010. Large-scale dynamics of the Meiyu-Baiu rainband: environmental forcing by the westerly jet. *Journal of Climate* **23**: 113–134.

Skamarock WC, Klemp JB, Dudhia J, Gill DO, Barker DM, Duda MG, Huang X-Y, Wang W, Powers JG. 2008. A description of the advanced research WRF version 3. NCAR Technical Note NCAR/TN-475+STR, National Center for Atmospheric Research: Boulder, CO; 113 pp.

Sugimoto S, Sato T, Nakamura K. 2013. Effects of synoptic-scale control on long-term declining trends of summer fog frequency over the Pacific side of Hokkaido Island. *Journal of Applied Meteorology and Climatology* **52**: 2226–2242.

van der Linden P, Mitchell JFB (eds). 2009. *ENSEMBLES: Climate Change and its Impacts: Summary of Research and Results from the ENSEMBLES Project*. Met Office Hadley Centre: Exeter, UK; 160 pp

Wang YQ, Leung LR, Mcgregor JL, Lee DK, Wang WC, Ding Y, Kimura F. 2004. Regional climate modeling: progress, challenges, and prospects. *Journal of the Meteorological Society of Japan* **82**: 1599–1628.

Wang SY, Gillies RR, Takle ES, Gutowski WJ. 2009. Evaluation of precipitation in the Intermountain Region as simulated by the NARCCAP regional climate models. *Geophysical Research Letters* **36**: L11704, doi: 10.1029/2009GL037930.

Yamada TJ, Sasaki J, Matsuoka N. 2012. Climatology of line-shaped rainband over northern Japan in boreal summer between 1990 and 2010. *Atmospheric Science Letters* **13**: 133–138.

Yatagai A, Kamiguchi K, Arakawa O, Hamada A, Yasutomi N, Kitoh A. 2012. APHRODITE: constructing a long-term daily gridded precipitation dataset for Asia based on a dense network of rain gauges. *Bulletin of the American Meteorological Society* **93**: 1401–1415.

14

Can marine cloud brightening reduce coral bleaching?

John Latham,[1,2] Joan Kleypas,[1] Rachel Hauser,[1,5] Ben Parkes[3] and Alan Gadian[4*]

[1] NCAR, Boulder, CO, USA
[2] SEAS, University of Manchester, Manchester, UK
[3] ICAS, University of Leeds, West Yorkshire, UK
[4] NCAS, ICAS, University of Leeds, West Yorkshire, UK
[5] Environmental Studies Program and Center for Science and Technology Policy Research, University of Colorado, Boulder, UK

*Correspondence to:
A. Gadian, NCAS, ICAS,
University of Leeds, West
Yorkshire, UK.
E-mail: alan@env.leeds.ac.uk

Abstract

Increases in coral bleaching events over the last few decades have been largely caused by rising sea surface temperatures (SST), and continued warming is expected to cause even greater increases through this century. We use a Global Climate Model to examine the potential of marine cloud brightening (MCB) to cool oceanic surface waters in three coral reef provinces. Our simulations indicate that under doubled CO_2 conditions, the substantial increases in coral bleaching conditions from current values in three reef regions (Caribbean, French Polynesia, and the Great Barrier Reef) were eliminated when MCB was applied, which reduced the SSTs at these sites roughly to their original values.

Keywords: coral bleaching; marine cloud brightening; sea surface temperature; cloud seeding; global climate modeling

1. Introduction

The continued increase in CO_2 emissions into the atmosphere has generated greater interest in engineering strategies to ameliorate the effects of climate change (Shepherd *et al.*, 2009; SRMGI, 2011). Some of these involve solar radiation management (SRM), i.e. 'shading' the planet from incoming sunlight and thus slowing the rate of warming at the Earth's surface. One such SRM technique is marine cloud brightening (MCB): (Latham, 1990, 2002; Bower *et al.*, 2006; Latham *et al.*, 2008; Jones *et al.*, 2009, 2011; Rasch *et al.*, 2009; Korhonen *et al.*, 2010; Bala *et al.*, 2011; Latham *et al.*, 2012a, 2012b). MCB geoengineering is designed to produce a cooling that in principle can maintain the Earth's average surface temperature and polar sea–ice cover at roughly current values in the face of increasing atmospheric CO_2 concentrations, at least up to the $2\times CO_2$ point. MCB involves seeding low-level marine stratocumulus (ice-free) clouds with submicrometer-sized seawater droplets. The particles have sufficiently high-salt mass to act as cloud condensation nuclei (CCN), thereby increasing the cloud droplet number concentration (CDNC) and the cloud optical thickness. The resulting effect is an increase in cloud albedo for incoming shortwave radiation (Twomey, 1977). Also, smaller cloud droplets coagulate more slowly, thus suppressing precipitation development and increasing cloud lifetime (Albrecht, 1989). The engineering design for MCB includes deployment of ocean-based vessels strategically placed to release seawater CCN into the turbulent boundary layer beneath marine stratocumulus clouds.

MCB was originally developed as a means of counteracting warming on a global scale. However, MCB in principle could also be used to target subglobal regions of particular interest. That is, seeding sites could be selected:

- to maximize a desired effect (e.g. reduction in warming);
- to minimize adverse effects (e.g. reduction in precipitation).

Some previous modeling studies have illustrated a cooling effect by seeding the three regions of most extensive marine stratocumulus clouds: off the west coasts of Africa, North America, and South America (Jones *et al.*, 2009, Latham *et al.*, 2012a; 2012b, 2011). Seeding these three regions, e.g. produced a significant reduction in sea surface temperatures (SST) where hurricanes traditionally develop in the Atlantic, raising the possibility of weakening them (Latham *et al.*, 2012b). Mostly, the three seeding regions covered less than 10% of the marine stratocumulus regions but some larger regions are considered in some of the articles.

The influence of SRM on marine ecosystems has recently been addressed by Russell *et al.* (2012). In this article, we evaluate computationally the potential for MCB seeding in these same three regions to significantly reduce tropical SSTs in three major coral reef provinces – the Caribbean, French Polynesia, and the Great Barrier Reef – and thereby reduce the rates of coral bleaching.

2. Methods

2.1. Simulations of SST with and without MCB

SST values were extracted from model calculations conducted with the Hadley Centre Global Environment

Model (HadGEM1), version 6.1 of the UK Meteorological Office Unified Model. Further details of this model are provided in Latham *et al.* (2012a, 2012b).

HadGEM1 was modified to have a fixed CDNC in the three aforementioned regions of low-level (about 1 km high) marine stratocumulus clouds. The normal (natural) value of CDNC in HadGEM1 is about $60\,cm^{-3}$. For the MCB simulations, the CDNC was given a value of $375\,cm^{-3}$ at all model levels between 0 and 3 km, which is consistent with the treatment used in Jones *et al.* (2009, 2011), Latham *et al.* (2008, 2012a, 2012b) and *Parkes et al.* (2012).

Three simulations were completed to determine the effects of MCB on SST values. They were (1) a Control simulation forced by an atmospheric CO_2 concentration of 440 ppm, (2) a $2\times CO_2$ simulation forced by an atmospheric CO_2 concentration of 560 ppm, and (3) a $2\times CO_2$ MCB simulation that includes MCB in the three regions of marine stratocumulus. For the $2\times CO_2$ simulations, the model was run from 2020 to 2045 with increasing CO_2 until 2045, and then held stable at double CO_2 values until 2090. For each 70-year simulation, the 10-day averaged SSTs of the final 20 years (2070–2090) were used for the coral bleaching analysis.

2.2. Coral bleaching calculations

For each simulation, the SSTs were used to calculate coral heat stress and bleaching. Heat stress in corals occurs once the temperature exceeds some predetermined threshold above the climatological maximum. This threshold is traditionally set at 1 °C (*Liu et al.*, 2003), or alternatively, is allowed to vary regionally as a function of the natural variability of the year-to-year maxima. In the latter case, the threshold is typically determined as $n \times SD_{max}$, where n is an empirically determined factor, and SD_{max} is the standard deviation of the annual maxima over the climatological period. We apply a threshold of $2.45 \times SD_{max}$ following Donner (2009).

Coral bleaching occurs when heat stress accumulates over a period of time. This is typically calculated as degree heating weeks (DHW), which is the accumulation of the heat stress over a 12-week period (Liu *et al.*, 2003). When weekly data are used, mild bleaching occurs once the DHW > 4, and severe bleaching occurs once DHW > 8.

SSTs of the final 20-year Control case were used to calculate the climatological SST maximum for every ocean cell within each region, including those that do not include reefs. In both the Caribbean and the Great Barrier Reef regions, many of the coastal grid cells were not resolved in the model; here, we assume that changes in offshore temperatures are largely representative of those affecting adjacent reefs. The climatological values derived from the Control run were used in the DHW calculations for each ocean cell for the three regions, for both the $2\times CO_2$ and $2\times CO_2 + MCB$ cases. To adjust for the

10-day average SSTs of the model output (vs weekly averages), a bleaching event was designated as mild when DHW = 3–5, and severe when DHW $\geq$ 6. These values are slightly lower than the 7-day DHW of 4 and 8, respectively. Bleaching events were calculated separately for the 1 °C and $2.45 \times SD_{max}$ heat-stress thresholds (Figure 1).

3. Results

3.1. Global SST

The changes in SST between the Control and $2\times CO_2$ simulations (Figure 2(a)) are consistent with patterns produced in other coupled GCM studies mentioned earlier. Temperature increases between 0.5 and 2 K occurred over much of the tropical area (Figure 2(a)). The effects of MCB on SSTs in the $2\times CO_2$ simulations more than cancel these temperature increases (Figure 2(b)). In case of three-patch MCB in the $2\times CO_2$ atmosphere, the global temperatures decrease by 0.12 K. Full-area seeding at $2\times CO_2$ leads to much greater cooling over the entire climate system and reduces tropical temperatures by more than 5 K, which is too large to be beneficial and could cause cold-temperature stress in corals.

3.2. Temperatures in coral reef regions

In all three coral reef regions, the doubling of atmospheric CO_2 raised the annual average temperatures by at least 0.5 K relative to the Control scenario, whereas MCB restored SSTs to near the Control values (Table I). The greater cooling in French Polynesia (Figure 2(b)) illustrates the cooling effect of MCB in waters off Peru that are then transported by ocean currents westward across the Pacific.

Coral bleaching events were rare based on SSTs from the Control case (Table II); only French Polynesia exhibited bleaching for either heat-stress threshold. The number of bleaching events in the $2\times CO_2$ case was dramatically higher for all three regions. For the $2\times CO_2 + MCB$ case, however, bleaching events were almost entirely eliminated in all three regions.

4. Discussion

The possible utilization of MCB for examining the subglobal scale topic of coral bleaching amelioration possesses a number of positive and negative attributes. On the positive side, our model-based analysis indicates that MCB seeding in key patches of marine stratocumulus would not only lower temperatures over all three major reef regions studied, but could also restore temperatures to the Control levels. Coral reef ecosystems may thus be beneficiaries of MCB geoengineering. Other subglobal locations for which calculations predict that cooling due to MCB can compensate for the warming produced by fossil

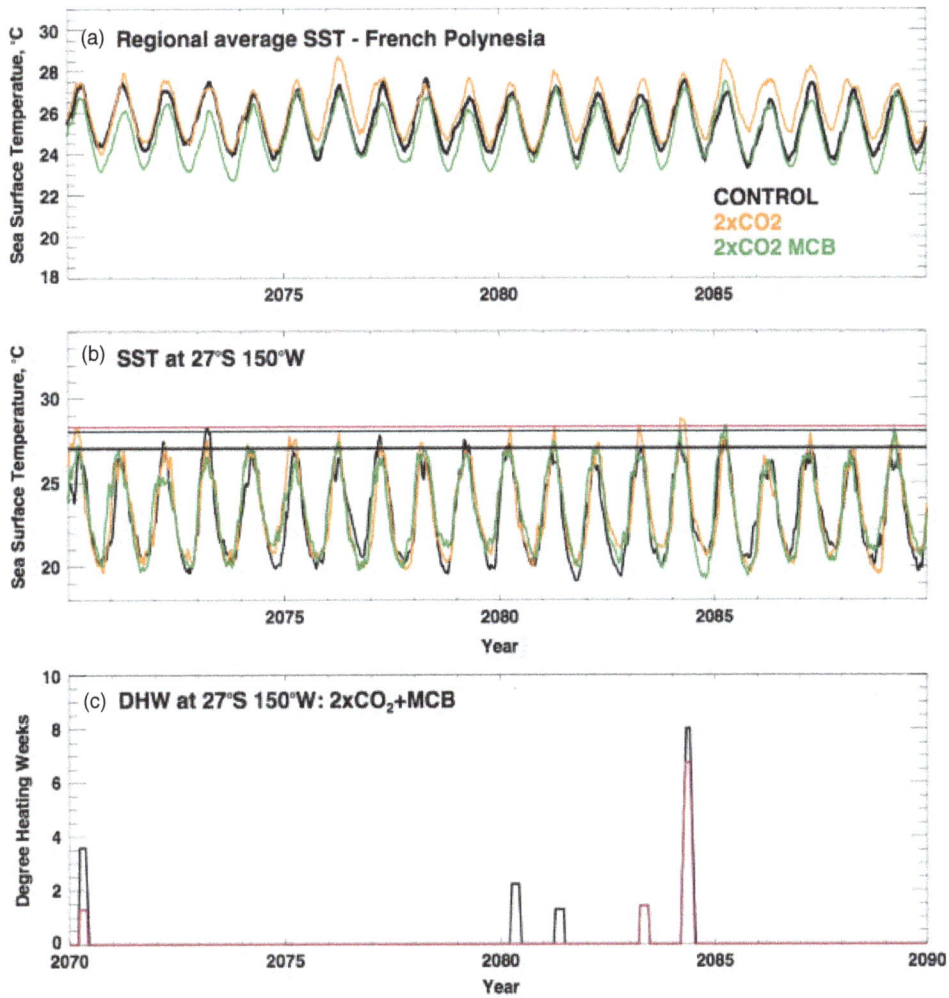

Figure 1. (a) Regional average sea surface temperature (SST) for French Polynesia, comparing SST for the three cases. (b) SST for a single cell within the French Polynesia region, showing the climatological maximum (thick black line), and the 1 °C (thin black line) and $2.45 \times SD_{max}$ (thin red line) thresholds above the climatological maximum. (c) The degree heating weeks (DHW) based on the 1 °C (thin black line) and $2.45 \times SD_{max}$ (thin red line) thresholds.

Table 1. Changes in sea surface temperatures between the $2 \times CO_2$ experiments and the Control, as described in Section 3.1 (all values in °C). The three regions, Caribbean, Great Barrier Reef, and Polynesia are shown in black rectangles in Figure 2.

Region	$2 \times CO_2$ – Control	$2 \times CO_2$ + MCB – Control
Global	0.67	−0.12
Caribbean	0.53	−0.04
Great Barrier Reef	0.55	0.13
Polynesia	0.64	−0.58

$2 \times CO_2$ + MCB all clouds – control: global temperature change = −6.5 °C

fuel burning are the Arctic (Parkes *et al.*, 2012) and the hurricane-generating waters of the tropical Atlantic (Latham *et al.*, 2012b). MCB has additional advantages such as: (1) it does not require a long spin-up time (about 1 year) to achieve a cooling effect, (2) once halted, the effects of MCB on clouds are easily and quickly reversed [the life-time of salt particles in the atmosphere is about 10 days (Salter *et al.*, 2008)], and (3) the utilization of mobile MCB spray vessels could

provide a high degree of control, given the relative ease of adjusting their positioning, as well as the flux of sea salt spray particles that act as CCN. The regions where MCB would be most effective are those where marine stratocumulus clouds are frequently present or where surface currents deliver waters that have been cooled by MCB. The flexibility in the location of MCB seeding may permit fine tuning to maximize its effectiveness and to minimize side effects.

Latham *et al.* (2012a) present a detailed three-stage plan for field-testing MCB, should this be authorized, on a spatial scale of about 100×100 km, which seems very likely to be too small to produce any significant climate effects. It would be based on – but on a much smaller scale than – the successful VOCALS field experiment (Wood *et al.*, 2011) which involved exhaustive studies of marine stratocumulus clouds. A field study of the impact of MCB on coral bleaching at a selected site would utilize some but by no means all of the equipment and procedures required for the three-stage field experiment.

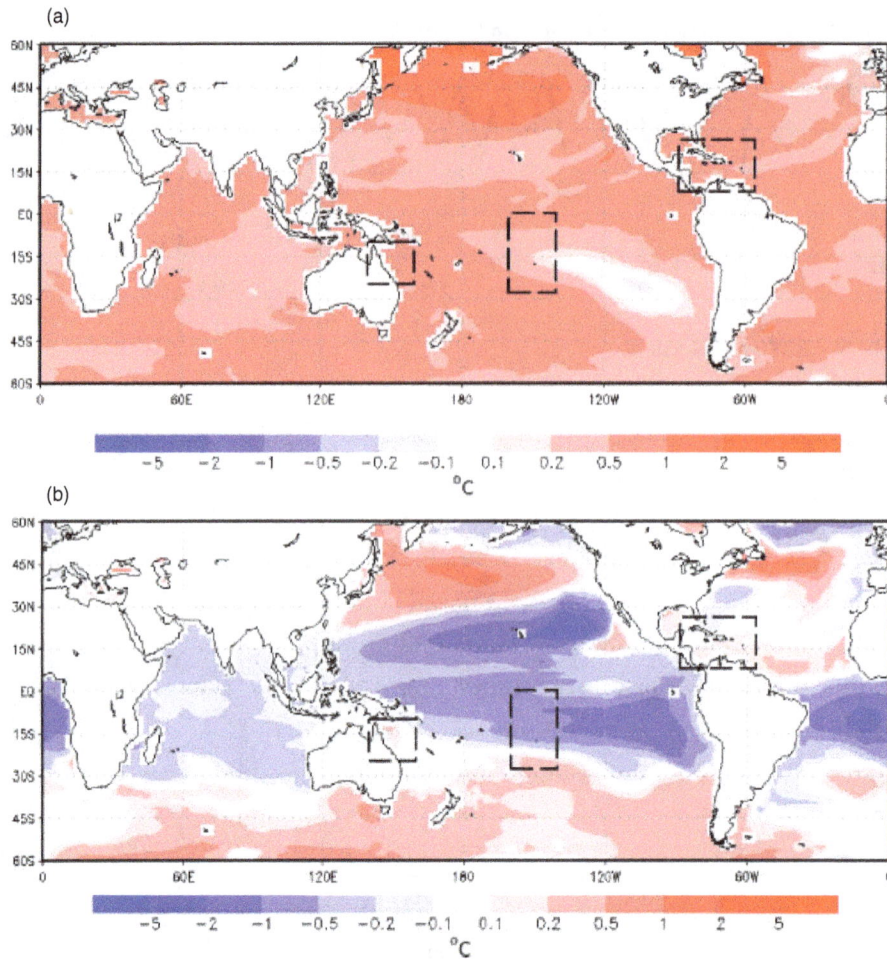

Figure 2. (a) Change in annual average sea surface temperature ($^{\circ}$C) between the $2\times CO_2$ and control simulations. (b) Change in annual average sea surface temperature ($^{\circ}$C) between the control and $2\times CO_2 + MCB$ simulations. The dashed black boxes in both panels represent the three coral reef regions.

Table II. The number of mild and severe bleaching events over a 20-year period for the three simulations: Control, $2\times CO_2$, and $2\times CO_2+MCB$. Bleaching events are calculated based on two heat-stress thresholds: 1 $^{\circ}$C above the climatological maximum, and $2.45 \times SD_{max}$ above the climatological maximum. Climatological maxima were calculated based on the Control. Note that the Control provides the background number of bleaching events expected under normal conditions. Numbers in parentheses are the number of ocean cells within each reef region.

	Control		$2\times CO_2$		$2\times CO_2 + MCB$	
Region	Mild	Severe	Mild	Severe	Mild	Severe
Caribbean (439)						
1 $^{\circ}$C	0	0	294	104	0	0
$2.45 \times SD_{max}$	0	0	655	177	3	0
Great Barrier Reef (179)						
1 $^{\circ}$C	0	0	323	71	0	0
$2.45 \times SD_{max}$	0	0	311	36	0	0
French Polynesia (899)						
1 $^{\circ}$C	95	2	1056	1884	32	0
$2.45 \times SD_{max}$	3	0	850	1502	2	0

However, there are a number of possible disadvantages and limitations of MCB seeding designed to reduce coral bleaching: (1) use of MCB may reduce the photosynthetically available radiation (PAR) reaching the surface, which would affect both the depth of light penetration (which affects the depth limits of corals and seagrass beds), as well as primary production in the water column, (2) Although blocking shortwave radiation leads to a reduction in ocean temperatures in these reef regions, it does not reduce the process of ocean acidification, which is a direct consequence of rising atmospheric CO_2, and is detrimental to coral growth and reef development, and (3) as with any SRM geoengineering technique, the

MCB process would have to be continued indefinitely. Cessation of MCB, particularly if atmospheric CO_2 concentrations continue to rise, could result in very rapid warming, which for coral reefs is the most common condition leading to coral bleaching. The estimated annual cost of deploying MCB to inhibit coral bleaching around the major reefs is about $40M. However, social, ethical, and political costs related to compensation are extremely hard to estimate, and may dominate the costs.

As mentioned earlier, a primary requirement of SRM research is to fully examine all possible adverse consequences of deployment, and to abandon this work if significant ones cannot be remedied. Previous modeling analyses of the potential adverse effects of MCB have largely been restricted to global applications of MCB and mostly to the possibility that rainfall would be reduced in particular regions. Jones *et al.* (2009), e.g. found that MCB reduced rainfall significantly in the Amazonian region, but later (Jones *et al.* 2011) found that altering the locations of seeding largely eliminated this problem. Bala *et al.* (2011) also found that MCB seeding would cause a substantial reduction in rainfall, but that virtually all of this loss was over the oceans, with no net loss over land. Whether or not appreciable rainfall reduction occurs specifically over Amazonia appears to depend on the choice of seeding site(s). Thus, there exists a useful element of flexibility regarding the unforeseen consequences issue. Latham *et al.* (2012a, 2012b) provide a more detailed account of research into this issue.

Another major consideration relates to governing the research associated with geoengineering, including MCB. This is not a trivial consideration, particularly for the large-scale deployments necessary to achieve the experimental ends suggested by the modeling in this article. A governance framework – the processes, mechanisms, institutions, and individuals guiding ordered rule and collective action (Folke *et al.*, 2005) – will be necessary to guide and inform MCB research and will almost certainly need to be developed in advance of field testing (Shepherd *et al.*, 2009; Keith *et al.*, 2010). Establishing such a framework will assist scientists, decision makers, and the public with evaluating and managing both known and as-yet-unknown risks and benefits of MCB.

Among the governance questions to consider are: How can ongoing, independently run assessments of geoengineering research and its outcomes be achieved? How may newly acquired knowledge be best incorporated into existing governance and research structures? Also, means of allowing public scrutiny and input, and research and decision transparency will be important to incorporate when developing SRM or MCB governance. Issues of liability and equity with regard to MCB decision-making and application will need to address how to attend to the needs of the voiceless and those left worse off from MCB use. In essence, effective governance will frame ways in which MCB and other geoengineering efforts are best regulated as a 'public good' (Rayner *et al.*, 2009). Governance will benefit from integrating existing research protocols and lessons learned with geoengineering-specific governance guidelines and ideas such as those generated by the Oxford Principles (Rayner *et al.*, 2009) and groups like the Solar Radiation Management Initiative (SRMGI, 2011).

We conclude that MCB seeding would likely lower tropical SSTs and thus lower the risk of coral reef bleaching for several decades. However, various technological and governance hurdles remain before a fully operational MCB system can be put in place. As with any SRM geoengineering technique, much more work is also required before we can fully gauge whether MCB will cause adverse consequences that cannot be remedied – in which case it should not be utilized. Given the current rate of CO_2 increase in the atmosphere, however, MCB should continue to be evaluated as a measure to prevent particularly dangerous aspects of climate change.

Acknowledgements

We are grateful for the use of NERC, NCAS, HECToR supercomputer resources. Support for elements of this research was provided by the Fund for Innovative Climate and Energy Research, FICER, at the University of Calgary. This does not constitute endorsement of deployment in any form of cloud albedo modification by the funding agency. Part of this work was undertaken while working on the End-to-end quantification of uncertainty for impacts prediction project (EQUIP) funded by NERC (Grant number NE/H003525/1). NCAR is sponsored by the National Science Foundation.

References

Albrecht BA. 1989. Aerosols, cloud microphysics, and fractional cloudiness. *Science* **245**(4923): 1227–1230, DOI: 10.1126/science.245.4923.1227.

Bala G, Caldeira K, Nemani R, Cao L, Ban-Weiss G, Shin HJ. 2011. Albedo enhancement of marine clouds to counteract global warming: impacts on the hydrological cycle. *Climate Dynamics* **37**(5–6): 915–931, DOI: 10.1007/s00382-010-0868-1.

Bower K, Choularton TW, Latham J, Sahraei J, Salter S. 2006. Computational assessment of a proposed technique for global warming mitigation via albedo-enhancement of marine stratocumulus clouds. *Atmospheric Research* **82**: 328–336.

Donner SD. 2009. Coping with commitment: projected thermal stress on coral reefs under different future scenarios. *PloS One* **4**(6), DOI: 10.1371/journal.pone.0005712.

Folke C, Hahn T, Olsson P, Norberg J. 2005. Adaptive governance of social-ecological systems. *Annual Review of Environment and Resources* **30**: 441–473.

Jones A, Haywood J, Boucher O. 2009. Climate impacts of geoengineering marine stratocumulus clouds. *Journal of Geophysical Research-Atmospheres* **114**, DOI: 10.1029/2008jd011450.

Jones A, Haywood J, Boucher O. 2011. A comparison of the climate impacts of geoengineering by stratospheric SO_2 injection and by brightening of marine stratocumulus cloud. *Atmospheric Science Letters* **12**(2): 176–183, DOI: 10.1002/asl2.291.

Keith DW, Parson E, Morgan MG. 2010. Research on global sun block needed now. *Nature* **463**(7280): 426–427, DOI: 10.1038/463426a.

Korhonen H, Carslaw KS, Romakkaniemi S. 2010. Enhancement of marine cloud albedo via controlled sea spray injections: a global model study of the influence of emission rates, microphysics and transport. *Atmospheric Chemistry and Physics* **10**(9): 4133–4143, DOI: 10.5194/acp-10-4133-2010.

Latham J. 1990. Control of global warming. *Nature* **347**(6291): 339–340, DOI: 10.1038/347339b0.

Latham J. 2002. Amelioration of global warming by controlled enhancement of the albedo and longevity of low-level maritime clouds. *Atmospheric Science Letters* **3**(2–4): 52–58, DOI: 10.1006/asl2e.2002.0048.

Latham J, Rasch P, Chen CC, Kettles L, Gadian A, Gettelman A, Morrison H, Bower K, Choularton T. 2008. Global temperature stabilization via controlled albedo enhancement of low-level maritime clouds. *Philosophical Transactions of the Royal Society A: Mathematical Physical and Engineering Sciences* **366**(1882): 3969–3987, DOI: 10.1098/rsta.2008.0137.

Latham J, Bower K, Choularton T, Coe H, Connolly P, Cooper G, Craft T, Foster J, Gadian A, Galbraith L, Iacovides H, Johnston D, Launder B, Leslie B, Meyer J, Neukermans A, Ormond B, Parkes B, Rasch P, Rush J, Salter S, Stevenson T, Wang H, Wang Q, Wood R. 2012a. Marine cloud brightening. *Philosophical Transactions of the Royal Society A: Mathematical Physical and Engineering Sciences* **370**: 4217–4262, DOI: 10.1098/rsta.2012.0086.

Latham J, Parkes B, Gadian A, Salter S. 2012b. Weakening of hurricanes via marine cloud brightening (MCB). *Atmospheric Science Letters* **13**(4): 231–237, DOI: 10.1002/asl2.402.

Liu G, Strong A, Skirving W. 2003. Remote sensing of sea surface temperature during 2002 Barrier Reef coral bleaching. *Eos, Transactions of the American Geophysical Union* **84**: 137–141.

Parkes B, Gadian A, Latham J. 2012. The effects of marine cloud brightening on seasonal polar temperatures and the meridional heat flux. *ISRN Geophysics* , DOI: 10.5402/2012/142872.

Rasch PJ, Latham J, Chen CC. 2009. Geoengineering by cloud seeding: influence on sea ice and climate system. *Environmental Research Letters* **4**(4), DOI: 10.1088/1748-9326/4/4/045112.

Rayner S, Redgwell C, Savulescu J, Pidgeon N, Kruger T (2009) Memorandum on draft priniciples for the conduct of geoengineering research. U.K. House of Commons Science and Technology Committee Enquiry into the Regulation of Geoengineering, 5 pp. http://www.sbs.ox.ac.uk/centres/insis/Documents/regulation-of-geoengineering.pdf.

Russell LM, Rasch PJ, Mace GM, Jackson RB, Shepherd J, Liss P, Leinen M, Schimel D, Vaughan NE, Janetos AC, Boyd PW, Norby RJ, Caldeira K, Merikanto J, Artaxo P, Melillo J, Morgan MG. 2012. Ecosystem impacts of geoengineering: a review for developing a science plan. *Ambio* **41**(4): 350–369, DOI: 10.1007/s13280-012-0258-5.

Salter S, Sortino G, Latham J. 2008. Sea-going hardware for the cloud albedo method of reversing global warming. *Philosophical Transactions of the Royal Society A: Mathematical Physical and Engineering Sciences* **366**(1882): 3989–4006, DOI: 10.1098/rsta.2008.0136.

Shepherd, J, Caldeira, K, Cox, P, Haigh, J, Keith, D, Launder, B, Mace, G, MacKerron, G, Pyle, J, Rayner, S, Redgwell, C, and Watson, A (2009) Geoengineering the Climate: Science, governance and uncertainty. Royal Society Policy document, 82 pp.

Solar Radiation Management Governance Initiative (SRMGI) (2011) Solar radiation management: the governance of research. Environmental Defense Fund edited, The Royal Society and TWAS; 70 pp.

Twomey S. 1977. Influence of internal scattering on the optical properties of particles and drops in the near-infrared. *Applied Optics* **26**(7): 1342–1347.

Wood R, Mechoso CR, Bretherton CS, Weller RA, Huebert B, Straneo F, Albrecht BA, Coe H, Allen G, Vaughan G, Daum P, Fairall C, Chand D, Klenner LG, Garreaud R, Grados C, Covert DS, Bates TS, Krejci R, Russell LM, de Szoeke S, Brewer A, Yuter SE, Springston SR, Chaigneau A, Toniazzo T, Minnis P, Palikonda R, Abel SJ, Brown WOJ, Williams S, Fochesatto J, Brioude J, Bower KN. 2011. The VAMOS Ocean–Cloud–Atmosphere–Land Study Regional experiment (VOCALS-REx): goals, platforms, and field operations. *Atmospheric Chemistry and Physics* **11**(2): 627–654, DOI: 10.5194/acp-11-627-2011.

15

Climatology of convective available potential energy (CAPE) in ERA-Interim reanalysis over West Africa

Cyrille Meukaleuni, André Lenouo* and David Monkam

Department of Physics, Faculty of Science, University of Douala, Cameroon

*Correspondence to:
A. Lenouo, Department of Physics, Faculty of Science, University of Douala, P.O Box 24157, Douala, Cameroon.
E-mail: lenouo@yahoo.fr*

Abstract

Seasonal study of convective available potential energy (CAPE) is done using 6-h ERA-Interim data over West Africa during 35 years (1979–2014). Climatology of CAPE presented in terms of seasonal means, variances and trends shows large values toward 12°–16°N with maxima during summer, according to higher relative humidity due to the arrival of monsoon in West Africa. Spectral analysis in the zone 10°–20°N/20°W–30°E, centered on the latitudes of maxima of CAPE trends at 12°–16°N toward inter-tropical convergence zone (ITCZ) mean position in summer, shows significant power in the 3–5 day within the regions of tropical deep convection in connection with African easterly waves.

Keywords: CAPE; West Africa; ITCZ; ERA-Interim; trends

1. Introduction

Convection plays a crucial role in the terrestrial climate with the formation of the clouds such as the cumulonimbus. The cumulus and other cumulonimbus are conditioned by the general circulation of the atmospheric air (orientation and force of the wind on ground) and the mode of precipitation (Lenouo *et al.*, 2010). The human activities are directly influenced by their effects: cloud cover, downpours, storm, gust of wind, etc. In order to better understand and envisage the episodes of convection's effects, international program as African Monsoon Multidisciplinary Analyses (AMMA) organized a series of measurements on the ground to study the atmospheric stability over West Africa.

The convective available potential energy (CAPE) has become an instability index widely used in the past few decades to evaluate the convective potential of the atmosphere. It is calculated by means of an integral of a vertical profile of cloud buoyancy and has been used for different kinds of studies. For example, CAPE was the appropriate parameter to analyze the conditional instability in the tropical atmosphere (Williams and Renno, 1993) for studies on tropical 'Hot Towers' (Williams *et al.*, 1992) and for other research projects on atmospheric convection (Renno and Ingersoll, 1996). The importance of the CAPE in all these projects has led to a more precise way of calculating their values, and to certain approximations and corrections that may be included in the general formula. Later research into the CAPE also inquired into the possible relationships between this and other parameters that characterize atmospheric conditions (Blanchard, 1998; Monkam, 2002). Brooks *et al.* (2003, 2007) presented global CAPE climatology derived from 7 years using National Centre for Atmospheric Research (NCAR)/National Centres for Environmental Prediction (NCEP) reanalysis. For the United States and Europe, many researches have been done, sometimes through the use of several coupled atmosphere and ocean models (Trapp *et al.*, 2007; Riemann-Campe *et al.*, 2009, 2010) or by using 30 year climatology of CAPE and Convective Inhibition (CIN; Romero *et al.*, 2007) based on ERA-40 from the European Centre for Medium-Range Weather Forecast (ECMWF). Riemann-Campe *et al.* (2010) also provided a global climatology of CAPE and CIN and their relation to convective precipitation, using 1979–2001 ERA-40 reanalyses data and 1979–2009 ECHAM5/MPI-OM model and analyzed both parameters in terms of trends and how they change in a warmer climate. Such statistical analyses which can characterize the climatology of a region, specifically the West Africa region that has a large zone of deep convection deserve to be done.

Therefore, we propose to complete the CAPE analyses from the West Africa climatology based on 35 years (1979–2014) of ERA-Interim data, to provide seasonal ensemble means, a trend analysis and its link to African easterly waves (AEWs). The outline of the article, is as follows: the data and methods of analysis are presented in the Section 2. In Section 3, we present the result of the climatology of CAPE, trends and relationship with AEW. Section 4 presents conclusions and the outlook on future research on higher order statistics.

2. Data and methods

In this document, we have calculated the different variables by using the 0.75° grid ERA-Interim data firstly because of its high quality even if differences sometimes occur due to the varying density of observations (as in the southern hemisphere, and over oceans), and secondly because it is the most recent reanalysis from the ECMWF started in 1979. By taking into account the satellite observations, the quality and quantity of data have improved considerably (Uppala *et al.*, 2005), even if the change in the measurement system leads to an artificial warming trend in the global mean temperature of the lower troposphere (Bengtsson *et al.*, 2004). The differences between observed and ERA-Interim temperature trends were also pointed out by Simmons *et al.* (2004). However, ERA-Interim data yield the general trend signal with an improving performance after 1979. Thus, trends in CAPE can be analyzed for the whole period of time and additionally before and after 1979.

CAPE is calculated between the level of free convection (LFC) and the Level of Neutral Buoyancy (LNB) as measures of the bottom and top of the cloud, respectively. As the effect of moisture on buoyancy is taken into account, the virtual temperature T_V is used and the CAPE is given by R_d gas constant for dry air (287.05 J kg^{-1} K^{-1}):

$$\text{CAPE} = \int_{\text{LFC}}^{\text{LNB}} R_d \left(T_{\text{vp}} - T_{\text{ve}} \right) d\ln(P) \qquad (1)$$

where P, T_{vp}, T_{ve}, LNB and LFC are, respectively, the pressure, the virtual temperature of the air parcel, the virtual temperature of the environment of the air parcel, the LNB and the LFC.

The air parcel rises dry adiabatically from the surface to the lifting condensation level (LCL). Above the LCL, the parcel raises pseudo adiabatically which means that any condensates will immediately fall out of the parcel as rain (Riemann-Campe *et al.*, 2009). Between cloud bottom and cloud top the parcel rises freely as the temperature of the parcel is higher than the temperature of its environment. The pseudo equivalent potential temperature θ_{ep} is used to calculate the temperature of the rising parcel (Emanuel, 1994).

$$\theta_{\text{ep}} = T \left(\frac{P_{sfc}}{P} \right)^{0.2854(1-0.28r)}$$
$$\times \exp \left[r(1 + 0.81r) \left(\frac{3376}{T_{\text{LCL}}} - 2.54 \right) \right] \qquad (2)$$

with the mixing ratio r of dry to moist air and the temperature T_{LCL} of the parcel at the LCL, where the ascent of the air parcel changes from dry adiabatic to pseudo adiabatic and P_{sfc} is the pressure at the sea surface (1000 hPa). The expression of the saturation temperature T_{LCL} can be approximated by (Bolton, 1980):

$$T_{\text{LCL}} = \frac{2840}{3.5 \log T - \log E - 4.805} + 55 \qquad (3)$$

with the vapor pressure E.

Local seasonal trends in CAPE are analyzed by the Mann–Kendall trend test which is a robust trend estimator applicable for any theoretical distribution. As noted by Wilks (2011), investigating the possible trend through time of the central tendency of a data series is of interest in the context of a changing underlying climate,

among other settings. The usual parametric approach to this kind of question is through regression analysis with a time index as the predictor, and the associated test for the null hypothesis that a regression slope is zero. This particular test is used here, as CAPE is not normally distributed in general. The trend in CAPE is calculated on a seasonal mean basis. Only positive CAPE values are considered in the calculation. Grid points with less than two positive CAPE values per season are neglected. The Mann–Kendall score indicates a given trend being positive or negative, which is supplemented by a two sided p-value to provide the probability of a detected trend (Riemann-Campe *et al.*, 2009). Trends are only included in the analysis if their probability exceeds the 95% significance level. The magnitude of a given trend is estimated by linear regression, although the error is rarely normally distributed in CAPE.

The wavelet analysis method is applied here to identify the dominant synoptic oscillation modes and to isolate the synoptic oscillation components. In this work, it is used on daily CAPE time series over West Africa for the period 1979–2009 in order to evaluate the seasonality of the variance of the synoptic time scale and of the related AEW signal. Previous studies have shown that dominant modes of monsoon synoptic variability are characterized by strong and reproducible 3–5 day oscillations (Sultan and Janicot, 2003). Hence, CAPE spectra are calculated on each individual June–September period using wavelet analysis method. A red noise background spectrum is computed from the formula of Gilman *et al.* (1963). The 95% confidence limits about this red noise spectrum are determined using F-statistic (Wilks, 2011).

3. Results and discussions

3.1. Means and variability

The global distribution of seasonally averaged CAPE (Figure 1) follows basically the distribution of the wind at 850 hPa over West Africa. CAPE generally increases from Sahara which is the arid zone with values of about 500 J kg^{-1} to the equator where they can reach 2000 J kg^{-1}. CAPE minima are observed in regions of cold water upwelling and where currents are colder than the ambient ocean temperatures, and in arid regions. CAPE also moves with the inter-tropical convergence zone (ITCZ) as show in Figure 1. However, the northward evolution of monsoon around 10°N during the months of April–May–June (AMJ, Figure 1(b)) to about 15°N in July–August–September (JAS, Figure 1(c)) before decrease southward at October–November–December (OND, Figure 1(d)) suggested that good correlation can exist between CAPE and the seasonal migration of the West Africa monsoon where sufficient moisture are available. The location and intensity of convective systems occurring during JAS are more frequently north of the ITCZ. This suggests the possible influence of AEW *versus* CAPE in their development during

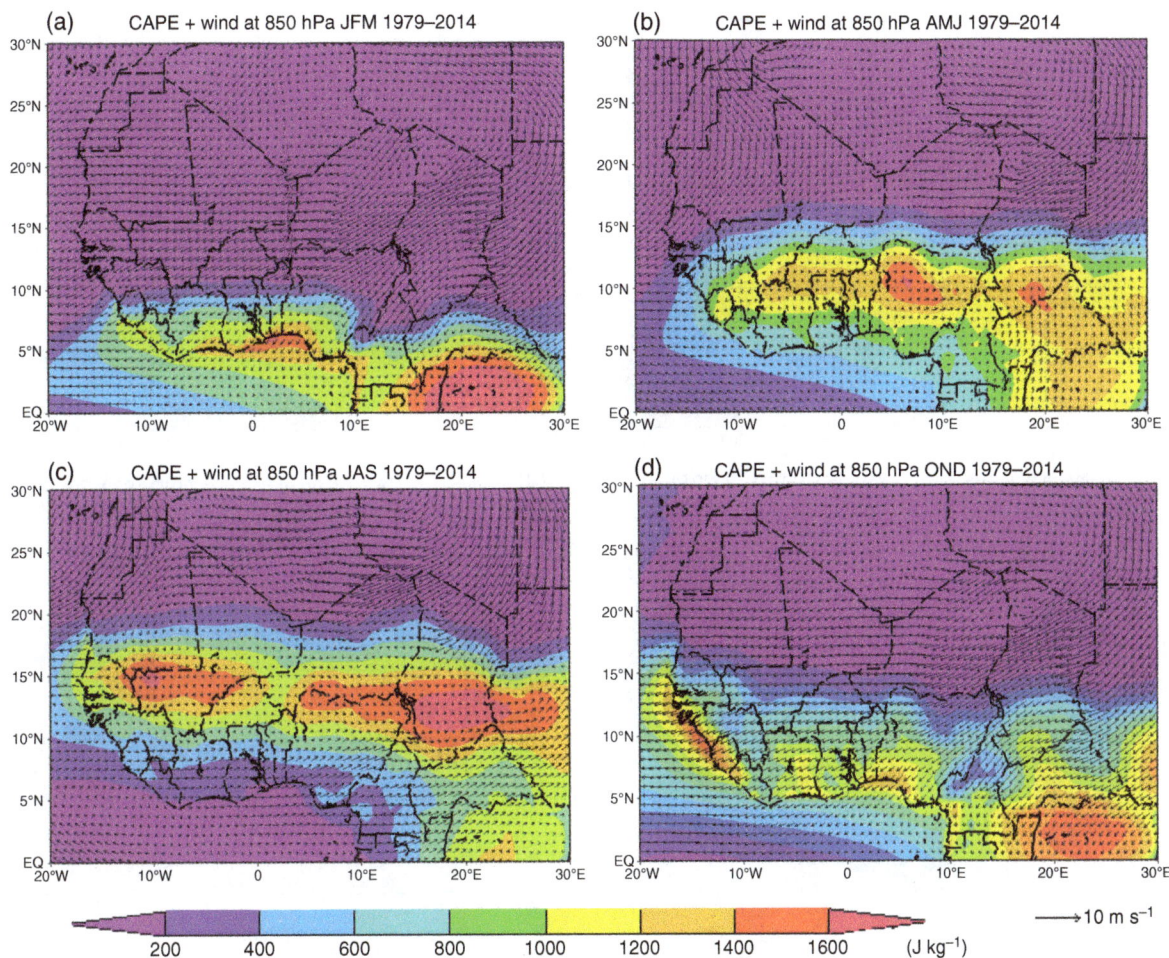

Figure 1. Climatological (1979–2014) seasonal mean of CAPE (in J kg^{-1}) and wind at 850 hPa (in m s^{-1}) averaged during JFM (a), AMJ (b), JAS (c) and OND (d); the country contours are represented by dashed lines. The interval of CAPE in the color scale is 200 J kg^{-1}.

AMJ and JAS. In other way, the maximum CAPE over Congo basin or in Guinea coast can be viewed as the development of Kelvin wave versus CAPE during JFM and OND.

Interannual variance of the seasonal CAPE means varied from 500 to about 6.5×10^7 (J kg^{-1})^2, with mean values around 1.3×10^5 (J kg^{-1})^2 (Figure 2). The zone of maximum variance corresponds to the growth in relative humidity. The geographical distribution of the seasonal differences is similar to those of the seasonally averaged mean. In general, larger values are observed in summer (AMJ, Figure 2(b) and JAS, Figure 2(c)) according to higher relative humidity due to the arrival of the monsoon over West Africa. In the dry season (JFM), high variance is present around the Congo basin and the Benin coast where relative humidity is 80 and 65%, respectively, and in OND (Figure 2(c)) around the Guinean coast where the relative humidity remains around 65%.

3.2. Trends

CAPE trend magnitudes are displayed (Figure 3) at the level of significance exceeding 95%. Trends are

shown for the full 35 years time series (1979–2014). Calculated trends before 1979 might correspond to changes in measurements instead of changes in climate (Bengtsson *et al.*, 2004). Significant trends in CAPE occur in most parts of the West Africa except in dry region. Regions with a positive trend outnumber the regions of negative trends considerably with magnitudes varying in the time periods considered. Trend magnitudes range from about −800 J kg^{-1} to about 1000 J kg^{-1} per decade during 1979–2014. The largest increase in CAPE of about 1600 J kg^{-1} per decade occurs in the Congo basin during JFM (Figure 1(a)) and around the AEW axis (15°N) in the south of Chad and at the borders of Mali and Senegal during JAS (Figure 1(c)). The change of sign yields in a net decrease by 200–600 J kg^{-1} per decade from the dry period (OND and JFM) to the wet season (AMJ and JAS) in the band of latitudes 10°–20°N can be due to the migration of the ITCZ which is also linked to the southward decrease of CAPE during the two periods.

3.3. Link with AEW

Figure 4 shows the relationship between CAPE and AEW through spectral analysis of CAPE during the

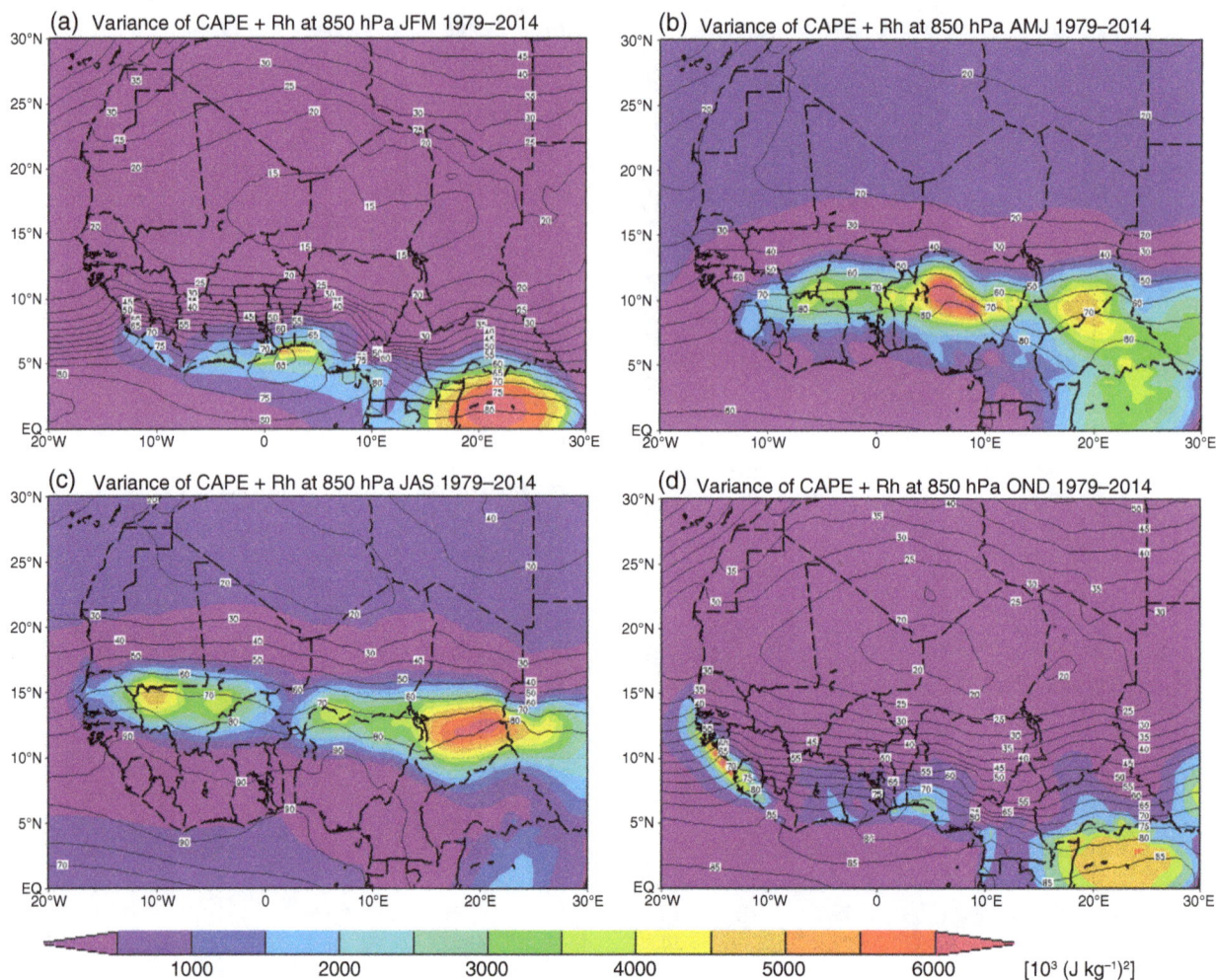

Figure 2. Climatological (1979–2014) inter annual variance of the seasonal CAPE in $(J\,kg^{-1})^2$ and relative humidity at 850 hPa (in %) averaged during JFM (a), AMJ (b), JAS (c) and OND (d); the country contours are represented by dashed lines.

period JAS of 1979–2014. Wavelet analysis is a common tool for decomposing a time series into a time-frequency space and detecting time-frequency variations. The wavelet transform allows the treatment of a signal as a wavelet function called a mother wavelet (here the Morlet wavelet is used; Sultan and Janicot, 2003; Mohr and Thorncroft, 2006). We have tested different mother wavelets and the results look similar. Because the wavelet transform is a band pass filter with a known response function (the wavelet function), it is also a powerful filtering technique. The wavelet analysis method is used here to identify the dominant synoptic oscillation modes and to isolate the synoptic oscillation components. The mean seasonal cycle and interannual variability were removed before computing the spectrum. At the right (Figure 4), the wavelet significance with the red noise background spectrum and the 95% confidence level are shown.

Spectral analysis using high resolution data shows significant power in the 3–5-day band period over tropical West and Central Africa in the box 10°–20°N and 20°W–30°E, where CAPE is high as shown in Figure 1. Within the regions of tropical deep convection, the 3–5-day time scale variance accounts for about

25–35% of the total variance. The 3–5-day convective variance has similar amplitudes from east to west in the band 10°–20°N, while dynamic measures of AEW activity show stronger amplitudes in the west. Weak AEW activity in the east is consistent with initial wave development there. AEWs are initiated by the convection triggered on the western sides of the Darfur Mountains (western Sudan) and Ethiopia. The subsequent development and growth of AEWs are associated with stronger coherence with convection there (Mohr and Thorncroft, 2006).

4. Summary and conclusions

By using the ERA-Interim data, with a 0.75° grid over West Africa, it was found that CAPE minima are observed in arid regions over West Africa and move with the ITCZ. The evolution of the monsoon around 10°N during the months of AMJ to about 15°N in JAS prior to a southward decrease in OND suggests good correlation between CAPE and the seasonal migration of West Africa monsoon where sufficient moisture are available. Larger values of CAPE variance are observed

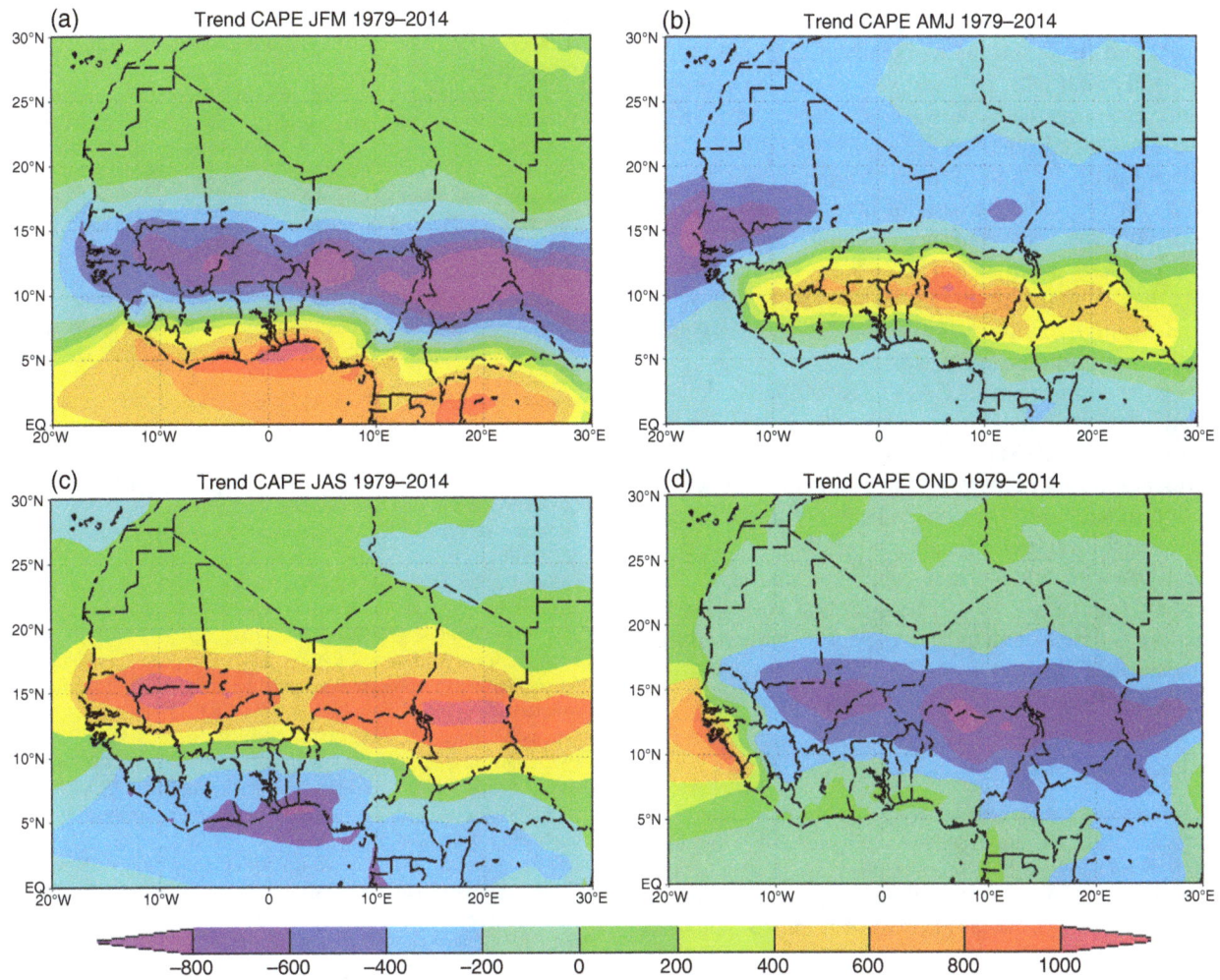

Figure 3. Seasonal trends of CAPE in J kg^{-1} per decade, significant at the 95% level over West Africa during JFM (a), AMJ (b), JAS (c) and OND (d) from 1979 to 2014. The country boundaries are represented by a dash.

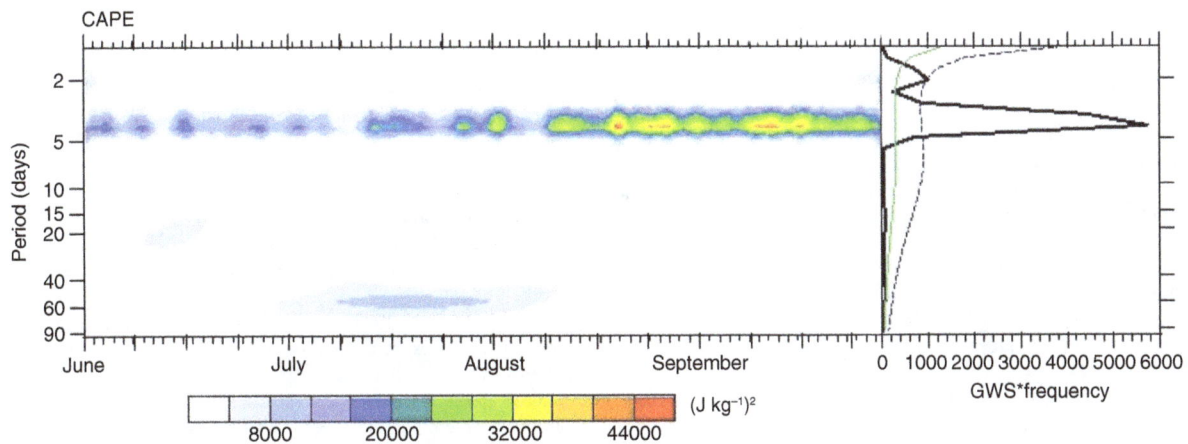

Figure 4. Diagrams of mean wavelet variance of daily unfiltered CAPE over West Africa for 10°–20°N/20°W–30°E during the summer (JAS) in the period 1979–2009. The mean seasonal cycle and interannual variability were removed before computing the spectrum. At the right, the wavelet significance with the red noise backgrounds spectrum and the 95% confidence level.

in the summer according to higher relative humidity due to the arrival of the monsoon in the continent. We found that significant trends in CAPE occur in most parts of the West Africa except in dry region. Trend magnitudes range from about −800 J kg^{-1} to 1000 J kg^{-1} per decade during 1979–2014. The largest increase in CAPE of about 1600 J kg^{-1} per decade occurs in the Congo basin during JFM and around the AEW axis (15°N) in the south of Chad and at the borders of Mali and Senegal during JAS. The change of sign yields decrease from

200 to 600 J kg^{-1} per decade between the dry period (OND and JFM) and the wet season (AMJ and JAS) in the band of latitudes 10°–20°N. Spectral analysis using high resolution data shows significant power in the 3–5-day range in the box 10°–20°N and 20°W–30°E, where CAPE is high. Within the regions of tropical deep convection in the band of latitude 10°–20°N, the 3–5-day time scale variance accounts for about 25–35% of the total variance. The 3–5-day convective variance has similar amplitudes from east to west of this band, while dynamic measures of AEW activity show stronger amplitudes in the west.

Similar intra-seasonal CAPE fluctuations occur also over the Congo basin, especially during the dry season (OND and JFM). This point must be investigated further. Another point would be to quantify the contribution of such intra-seasonal variability to the interannual variability. It will be also interesting to investigate the links with atmospheric modes of this zone since Kamsu-Tamo *et al.* (2014) showed that corresponding regressed deseasonalized atmospheric fields highlight an eastward propagation of patterns consistent with convectively coupled equatorial Kelvin wave dynamics.

Acknowledgements

This research is supported by ICTP, Trieste Italy through the Associate and Federation Schemes Program and 'Programme Pilote Régional Forêts Tropicales Humides d'Afrique Centrale (PPR FTH-AC)' of IRD. We also appreciated the comments of the anonymous reviewers which helped to improve the manuscript.

References

Bengtsson L, Hagemann S, Hodges KI. 2004. Can climate trends be calculated from reanalysis data? *Journal of Geophysical Research* **109**: D11111.

Blanchard DO. 1998. Assessing the vertical distribution of convective available potential energy. *Weather and Forecasting* **13**: 870–877.

Bolton D. 1980. The computation of equivalent potential temperature. *Monthly Weather Review* **108**: 1046–1053.

Brooks HE, Lee JW, Craven JP. 2003. The spatial distribution of severe thunderstorm and tornado environments from global reanalysis data. *Atmospheric Research* **67**: 73–94.

Brooks HE, Anderson AR, Riemann K, Ebbers I, Flachs H. 2007. Climatological aspects of convective parameters from the NCAR/NCEP reanalysis. *Atmospheric Research* **83**: 294–305.

Emanuel KA. 1994. *Atmospheric Convection*. Oxford University Press: New York, NY.

Gilman DL, Fuglister FJ, Mitchel JM Jr. 1963. On the power spectrum of red noise. *Journal of the Atmospheric Sciences* **20**: 182–184.

Lenouo A, Monkam D, Mkankam Kamga F. 2010. Variability of static stability over West Africa during northern summer 1979-2005. *Atmospheric Research* **98**: 353–362.

Mohr KJ, Thorncroft CD. 2006. Intensive convective systems in West Africa and their relationship to the African easterly jet. *The Quarterly Journal of the Royal Meteorological Society* **132**: 163–176.

Monkam D. 2002. Convective available potential energy (CAPE) in Northern Africa and Tropical Atlantic and study of its connections with rainfall in Central and West Africa during summer 1985. *Atmospheric Research* **62**: 125–147.

Kamsu-tamo P-H, Janicot S, Monkam D, Lenouo A. 2014. Convection activity over the Guinean coast and Central Africa during northern spring from synoptic to intra-seasonal timescales. *Climate Dynamics* **43**: 3377–3401, doi: 10.1007/s00382-014-2111-y.

Renno NO, Ingersoll AP. 1996. Natural convection as a heat engine: a theory for CAPE. *Journal of Atmospheric Sciences* **53**: 572–585.

Riemann-Campe K, Fraedrich K, Lunkeit F. 2009. Global climatology of convective available potential energy (CAPE) and convective inhibition (CIN) in ERA-40 reanalysis. *Atmospheric Research* **93**: 534–545.

Riemann-Campe K, Blender R, Fraedrich K. 2010. Global memory analysis in observed and simulated CAPE and CIN. *International Journal of Climatology* **31**(8): 1099–1107.

Romero R, Gaya M, Doswell CA III. 2007. European climatology of severe convective storm environmental parameters: a test for significant tornado events. *Atmospheric Research* **83**: 389–404.

Simmons AJ, Jones PD, da Costa Bechtold V, Beljaars ACM, Kllberg PW, Saarinen S, Uppala SM, Viterbo P, Wedi N. 2004. Comparison of trends and low-frequency variability in CRU, ERA-40, and NCEP/NCAR analyses of surface air temperature. *Journal of Geophysical Research* **109**: D24115.

Sultan B, Janicot S. 2003. The West African monsoon dynamics. Part II: the "preonset" and "onset" of the summer monsoon. *Journal of Climate* **16**: 3407–3427.

Trapp RJ, Diffenbaugh NS, Brooks HE, Baldwin ME, Robinson ED, Pal JS. 2007. Changes in severe thunderstorm environment frequency during the 21st century caused by anthropogenically enhanced global radiative forcing. *Proceedings of the National Academy of Sciences* **104**: 19719–19723.

Uppala SM, Kållberg PW, Simmons AJ, Andrae U, da Costa Bechtold V, Fiorino M, Gibson JK, Haseler J, Hernandez A, Kelly GA, Li X, Onogi K, Saarinen S, Sokka N, Allan RP, Andersson E, Arpe K, Balmaseda MA, Beljaars ACM, van de Berg L, Bidlot J, Bormann N, Caires S, Chevallier F, Dethof A, Dragosavac M, Fisher M, Fuentes M, Hagemann S, Holm E, Hoskins BJ, Isaksen L, Janssen PAEM, Jenne R, McNally AP, Mahfouf J-F, Morcrette J-J, Rayner NA, Saunders RW, Simon P, Sterl A, Trenberth KE, Untch A, Vasiljevic D, Viterbo P, Woollen J. 2005. The ERA-40 reanalysis. *The Quarterly Journal of the Royal Meteorological Society* **131**: 2961–3012.

Wilks DS. 2011. *Statistical Methods in the Atmospheric Sciences*. International Geophysics Series, 3rd ed, Vol. **100**. Academic Press: New York, NY; 669 pp.

Williams E, Renno N. 1993. An analysis of the conditional instability of the tropical atmosphere. *Monthly Weather Review* **121**: 21–36.

Williams ER, Geotis SG, Renno N, Rutledge SA, Rasmussen E, Rickenback T. 1992. A radar and electrical study of tropicalAhot towerB. *Journal of the Atmospheric Sciences* **49**: 1386–1395.

16

Tibetan ice core evidence for an intensification of the East Asian jet stream since the 1870s

G. W. K. Moore*

Department of Physics, University of Toronto, Toronto, Ontario, Canada

*Correspondence to:
G. W. K. Moore, Department of Physics, University of Toronto, Toronto, Ontario, Canada
E-mail: gwk.moore@utoronto.ca*

Abstract

The subtropical jet stream plays a significant role in the climate system and there exists evidence that it has intensified since the 1950s. However, the lack of upper-air observations before this time makes it difficult to place this trend in a longer term context. Here, we show that a Tibetan ice core's snow accumulation and dust concentration records are respectively correlated and anti-correlated with wind speeds in the jet stream over Asia as represented in both the NCEP and the 20th Century Reanalyses. This behavior is consistent with an intensification of this jet stream that began in the 1870s.

Keywords: paleoclimate; East Asian jet stream; climate change; ice core

1. Introduction

Strong upper-level westerly winds were first identified over Japan in the 1920s (Lewis, 2003). Subsequent analysis indicated that these winds are a global phenomenon associated with the so-called subtropical jet stream that results from the meridional temperature gradient of opposing sign in the troposphere and stratosphere (Krishnamurti, 1961; Palmen and Newton, 1969). The high winds first discovered over Japan are associated with a regional manifestation known as the East Asian jet stream (EAJS) that plays a significant role in the meteorology and climate of the Asia-Pacific region (Yang *et al.*, 2002; Lin and Lu, 2005).

Since the 1950s, there has been an intensification of the subtropical jet stream (Fu *et al.*, 2006; Strong and Davis, 2007). This change has occurred in association with an increase in the height of the tropopause (Santer *et al.*, 2003; Wilcox *et al.*, 2012). All of these changes have been attributed to differential tropospheric warming and stratospheric cooling arising from the increasing concentration of greenhouse gases in the atmosphere (Santer *et al.*, 2003). Climate models predict that this intensification will continue throughout the 21st century under increasing greenhouse gas concentration scenarios (Yin, 2005; Lorenz and DeWeaver, 2007).

However, the lack of upper-tropospheric data prior to the establishment of the synoptic network of upper-air observations in the late 1940s makes it difficult to place these recent changes in a longer term perspective. High altitude ice cores have been shown to contain information on atmospheric processes in the upper-troposphere (Thompson, 2000; Moore *et al.*, 2002; Zhao and Moore, 2006; Wu *et al.*, 2013) and so offer the *potential* to provide such a perspective. However, this potential is for the most part unrealized because of the lack of information on upper-tropospheric climate variability prior to the late 1940s. The recent availability of such data from the 1870s onward, as contained in the 20th Century Reanalysis (20CR) (Compo *et al.*, 2011), provides the opportunity for a significant expansion in our knowledge of the climate signals in high-elevation ice cores.

In this article, we use both the National Centers for Environmental Prediction Reanalysis (NCEP) and the 20CR to show that the inter-annual variability in an ice core extracted from 7200 m asl2 on the Dasuopu Glacier along the southern boundary of the Tibetan Plateau (Thompson *et al.*, 2000; Davis *et al.*, 2005) is correlated with the variability in the EAJS. Furthermore, the statistically significant trends in snow accumulation (DSA) and dust concentration (DSD) in the ice core that began in the middle of the 19th century are consistent with an intensification in the EAJS over the same period.

2. Data

The long-term atmospheric data sets that are the result of retrospective analyses or 'reanalyses' have become an essential tool in the diagnostic studies of the climate system. Conventional reanalyses require assimilation of upper-air observations and so are restricted to the period after the late 1940s (Kalnay *et al.*, 1996; Compo *et al.*, 2011). This limits their ability to look at variability that predates the establishment of the synoptic upper-air network. This is a particular problem for upper-tropospheric variability that may only have a muted expression at the surface, where longer term observational data sets are available.

The 20CR assimilates only surface pressure observations and uses an ensemble of 56 short-term parallel integrations of the atmospheric component of the

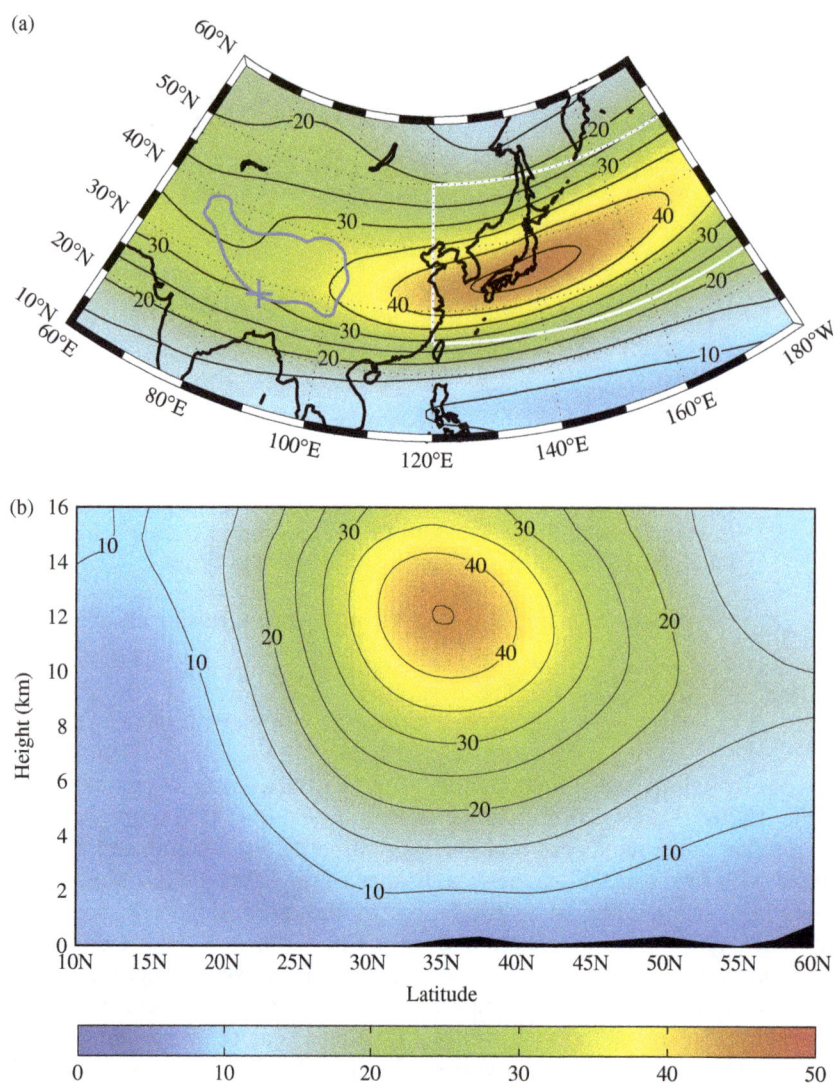

Figure 1. The annual mean wind speed (m s^{-1}) from the NCEP Reanalysis for the period 1948–2012: (a) at 200 mb and (b) along 140°E. In (a) the areal extent of the Tibetan Plateau is indicated by the thick blue line, the Dasuopu ice core site is indicated by the '+' and the region over which the EAJSI is calculated is bounded by the thick white lines.

National Centers for Environmental Prediction's Climate Forecast System model, forced by prescribed sea surface temperatures and sea ice concentrations, to achieve the first three-dimensional representation of the state of the troposphere prior to the establishment of the synoptic upper-air network (Compo *et al.*, 2006, 2011). For this article, we will use monthly mean data from the second version of the 20CR available for the period from 1871 to 2010. Comparisons between the 20CR with independent mid- and upper-tropospheric temperature and pressure data throughout the 20th century show good agreement indicating that the 20CR has some skill in representing tropospheric climate variability (Compo *et al.*, 2011; Stachnik and Schumacher, 2011). The 20CR has a horizontal resolution of 2° and contains 24 vertical levels from 1000 to 10 mb (Compo *et al.*, 2011).

Results from the 20CR will be compared to those from the more conventional NCEP Reanalysis that assimilates upper-air and satellite observations in addition to surface data (Kalnay *et al.* 1996). The NCEP Reanalysis is available from 1948 onward. The NCEP has a horizontal resolution of 2.5° and contains 17 vertical levels from 1000 to 10 mb (Kalnay *et al.* 1996)

In 1997, three ice cores were drilled at a site 7200 m above sea level on the Dasuopu Glacier (28.38°N, 85.72°E) in southern Tibet (Thompson 2000; Davis *et al.* 2005). The annually resolved DSA and DSD time series averaged over the cores extend from 1450 to 1996. The DSA time series showed an increase during the early part of the 19th century that was followed by a decrease that began during the latter half of that century (Thompson *et al.*, 2000). This behavior was argued to the result of modulations in the intensity of the summer precipitation over north central India (Sontakke and Singh, 1996; Thompson *et al.*, 2000). Zhao and Moore (2006) argued that the DSA was associated with anomalous easterly moisture transport at the height of the ice core site. This connection with

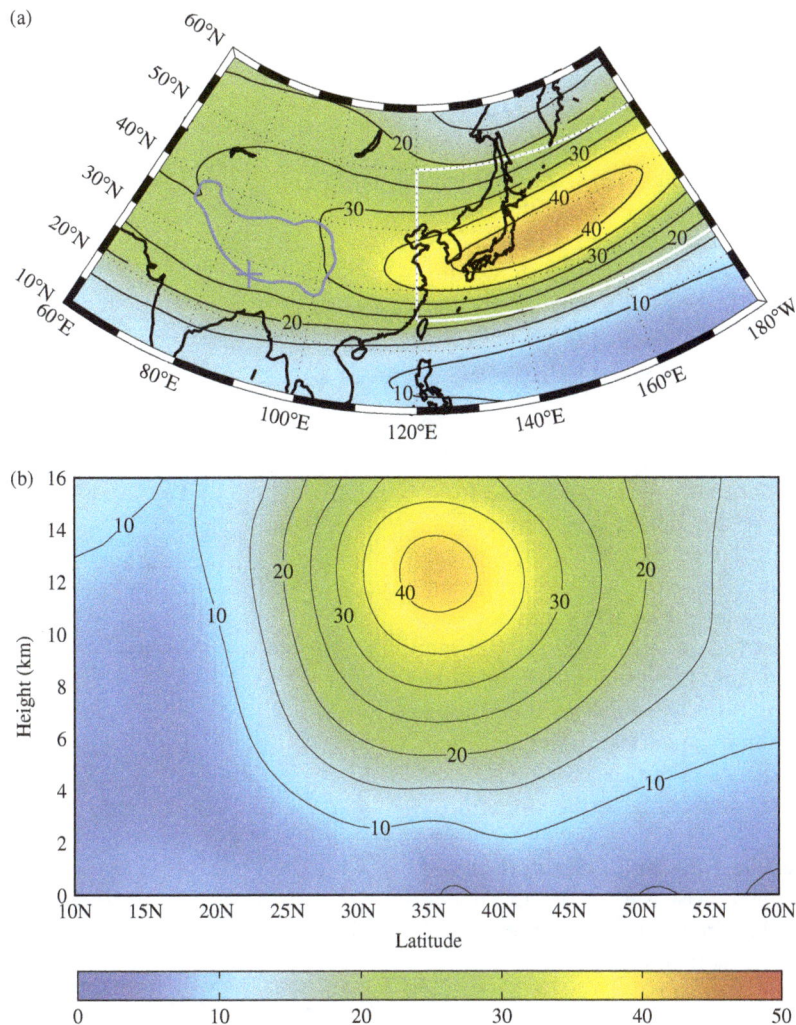

Figure 2. The annual mean wind speed (m s^{-1}) from the 20CR Reanalysis for the period 1871–2010: (a) at 200 mb and (b) along 140°E. In (a) the areal extent of the Tibetan Plateau is indicated by the thick blue line, the Dasuopu ice core site is indicated by the '+' and the region over which the EAJSI is calculated is bounded by the thick white lines.

the upper-tropospheric circulation is supported by an analysis of stable isotope data of snow and melt water from the Dasuopu Glacier that indicates there is a significant contribution from winter snowfall arising from westerly wind transport associated with the EAJS (Tian *et al.*, 2005).

With respect to the DSD in the ice cores, Thompson *et al.* (2000) showed that there was a fourfold increase in the DSD time series since 1860 and proposed this to the result of a reduction in snow cover and land-use changes in the upwind source region. Wu *et al.* (2013) reported that the dust concentration in an ice core extracted from Tanggula in central Tibet has increased over the past 4 decades and attributed it to a strengthening of the upper-level westerlies over the Tibetan Plateau.

3. Results

Figure 1 shows the structure of the annual mean EAJS from the NCEP Reanalysis for the period 1948–2012.

At 200 mb, the jet stream is characterized by a region of high wind speed extending eastward from the Tibetan Plateau with a core situated over southern Japan along 140°E (Figure 1(a)). A meridional cross-section along 140°E, i.e. through the jet core, shows that the EAJS has its highest wind speeds near 12 km (~200 mb) and is bounded below by a region of positive vertical shear and above by a region of negative shear that are associated with the opposing tropospheric and stratospheric meridional temperature gradients (not shown).

In Figure 2, we show the structure of the annual mean EAJS from the 20CR for the period 1871–2010. A comparison with Figure 1 shows good agreement with regard to the overall structure of the jet. There is however a reduction in the magnitude of the maximum wind speed over Japan in the 20CR. The structure of the EAJS in the 20CR is also poorer over the Tibetan Plateau and this may represent a difficulty of the 20CR in representing the flow over the complex topography of this region. There is also a difference in the time period used for the NCEP mean, i.e. 1948–2012, and the 20CR mean, i.e. 1871–2010. The mean structure of the EAJS from the 20CR for the period of overlap with the NCEP Reanalysis, i.e. 1948–2010, shows better agreement, however the 20CR is still biased low (not shown).

An index of the strength of the EAJS (EAJSI) was calculated by a latitude-weighted average of the annual mean 200 mb wind speed over the region from $25°N$ to $50°N$ and from $120°E$ to $180°E$. Figure 3 shows this index as contained in both the NCEP Reanalysis and the 20CR. For the period of overlap, 1948–2010, the correlation between the two indices was 0.75 with a root-mean-square error of $0.97\,m\,s^{-1}$. Also shown are the linear least squares fits to the two indices. Over the period from 1871–2010, the trend in the 20CR EAJSI was $0.33\,m\,s^{-1}\,decade^{-1}$. Over the period from 1948–2010, the trend in the EAJSI from the NCEP Reanalysis was $0.4\,m\,s^{-1}\,decade^{-1}$. Both trends are statistically significant at the 99% confidence level using a Monte Carlo technique that takes into account the spectral characteristics of the underlying time series, thereby capturing any temporal autocorrelation that reduces the degrees of freedom resulting in a bias in conventional significance tests (Rudnick and Davis, 2003; Moore, 2012). There is an evidence of a recent, i.e. post 2000, weakening in the strength of both indices.

Figure 3 also shows the DSA and DSD time series for the period 1870–1996. Also shown are the linear least squares fits to the time series. As previously reported, the DSA time series has a trend toward lower values, while the DSD time series exhibits an opposite trend. Both trends are statistically significant at the 99% level using the same significance test as used above. Despite the low frequency anti-correlation between the DSA and DSD time series evident in Figure 3, there is a reduction in the magnitude of the correlation as higher frequency variability is included. For example, if one includes variability on time scales greater than 40 years, the correlation between the two time series is −0.71. The correlation is reduced to −0.13 if the full annually resolved time series are used.

Figure 4 shows the spatial correlation field between the annual mean wind speed from the NCEP Reanalysis and the DSA for the period of overlap between the two data sets, i.e. 1948–1996. The statistical significance of the correlation at each grid point was assessed using the same approach as used in Figure 3. At 200 mb (Figure 4(a)), the wind speeds are anti-correlated with the DSA along a broad and spatially coherent zonal band centered on the latitude of the ice core. Embedded with this band, there are regions, including one near $140°E$, where the anti-correlation is statistically significant at the 95% confidence level. There is also a region with a statistically significant positive correlation to the north of the Tibetan Plateau. Along $140°E$ (Figure 4(b)), there is vertical coherence to the region with a statistically significant anti-correlation that extends from ∼2 to ∼14 km with a maximum near the 8 km.

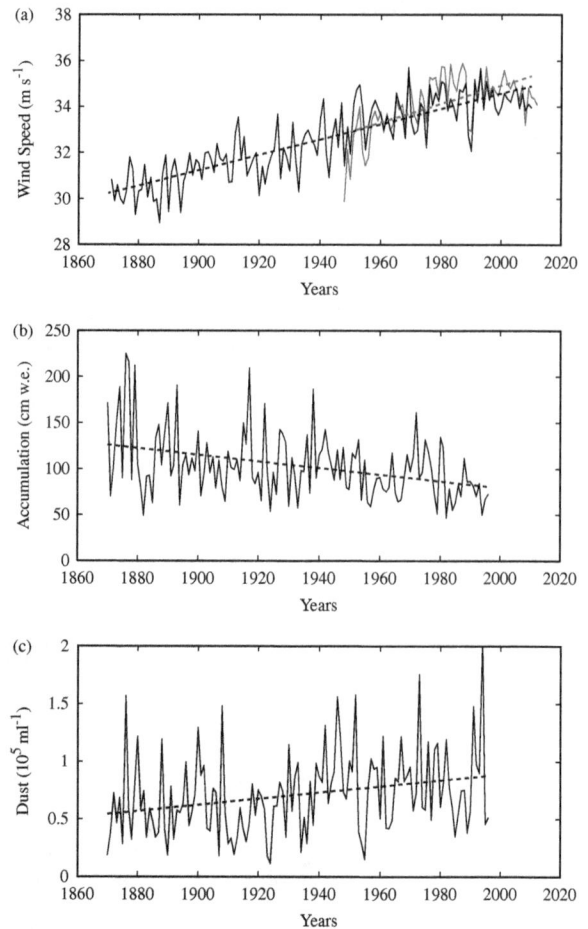

Figure 3. (a) The EAJSI from the NCEP Reanalysis 1948–2012 (blue line) and the 20CR 1871–2010 (black line). The annual: (b) DSA (cm w.e.) and (c) DSD ($10^5\,ml^{-1}$) time series from the Dasuopu ice core 1870–1996. The respective linear least squares fit are indicated by the dashed lines.

Figure 5 shows the same spatial correlation fields except from the 20CR for the period of the overlap of the two data sets, i.e. 1871–1996. At 200 mb (Figure 5(a)), there is a broad and spatially coherent zonal band of statistically significant anti-correlation along the latitude of the ice core that extends across the domain of interest with the largest magnitude of the anti-correlation in the vicinity of $140°E$. Along $60°N$, there is another broad region of statistically anti-correlation. Along $140°E$ (Figure 5(b)), there is again vertical coherence to the region of anti-correlation along the latitude of the ice core. There is also vertical coherence to the region of anti-correlation along $60°N$.

Figure 6 shows the spatial correlation field between the annual mean wind speed at 200 mb from both the NCEP Reanalysis and the 20CR and the DSD for the respective periods of overlap, i.e. 1948–1996 and 1871–1996. With respect to the 20CR, there is an evidence of downstream region of positive correlation but it does not reach the magnitude or statistical significance that was attained with the DSA (Figure 5(a)). In addition, both correlation fields contain a meridional

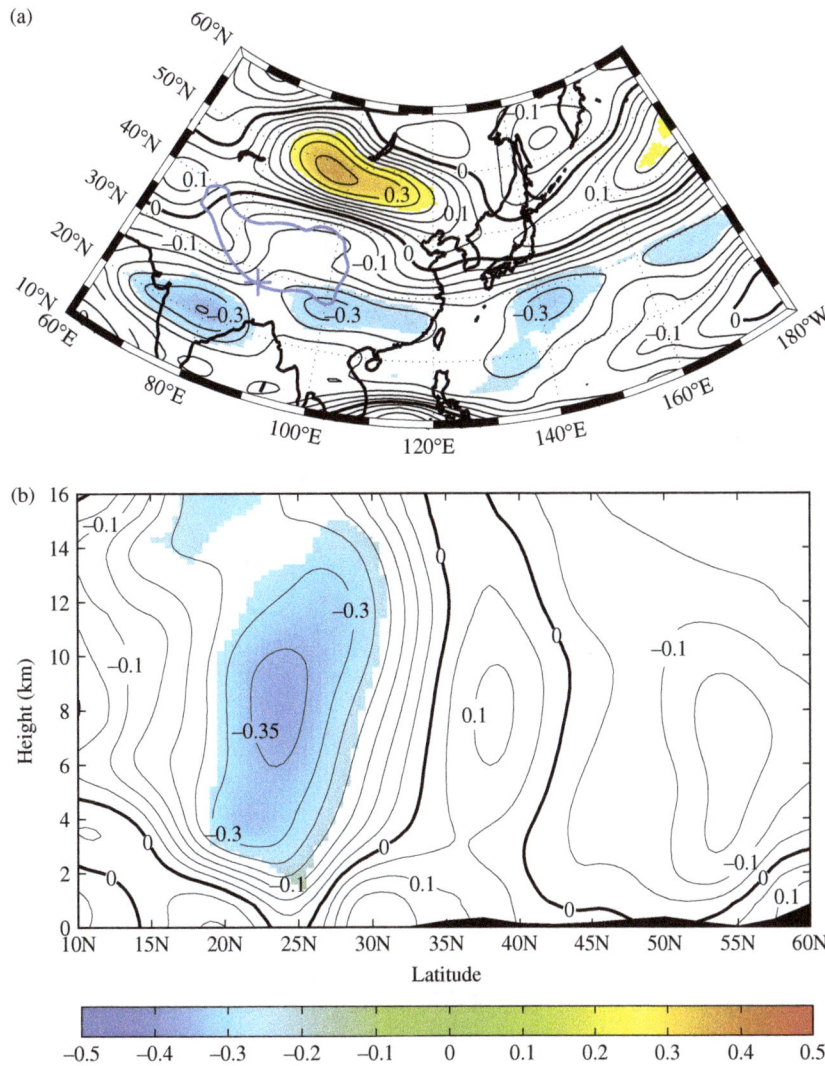

Figure 4. The spatial correlation between the annual mean wind speed from the NCEP Reanalysis and the DSA for the period 1948–1996: (a) at 200 mb and (b) along 140°E. The correlation is statistically significant at the 95% confidence level in the shaded regions. In (a) the areal extent of the Tibetan Plateau is indicated by the thick blue line whereas the Dasuopu ice core site is indicated by the '+'.

dipole across the Tibetan Plateau with regions of positive correlation in the vicinity of the ice core site.

4. Discussion

In this article, we have also shown that the 20CR is able to capture the mean structure of the EAJS over the period 1871–2010 and agrees with the NCEP Reanalysis regarding its inter-annual variability since 1948. In agreement with previous work (Strong and Davis 2007), both reanalyses indicate a recent tendency toward an intensification of the EAJS. Using the 20CR, the period over which this intensification has occurred has been extended back to 1871. The wind speed in the core of the EAJS has increased by approximately $4 \, \mathrm{m \, s^{-1}}$ or 10% over the 20th Century.

For both the NCEP Reanalysis and the 20CR, the annual mean wind speed at 200 mb shows a broad and spatially coherent zonal band of anti-correlation with

the DSA along the southern boundary of the EAJS. For the shorter period of overlap with the NCEP Reanalysis, there are only scattered regions within this band where the correlation is statistically significant. In contrast, the entire band is statistically significant for the period of the overlap with the 20CR. It should be emphasized that the significance test was applied independently at each grid point and so it does not take into account the spatial coherence of the correlation field. This suggests that the statistical significance of the correlation with the NCEP Reanalysis is underestimated. The correlation with the 20CR also had a region of anti-correlation centered on 60°N that was absent from the correlation with the NCEP Reanalysis. This is most likely associated with issues at high northern latitudes in the 20CR that result from issues with the specification of sea ice field (Compo *et al.*, 2011).

For both reanalyses, the correlation with the DSD is positive in the vicinity of the ice core site and is negative to north. Although the magnitude of the

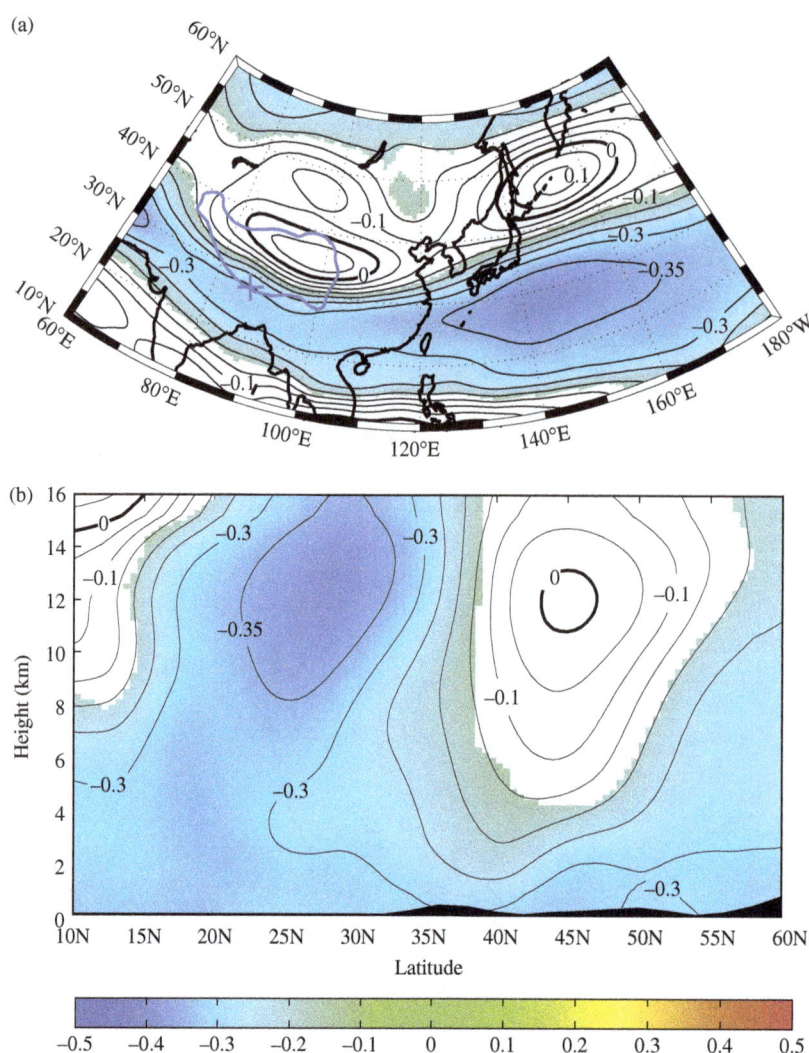

Figure 5. The spatial correlation between the annual mean wind speed from the 20CR and the DSA for the period 1871–1996: (a) at 200 mb and (b) along 140°E. The correlation is statistically significant at the 95% confidence level in the shaded regions. In (a) the areal extent of the Tibetan Plateau is indicated by the thick blue line whereas the Dasuopu ice core site is indicated by the '+'.

correlation is low, it is spatially coherent in both reanalyses suggesting that it represents an actual climate signal that may be related to flow splitting of the EAJS by the high topography of the Tibetan Plateau (Ren *et al.*, 2011). Furthermore, the association of higher wind speeds in the vicinity of the ice core site with higher DSD in the ice core is similar to that observed at Tanggula in central Tibet (Wu *et al.*, 2013).

Despite the differences in the overall structure of the 200 mb annual mean wind speed spatial correlation field with the DSA and DSD, both are consistent with an increase in wind speed along the southern flank of the Tibetan Plateau since the 1870s. The correlation with the DSA and the spatial coherence of the 200 mb wind field suggests that this intensification has also been occurring in the core of the EAJS. A result that is consistent with the observed increase in the EAJSI as represented in the 20CR.

With regard to regional climate variability, previous work has shown that winter variability in the strength of the EAJS is anti-correlated with precipitation in the region upstream of the ice core site over the period 1968–2000 (Yang *et al.*, 2002). This association was shown to be the result of large-scale changes in the atmospheric mass field that favored an intensification of the planetary wave pattern across the Northern Hemisphere; including a strengthening of the Siberian High (Yang *et al.*, 2002). There has also been a trend toward higher surface pressures over the Tibetan Plateau that is associated with regional warming and that would also contribute to a reduction in precipitation in the vicinity of the ice core site (Moore, 2012). The relationships identified in this article are therefore consistent with and extend back in to time to the 1870s the view that the EAJS modulates precipitation in the vicinity of the ice core site. It also provides a mechanism, i.e. variability in the Northern Hemisphere planetary wave pattern, for the observed correlation between the Dasuopu ice core record and the North Atlantic Oscillation (Davis *et al.*, 2005).

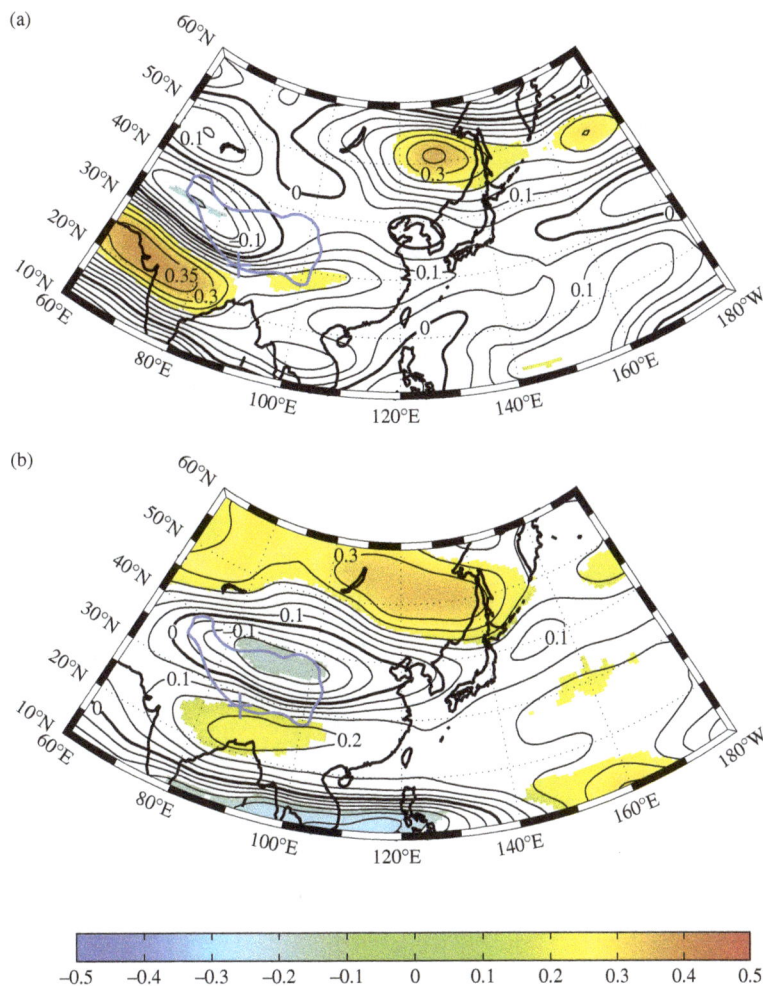

Figure 6. The spatial correlation between the annual mean wind speed from: (a) the NCEP Reanalysis and (b) the 20CR with the DSD at 200 mb. In (a) the period of correlation is 1948–1996, whereas in (b) the period of correlation is 1871–1996. The correlation is statistically significant at the 95% confidence level in the shaded regions. The areal extent of the Tibetan Plateau is indicated by the thick blue line whereas the Dasuopu ice core site is indicated by the '+'.

Finally, this work serves to show the potential of using the 20CR to understand the long-term climate signal in high-elevation ice cores.

Acknowledgements

The author would like to thank the NOAA Earth System Research Laboratory for access to the NCEP Reanalysis and the 20CR data as well as Professor L. Thompson and the NOAA Paleoclimatology Program for access to the Dasuopu ice core data. In addition, comments made during the review process contributed to the article. The author was supported by the Natural Science and Engineering Research Council of Canada.

References

Compo GP, Whitaker JS, Sardeshmukh PD. 2006. Feasibility of a 100-year reanalysis using only surface pressure data. *Bulletin of the American Meteorological Society* **87**: 175–190.

Compo GP, Whitaker JS, Sardeshmukh PD, Matsui N, Allan RJ, Yin X, Jr Gleason BE, Vose RS, Rutledge G, Bessemoulin P, Broennimann S, Brunet M, Crouthamel RI, Grant AN, Groisman PY, Jones PD, Kruk MC, Kruger AC, Marshall GJ, Maugeri M, Mok HY, Nordli O, Ross TF, Trigo RM, Wang XL, Woodruff SD, Worley SJ. 2011.

The Twentieth Century Reanalysis Project. *Quarterly Journal of the Royal Meteorological Society* **137**: 1–28.

Davis ME *et al.* 2005. Forcing of the Asian monsoon on the Tibetan Plateau: evidence from high-resolution ice core and tropical coral records. *Journal of Geophysical Research-Atmospheres* **110**: 1–13.

Fu Q, Johanson CM, Wallace JM, Reichler T. 2006. Enhanced mid-latitude tropospheric warming in satellite measurements. *Science* **312**: 1179.

Kalnay E, Kanamitsu M, Kistler R, Collins W, Deaven D, Gandin L, Iredell M, Saha S, White G, Woollen J, Zhu Y, Chelliah M, Ebisuzaki W, Higgins W, Janowiak J, Mo KC, Ropelewski C, Wang J, Leetmaa A, Reynolds R, Jenne R, Joseph D. 1996. The NCEP/NCAR 40-year reanalysis project. *Bulletin of the American Meteorological Society* **77**: 437–471.

Krishnamurti TN. 1961. The subtropical jet stream of winter. *Journal of Meteorology* **18**: 172–191.

Lewis JM. 2003. Ooishi's observation - viewed in the context of jet stream discovery. *Bulletin of the American Meteorological Society* **84**: 357–369.

Lin ZD, Lu RY. 2005. Interannual meridional displacement of the east Asian upper-tropospheric jet stream in summer. *Advances in Atmospheric Sciences* **22**: 199–211.

Lorenz DJ, DeWeaver ET. 2007. Tropopause height and zonal wind response to global warming in the IPCC scenario integrations. *Journal of Geophysical Research-Atmospheres* **112**.

Moore GWK. 2012. Surface pressure record of Tibetan Plateau warming since the 1870s. *Quarterly Journal of the Royal Meteorological Society* **138**: 1999–2008.

Moore GWK, Holdsworth G, Alverson K. 2002. Climate change in the North Pacific region over the past three centuries. *Nature* **420**: 401–403.

Palmen EH, Newton CW. 1969. *Atmospheric Circulation Systems: Their Structure and Physical Interpretation*, Vol. **xvii**. Academic Press: New York, NY; 603.

Ren X, Yang X, Zhou T, Fang J. 2011. Diagnostic comparison of wintertime East Asian subtropical jet and polar-front jet: large-scale characteristics and transient eddy activities. *Acta Meteorologica Sinica* **25**: 21–33.

Rudnick DL, Davis RE. 2003. Red noise and regime shifts. *Deep Sea Research Part I: Oceanographic Research Papers* **50**: 691–699.

Santer BD, Sausen R, Wigley TML, Boyle JS, AchutaRao K, Doutriaux C, Hansen JE, Meehl GA, Roeckner E, Ruedy R, Schmidt G, Taylor KE. 2003. Behavior of tropopause height and atmospheric temperature in models, reanalyses, and observations: decadal changes. *Journal of Geophysical Research-Atmospheres* **108**: 1–22.

Sontakke NA, Singh N. 1996. Longest instrumental regional and all-India summer monsoon rainfall series using optimum observations: reconstruction and update. *The Holocene* **6**: 315–331.

Stachnik JP, Schumacher C. 2011. A comparison of the Hadley circulation in modern reanalyses. *Journal of Geophysical Research-Atmospheres* **116**: 1–14.

Strong C, Davis RE. 2007. Winter jet stream trends over the Northern Hemisphere. *Quarterly Journal of the Royal Meteorological Society* **133**: 2109–2115.

Thompson LG. 2000. Ice core evidence for climate change in the Tropics: implications for our future. *Quaternary Science Reviews* **19**: 19–35.

Thompson LG, Yao T, Mosley-Thompson E, Davis ME, Henderson KA, Lin PN. 2000. A high-resolution millennial record of the South Asian Monsoon from Himalayan ice cores. *Science* **289**: 1916–1919.

Tian LD, Yao TD, White JWC, Yu WS, Wang NL. 2005. Westerly moisture transport to the middle of Himalayas revealed from the high deuterium excess. *Chinese Science Bulletin* **50**: 1026–1030.

Wilcox LJ, Hoskins BJ, Shine KP. 2012. A global blended tropopause based on ERA data. Part II: Trends and tropical broadening. *Quarterly Journal of the Royal Meteorological Society* **138**: 576–584.

Wu G, Zhang C, Xu B, Mao R, Joswiak D, Wang N, Yao T. 2013. Atmospheric dust from a shallow ice core from Tanggula: implications for drought in the central Tibetan Plateau over the past 155 years. *Quaternary Science Reviews* **59**: 57–66.

Yang S, Lau KM, Kim KM. 2002. Variations of the East Asian jet stream and Asian-Pacific-American winter climate anomalies. *Journal of Climate* **15**: 306–325.

Yin JH. 2005. A consistent poleward shift of the storm tracks in simulations of 21st century climate. *Geophysical Research Letters* **32**: 1–4.

Zhao H, Moore GWK. 2006. Reduction in Himalayan snow accumulation and weakening of the trade winds over the Pacific since the 1840s. *Geophysical Research Letters* **33**: 1–5.

17

Impact of Congo Basin deforestation on the African monsoon

R. Nogherotto,*E. Coppola, F. Giorgi and L. Mariotti

Earth System Physics Section, The Abdus Salam International Centre for Theoretical Physics, Trieste, Italy

*Correspondence to:
R. Nogherotto, Earth System
Physics Section, The Abdus Salam
International Centre for
Theoretical Physics, Strada
Costiera 11, 34151 Trieste, Italy.
E-mail: moghero@ictp.it*

Abstract

The interactions between land cover change in Central Africa and the Africa monsoon is investigated using a regional climate model simulation (2000–2007) under a Congo Basin deforestation scenario. Decreased evaporation over the deforested area locally produces a heat low and reduced precipitation. In JJA, this low strengthens the West Africa monsoon causing increased precipitation over the Sahel and decreased precipitation over the Guinea coast. In DJF, it strengthens the south-equatorial African monsoon causing an increase of precipitation over south-equatorial Africa. Therefore our simulations indicate that the biosphere-state in Central Africa may play an important role locally but also remotely via interactions with regional monsoon dynamics.

Keywords: regional climate model; deforestation; Africa; monsoon

1. Introduction

The Congo Basin forests of Central Africa cover a large region of over 180 million ha, spreading across the Democratic Republic of Congo (DRC), most of Congo-Brazzaville, the southeastern reaches of Cameroon, the southern Central African Republic (CAR), Gabon and Equatorial Guinea. According to the UN Food and Agriculture Organization (FAO), Central Africa lost ~721.000 ha per year to deforestation between 2000 and 2010. Current estimates place the region's annual deforestation rate at 0.23% per year for the past 10 years (FAO, 2011), but an increase to 0.3–0.5% per year is predicted by 2020–2030 (Nayar, 2009).

Deforestation causes an increase of surface albedo and a decrease of surface roughness, in addition to changing the potential evapotranspiration properties of the surface. The increase in albedo leads to a reduction of radiation absorbed by surface and thus to a reduction of energy available for sensible and latent heat fluxes, while the decrease in leaf area index leads to a decrease of surface evapotranspiration. These changes in surface fluxes may exert a forcing on climate not only locally but also remotely through teleconnection patterns (Werth and Avissar, 2002; Avissar and Werth, 2005; Hasler *et al*., 2009) and, for example, the equatorial and sub-equatorial Africa regions have been identified as being among the regions of the globe where the effects of land surface conditions on regional climate and dynamics are most pronounced (Koster *et al*., 2004).

A number of numerical experiments on deforestation have been carried out to date focusing mostly on the Amazon basin (Henderson-Sellers and

Gornitz, 1984; Henderson-Sellers *et al*., 1988; Sud *et al*., 1996; Lean and Rowntree, 1997; Hasler *et al*., 2009). Some experiments carried out for both the Amazon basin and equatorial Africa (Polcher and Laval, 1994; Zhang *et al*., 2001) found a prevailing local decrease of rainfall, along with cases of increase over equatorial Africa in response to increases of moisture convergence. Nevertheless in most cases, these experiments showed a decrease in precipitation and an increase in temperature over the region as a result of deforestation. Another set of experiments have investigated the issue of the effects of sub-tropical African deforestation and land degradation on the West Africa monsoon, using both global and regional climate models (RCMs) (Charney, 1975; Zheng and Eltahir, 1998; Clark *et al*., 2001; Abiodun *et al*., 2007; Zeng and Neelin, 1999; Dirmeyer and Shukla, 1994; Henderson-Sellers and Gornitz, 1984; Werth and Avissar, 2005). They showed that the land surface conditions can affect the development of the West Africa Monsoon through the modification of moist static energy gradients.

To date, however, only few studies have focused specifically on the deforestation of the Congo Basin forests. They show a local reduction in rainfall and suggest a modification in the monsoon circulation (Semazzi and Song, 2001; Shem and Dickinson, 2006). Being one of the largest forested basins in Africa, and being degraded at a fast rate, it can indeed be expected that the Congo Basin deforestation might be an important factor affecting future climate conditions of Africa. On the basis of this premise we here investigate the impact of the full deforestation of the Congo Basin on the climate of Africa by using a regional climate modeling approach. This is achieved

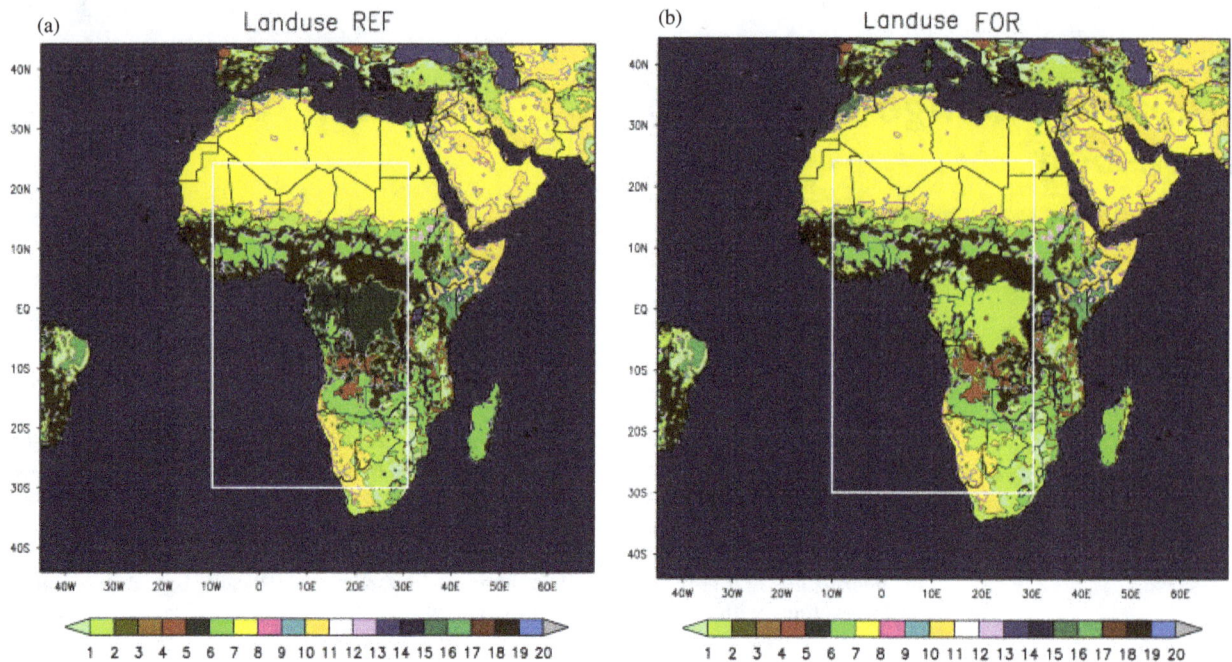

Figure 1. Landuse map for the REF (a) and FOR (b) experiments. In the Congo Basin the landuse is changed from 'Evergreen broadleaf tree' (a) to 'Short grass' (b). White boxes indicate the area over which the radiation and surface fluxes have been averaged in Figure 3.

by intercomparing simulations driven by analyses of observations with and without deforestation over a domain encompassing the entire African continent, and by analyzing the effects of deforestation on the dynamics of the African monsoon systems and the associated precipitation patterns.

2. Model and experiment design

We use the International Centre for Theoretical Physics (ICTP) RCM version 3 (RegCM3), which is extensively described in Giorgi and Mearns (1999) and Pal *et al*. (2007). The model configuration is the same as in Sylla *et al*. (2010). The domain covers the entire African continent and surrounding areas at 50 km grid spacing (Figure 1) and 18 vertical sigma levels. Of specific relevance to this study is the use of the mass flux convection scheme by Grell (1993), the boundary layer scheme of Holtslag *et al*. (1990) and the Biosphere-Atmosphere Transfer Scheme (BATS, Dickinson *et al*., 1993) for land surface processes. In particular, BATS, which has been used by a wide community for many years, includes a three-layer soil moisture module, a force-restore soil temperature model, and one layer vegetation module including transpiration and rain interception and a simple surface runoff scheme.

The integration period is January 2000 through December 2007 (8 years) with lateral boundary conditions provided by the ERA-Interim re-analysis (Simmons *et al*., 2007) which, as shown by Uppala *et al*., 2008, corrects some of the errors of the ERA-40 reanalysis regarding the hydrologic cycle of

the tropics. Two numerical experiments are performed using two different land cover patterns (Figure 1). In the first experiment, denoted with REF, we use 'present-day' vegetation cover as characterized by United States Geological Survey (USGS) Global Land Cover Characterization (GLCC) (Loveland *et al*., 2000). In the second, denoted with FOR, the land cover between 10E and 30E and between 7S and 7N (evergreen broadleaf trees) is replaced by short grass (Figure 1). Compared to evergreen broadleaf forest, short grass has lower leaf area index, lower roughness and higher albedo as reported in Table I, which contains BATS vegetation/land-cover values (BATS, Dickinson *et al*., 1993).

Initial soil moisture is also reduced over the grass area according to the approach of Giorgi and Bates (1989), however, this initialization is not important in view of the length of the simulation. We analyze results for two seasons, a boreal warm (austral cold) season, composed by June, July, August and September (JJAS), and a boreal cold (austral warm) season, composed by October, November, December, January and February (ONDJF), corresponding to distinctly different climates over different African regions. In JJAS, the monsoon produces maximum precipitation amounts north of the equator up to the Sahel region, while in ONDJF the ITCZ moves southward and precipitation peaks over sub-equatorial Africa down to about 30S.

We do not present here a validation of the REF simulation because this is already extensively discussed in Sylla *et al*. (2010). As shown in that study, comparison with different station and satellite-based observation datasets showed that RegCM3

Table I. BATS parameters used for the two vegetation/land cover types used in the reference and deforestation experiments.

Parameter	(2) Short grass	(6) Tropical forest
Maximum fractional vegetation cover	0.80	0.90
Difference between max fractional vegetation cover and cover at 269 K	0.1	0.3
Roughness length (m)	0.05	2.00
Displacement height (m)	0.0	18.0
Minimum stomatal resistance (s/m)	60	60
Maximum Leaf Area Index	2	6
Minimum Leaf Area Index	0.5	5
Stem (dead matter area index)	4.0	2.0
Inverse square root of leaf dimension ($m^{(-1/2)}$)	5	5
Light sensitivity factor ($m^2 W^{(-1)}$)	0.02	0.06
Upper soil layer depth (mm)	100	100
Root zone soil layer depth (mm)	1000	1500
Depth of total soil (mm)	3000	3000
Soil texture type	6	8
Soil color type	3	4
Vegetation albedo for wavelenghts < 0.7 μm	0.10	0.04
Vegetation albedo for wavelenghts > 0.7 μm	0.30	0.20

captures all the main climatic features regulating the Continental Africa monsoon climates in the two contrasting seasons, both in terms of precipitation and main circulation features (African Easterly Jet, AEJ; Tropical Easterly Jet, TEJ; African Easterly Wave activity; Intertropical Convergence Zone, ITCZ seasonal migration; Figure 2(a) and (b)). The model reproduces the seasonal cycle of precipitation over different regions, although in some cases the amounts of precipitation differ from observed. The West Africa monsoon and associated rain band is well captured in its interaseasonal evolution, although it shows a slight shift to the north compared to observations. The reader is referred to Sylla *et al.* (2010) for more details.

3. Results

Figures 2(c) and (d) and 4(a) and (b) show the difference in precipitation and temperature, respectively, between the FOR and REF simulations in the two seasons analyzed. The deforestation of the Congo Basin produces a statistically significant warming (Supporting information Figure S4) over the region of up to 2–4 °C, with this warming being somewhat more pronounced and extended during the southern hemisphere summer season. This warming is triggered by a decrease in surface evapotranspiration and

latent heat flux associated with a decrease in leaf area index (Figure 3). The warming is then amplified by a feedback process by which the reduced evaporation causes a statistically significant (Supporting information Figure S4) reduction in precipitation over the Congo Basin (Figure 2(c) and (d)) and thus a reduction of soil moisture and cloud cover (Supporting information Figure S1). In fact, a strong reduction of precipitation of up to 50% over the Congo Basin is produced by the deforestation (Figure 2).

It should be noted that deforestation not only affects 'potential evaporation' but also changes the relationship between soil moisture and evaporation. Both soil moisture and evapotranspiration decrease under the deforestation conditions, however, the deforested surface may be less able to extract and use the soil moisture than the forest. To verify this hypothesis, we calculated the ratio of evapotranspiration over soil moisture in the root zone and indeed found that, when averaged over the entire deforested area, this ratio decreased by ~7.7% in JJAS and 4.5% in ONDJF. This result implies that the deforested area might feel an additional water stress due to its decreased ability to extract water from the soil.

To achieve a better understanding of the thermodynamic processes underlying the model response to the deforestation, Figure 3 shows the FOR–REF difference in sensible heat, latent heat and net radiation fluxes over the deforested basin. In general, sensible heat increases with the warmer and drier conditions caused by the deforestation, while latent heat shows a corresponding decrease. These changes are directly tied to the changes in precipitation (Figure 1). The net radiation shows a more complex response. Overall, the net radiation absorbed at the surface decreases (Figure 3) due to the greater albedo of grass compared to forest and the reduced downward infrared flux caused by lower cloudiness, which overcome the corresponding increase in incoming solar radiation (Supporting information Figure S1). However, in both seasons there is a latitudinal shift between the maximum in sensible/latent heat difference and the maximum in net radiation difference. Comparison of Figures 2 and 3 and Supporting information Figure S1 indicates that this is because, while the changes in sensible and latent heat fluxes are essentially driven by the changes in precipitation, the change in net radiation is more tied to cloudiness and the peak in cloud change is not co-located with the peak in precipitation change (most noticeably in ONDJF). In addition, Supporting information Figure S2 shows that the decrease in evaporation for the deforested case is dynamically compensated by an increase in atmospheric moisture convergence, a result consistent with the findings of Polcher and Laval (1994).

The important aspect of the precipitation changes shown in Figure 2(b)–(e) is that, in addition to the local drying over the basin, the Congo Basin deforestation has remote effects on precipitation in other areas of Africa. During the boreal summer (JJAS), we

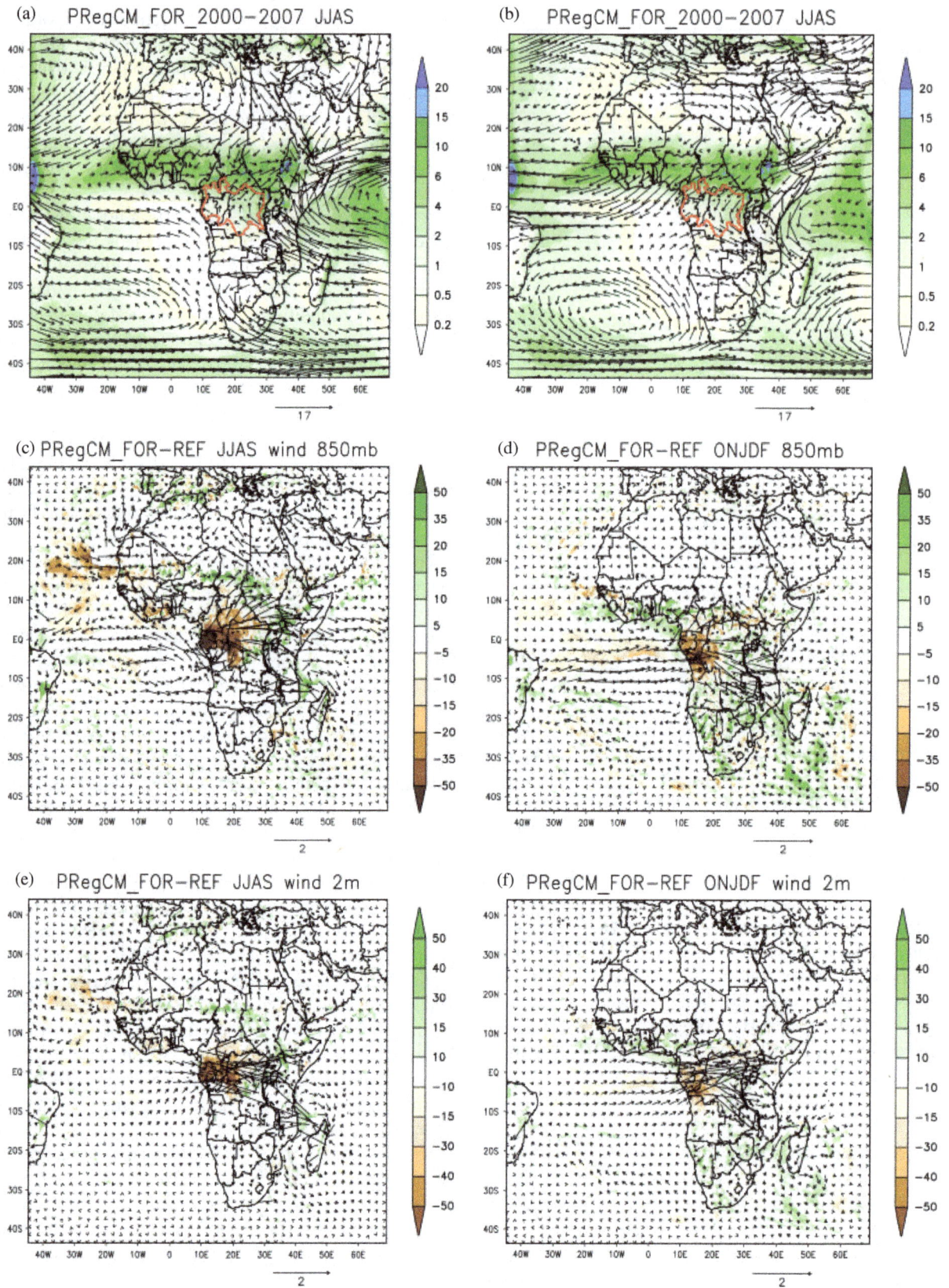

Figure 2. Mean precipitation (mm) and 850 mb wind for the 2000–2007 JJAS (a) and ONDJF (b) season in FOR. Precipitation differences (in percentage) and wind differences at 850 mb between FOR and REF in the JJAS (c) and ONDJF season (d). Precipitation differences (in percentage) and wind differences at 2 m between FOR and REF in the JJAS (e) and ONDJF season (f). In Fig. 2 a–b the deforested area is highlighted in red.

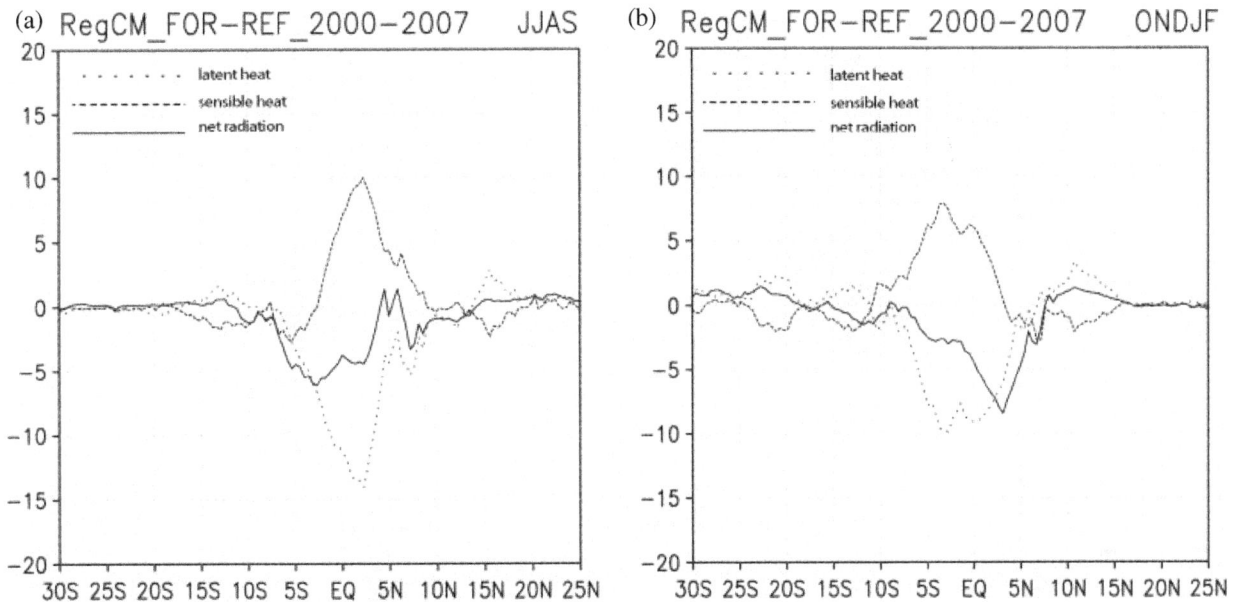

Figure 3. Difference of latent heat, sensible heat and net radiation between FOR and REF in the JJAS (a) and in the ONDJF (b) season, averaged for the years 2000–2007 between 10W and 30E longitude.

find an increase in precipitation over the Sahel region and over some areas of the Ethiopian highlands and a decrease of precipitation over the Guinea coast. The 850 hPa wind patterns of Figure 2(c) and (d) show that this is due to an intensification of the West Africa monsoon, which causes a northward shift of the monsoon rain band (Lin *et al.* (2008)). This intensification is in turn associated with the formation of a heat low over the Congo Basin (Figure 4(a)) that causes relatively dry converging flow at the surface (Figure 2(e)) and diverging flow at 850 hPa (Figure 2(c)). Similarly, in the austral summer season the heat low (Figure 4(b)) causes relatively dry converging flow at the surface (Figure 2(f)) and diverging flow at 850 hPa (Figure 2(d)), which reinforces the northwesterly monsoon flow over southern equatorial Africa, causing increased precipitation there. All these changes are mostly statistically significant with a confidence level of 95% (Supporting information Figures S3 and S4).

Our study provides the first regional model-based specific investigation of the climatic impacts of the Congo Basin deforestation and shows that this deforestation can have far-reaching climatic effects over the African continent.

This result confirms the critical role of land surface processes for African Climates (Koster *et al.*, 2004). Since the anthropogenic pressure on the Congo Basin forest ecosystem is large and increasing, it might be expected that it will interact with other environmental pressures, such as due to increasing GHG (Greenhouse gas) forcing, resulting in possibly substantial impacts on the climate of different African regions. It is thus important that more in depth studies are carried out focusing on the Congo Basin and its interactions with other regional climates.

4. Conclusions

Using a set of experiments with the RCM RegCM3 we find both local and remote effects of the Congo Basin deforestation. Locally, decreased evapotranspiration determines substantial warming (up to 3–4 °C) and reduced precipitation (up to 50%) in both seasons. The associated heat low generates thermally driven organized overturning circulations similar to those found in early studies of land surface modifications at the regional scale (Seth and Giorgi, 1996; Giorgi and Avissar, 1997; Pielke *et al.*, 2002). The regional circulations interact with large scale flows resulting in an intensification of the West Africa monsoon flow in the Boreal Summer and an intensification of the Southern Equatorial monsoon flow in the Austral summer. These in turn cause increased (decreased) precipitation over the Sahel (Guinea Coast) in the boreal summer and increased precipitation over Southern Equatorial Africa.

Our study provides the first regional model-based specific investigation of the climatic impacts of the Congo Basin deforestation and shows that this deforestation can have far-reaching climatic effects over the African continent.

This result confirms the critical role of land surface processes for African Climates (Koster *et al.*, 2004). Since the anthropogenic pressure on the Congo Basin forest ecosystem is large and increasing, it might be expected that it will interact with other environmental pressures, such as due to increasing GHG forcing, resulting in possibly substantial impacts on the climate of different African regions. It is thus important that more in depth studies are carried out focusing on the Congo Basin and its interactions with other regional climates to better understand the development

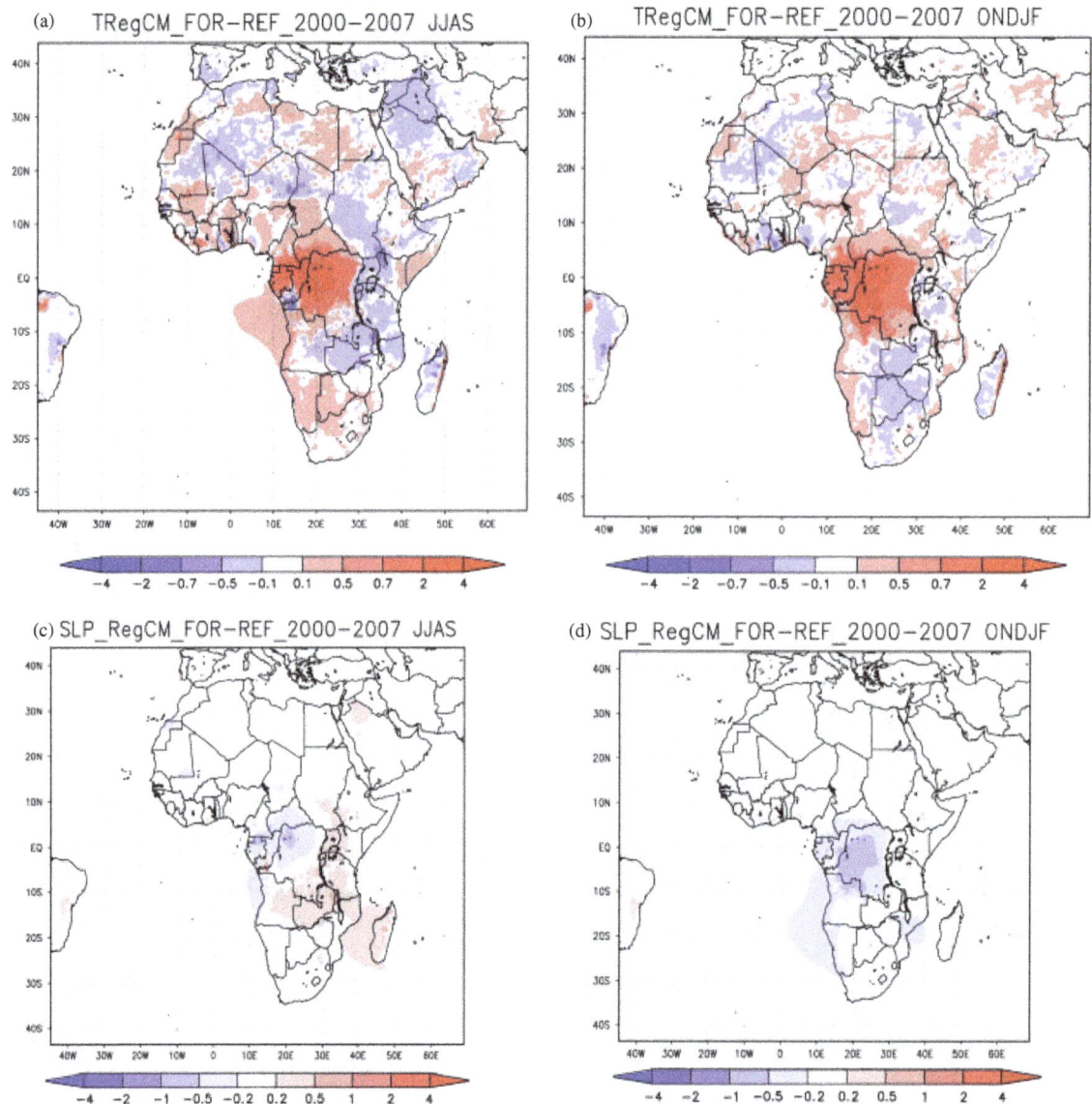

Figure 4. Differences of surface temperature (degree) between FOR and REF in the JJAS (a) and ONDJF (b) seasons for the years 2000–2007. Differences of sea level pressure (mb) between FOR and REF in the JJAS (c) and ONDJF (d) season for the years 2000–2007.

of the impacts of the deforestation also in the perspective of promoting policies of conservation [like for example the Reducing Emissions from Deforestation and Degradation (REDD)], Angelsen *et al*., 2009).

Supporting information

The following supporting information is available:

Figure S1. (a–b) Cloud cover difference between FOR and REF (in percentage). (c–d) Difference of net radiation, short wave incident, long wave downward, net short wave and net long wave between FOR and REF in the JJAS and in the ONDJF season, averaged for the years 2000–2007 between 10 W and 30 E longitude.

Figure S2. Difference of moisture convergence (mg/s) between the FOR and the REF for the JJAS (a) and the ONDJF (b) season, averaged for the years 2000–2007.

Figure S3. T-test of the U-wind component (upper panels) and of the V-wind component (lower panels) at 850 hPa, for the JJAS and ONDJF seasons for the years 2000–2007.

Figure S4. T-test of the temperature (upper panels) and precipitation (lower panels), for the JJAS and ONDJF seasons for the years 2000–2007.

References

Abiodun BJ, Pal JS, Afiesamama EA, Gutowski WJ, Adedoyin A. 2007. Simulation of West Africa monsoon using RegCM3 Part II: impacts of deforestation and desertification. *Theoretical and Applied Climatology* **93**: 245–261.

Angelsen A, Brown S, Loisel C, Peskett L, Streck C, Zarin D. 2009. Reducing emissions from deforestation and forest degradation (REDD): An Options Assessment Report. Prepared for the Government of Norway, Meridian Institute.

Avissar R, Werth D. 2005. Global hydroclimatological teleconnections resulting from tropical deforestation. *Journal of Hydrometeorology* **6**: 134–145.

Charney JG. 1975. Dynamics of deserts and drought in the Sahel. *The Quarterly Journal of the Royal Meteorological Society* **101**: 193–202.

Clark DB, Xue Y, Harding RJ, Valdes PJ. 2001. Modeling the impact of land surface degradation on the climate of tropical North Africa. *Journal of Climate* **14**: 1809–1822.

Dickinson RE, Henderson-Sellers A, Kennedy PJ. 1993. Biosphere Transfer Scheme (BATS) Version le as coupled to the NCAR Community Climate Model. *NCAR Technical Note*, TN-387 + STR, 72 pp.

Dirmeyer PA, Shukla J. 1994. Albedo as a modulator of climate response to tropical deforestation. *Journal of Geophysical Research* **99**: 20863–20877.

FAO. 2011. *The State of Forests in the Amazon Basin, Congo Basin and Southeast Asia*. Food and Agriculture Organization of the United Nations, Rome; 1–83.

Giorgi F, Avissar R. 1997. Representation of heterogeneity effects in earth system modeling: experience from land surface modeling. *Reviews of Geophysics* **35**: 413–437.

Giorgi F, Bates GT. 1989. The climatological skill of a regional model over complex terrain. *Monthly Weather Review* **117**: 2325–2347.

Giorgi F, Mearns LO. 1999. Introduction to special section: regional climate modeling revisited. *Journal of Geophysical Research* **104**: 6335–6352.

Grell GA. 1993. Prognostic evaluation of assumptions used by cumulus parametrization. *Monthly Weather Review* **121**: 764–787.

Hasler N, Werth D, Avissar R. 2009. Effects of tropical deforestation on global hydroclimate: a multimodel ensemble analysis. *Journal of Climate* **22**: 1124–1141.

Henderson-Sellers A, Gornitz V. 1984. Possible climatic impacts of land cover transformations, with particular emphasis on tropical deforestation. *Climatic Change* **6**: 231–257.

Henderson-Sellers A, Dickinson RE, Wilson MF. 1988. Tropical deforestation – important processes for climate models. *Climatic Change* **13**: 43–67.

Holtslag AAM, de Bruijn EIF, Pan HL. 1990. A high resolution air mass transformation model for short-range weather forecasting. *Monthly Weather Review* **118**: 1561–1575.

Koster RD, Dirmeyer PA, Guo Z, Bonan G, Chan E, Cox P, Gordon CT, Kanae S, Kowalczyk E, Lawrence D, Liu P, Lu C-H, Malyshev S, McAvaney B, Mitchell K, Mocko D, Oki T, Oleson K, Pitman A, Sud YC, Taylor CM, Verseghy D, Vasic R, Xue Y, Yamada T. 2004. Regions of strong coupling between soil moisture and precipitation. *Science* **305**: 1138–1140.

Lean J, Rowntree PR. 1997. Understanding the sensitivity of a GCM simulation of Amazonian deforestation to the specification of vegetation and soil characteristics. *Journal of Climate* **10**: 1216–1235.

Lin M, Fan K, Wang H. 2008. Somali jet changes under the global warming. *Acta Meteorologica Sinica* **22**: 502–510.

Loveland TR, Reed BC, Brown JF, Ohlen DO, Zhu Z, Yang L, Merchant JW. 2000. Development of a global land cover characteristic database and IGBP DISCover from 1 km AVHRR data. *International Journal of Remote Sensing* **21**: 1303–1330.

Nayar A. 2009. Model predicts future deforestation. *Nature News*. DOI: 10.1038/news.2009.1100.

Pal JS, Giorgi F, Bi XQ, Elguindi N, Solomon F, Gao XJ, Rauscher SA, Francisco R, Zakey A, Winter J, Ashfaq M, Syed FS, da Rocha RP, Sloan LC, Steiner AL. 2007. Regional climate modeling for the developing world - The ICTP RegCM3 and RegCNET. *Bulletin of American Meteorological Society* **88**: 1395–1409.

Pielke RA, Marland G, Betts RA, Chase TN, Eastman JL, Niles JO, Niyogi DS, Running SW. 2002. The influence of land-use change and landscape dynamics on the climate system: relevance to climate-change policy beyond the radiative effect of greenhouse gases. *Philosophical Transactions of the Royal Society of London* **A360**: 1705–1719.

Polcher J, Laval K. 1994. The impact of African and Amazonian deforestation on tropical climate. *Journal of Hydrology* **155**: 389–405.

Semazzi FH, Song Y. 2001. A GCM study of climate change induced by deforestation in Africa. *Climate Research* **17**: 169–182.

Seth A, Giorgi F. 1996. Three-dimensional model study of organized mesoscale circulations induced by vegetation. *Journal of Geophysical Research-Atmospheres* **101**: 7371–7391.

Shem OW, Dickinson RE. 2006. How the Congo basin deforestation and the equatorial monsoonal circulation influences the regional hydrological cycle, Paper Presented at the 86th Annual American Meteorological Society Meeting, 2006.

Simmons A, Uppala S, Dee D, Kobayashi S. 2007. ERA-Interim: new ECMWF reanalysis products from 1989 onwards. *ECMWF Newsletter 110 – Winter 2006/07*, 2961–3012.

Sud YC, Yang R, Walker GK. 1996. Impact in situ deforestation in Amazonia on the regional climate: general circulation model simulation study. *Journal of Geophysical Research-Atmospheres* **101**: 7095–7109.

Sylla MB, Coppola E, Mariotti L, Giorgi F, Ruti PM, Dell'Aquila A, Bi X. 2010. Multiyear simulation of the African climate using a regional climate model (RegCM3) driven by the high resolution ERA-Interim reanalysis. *Climate Dynamics*. **35**: 231–247 DOI: 10.1007/s00382-009-0613-9.

Uppala S, Dee D, Kobayashi S, Berrisford P, Simmons A. 2008. Towards a climate data assimilation system: status update of ERA-Interim. *ECMWF Newsl* **115**: 12–18.

Werth D, Avissar R. 2002. The local and global effects of Amazon deforestation. *Journal of Geophysical Research* **107**, 8087, 8 pp. DOI: 10.1029/2001JD000717.

Werth D, Avissar R. 2005. The local and global effects of African deforestation. *Geophysical Research Letters* **32**, L12704, 4 pp. DOI: 10.1029/2005GL022969.

Zeng N, Neelin JD. 1999. A land-atmosphere interaction theory for the tropical deforestation problem. *Journal of Climate* **12**: 857–872.

Zhang H, Henderson-Sellers A, McGuffie K. 2001. The compounding effects of tropical deforestation and greenhouse warming on climate. *Climatic Change* **49**: 309–338. DOI: 10.1023/A:1010662425950.

Zheng X, Eltahir EAB. 1998. The role of vegetation in the dynamics of West African monsoons. *Journal of Climate* **11**: 2078–2096.

18

Validation of CMIP5 models for the contiguous United States

Victor Privalsky[1]* and Vladislav Yushkov[2]

[1]Space Dynamics Laboratory (ret.), Logan, UT, USA
[2]Physics Department, Moscow State University, Russia

*Correspondence to:
V. Privalsky, 1272 Eastridge
Drive, Logan, UT 84321, USA.
E-mail: vprivalsky@gmail.com

Abstract

Major statistical characteristics – trend rates, mean values, standard deviations, probability densities, autoregressive model orders, persistence criteria, and spectra – of annual surface temperature over the contiguous United States from 1889 through 2005 are compared with respective characteristics of 47 time series generated within the CMIP5 historical experiment. The observed and most simulated time series are Gaussian. Most autoregressive orders, persistence criteria, and spectra of simulated time series are close to what is found in nature. Although the multi-model mean value is not biased, individual models can err by almost 3.5 °C. In addition, the models exaggerate linear trend rates and temperature variance is overestimated.

Keywords: climate models; statistical properties; verification; CONUS

1. Introduction

Efforts to show that numerical models correctly reproduce climatic variability including its stochastic components present a traditional research area in climatology (e.g. Jones *et al.*, 2013). A full-scale validation of climate models constitutes a statistically unmanageable task because it requires quantitative comparisons between time-dependent sequences of observed and simulated multidimensional random fields of climatic data. However, the task becomes doable if the data are spatially averaged so that multidimensional random fields become scalar time series. Comparing time series to each other is relatively easy. If a simulated time series obtained in this manner has statistically the same basic properties – such as trend rates, mean values and variances, probability density functions, *etc.* – as respective observed time series, one may regard the climate model as reliable at the given scale of spatial averaging. This is an obvious necessary condition for recognizing models as being in agreement with observations. If, on the contrary, a model generates data whose major statistical properties differ significantly from those obtained from observations, the model is inadequate.

This spatial averaging approach, first suggested in Privalsky and Croley (1992), is used here to compare basic statistical properties of the observed annual surface temperature (AST) averaged over the 48 states of the contiguous United States (CONUS) with respective properties of AST generated with CMIP5 models within the framework of CMIP5 historical experiment at the same scale of spatial averaging. The choice of CONUS is by no means random because the observation data over the CONUS territory since the end of the

19th century are probably more reliable than respective data over any other region of similar or bigger size.

The statistical properties of AST to be analyzed here include:

1. linear trend rates,
2. mean values and standard deviations,
3. type of probability density functions (PDFs),
4. orders of optimal autoregressive approximations and time series persistence (statistical predictability),
5. spectral densities.

2. Data and methods

The initial monthly observation data were taken from the HadCRUT4 file from the website of the University of East Anglia (Morice *et al.*, 2012) and averaged over the CONUS and over 12 months to obtain a time series of CONUS annual surface temperature. The observation data are available for the entire time interval of the CMIP5 historical experiment from 1850 through 2005 but we selected a shorter interval from 1889 through 2005 during which the coverage with observations was, according to the data set, never below 94%. The simulation data for the same time interval were obtained from the CMIP5 project data site (see Taylor *et al.*, 2012) and averaged in the same manner to obtain 47 time series of simulated AST. With just a few exceptions, comparisons were conducted for run 1 of each model.

The first four statistical characteristics were calculated in the usual manner while the spectra were estimated through the autoregressive (AR) time series modeling; the time domain models with properly selected AR orders served as a basis for the maximum

Table I. List of models and statistical properties of observed and simulated temperature.

Model	Trend, °C (100 years)$^{-1}$	Mean, °C	Standard deviation, °C	p*	d(1)*	$\tau_{0.9}$, yrs*
HadCRUT4 (observed)	0.56	11.16	0.46	1	0.97	0
ACCESS1-0	0.48	12.97	0.54	2	0.95	0
ACCESS1-3	0.29	14.59	0.46	1	0.93	1
BCC-CSM1-1	0.94	10.68	0.57	1	0.96	0
BCC-CSM1-1-M	0.93	11.65	0.58	0	1	0
BNU-ESM	1.24	9.86	0.64	0	0.95	1
CanESM2	0.53	10.26	0.62	1	0.92	1
CCSM4	0.23	11.46	0.57	0	1	0
CESM1-BGC	0.78	11.83	0.60	1	0.94	1
CESM1-CAM5	0.75	11.80	0.49	1	0.96	0
CESM1-FASTCHEM	1.04	12.11	0.57	1	0.94	1
CESM1-WACCM	0.97	11.61	0.55	0	1	0
CMCC-CESM	0.29	10.83	0.53	2	0.95	0
CMCC-CM	0.56	9.83	0.48	2	0.95	0
CMCC-CMS	0.42	10.39	0.54	2	0.95	0
CNRM-CM5	0.46	10.72	0.64	0	1	0
CNRM-CM5-2	0.40	10.52	0.53	0	1	0
CSIRO-MK3-6-0	0.35	12.20	0.54	1	0.95	1
CSIRO-MK3L-1-2	0.81	12.94	0.46	0	1	0
EC-EARTH	1.19	11.73	0.43	1	0.97	0
FGOALS-G2	0.90	7.80	0.52	1	0.91	1
FIO-ESM	0.84	11.63	0.43	2	0.97	0
GFDL-CM21	1.09	10.10	0.74	1	0.98	0
GFDL-CM3	0.38	10.64	0.52	1	0.91	1
GFDL-ESM2G	0.63	10.06	0.56	0	1	0
GFDL-ESM2M	0.67	10.62	0.63	2	0.84	2
GISS-E2-H	0.54	9.90	0.41	1	0.97	0
GISS-E2-H-CC	0.49	10.51	0.38	0	1	0
GISS-E2-R	0.27	9.66	0.45	1	0.97	0
GISS-E2-R-CC	0.27	9.40	0.44	1	0.92	1
HADCM3	0.68	10.00	0.71	1	0.94	1
HADGEM2-AO	0.34	12.49	0.64	3	0.84	3
HADGEM2-CC	−0.03	11.87	0.61	0	1	1
HADGEM2-ES	0.67	11.83	0.63	3	0.78	4
INM-CM4	0.96	10.02	0.42	0	1	0
IPSL-CM5A-LR	1.49	9.92	0.47	1	0.96	0
IPSL-CM5A-MR	0.88	10.97	0.47	0	1	0
IPSL-CM5B-LR	0.89	9.67	0.52	1	0.89	1
MIROC5	0.58	13.00	0.54	1	0.92	1
MIROC-ESM	0.74	12.24	0.48	3	0.88	2
MIROC-ESM-CHEM	0.42	11.92	0.45	1	0.94	1
MPI-ESM-LR	1.29	10.90	0.55	1	0.96	0
MPI-ESM-MR	1.37	11.33	0.58	1	0.98	0
MPI-ESM-P	1.21	11.65	0.65	1	0.95	1
MRI-CGCM3	0.58	9.66	0.44	0	1	0
MRI-ESM1	0.78	9.87	0.39	0	1	0
NORESM1-M	0.65	10.84	0.50	1	0.96	0
NORESM1-ME	0.62	10.48	0.55	1	0.95	0
AVERAGE	0.70	11.00	0.53	1	0.95	1

*p, optimal autoregressive order; d(1), relative prediction error variance within the Kolmogorov–Wiener theory at lead time $\tau = 1$ year; $\tau_{0.9}$, lead time at which the relative prediction error variance attains 0.9 or more.

entropy spectral estimates and estimates of the time series' persistence.

3. Results

The list of models analyzed here is given in the first column of Table 1 with graphs of observed and simulated time series shown in Figure 1.

3.1. Linear trend

As seen from Figure 1(a), both observed and simulated temperature anomalies reveal a positive trend. The trend

rates have been estimated under the assumption that every time series x_n, $n = 1, 2, \ldots, N$, presents a sum of a zero mean white noise sequence ε_n with a linear function of time: $x_n = \alpha n + \varepsilon_n$. As will be seen later, the statistical models of the observed and simulated time series of AST do not differ much from such a model. Estimates of the coefficient α are shown in Table 1 for the entire interval from 1889 through 2005. As seen from the table, the trend rate in the observed data amounts to $0.56 °C (100 \text{ years})^{-1}$ with the estimate's standard error $0.13 °C (100 \text{ years})^{-1}$. The average trend rate estimate for the simulated data is $0.70 °C (100 \text{ years})^{-1}$. Only 22 trend rate estimates for the simulated

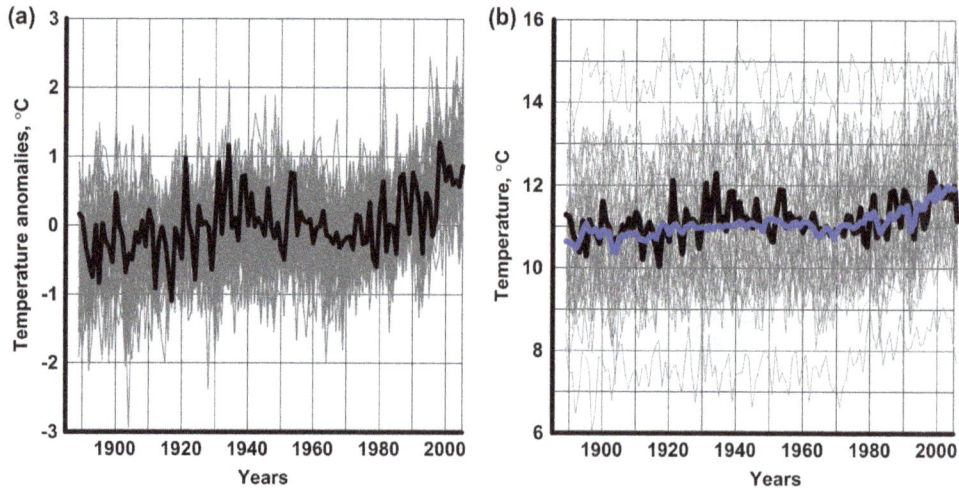

Figure 1. Observed (black) and simulated (grey) AST over CONUS: (a) anomalies; (b) temperature; the blue line shows the average simulated temperature.

data lie within the 95% confidence interval for the estimate of the trend rate in the observed time series. The range of trend rate estimates in simulated data is $1.5°C (100 \text{ years})^{-1}$ [from 0 to $1.5°C (100 \text{ years})^{-1}$]. In other words, the trend rate in the simulated data is seriously overestimated. This result is unsatisfactory. Note also that the positive bias in trend rate estimates can hardly be caused by observation errors because, according to Menne *et al.* (2010), there is 'no evidence that the CONUS average temperature trends are inflated due to poor station siting'.

As the linear trend is supposed to have been caused by external forcing, it has been removed from all time series prior to further analysis.

3.2. Mean values and standard deviations

The observed mean value for 1889–2005 is 11.16 °C. The multi-model average mean value estimate for the simulated data is 11.00 °C (Figure 1(b)). However, the mean value estimates for the simulated data lie between 7.8 and 14.6 °C (a 6.8 °C range) and everyone of them differs statistically significantly from the observed value. These results are unsatisfactory.

As seen from Figure 1(a), the simulated time series have a higher variance than observations. The observed standard deviation (s.d.) of AST is 0.42 °C with a 95% confidence interval between 0.37 and 0.48 °C. The average simulated s.d. is 0.53 °C – a statistically significant difference with observations. The difference between the standard deviation estimates of the observed and simulated data is statistically significant for 30 models at a 95% confidence level. On the whole, the standard deviations estimates of the CMIP5 data are positively biased. These results are unsatisfactory.

3.3. Probability density

The observed data have a probability density function that can be regarded as Gaussian according to

several criteria, including Kolmogorov–Smirnov's and chi-square. The simulated time series have the same Gaussian type of PDF with just two exceptions (HADGEM2-AO and IPSL-CM5B-LR). This is definitely a positive result.

3.4. AR model orders and persistence

The optimal model in this study was chosen for each time series with four order-selection criteria: Akaike's AIC, Parzen's CAT, Schwarz-Rissanen's BIC, and Hennan-Quinn's Ψ (see Bhansali, 1986; Broersen, 2000). After the linear trend removal, the observed time series x_n, $n = 1, \ldots, N$, where $N = 117$, is best approximated with an AR model of order $p = 1$ (a Gaussian Markov chain):

$$x_n \approx 0.21 x_{n-1} + a_n \qquad (1)$$

where a_n is a zero mean innovation sequence (white noise) with the variance $\sigma_a^2 = 0.17 \, (°C)^2$. The RMS error of the coefficient estimate in Equation (1) is 0.09.

The persistence of the stochastic model (1) is very low: the relative error of a one-year lead time prediction within the Kolmogorov–Wiener theory of extrapolation is $d(1) = \sigma_a^2/\sigma_x^2 \approx 0.97$. In other words, the observed time series is very close to a white noise. Most simulated time series behave in the same manner.

The limit of statistical predictability is defined here as the lead time of a Kolmogorov–Wiener linear extrapolation at which the error variance becomes equal to 0.90 or higher. Obviously, the observed time series has a zero limit of statistical predictability (zero persistence). Most simulated time series behave in the same manner (see Table 1). This is again a positive result.

3.5. Spectral densities

The spectra of the observed and simulated time series of annual surface temperature over the CONUS were

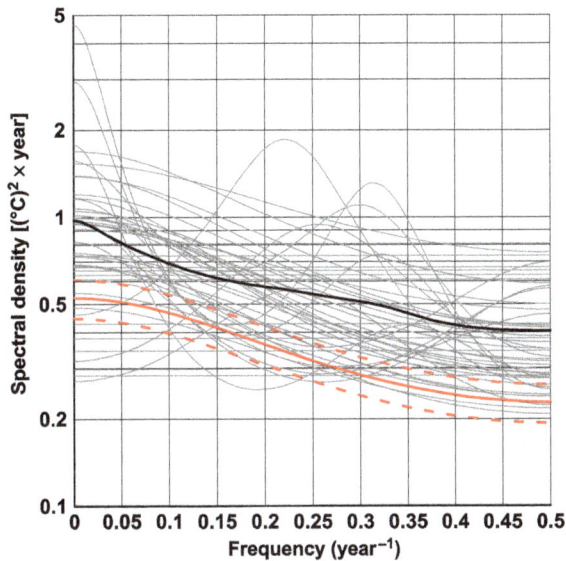

Figure 2. Maximum entropy estimates of observed (red, with 90% confidence bounds) and simulated (grey) AST spectra. The average simulated spectrum is shown with a thick black line.

obtained with the maximum entropy method (Burg, 1967; Jaynes, 1982). The algorithm includes fitting an optimal AR model to the time series and then calculating the spectral density on the basis of the selected time domain model. The approximate number of degrees of freedom v for the spectral estimates is determined as $v = N/p$, where $p > 0$ is the order of the selected AR model (Ulrich and Bishop, 1975).

The spectral estimates obtained for the 47 simulated time series are shown in Figure 2 along with the spectrum of observations (the red curves) and the average of the simulated spectra (the black curve). The higher position of the average simulated spectrum happens because of the positive bias in the variance estimates. Only nine spectra of simulated data have AR orders $p = 2$ or 3 and therefore can have one or two extrema or inflection points. All other spectra are constant (white noise, 13 cases) or monotonic (Markov chain, 25 cases). With few exceptions (BNU-ESM, CMCC-CMS, GFDL-ESM2M, HADGEM2-AO, MIROC-ESM), all spectra are close to the spectrum of observations. These results should be regarded as positive.

4. Conclusions

An agreement between basic statistical moments of the observed and simulated climates is a necessary condition for recognizing the validity of numerical models of climate. The CONUS cover a sizable part of the Earth's territory – over 5% – and present a region with a practically complete coverage with instrumental

observations of surface temperature since the end of the 19th century. These features make CONUS an object particularly suitable for trustworthy validations of numerical models of climate. Our analysis of time series of the average over the CONUS annual surface temperature generated with CMIP5 models showed that the mean value averaged over the ensemble is unbiased, most simulated time series have the same Gaussian PDF as the observed time series. Most CMIP5 models correctly reproduce the time and frequency domains behavior of the observed time series (including probability distribution type, persistence parameters and spectrum), which is close to the behavior of a white noise sequence. These results are satisfactory.

Yet, the linear trend rates and standard deviations of simulated annual surface temperature differ significantly from respective characteristics of the observed time series while the mean value estimates vary by as much as 6.8 °C. The trend rates are positively biased and can exceed the observed rate by as much as 220%. All mean value estimates for simulated data are statistically different from the observed mean annual temperature; they can differ from the observed mean value by more than 3 °C. The standard deviation estimates are positively biased and can exceed the observed value by 60%. These results are unsatisfactory.

References

Bhansali RJ. 1986. The criterion autoregressive transfer function of Parzen. *Journal of Time Series Analysis* **7**: 79–104.

Broersen MT. 2000. Finite sample criteria for autoregressive order selection. *IEEE Transactions of Signal Processing* **48**: 3550–3558.

Burg JP. 1967. Maximum entropy spectral analysis. In Paper presented at the 37th Meeting of Society for Exploration Geophysics, Oklahoma City, OK, Oct. 31, 1967.

Jaynes ET. 1982. On the rational of the maximum-entropy methods. *Proceedings of IEEE* **70**: 939–952.

Jones GS, Stott PA, Christidis N. 2013. Attribution of observed historical near–surface temperature variations to anthropogenic and natural causes using CMIP5 simulations. *Journal of Geophysical Research, [Atmospheres]* **118**: 4001–4024.

Menne MJ, Williams CN Jr, Palecki MA. 2010. On the reliability of the U.S. surface temperature record. *Journal of Geophysical Research* **115**: D11108, doi: 10.1029/2009JD013094.

Morice CP, Kennedy JJ, Rayner AA, Jones PD. 2012. Quantifying uncertainties in global and regional temperature change using an ensemble of observational estimates: the HadCRUT4 data set. *Journal of Geophysical Research* **117**: D08101, doi: 10.1029/2011JD017187.

Privalsky V, Croley TE. 1992. Statistical validation of GCN-simulated climates for the U.S. Great Lakes and C.I.S. Emba and Ural River basins. *Stochastic Hydrology and Hydraulics* **6**: 69–80.

Taylor KE, Stouffer RJ, Meehl GA. 2012. An overview of CMIP5 and the experiment design. *Bulletin of the American Meteorological Society* **93**: 485–498.

Ulrich TJ, Bishop TN. 1975. Maximum entropy spectral analysis and autoregressive decomposition. *Reviews of Geophysics and Space Physics* **13**: 183–200.

19

Land surface–atmosphere interaction in future South American climate using a multi-model ensemble

R. C. Ruscica,[1] C. G. Menéndez[1,2] and A. A. Sörensson[1]*

[1]Centro de Investigaciones del Mar y la Atmósfera, Consejo Nacional de Investigaciones Científicas y Técnicas, Universidad de Buenos Aires, Argentina
[2]Departamento de Ciencias de la Atmósfera y los Océanos, FCEN, Universidad de Buenos Aires, Argentina

*Correspondence to:
A. A. Sörensson, Centro de Investigaciones del Mar y la Atmósfera, Consejo Nacional de Investigaciones Científicas y Técnicas, Universidad de Buenos Aires, Ciudad Universitaria, Int. Guiraldes 2160, Pabellón 2, Piso 2, Ciudad Autónoma de Buenos Aires C1428EGA, Argentina.
E-mail: sorensson@cima.fcen.uba.ar

Abstract

The land–atmosphere interaction for reference and future climate is estimated with a regional climate model ensemble. In reference climate, more than 50% of the models show interaction in southeastern South America during austral spring, summer and autumn. In future climate, the region remains a strong hotspot although somewhat weakened due to the wet response that enhance energy limitation on the evapotranspiration. The region of the Brazilian Highlands and Matto Grosso appears as a new extensive hotspot during austral spring. This is related to a dry response which is probably accentuated by land surface feedbacks.

Keywords: land–atmosphere interaction; soil moisture; precipitation; coupling; South America; regional climate modeling

1. Introduction

On a monthly to seasonal time scale, soil moisture (SM) has potential to be a low-frequency modulator of climate because of its influence on the partitioning of heat fluxes and its long memory (e.g. Eltahir, 1998). Therefore, forecasts could be improved by including estimates of SM for the regions and seasons where this variable exerts a control on the atmosphere.

For SM to control precipitation (PP), the evapotranspiration (ET) regime has to be limited by SM rather than by radiation. The radiation (or energy)-limited ET regime dominates in wet regions where the moisture stress is low and variability of SM does not affect ET. In dry climates, SM availability is a first-order constraint on ET (e.g. Koster et al., 2004), but in very dry areas, however, SM does not affect PP because ET variability is too weak to induce PP generation. Therefore, regions where SM exerts control on PP tend to appear over transitions zones between dry and wet climates (Koster et al., 2004). A changing climate could alter these regimes due to increasing temperatures and altered PP amounts and temporal distribution.

South America embraces vast areas of both SM-limited regions, such as the Patagonian semi-arid steppe, and energy-limited regions, such as the Amazon rainforest. Southeastern South America (SESA) is one of the regions that have been identified as SM limited (Jung et al., 2010). SESA has also been identified as a region with strong SM–ET and SM–PP coupling during austral summer (DJF) in climate model studies (Wang et al., 2007; Sörensson and Menéndez, 2011; Dirmeyer et al., 2012; Ruscica et al., 2015) and as a region of SM–ET coupling during austral autumn (MAM) and spring (SON, Ruscica et al., 2015). In a reanalysis study by Zeng et al. (2010), subtropical South America was found to be a hotspot of interaction between SM and PP for DJF. Somewhat contrasting to these results, Dirmeyer et al. (2009) used data from several land surface models and identified SESA and the southern Amazon basin in boreal summer (JJA) as well as northeastern Brazil (NeB) and the Altiplano in MAM as regions that combine both land–atmosphere interactions and memory longer than 2 weeks, i.e. regions where forecasts could be improved by including SM information. On the other hand, a great number of studies on the global scale has focused on JJA (Koster et al., 2004; Notaro, 2008; Orlowsky and Seneviratne, 2010; Wei and Dirmeyer, 2010; Zeng et al., 2010; Zhang et al., 2011), showing different locations of high interaction in the South American tropics and agreeing on the absence of SM–ET and SM–PP interactions in subtropical areas.

The purpose of this work is to identify regions of land–atmosphere interaction over South America for all seasons and to identify changes in location and strength of these hotspots in future climate. The study is original in the sense that we use an ensemble of regional climate models (RCMs) simulations, which enhances the spatial resolution in comparison with global climate model (GCM) studies and spans a wider range of uncertainty than single-model studies.

Table 1. Basic information on the ensemble members.

	RCA	RegCM3	PROMES	LMDZ
GCM forcing at lateral boundaries and SST	EC5OM-R1, EC5OM-R2 and EC5OM-R3 (Roeckner et al., 2006)	EC5OM-R1 and HadCM3-Q0 (Gordon et al., 2000)	HadCM3-Q0	EC5OM-R3 and IPSL (Hourdin et al., 2006)
Reference	Samuelsson et al. (2011)	da Rocha et al. (2009)	Domínguez et al. (2010)	Hourdin et al. (2006)
Number of vertical atmospheric levels	40	18	37	19
Number of soil moisture levels	2	2	2	2

2. Models and methodology

The CLARIS-LPB project (Boulanger et al., 2011) has provided the first coordinated ensemble of RCM climate change simulations over the South American continent. The ensemble, which covers the period 1961–2100, is designed to span uncertainty of climate change scenarios using several RCMs forced by lateral boundary conditions and sea surface temperature from different GCMs, for the emission scenario A1B (Sánchez et al., 2015). In this study, we use time series of seasonal values of ET and PP for a period of 30 years of reference climate (1961–1990) and future climate (2071–2100) of eight ensemble members based on the combinations of four RCMs forced by three GCMs (Table 1). The land surface schemes of all four RCMs are second generation schemes where stomatal control on ET is parameterized (Sellers et al., 1997). The domains differ slightly among the models due to different types of grids, but all are of continental scale with borders over the surrounding oceans and a horizontal resolution of $0.5° \times 0.5°$. All models are hydrostatic.

The regions of SM–PP interaction are estimated with the index Γ, described by Zeng et al. (2010). Γ is defined as:

$$\Gamma = r_{pp,ET} \cdot \frac{\sigma_{ET}}{\sigma_{PP}}$$

where $r_{PP,ET}$ is the correlation between the time series of seasonal means of PP and ET, and the standard deviations σ_{ET} and σ_{PP} represent their temporal variability. For regions and seasons where the runoff is small in comparison to infiltration, PP can be interpreted as a proxy for SM. If $r_{PP,ET}$ is positive (negative), ET is SM (energy) limited. A necessary, although not sufficient, condition for SM to influence on PP is that the variability of ET is high enough. This criterion surges since, over dry regions, although ET is controlled by SM, the ET anomaly generated by a SM anomaly could be too small to generate a PP anomaly (Guo et al., 2006). In the Γ index, this criterion is represented by the variability of ET normalized by the variability of PP. To avoid spuriously high values of Γ, regions where interannual rainfall variability for each season is less than $0.5 \, \text{mm day}^{-1}$ are not included in the analysis.

3. Results

The observed and ensemble seasonal mean PP for the reference period are shown in Figure 1(a)–(d) (CRU TS3.1, Mitchell and Jones, 2005) and Figure 1(e)–(h), respectively. In the northeastern corner of Brazil during DJF, the ensemble simulates rainfall of $8 \, \text{mm day}^{-1}$ in a semiarid region of $1–2 \, \text{mm day}^{-1}$. This bias is the result of an artifact of the driving GCMs which, instead of the intertropical and the South Atlantic convergence zones, simulate only one convergence zone, with a maximum over this very dry region (Lin, 2007). When the RCMs of the ensemble are driven by reanalysis (Solman et al., 2013), this bias is not present. In SESA, there is an underestimation of rainfall for all seasons. This dry bias is common in both GCMs and RCMs (Vera et al., 2006; Menéndez et al., 2010) and is also present when the RCMs of the ensemble are driven by reanalysis (Solman et al., 2013).

The ensemble ET is shown in Figure 1(i)–(l). The spatial pattern resembles that of the PP with maximum located approximately over the same regions, except for in very wet situations such as in DJF over central Brazil or in JJA north of the equator. Over dry regions, the magnitude of the two variables is similar, while in wet regions such as the Amazonia, more than half of the PP goes to runoff.

The ensemble mean response to climate change is shown in Figure 2. The response of PP (Figure 2(a)–(d)) shows a dipole pattern of wet response over SESA and dry response north of it, similar to the study by Dirmeyer et al. (2014). Over SESA, future PP increases in DJF, MAM and SON with up to $1 \, \text{mm day}^{-1}$ (10–30% of reference PP), and around the border between Argentina, Paraguay and Brazil in SON, the response is larger than $1 \, \text{mm day}^{-1}$. The dry response is strongest in SON and extends over the entire continent north of 20°S. Figure 2(e)–(h) shows that the mean ET response has a similar pattern, although of less magnitude and extension. Over permanent wet regions such as the western Amazon basin or southern Chile, the mean PP decreases without any remarkable changes in mean ET, leading to potential decreased river discharge, similar to the results by Pokhrel et al. (2014) and Dirmeyer et al. (2014). Worth to notice is that even though the annual temperature of the Amazon region increases with 3°–6° (Sánchez et al., 2015), which

Reference climate (1961–1990)

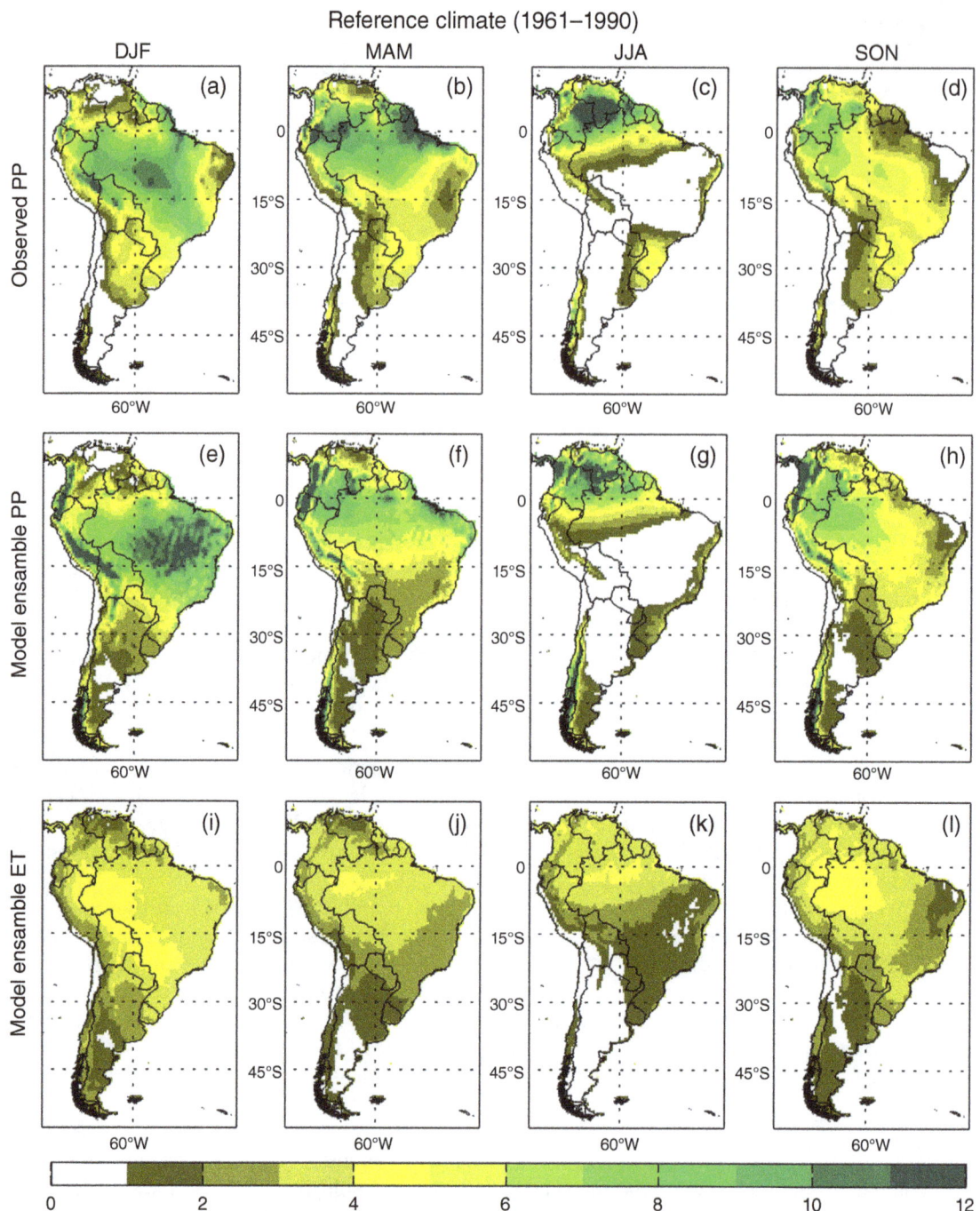

Figure 1. Seasonal means of continental precipitation (PP) and evapotranspiration (ET) during austral summer (DJF), autumn (MAM), winter (JJA) and spring (SON) in the reference climate period (1961–1990). Upper panels (a)–(d): observed PP of the CRU database. Middle panels (e)–(h): ensemble mean PP. Lower panels (i)–(l): ensemble mean ET. Units: mm day^{-1}.

should increase atmospheric demand, the ET does not increase. This could be due to other factors such as decreased wind speed, less insolation due to more clouds or stomata closure due to high temperatures.

Consensus results of land–atmosphere interaction were localized by calculating the number of ensemble members that had a Γ index equal or higher than 0.25. This is shown for the reference climate in Figure 3(a)–(d). There is a clear agreement among members that SESA is a region of land–atmosphere interaction (hotspot) for all seasons, except JJA. The

hotspot is strongest in DJF, where it extends from northern Patagonia in the south to Paraguay in the north and includes Uruguay and southern Brazil. In MAM and SON, the level of inter-member agreement is lower than in DJF, and the hotspot is located more inlands and further north. Overall, this is consistent with Wang *et al.* (2007); Sörensson and Menéndez (2011) and Ruscica *et al.* (2015). In SON, more than 50% of the ensemble members agree on land–atmosphere interaction over NeB. This is reasonable because during the dry JJA (Figure 1(g)), the soil water is depleted over this region,

Response to climate change (2071–2100, 1961–1990)

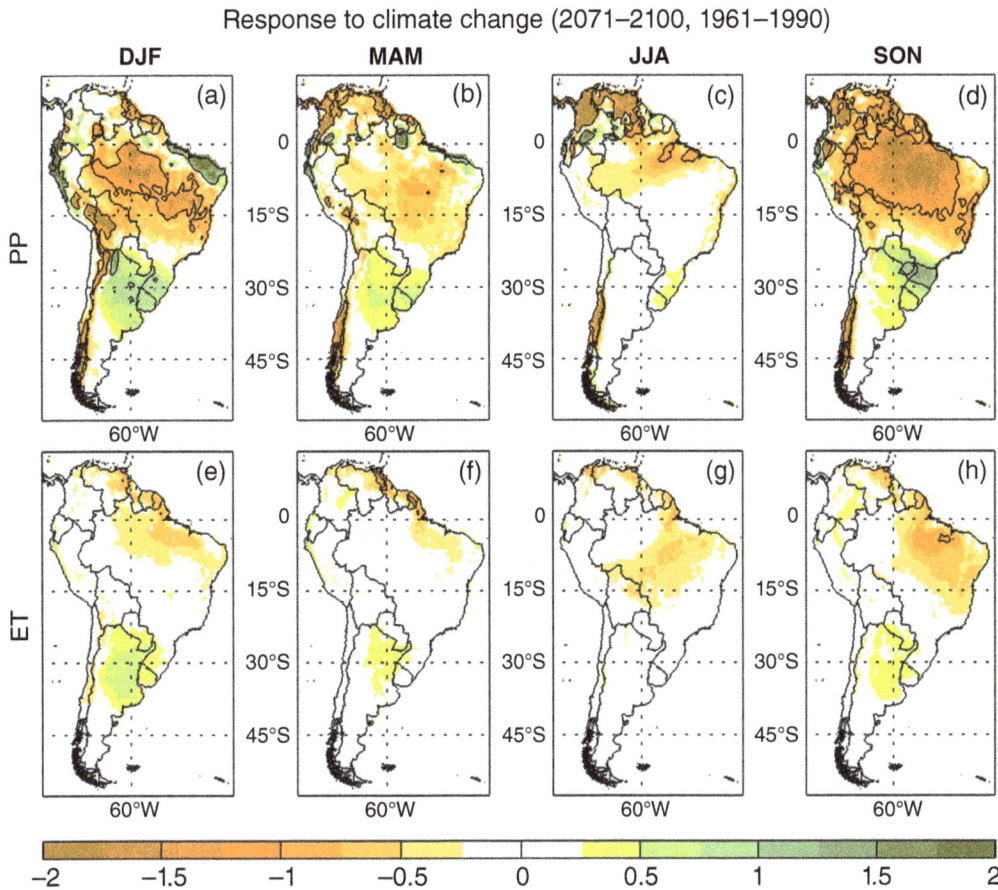

Figure 2. Ensemble response to climate change, defined as the difference of the future and reference climate mean fields (2071–2100 minus 1961–1990). Upper panels (a)–(d): precipitation (PP). Lower panels (e)–(h): evapotranspiration (ET). Units: $mm\,day^{-1}$. Contour lines delimit regions with a climate response of more (less) than $1\,mm\,day^{-1}$ ($-1\,mm\,day^{-1}$).

and when the wet season starts in SON (Figure 1(h)), the SM responds to the PP and ET responds to SM. It should be recalled that the very eastern corner of NeB has a wet PP bias in SON due to the GCM forcing (Figure 1(d) and (h)). During the same season, SON, half of the members show land–atmosphere interaction over eastern Amazonia.

Figure 3(e)–(h) summarizes the size and location of the hotspots in both periods of analysis. In future climate, SESA is still a region of high interaction (orange zones). This result is similar to the study by Dirmeyer et al. (2012), who analyzed the land–atmosphere interaction with a very high-resolution GCM for DJF and JJA for present and future climate. The most notable response to climate change (violet zones) on the continental scale occurs over the Brazilian Highlands and Matto Grosso, where many members agree on strong interaction during SON. This is a consequence of the dryer future climate during this season (Figure 2(h)), transforming the region in a transition zone with better potential conditions for the SM–PP coupling. Over the eastern corner of NeB, the hotspot is lost (green zone) in future climate. This is because Γ is not defined for regions where PP variability is less than $0.5\,mm\,day^{-1}$, and this region loses variability as the mean value decreases (Figure 2(d)). Over the Amazon region in JJA,

it can be seen how the double dry–wet transition zone line (green and orange grid points) migrates to a northern single one (violet grid points). This is consistent with the northward shift of the maximum PP gradient during this season (see Figures 1(g) and 2(c)). Central and eastern Amazonia maintain their energy-limited ET regime in spite of decreased PP (Figure 2(a)–(d)), which is in agreement with the Pokhrel et al. (2014) study that drive a land hydrology model with future scenario from a GCM.

As SESA in DJF is the most important and well-defined hotspot on the continental scale in both climates, Figure 4 summarizes the response of its size and strength over this region. The red region in the Buenos Aires province is a region where seven to eight members agree on interaction in both reference and future climate (strong hotspot). The violet color over northeastern Argentina and Uruguay and the dark blue over southern Brazil mark regions where the land–atmosphere interaction loses in importance. This is the region of strongest wet response in SESA (Figure 2(a)), and as the PP response is higher than the ET response, the SM should also be higher in the future, being coherent with more energy-limited regime. The green color of the periphery of the hotspot is a moderate hotspot in both climates. This region has

Ensemble consensus of land–atmosphere interaction

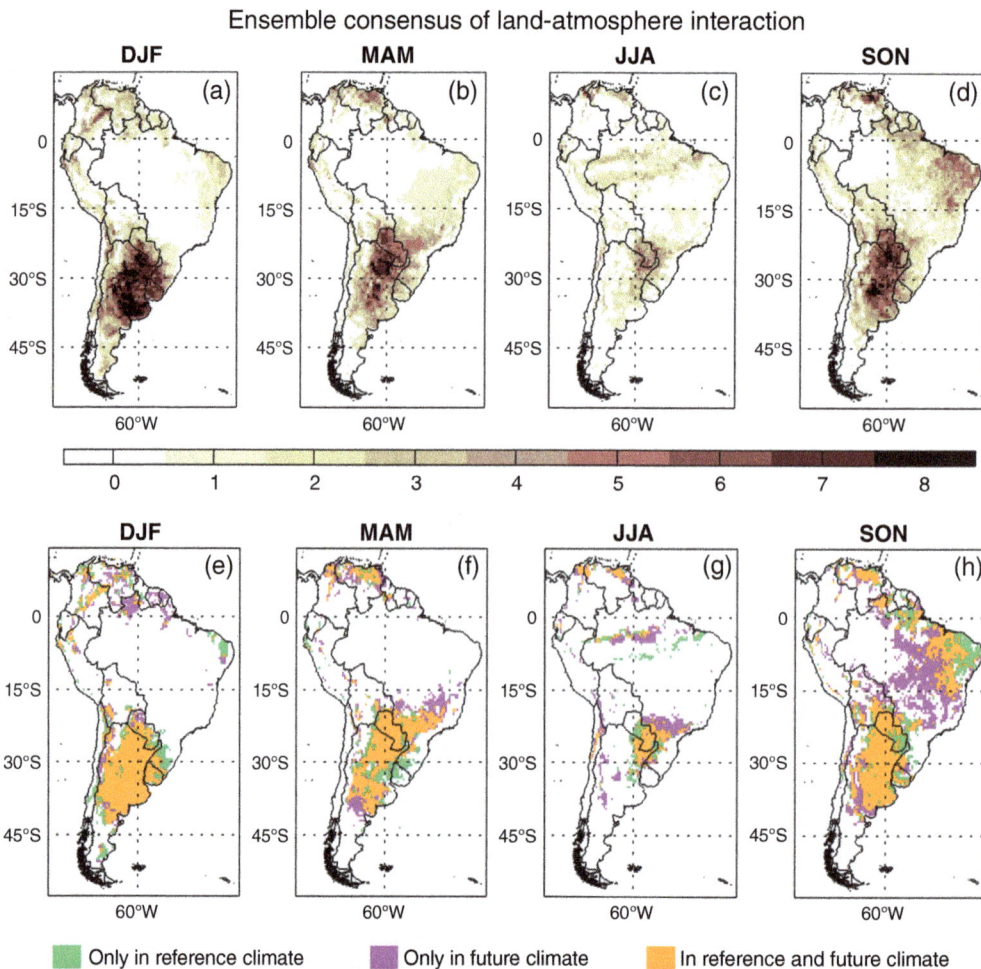

Figure 3. Upper panels (a)–(d): number of ensemble members with land–atmosphere interaction in the reference climate (1961–1990). White land grid points indicate regions without interaction or regions where Γ is not defined (see text for detailed description). Lower panels (e)–(h): extension and location of hotspots of land–atmosphere interaction that exist only in reference climate (1961–1990, green), only in future climate (2071–2100, violet) and that coexist in both climate periods (orange). Hotspot is defined as a region where three or more ensemble members coincide on having land–atmosphere interaction. White continental grid points indicate regions without hotspots in neither period.

a mean response of PP and ET of similar magnitude [less than 0.25 mm day^{-1}, Figure 2(a) and (e)], which is consistent with that the level of SM limitation does not change in the future.

4. Discussion and conclusions

The objectives of this study were to (1) achieve an estimation of land–atmosphere interaction over South America for all seasons and (2) evaluate the response of the interaction to future climate. For this purpose, an index representing land–atmosphere interaction was evaluated for an ensemble of eight RCM–GCM members for both reference and future climate.

The Γ index is advantageous in comparison with others since it is easily computed, does not require ad hoc experiments with climate models and can be used for any consistent ET and PP datasets. However, its main disadvantage is that it does not isolate the causality between two variables (e.g. SM anomalies and

PP anomalies). High Γ values identify regions where land–atmosphere interactions are possible and likely, but, a high correlation between two variables could also be caused by a third one, i.e. a common forcing as SST (Orlowsky and Seneviratne, 2010). However, the results obtained here agree with other modeling studies whose methodologies isolate SM variability as a cause of ET and PP variabilities (Wang *et al.*, 2007; Sörensson and Menéndez, 2011; Ruscica *et al.*, 2014, 2015).

For reference climate (1961–1990), SESA appears as a region of strong land–atmosphere interaction, in particular during austral summer (DJF). This is consistent with other studies (Sörensson and Menéndez, 2011; Dirmeyer *et al.*, 2012; Ruscica *et al.*, 2015), where the most robust hotspot is found during DJF when the variability of ET is high enough to get an effective land–atmosphere interaction. In austral winter (JJA), which is the coldest and driest season in SESA, the region of interaction has a smaller extension and there is less coincidence among ensemble members. However, SESA is simulated too dry in most ensemble

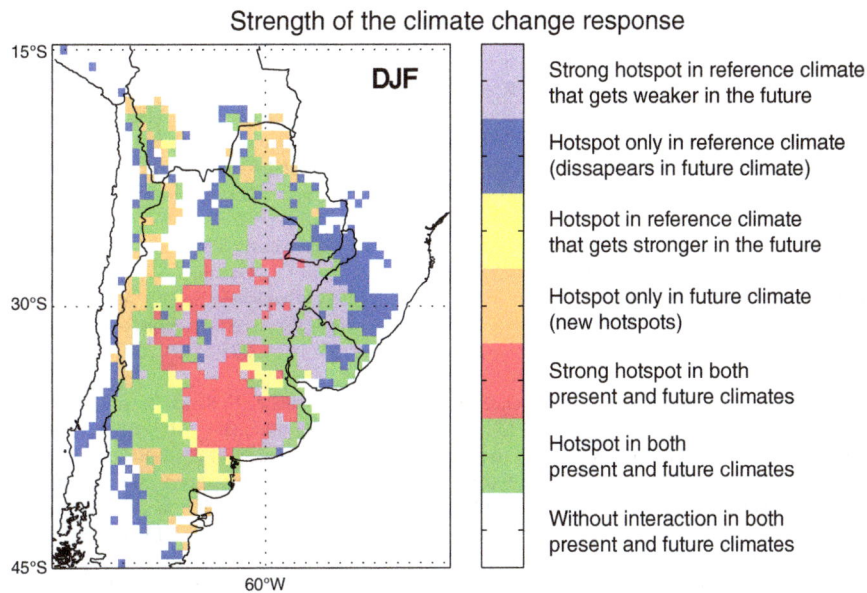

Figure 4. Response of extension and strength of the hotspot over SESA and surroundings in DJF. The definition of hotspot is the same as in Figure 3(e)–(h), and strong hotspot is defined as the agreement of seven to eight ensemble members.

members/seasons, which probably will accentuate the interaction (Ruscica *et al.*, 2015).

More than half of the members agree on strong interaction over NeB and eastern Amazonia during austral spring (SON). In DJF and austral autumn (MAM), the ensemble has a large wet bias over NeB, and therefore it is possible that this region has a strong interaction also in these seasons without being captured by this ensemble.

In future climate, SESA maintains its status as a hotspot, with in general less agreement between members. In regions where PP increases more than ET in DJF, the hotspot loses in strength and extension, while in regions where the response of the two variables is similar, the hotspot maintains its strength. This could be explained by the fact that in the latter case, the region maintains the same level of SM limitation, while in the first case, the ET depends more on atmospheric demand in future climate. In SON, the Brazilian Highlands and Matto Grosso appear as a new, extensive hotspot. This is due to the strong negative response of PP and ET, converting the energy-limited region to a SM-limited region. The dry trend over this region is consistent with the single model results by Sörensson *et al.* (2010). They found that the dry response of SON is associated with a longer dry season in this region and a delay of the monsoon onset. The results here are important because they suggest that this trend can in part be due to, or be amplified by, positive land–atmosphere feedbacks.

Acknowledgements

We acknowledge the European Community's Seventh Framework Program (CLARIS-LPB, Grant Agreement No 212492) and projects LEFE/INSU AO2015-876370 (CNRS, France), PIP No 11220110100932 (CONICET, Argentina) and PICT 2014-0887 (ANPCyT, Argentina).

References

Boulanger JP, Schlindwein S, Gentile E. 2011. CLARIS LPB WP1: metamorphosis of the CLARIS LPB European project: from a mechanistic to a systemic approach. *CLIVAR Exchanges* **16–57**: 7–10.

Dirmeyer PA, Schlosser CA, Brubaker KL. 2009. Precipitation, recycling, and land memory: an integrated analysis. *Journal of Hydrometeorology* **10**: 278–288, doi: 10.1175/2008JHM1016.1.

Dirmeyer PA, Cash BA, Kinter III JL, Stan C, Jung T, Marx L, Towers P, Wedi N, Adams JM, Altshuler EL, Huang B, Jin EK, Manganello J. 2012. Evidence for enhanced land–atmosphere feedback in a warming climate. *Journal of Hydrometeorology* **13**: 981–995, doi: 10.1175/JHM-D-11-0104.1.

Dirmeyer PA, Fang G, Wang Z, Yadav P, Milton A. 2014. Climate change and sectors of the surface water cycle in CMIP5 projections. *Hydrology and Earth System Sciences* **18**: 5317–5329, doi: 10.5194/hess-18-5317-2014.

Domínguez M, Gaertner MA, de Rosnay P, Losada T. 2010. A regional climate model simulation over West Africa: parameterization tests and analysis of land-surface fields. *Climate Dynamics* **35**: 249–265, doi: 10.1007/s00382-010-0769-3.

Eltahir EAB. 1998. A soil moisture-rainfall feedback mechanism. 1.Theory and observations. *Water Resources Research* **34**: 765–776.

Gordon C, Cooper C, Senior CA, Banks H, Gregory JM, Johns TC, Mitchell JFB, Wood RA. 2000. The simulation of SST, sea ice extents and ocean heat transports in a version of the Hadley Centre coupled model without flux adjustments. *Climate Dynamics* **16**: 147–168, doi: 10.1007/s003820050010.

Guo Z, Dirmeyer PA, Koster RD, Sud YC, Bonan G, Oleson KW, Chan E, Verseghy D, Cox P, Gordon CT, McGregor JL, Kanae S, Kowalczyk E, Lawrence D, Liu P, Mocko D, Lu C-H, Mitchell K, Malyshev S, McAvaney B, Oki T, Yamada T, Pitman A, Taylor CM, Vasic R, Xue Y. 2006. GLACE: the global land–atmosphere coupling experiment. Part II: analysis. *Journal of Hydrometeorology* **7**: 611–625, doi: 10.1175/JHM511.1.

Hourdin F, Musat I, Bony S, Braconnot P, Codron F, Dufresne J-L, Fairhead L, Filiberti M-A, Friedlingstein P, Grandpeix J-Y, Krinner K, LeVan P, Li Z-X, Lott F. 2006. The LMDZ4 general circulation model: climate performance and sensitivity to parametrized physics with emphasis on tropical convection. *Climate Dynamics* **27**: 787–813, doi: 10.1007/s00382-006-0158-0.

Jung M, Reichstein M, Ciais P, Seneviratne SI, Sheffield J, Goulden ML, Bonan G, Cescatti A, Chen J, de Jeu R, Dolman JA, Eugster W,

Gerten D, Gianelle D, Gobron N, Heinke J, Kimball J, Law BE, Montagnani L, Mu Q, Mueller B, Oleson K, Papale D, Richardson AD, Roupsard O, Running S, Tomelleri E, Viovy N, Weber U, Williams C, Wood E, Zaehle S, Zhang K. 2010. Recent decline in the global land evapotranspiration trend due to limited moisture supply. *Nature* **467**: 951–954, doi: 10.1038/nature09396.

Koster RD, Dirmeyer PA, Guo Z, Bonan G, Chan E, Cox P, Gordon CT, Kanae S, Kowalczyk E, Lawrence D, Liu P, Lu C-H, Malyshev S, McAvaney B, Mitchell K, Mocko D, Oki T, Oleson K, Pitman A, Sud YC, Taylor CM, Verseghy D, Vasic R, Xue Y, Yamada T. 2004. Regions of coupling between soil moisture and precipitation. *Science* **305**: 1138–1140, doi: 10.1126/science.1100217.

Lin JL. 2007. The double-ITCZ problem in IPCC AR4 coupled GCMs: ocean–atmosphere feedback analysis. *Journal of Climate* **20**: 4497–4525, doi: 10.1175/JCLI4272.1.

Menéndez CG, de Castro M, Boulanger J-P, D'Onofrio A, Sanchez E, Sörensson AA, Blazquez J, Elizalde A, Jacob D, Le Treut H, Li ZX, Núñez MN, Pessacg N, Pfeiffer S, Rojas M, Rolla A, Samuelsson P, Solman SA, Teichmann C. 2010. Downscaling extreme month-long anomalies in southern South America. *Climatic Change* **98**: 379–403, doi: 10.1007/s10584-009-9739-3.

Mitchell TD, Jones PD. 2005. An improved method of constructing a database of monthly climate observations and associated high resolution grids. *International Journal of Climatology* **25**: 693–712, doi: 10.1002/joc.118.

Notaro M. 2008. Statistical identification of global hot spots in soil moisture feedbacks among IPCC AR4 models. *Journal of Geophysical Research* **113**: D09101, doi: 10.1029/2007JD009199.

Orlowsky B, Seneviratne SI. 2010. Statistical analyses of land atmosphere feedbacks and their possible pitfalls. *Journal of Climate* **23**: 3918–3932, doi: 10.1175/2010JCLI3366.

Pokhrel YN, Fan Y, Miguez-Macho G. 2014. Potential hydrologic changes in the Amazon by the end of the 21st century and the groundwater buffer. *Environmental Research Letters* **9**: 084004, doi: 10.1088/1748-9326/9/8/084004.

Roeckner E, Brokopf R, Esch M, Giorgetta M, Hagemann S, Kornblueh L, Manzini E, Schlese U, Schulzweida U. 2006. Sensitivity of simulated climate to horizontal and vertical resolution in the ECHAM5 atmosphere model. *Journal of Climate* **19**: 3771–3791, doi: 10.1175/JCLI3824.1.

Ruscica RC, Sörensson AA, Menéndez CG. 2014. Hydrological links in southeastern South America: soil moisture memory and coupling within a hot spot. *International Journal of Climatology* **34**: 3641–3653, doi: 10.1002/joc.3930.

Ruscica RC, Sörensson AA, Menéndez CG. 2015. Pathways between soil moisture and precipitation in southeastern South America. *Atmospheric Science Letters* **16**: 267–272, doi: 10.1002/asl2.552.

Samuelsson P, Jones CG, Willén U, Ullerstig A, Gollvik S, Hansson U, Jansson C, Kjellström E, Nikulin G, Wyser K.

2011. The Rossby centre regional climate model RCA3: model description and performance. *Tellus A* **63**: 4–23, doi: 10.1111/j.1600-0870.2010.00478.x.

Sánchez E, Solman S, Remedio ARC, Berbery H, Samuelsson P, Da Rocha RP, Mourão C, Li L, Marengo J, de Castro M, Jacob D. 2015. Regional climate modelling in CLARIS-LPB: a concerted approach towards twentyfirst century projections of regional temperature and precipitation over South America. *Climate Dynamics* **45**: 2193–2212, doi: 10.1007/s00382-014-2466-0.

Sellers PJ, Dickinson RE, Randall DA, Betts AK, Hall FG, Berry JA, Collatz GJ, Denning AS, Mooney HA, Nobre CA, Sato N, Field CB, Henderson-Sellers A. 1997. Modeling the exchanges of energy, water, and carbon between continents and the atmosphere. *Science* **275**(5299): 502–509, doi: 10.1126/science.275.5299.502.

Solman SA, Sanchez E, Samuelsson P, da Rocha RP, Li L, Marengo J, Pessacg NL, Remedio ARC, Chou SC, Berbery H, Le Treut H, de Castro M, Jacob D. 2013. Evaluation of an ensemble of regional climate model simulations over South America driven by the ERA-interim reanalysis: model performance and uncertainties. *Climate Dynamics* **41**: 1139–1159, doi: 10.1007/s00382-013-1667-2.

Sörensson AA, Menéndez CG. 2011. Summer soil-precipitation coupling in South America. *Tellus A* **63**: 56–68, doi: 10.1111/j.1600-0870.2010.00468.x.

Sörensson AA, Menéndez CG, Ruscica R, Alexander P, Samuelsson P, Willén U. 2010. Projected precipitation changes in South America: a dynamical downscaling within CLARIS. *Meteorologische Zeitschrift* **19**: 347–355, doi: 10.1127/0941-2948/2010/0467.

Vera C, Silvestri G, Liebmann B, Gonzalez P. 2006. Climate change scenarios for seasonal precipitation in South America from IPCC-AR4 models. *Geophysical Research Letters* **33**: L13707, doi: 10.1029/2006GL025759.

Wang G, Yeonjoo K, Wang D. 2007. Quantifying the strength of soil moisture–precipitation coupling and its sensitivity to changes in surface water budget. *Journal of Hydrometeorology* **8**: 551–570, doi: 10.1175/JHM573.1.

Wei J, Dirmeyer PA. 2010. Toward understanding the large-scale land-atmosphere coupling in the models: roles of different processes. *Geophysical Research Letters* **37**: L19707, doi: 10.1029/2010GL044769.

Zeng X, Barlage M, Castro C, Fling K. 2010. Comparison of land–precipitation coupling strength using observations and models. *Journal of Hydrometeorology* **11**: 979–994, doi: 10.1175/2010JHM1226.1.

Zhang L, Dirmeyer PA, Wei J, Guo Z, Cheng-Hsuan L. 2011. Land–atmosphere coupling strength in the global forecast system. *Journal of Hydrometeorology* **12**: 147–156, doi: 10.1175/2010JHM1319.1.

20

On the link between cold fronts and hail in Switzerland

Sebastian Schemm,[1,2,3]* Luca Nisi,[2,3,4] Andrey Martinov,[2,3,4] Daniel Leuenberger[5] and Olivia Martius[2,3,4]

[1] Geophysical Institute and Bjerknes Centre for Climate Research, University of Bergen, Bergen, Norway
[2] Institute of Geography, University of Bern, Bern, Switzerland
[3] Oeschger Centre for Climate Change Research, University of Bern, Bern, Switzerland
[4] Mobiliar Lab for Natural Risks, University of Bern, Bern, Switzerland
[5] Federal Office of Meteorology and Climatology MeteoSwiss, Zurich, Switzerland

*Correspondence to:
S. Schemm, Geophysical Institute
and Bjerknes Centre for Climate
Research, University of Bergen,
NO-5020 Bergen, Norway.
E-mail:
sebastian.schemm@uib.no

Abstract

Hail is the costliest atmospheric hazard in Switzerland, causing substantial damage to agriculture, cars and buildings every year. In this study, a 12-year statistic of objectively identified cold fronts and a radar-based hail statistic are combined to investigate the co-occurrence of cold fronts and hail in Switzerland. In a first step, an automated front identification scheme, which has previously been designed for and applied to global reanalysis data, is modified for a high-resolution regional analysis data set. This front detection method is then adapted, tested and applied to the Consortium for Small Scale Modelling (COSMO) analysis data for the extended hail season (May to September) in the years 2002–2013. The resulting cold front statistic is presented and discussed. In a second step, the frequency of cold fronts is linked to a high-resolution radar-based hail statistic to determine the relative fraction of hail initiation events in pre-frontal environments. Up to 45% of all detected hail events in north-eastern and southern Switzerland form in pre-frontal zones. Similar fractions are identified upstream of the Jura and the Black Forest mountains. The percentage of front-related hail formation is highest in regions where hail is statistically less frequent, with the exception of southern Switzerland. Furthermore, it is shown that fronts create wind-sheared environments, which are favourable for hail cells.

Keywords: hail; front; climatology; thunderstorm; high impact weather; severe weather

1. Introduction

Hail-producing thunderstorms are a frequent natural hazard in Switzerland during the summer months and pose a challenge to numerical weather models and forecasters. Many hail storms are accompanied by large-scale weather fronts but the precise number of pre-frontal hail storms and its regional variability is unknown. In this study, we use a high-resolution regional analysis to quantify the pre-frontal hail fraction of a 12-year radar-observed hail statistic over Switzerland for the 'extended hail season' May to September (Nisi *et al.*, 2016). Recent studies focused on hail occurrence over central Europe (Suwala and Bednorz, 2013; Punge *et al.*, 2014) but yet the number of pre-frontal hail events has not been quantified.

Research on the ambient state of the atmosphere that is favourable for the formation of severe thunderstorms that can produce hail, convective wind gusts or tornadoes has a long tradition, e.g. Fawbush and Miller (1954); Houze *et al.* (1993); Willemse (1995); Schiesser *et al.* (1997); Huntrieser *et al.* (1997); Rasmussen and Blanchard (1998); Brooks *et al.* (2003); Groenemeijer and Delden (2007); Naylor and Gilmore (2012); Grams *et al.* (2012); Manzato (2012); Mohr and Kunz (2013); Gensini *et al.* (2014); Allen and Karoly (2014); Merino *et al.* (2014). Such studies often involve the analysis of proximity-soundings, i.e. soundings obtained in

close spatial and temporal proximity to such an event. Proximity soundings allow for the identification of the ambient conditions conducive to the development of severe thunderstorms (Brooks, 2009). These include high values of convective available potential energy (CAPE; Groenemeijer and Delden, 2007), low convective inhibition (CIN; Davies, 2004) and strong vertical wind shear between approximately 0–6 km (Weisman and Klemp, 1982; Doswell and Evans, 2003; Kaltenboeck and Steinheimer, 2015). Also important are a reduced mid-tropospheric lapse rate, a high mixing ratio and high storm relative helicity (Droegemeier *et al.*, 1993). A variety of derived indices, e.g. the significant hail parameter (SHIP) (details are provided at http://www.spc.noaa.gov/sfctest/help/help_sigh.html) or the SWISS index (Huntrieser *et al.*, 1997), have been developed based on empirically or statistically derived combinations of the above-mentioned quantities and others (Manzato, 2013). However, proximity soundings are only representative for the state of the lower troposphere in close vicinity (50–100 km; Haklander and Delden, 2003) to the measurement site and can only provide limited information about the synoptic-scale setting.

The synoptic-scale flow situation can be studied using surface weather charts, radar information, satellite data and analysis data (Towery and Changnon, 1970; Doswell, 1987). Indeed, the role of synoptic-scale forcing and especially of cold-fronts is important for hail storms, as early investigations of radar data have demonstrated. Studies such as Towery and Changnon (1970) showed that hail cells associated with fronts live longer and move faster, and that a substantial number of hail cases (more than 50%) are linked to frontal features. In a number of case studies, the role of cold fronts for the initiation and life-cycle of severe convection in North America was discussed in detail (e.g. Ogura and Portis, 1982; Heymsfield and Schotz, 1985; Doswell, 1987; Locatelli *et al.*, 1995; Neiman and Wakimoto, 1999). The role of synoptic-scale forcing is mainly to 'set the scene' for severe convection by contributing to a moistening of the lower troposphere through low-level moisture advection and by reducing existing capping inversions through upper-level cold air advection (Doswell, 1987). In addition, cold fronts are important for (1) the triggering of convection (Ogura and Portis, 1982), (2) the upward advection of moisture by the vertical frontal circulation and the initiation of convection through the vertical circulation (Neiman and Wakimoto, 1999) and (3) the formation of convergence lines that trigger the convection (Heymsfield and Schotz, 1985). Alberoni *et al.* (2000); Costa *et al.* (2001) and Giaiotti *et al.* (2003) discuss the role of synoptic-scale forcing and in particular the role of cold fronts associated with severe convection and hail in Northern Italy. Both Costa *et al.* (2001) and Giaiotti *et al.* (2003) point out the importance of the interaction of the large-scale driven flow with the local orography for the formation of hail-producing thunderstorms. This interaction results in important triggers for convection, namely convergence zones associated with cold fronts and the destabilization by the cold advection aloft.

2. Outline and objectives

In this study, we extend the case studies cited above that focused on a limited number of cases (Ogura and Portis, 1982; Heymsfield and Schotz, 1985; Doswell, 1987; Locatelli *et al.*, 1995; Neiman and Wakimoto, 1999) by investigating a larger sample of hail events (~14 000) that occurred in Switzerland and adjacent regions during the extended hail season between 2002 and 2013. The main objectives of this study are:

1 To apply an automated front recognition algorithm to an analysis data set with high spatial and temporal resolution (Steppler *et al.*, 2003; Jenkner *et al.*, 2010). The front algorithm and radar-based hail detection are discussed in Section 3; the front frequency for Switzerland is presented in Section 4.

2 To link the front data to a radar-based data set of hail initiations, *defined as the time when hail is first detected in a thunderstorm cell by the radar*, and to quantify the fraction of pre-frontal hail formations in different regions of Switzerland (Section 5).

3 To characterize the temporal evolution of the vertical wind shear in pre-frontal zones prior to hail formation (Section 6).

3. Methods

3.1. Front detection

The automated detection algorithm for synoptic-scale fronts in the Consortium for Small Scale Modelling (COSMO) analysis data set is based on the work of Jenkner *et al.* (2010) and Schemm *et al.* (2015) and has been shown to be useful for the automated recognition of synoptic-scale fronts in a global reanalysis (Schemm and Sprenger, 2015; Schemm *et al.*, 2015) and in the high resolution COSMO-7 reanalysis data (Jenkner *et al.*, 2010). In this study, the methodology and specific complications arising from its application to a 2 km data set are outlined. The COSMO analysis data set is operationally produced by Meteo Swiss and encompasses approximately 2°–17°E, 42°–50°N (more details at: http://www. cosmo-model.org/content/tasks/operational/meteo Swiss/). The hourly model output is provided with a 6.6 km grid-spacing until 2007 (COSMO-7) and with a 2.2 km grid-spacing (COSMO-2) since 2007 (Steppler *et al.*, 2003; Jenkner *et al.*, 2010). Because the frequency of large-scale weather fronts in the limited area weather predication COSMO model (which is driven by ECMWF forecast fields) is not affected by the increase in the spatial resolution in 2007, and because we do not make use of more sensitive information such as frontal gradients, we decided to use the full time period 2002–2012.

The identification strategy focuses on the localization of the thermal front parameter (TFP; Hewson, 1998), which is defined as

$$ \mathrm{TFP} = -\nabla |\nabla \theta_e| \cdot \frac{\nabla \theta_e}{|\nabla \theta_e|}, \qquad (1) $$

where θ_e denotes the equivalent potential temperature. The parameter describes 'the gradient of the magnitude of the gradient of a thermodynamic scalar quantity (here θ_e), resolved into the direction of the gradient of that quantity' (Renard and Clarke, 1965). In other words, the change in the gradient of θ_e is projected along the normal component of the front (perpendicular to the frontal zone) in order to take the curvature of a front into account. The TFP is used to identify the location of the maximum of $|\nabla \theta_e|$ in narrow zones of strong thermal gradients, which are referred to as frontal zones or frontal areas. After a series of manual tests in the COSMO-2 data, we set the threshold for the frontal zone to $|\nabla \theta_e| > 6$ K $(100 \, \mathrm{km})^{-1}$ at 700 hPa. Note that previous studies use weaker thresholds, e.g. 4 K $(100 \, \mathrm{km})^{-1}$ by Schemm *et al.* (2015), which is primarily due to the lower resolution of the data sets used in their analyses. Inside a frontal zone, the actual front is located where the TFP is zero (see Figure 1). To test the sensitivity of the frontal frequencies to the choice of the minimum frontal length, the frontal frequencies obtained with a minimum length of 300 and 500 km are discussed (Section 4). The TFP in this study is evaluated at 700 hPa, because the traditional altitude of

850 hPa (Hewson, 1998; Berry *et al.*, 2011, and references therein) lies below most of Switzerland's high topography. For further details and technical aspects regarding the implementation, testing and validation of the algorithm, the reader is referred to Schemm *et al.* (2015) and Jenkner *et al.* (2010).

The high-resolution data set used in this study poses some additional challenges compared to the previous application of the front detection algorithm:

- To overcome the large gridpoint-to-gridpoint variability in a high-resolution data set, and to avoid the misidentification of localized noise as fronts, a diffusive smoothing is applied to θ_e using a Laplacian filter (see Jenkner *et al.*, 2010).
- In general, the fronts are very thin lines (where TFP = 0, see also Figure 1); the slightly broader frontal zones ($|\nabla\theta_e| > 6\,\mathrm{K}\,(100\,\mathrm{km})^{-1}$) around the detected fronts are therefore used for the calculation of the frontal frequencies.

Finally, we note that frontal identification strategies based on thermal gradients frequently identify thermal boundaries over the Alps, especially during the late evening, that are formed by Alpine pumping (Weissmann *et al.*, 2005). As we are mainly interested in synoptic-scale fronts, we restrict our study to cold fronts and require all identified fronts to live for at least 3 h. Due to the hourly resolution of the data set, an overlap criterion between consecutive time steps is sufficient to determine the lifetime of every front. No frontal tracking algorithm is required. The front type is determined by defining a wind component v_f perpendicular to the front:

$$v_f = \mathbf{v} \cdot \frac{\nabla \mathrm{TFP}}{|\nabla \mathrm{TFP}|}, \qquad (2)$$

This advection criterion allows the scheme to distinguish cold ($v_f > 0$) from warm ($v_f < 0$) fronts (where $\mathbf{v}$ denotes the horizontal wind). The requirements introduced above (i.e. a minimum lifetime of 3 h, limitation to cold fronts and minimum length) remove the majority of thermal fronts induced by Alpine pumping. The introduction of additional criteria to remove thermal fronts, e.g. a minimum advection threshold, the time of detection or the location above the Alpine ridge, did not improve the algorithm.

The main differences between the fronts used in this study and the fronts used by Jenkner *et al.* (2010) are the application of the length criterion, the requisite minimum lifetime and the restriction to cold fronts. In addition, the frontal threshold used in this study is slightly higher due to the increased spatial resolution of the data.

3.1.1. Front detection – an illustrative example

To depict a typical synoptic situation with a cold front and pre-frontal hail formation on the northern side of the Alps, an example from the first of July 2012 at 0600 UTC is presented in Figure 1. The synoptic situation was dominated by a long-wave quasi-stationary trough over the East Atlantic and Western Europe. The Alpine area was located in a warm sector ahead of a cold front that moved eastward across Central Europe. Because of Foehn winds from the south, the northern side of the Alps was initially rather dry. However, a moderate mid-level south-westerly flow brought very warm, moist and unstable air over Central Europe, creating an environment with high potential instability and therefore conditions favourable to severe thunderstorm development. The first thunderstorms developed during the late evening hours, and the northern Alpine area was affected later that night by severe hailstorms (not shown).

The flow at 700 hPa is south-westerly (Figure 1(a)) and the cold front is detected in the north-western part of the domain over France. The propagation of the front is slow: over the next 2 days, the cold front remains almost stationary. The radar reflectivity pattern indicates a precipitation band with an orientation similar to that of the cold front located approximately 400 km downstream of the cold front. Noticeable is the hail cell that is detected in north-eastern Switzerland at 47.2°N and 8.7°E with a radar reflectivity of ~60 dBZ.

In Figure 1(b), the same situation is shown at 850 hPa. The cold front is again identified in the western part of the domain but is located slightly more to the east. This can be expected from the characteristic vertical tilt of the front. In addition to the mature cold front, a second front located at the centre of the main reflectivity pattern is identified with strong wind convergence. Such a pre-frontal convergence line is a characteristic that is frequently observed in conjunction with strong pre-frontal thunderstorm activity (Newton, 1950; Rotunno *et al.*, 1988; Meischner *et al.*, 1991; Delden, 1998; Schultz, 2005). From the kinematic perspective of temperature advection, the convergence line is identified as a cold front by our front identification method.

The shape of the frontal areas at 850 hPa is strongly affected by the latent heat released during the formation of cloud droplets. In addition, the 850 hPa surface intersects the ground in many areas, as illustrated in Figure 1(b) by areas with no wind vectors. This underpins the decision to use the fronts as detected at 700 hPa, even though they might be located relatively far from the centre of precipitation. Furthermore, convergence lines are detected more rarely at 700 hPa than at 850 hPa.

For the linkage between hail initiations and approaching cold fronts to be dynamically meaningful, the frontal zone must be located upstream of the hail cell (see Figure 1). Due to the small domain size, all fronts upstream of hail cells are considered in the analysis and no maximum distance criterion is used (the average distance between hail formation and a front is around 200 km; the median distance is 130 km).

3.2. Radar-based hail detection

Radar-based detection of the probability of hail (POH) provides an indirect measure of the occurrence of hail

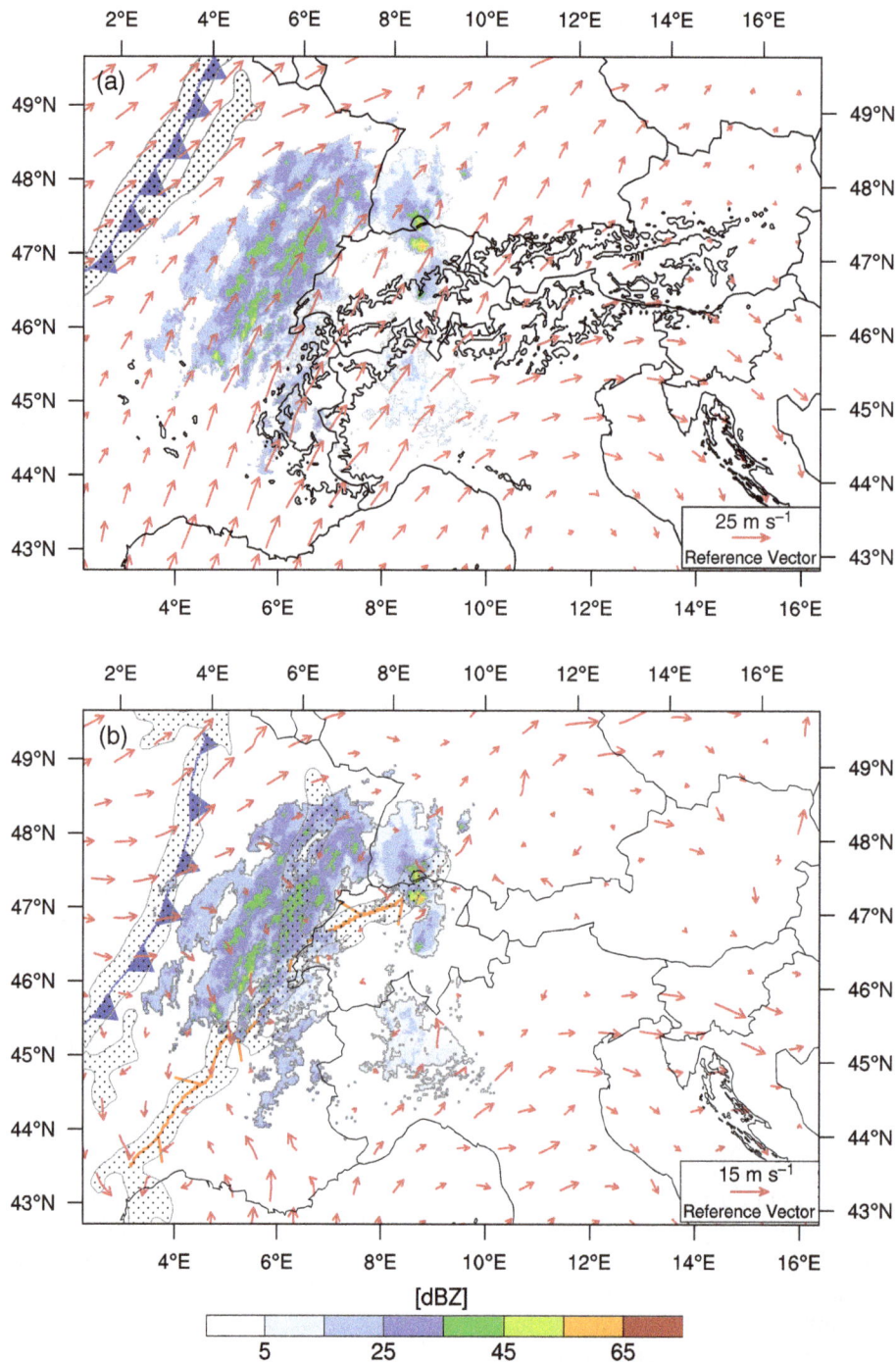

Figure 1. An illustrative example of fronts detected at two different altitudes in COSMO-2 and radar reflectivity at 0600 UTC 1 July 2012. Cold fronts (blue) and wind vectors are shown at (a) 700 hPa and (b) 850 hPa. A pre-frontal convergence line (orange) at 850 hPa aligns with the area of highest reflectivity. Also shown are the frontal zones around each front (stippled lines). Note that the hail cell (~60 dBZ) in north-eastern Switzerland is located on the north-eastern tip of the 850 hPa convergence line. The 1500 m (a.s.l.) topography is also shown in (a).

at the surface. The 3-D radar-based POH statistic of Nisi *et al.* (2016) is based on the difference between the 45-dBZ radar echo top height, i.e. the maximum altitude of the 45-dBZ echo (ET^{45}), and the height of the melting layer (H^0). In this study, H^0 is taken from the COSMO analysis. An empirical relationship, originally derived by Waldvogel *et al.* (1979) from field experiments, relates the height difference between ET^{45} and H^0 in a cloud to the POH. We use the modified

empirical relationship proposed by Foote *et al.* (2005), which has been in operational use at MeteoSwiss since 2008. For values below 1.65 km, POH is assumed to be zero. For values greater than 5.8 km, POH is assumed to be 100%. The fit for intermediate values is given following Foote *et al.* (2005). The relationship was used to reprocess POH for the extended hail season over Switzerland for the period between 2002 and 2013. For every 5-min time step, POH is calculated with a spatial resolution of 1×1 km. Refer to Chapter 3 in the article of Nisi *et al.* (2016) for more details regarding the quality of the hail detection over complex terrain, high-resolution hail statistics and the calibration data. For a detailed description of the Swiss radar network, which consists of three radar stations in western (La Dôle; 6.1°E, 46.4°N; 1680 m a.s.l), north-eastern (Albis; 8.5°E, 47.2°N; 928 m a.s.l) and southern (Monte Lema; 8.8°E, 46.0°N; 1625 m a.s.l.) Switzerland, the reader is referred to the study of Germann *et al.* (2006) and Joss *et al.* (1998). In Figure 2(a), the radar stations are located near Geneva, Zurich and Locarno.

The life cycle of individual thunderstorms and hail cells can be investigated using the radar tracking algorithm Thunderstorm Radar Tracking (TRT) by Hering *et al.* (2004). The underlying criteria for the detection of a convective hail cell in the data set of Nisi *et al.* (2016) are

- a minimum convective cell lifetime of 15 min and
- at least 5 pixels in each cell with POH values greater than 90% for at least 5 min.

A total of 14 000 convective hail cells are identified in the 12-year study period. The locations of hail initiation in the convective cell represent the data base of this study. We henceforth refer to the location of the first detection of a hail signal as the location of hail initiation.

4. Frequency of front detection

Switzerland is commonly affected by eastward propagating synoptic-scale disturbances that are embedded in the mean westerly flow. Cold fronts typically approach Switzerland from the northwest (75%) and only rarely from the southeast (Jenkner *et al.*, 2010). During summer, the largest fraction of fronts that do not propagate into Switzerland from the northwest instead approach the Alps from the southwest (Jenkner *et al.*, 2010). The Alps strongly affect the speed and direction of travel. The distortion of the flow field leads to the slowing and deformation of the fronts (Davies, 1984; Schumann, 1987).

In this section, the frequencies of cold fronts with two different minimum lengths are discussed. The climatology is presented in terms of an occurrence rate (%) of the frontal zones at every gridpoint, which is obtained by a time average of binary frontal fields. To obtain this average, gridpoints containing fronts are labelled with a one; those without are labelled with a zero. A time average over all time steps results in a front frequency. The results presented here show the entire COSMO-2 domain; in all later sections, only a sub-domain around Switzerland is shown to account for the decrease in the quality of radar detection.

The detection frequencies of fronts with minimum lengths of 500 km are shown in Figure 2(a). The obtained frontal frequency highlights regions where fronts typically slow down or become stationary, i.e.

upstream of mountain crests, before they become deformed or split. Fronts are frequently identified upstream of the main Alpine ridge (up to ~6%, Figure 2(a)), as well as upstream of the Jura Mountains in north-western Switzerland (~4%) and on the southern side of the Alps (~5%). The detection rates decrease towards the boundaries of the domain due to the fact that fronts that have just entered the domain or are leaving the domain do not fulfil the length criterion.

The front frequencies for fronts with a minimum length of 300 km (Figure 2(b)) show higher detection rates over the Alpine foothills (~13%). This can be explained by a type of quasi-stationary front that is mainly due to diurnal circulations and that is less frequently detected with the 500 km length criterion. For example, a strong increase in frontal detection rates is found in the south-western part of the Alps, forming a bow-shaped pattern that extends southward towards the Mediterranean, that is much less pronounced in Figure 2(a). This finding agrees with the results of Jenkner *et al.* (2010) and their climatology of meso-scale fronts. They find enhanced detection rates of thermally induced meso-scale fronts along the border between France and Italy, along the upper Rhône Valley and along the French Alps towards the Mediterranean coastline. These fronts are mainly driven by the diurnal evolution of the planetary boundary layer in this region. Moreover, Jenkner *et al.* (2010) point out the pronounced effect of the thermal circulation on the detection of meso-scale fronts along the Mediterranean coastline. In this area, sea breezes contribute to the break up of the inversion layer and can hence assist the growth of localized thermal fronts (Bastin *et al.*, 2005). From the kinematic perspective, these fronts may, according to Equation (2), be classified as cold fronts and hence become part of our front data set, if they also meet the minimum length criterion of 300 km. Based on these results, we decided to require the cold fronts in this study to have a minimum length of 500 km.

5. Statistics of pre-frontal hail formation

Turning to the fraction of pre-frontal hail initiations, a spatially smoothed hail cell initiation statistic is presented in Figure 3(a). The southern parts of Switzerland and the southern Prealps, i.e. the Alpine foothills, are hail initiation hotspots (Figure 3(a); see Punge *et al.*, 2014; Mohr *et al.*, 2015). North of the Alps, western Switzerland, especially along the Jura mountains and the pre-Alpine areas south of Bern, shows an increased occurrence of hail cell initiations. The hotspot region in south-western Germany is a well-known area of frequent thunderstorm activity (Kunz and Puskeiler, 2010). Note that areas of high topography (>1500 m a.s.l.), where the total number of detected hail cells is low, are masked. A more detailed discussion can be found in the study of Nisi *et al.* (2016).

Figure 3(b) presents the relative percentage of all hail initiations occurring downstream of a synoptic-scale

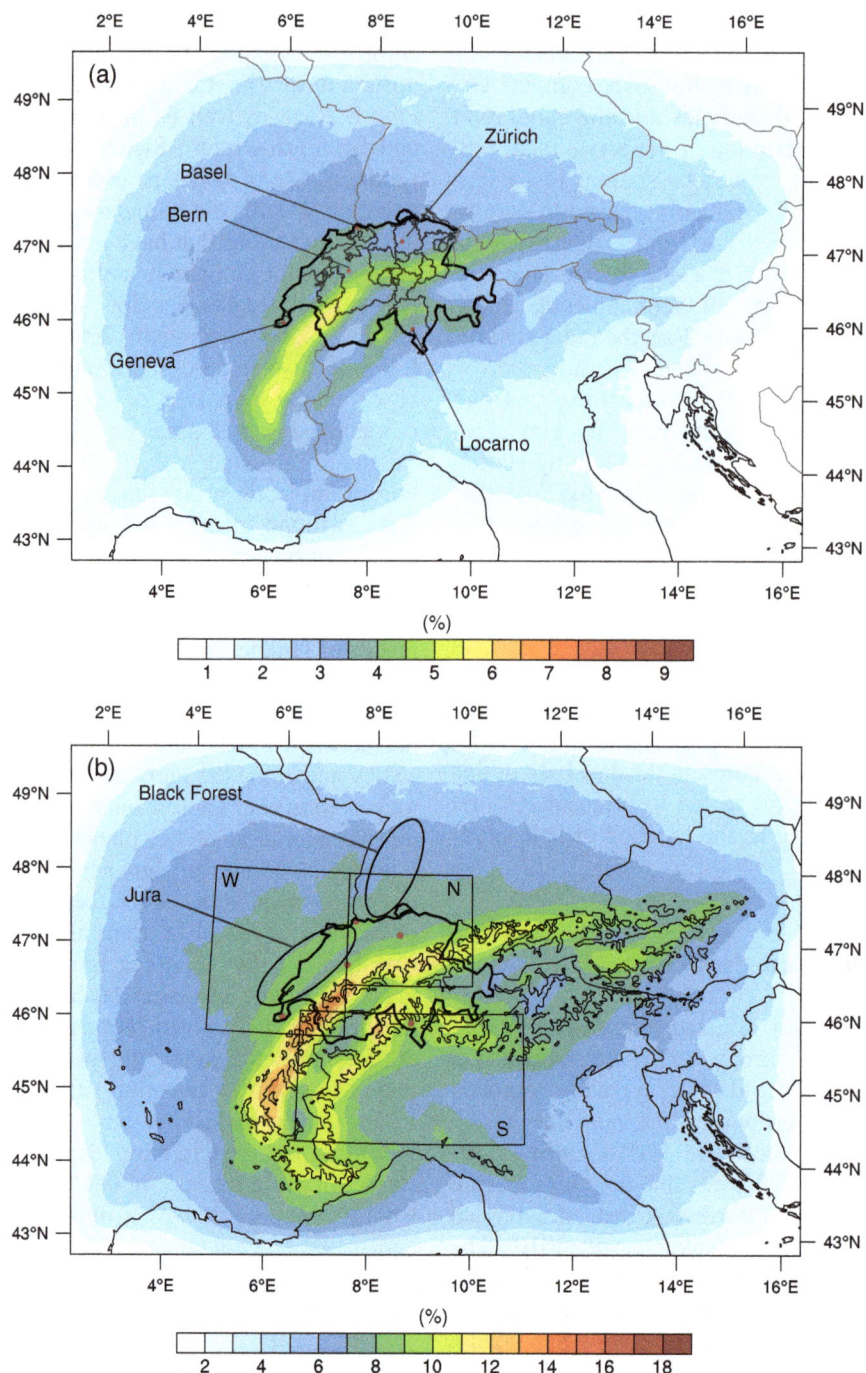

Figure 2. Detection rates of frontal zones with colds fronts with minimum lengths of (a) 500 km and (b) 300 km for the period between 2002–2013 (May to September). The entire COSMO-2 domain is shown, as well as (a) the Swiss administrative areas and five major cities and (b) the 1500 m COSMO-2 model topography (a.s.l).

cold front, i.e. the frontal gridpoint nearest to the hail initiation must be located to its west. Due to the limited domain size, all hail initiations downstream of a cold front are taken into account and no maximum distance limit is set. In the region upstream (windward side) of the Jura Mountains, we find that up to 40–45% of all hail initiations are associated with a cold front to their west. In the Jura Mountains, however, where a high fraction of pre-frontal hail formations are found upstream, a local minimum is identified (<20%). This dipole pattern around an orographic obstacle

is also found in south-western Germany, where a minimum in the pre-frontal hail fraction is identified over the Black Forest (<25%) and 30–40% of all pre-frontal hail initiations are formed to the west of the Rhine valley. In north-eastern Switzerland, around Basel and Zurich, more hail initiations are associated with fronts (30–40%) as compared to western Switzerland (20–30%). In western Switzerland, hail formation is frequently due to air mass convection in south-westerly flow situations (Nisi et al., 2016). For southern Switzerland, we identify fronts in up to

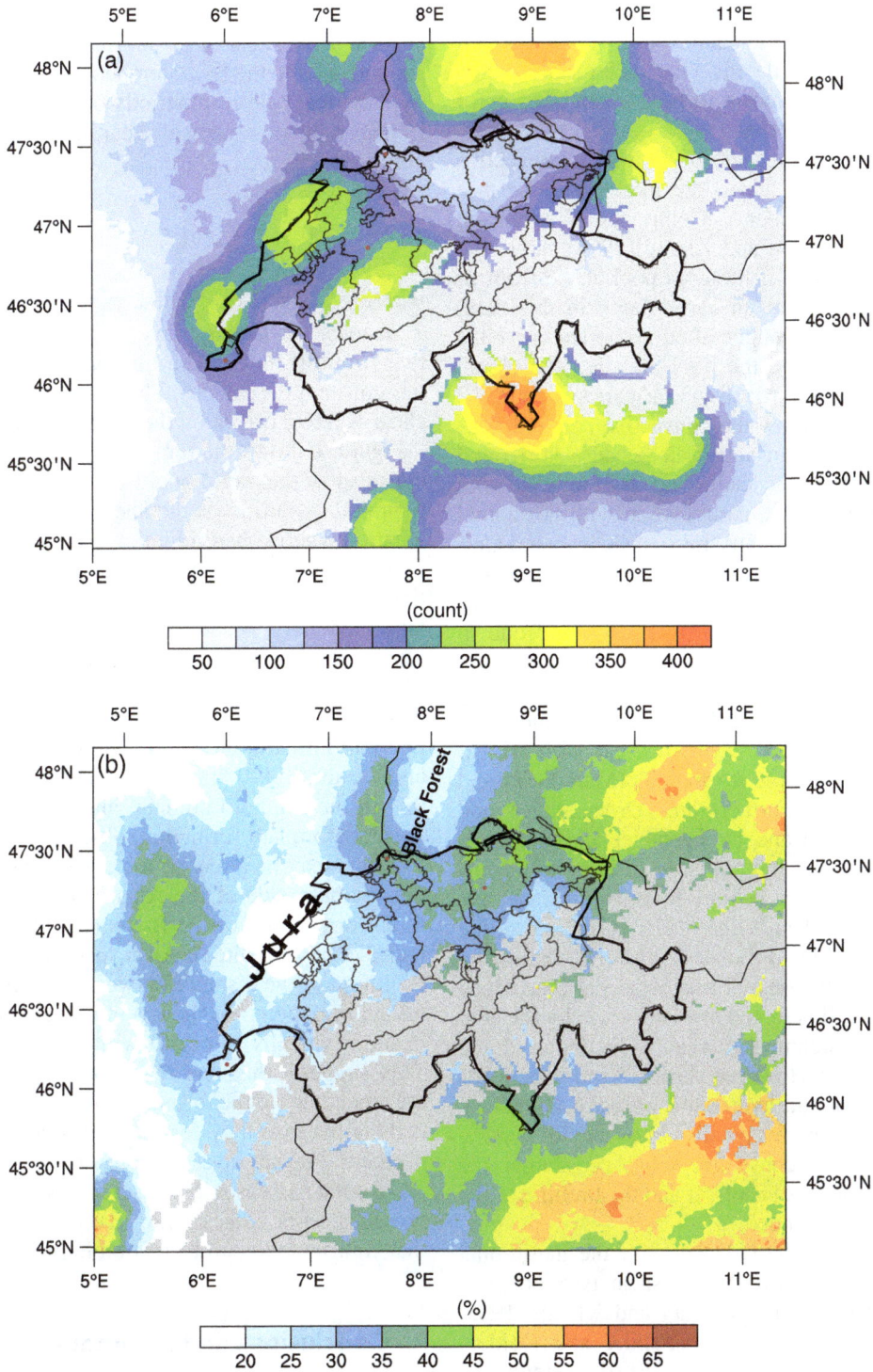

Figure 3. Shown is (a) the smoothed geographical distribution of detected hail cell initiations between 2002 and 2013 (May and September) and (b) the relative fraction of events with a synoptic-scale cold front located to the west. Regions above 1500 m (a.s.l) are masked. The Jura and Black Forest sub-Alpine mountains are labelled.

40% of all hail cases. Although there appears to be a west–east gradient in the percentage of hail initiations with associated synoptic-scale fronts, the local variability over the domain suggests that the eastward increase is not due to the influence of the limited domain size. Although hail occurs statistically less frequently in pre-frontal zones, these findings suggest that hail occurs primarily in pre-frontal zones, with the

exception of Ticino on the southern side of the Alps (see Figure 3(a) and (b)).

6. Vertical shear during pre-frontal hail formation

So far, we have focused on the statistical relationship between fronts and the formation of hail. In the following section, we touch briefly on the potential physical mechanisms that make fronts a key ingredient for the formation of hail cells. One such mechanism is wind shear, which is known to be a crucial factor for long-lived convective cells.

Shear is able to extend the lifetime of a convective cell and multiple-cell storms are likely to occur in a wind-sheared environment. Klemp and Wilhelmson (1978) and Wilhelmson and Klemp (1978) show that convective cells decay if the cold pool at the front of the cell moves away at an early stage. In this case, moist warm air is no longer lifted at the leading edge of the cold pool above the pool and into the cloud. If there is sufficient vertical wind shear; however, the cold pool remains below the convective cell and adds to the vertical lifting of warm air because horizontal vorticity generated at the leading edges of the cold pool forces vertical motion in combination with horizontal ambient vorticity of the opposite sign (Knupp and Cotton, 1982; Rotunno and Klemp, 1985; Rotunno et al., 1988; Weisman et al., 1988; Weisman and Rotunno, 2004; Schlemmer and Hohenegger, 2014) or due to buoyancy forcing by evaporation at the cold pool boundaries (Khairoutdinov and Randall, 2006; Böing et al., 2012; Torris et al., 2015). These mechanical and thermodynamic cold-pool dynamics can also trigger the formation of multiple storm cells (Markowski and Richardson, 2010). Furthermore, Droegemeier and Wilhelmson (1985a) and Droegemeier and Wilhelmson (1985b) describe the collision of two cold pools as driving mechanism behind long-lived convective clouds. For further details, the reader is referred to the study of Markowski and Richardson (2010). Following the arguments presented by Emanuel (1994), the decay of a convective cell is forced at a later stage by the cold pool if updrafts become too weak to lift the cold and dense air into cloud. However, vertical wind shear leads to the formation of non-hydrostatic pressure anomalies, which are a surrogate for positive buoyancy or heating anomalies inside the convective cell. Accordingly, the forcing by the cold pool is offset and the lifetime of the convective cell is extended.

Here we focus on the bulk shear of the meridional wind component (Figure 4). The shear is often measured between 0–6 km (Weisman and Klemp, 1982; Doswell and Evans, 2003; Haklander and Delden, 2003; Groenemeijer and Delden, 2007; Kaltenboeck and Steinheimer, 2015), but because our focus lies not on the actual quantification of the forced lifting but rather on a qualitative assessment of the shear conditions in a broader pre-frontal zone, we consider 700 hPa (which is the pressure level at which fronts are identified) and the 850 hPa level. In the *semi-geostrophic* approximation, the thermal wind balance implies a vertical wind shear of the *along-frontal* wind component proportional to the *across-frontal* temperature gradient (Holton, 2004). If we consider a broader zone around the frontal lines (±500 km) that encompasses the pre-frontal and post-frontal areas, we may argue that the geostrophic approximation is also justified. In

the *geostrophic* case, thermal wind balance implies a vertical wind shear of the *meridional* wind component proportional to the *zonal* temperature gradient, i.e. for a hypothetical cold front strictly oriented north–south, it implies a positive vertical shear of the meridional wind component:

$$ v_T = v\left(p_1\right) - v\left(p_0\right) = \frac{R}{f} \ln\left(\frac{p_0}{p_1}\right) \left(\frac{\partial \overline{T}}{\partial x}\right)_p . \quad (3) $$

where p_0 is set to 850 hPa, p_1 is set to 700 hPa and $\overline{T}$ denotes the average layer temperature between the two limits (f is the Coriolis parameter and R is the gas constant for dry air). The temporal evolution of the shear is likely driven by the approaching cold fronts.

Figure 4 shows the temporal evolution of the vertical shear of the meridional wind at every hail site. The temporal evolution of the shear is calculated at every location where a hail initiation is detected by the radar. Afterwards we average over all time series with hail initiation in western, north-eastern and southern Switzerland separately. We additionally separate cases in western, north-eastern (northern) and southern Switzerland (domains are shown in Figure 2(b)). However, the vertical shear of the meridional wind increases prior to the first hail initiation in all three domains. In the northern and western sub-domain, this increase is stronger than that in the southern domain, and reaches a maximum 2–4 h before hail formation. Shielding of the fronts by the Alps may be a reason for a weaker shear signal on the southern side of the Alps or a systematically different orientation of the fronts (i.e. more zonally than meridionally oriented). An inspection of the temporal evolution of the distance between the front and the site of hail initiation (not shown) indicates that the increase in shear typically co-occurs with the approaching front. This suggests that one reason why pre-frontal zones are amenable for the formation of long-lived convective cells is because they form a large-scale sheared environment. Further investigations are needed to compare pre-frontal hail cases to front-free cases, but the analysis provides a first indication of the physical mechanisms that contribute to hail formation in pre-frontal zones.

7. Conclusions and summary

The main objective of this study was to investigate the co-occurrence of cold fronts and hail over Switzerland using a large sample (14 000) of hail events for the period between 2002 and 2012 (May–September). To this end, an automated front identification algorithm is applied to a high-resolution regional analysis data set (COSMO-2/-7) covering the Alpine ridge and the front data are linked with radar-based hail signals. Based on this information, we present a statistic of pre-frontal hail initiations.

The front algorithm was successfully adapted to operate on a grid spacing of 2.2 km and performs well in the identification of synoptic-scale cold fronts on 700 hPa.

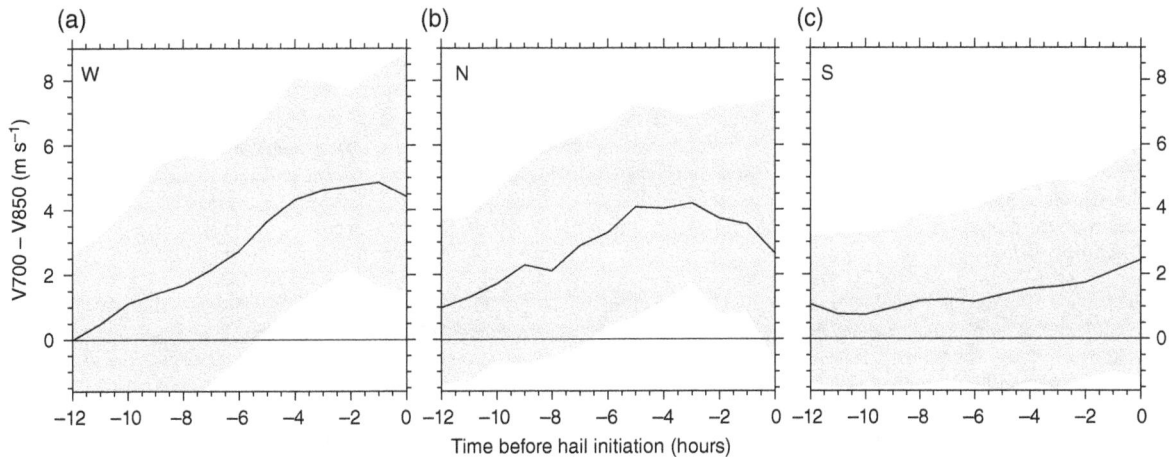

Figure 4. Temporal evolution of meridional bulk wind shear between the 700 and 850 hPa levels (m s^{-1}) prior to hail initiation in (a) western, (b) north-eastern and (c) southern Switzerland. Shown is the median (black) and the interquartile range (25–75% percentile, grey shading).

The obtained 12-year frontal frequencies show maxima in areas where fronts typically become stationary along the Alpine foothills.

We identify regions where the fraction of pre-frontal hail initiations is particularly high. Locally, the fraction increases to up to 40–45%, *particularly in areas where the total number of hail initiations is small*, i.e. in north-eastern Switzerland. This finding suggests that hail initiations in these areas require fronts as an additional agent to force intense convection. Elsewhere orographic forcing is potentially strong enough to provide enough lifting to trigger hail storms. On the southern side of the Alps, the Ticino region is an exception. There, hail occurs frequently and is associated with pre-frontal conditions in 40% of all cases. On average, the distance between the front line and the point of hail initiation is around 200 km, with a median value of ~130 km.

Finally, we examined the temporal evolution of the vertical shear of the meridional wind (between 850 and 700 hPa) at the sites of hail initiation. We find an increase in shear until around 1 h prior to the detection of hail for cases north of the Alps, and a continued increase for cases south of the Alps (note that the temporal resolution of the data set is 1 h). Because cold fronts are associated with vertical wind shear, and because wind shear is known to be a favourable condition for the formation of long-lived convective cells, we argue that frontal zones are amenable for the formation of hail (among other effects, not discussed in this study; such as along-frontal moisture transport (see Doswell, 2001)).

The obtained data set of pre-frontal hail initiations over Switzerland can be further used to, e.g. directly compare the characteristics of hail cases associated with fronts to cases without fronts and to quantify the role of pre-frontal moisture advection, lifting and destabilization. The data set might also be used to explore the relationship between hail and other significant weather events such as wind gusts or lightning.

Acknowledgements

We would like to thank MeteoSwiss for granting access to the COSMO-2/-7 analysis and radar data set.

References

Alberoni PP, Levizzani V, Watson RJ, Holt AR, Costa S, Mezzasalma P, Nanni S. 2000. The 18 June 1997 companion supercells: multiparametric doppler radar analysis. *Meteorology and Atmospheric Physics* **75**: 101–120, doi: 10.1007/s007030070018.

Allen JT, Karoly DJ. 2014. A climatology of Australian severe thunderstorm environments 1979–2011: inter-annual variability and ENSO influence. *International Journal of Climatology* **34**: 81–97, doi: 10.1002/joc.3667.

Bastin S, Dobrinski P, Dabas A, Delville P, Reitebuch O, Werner C. 2005. Impact of the Rhône and Durance valley on sea-breeze circulation in the Marseille area. *Atmospheric Research* **74**: 303–328, doi: 10.1016/j.atmosres.2004.04.014.

Berry G, Reeder MJ, Jakob C. 2011. A global climatology of atmospheric fronts. *Geophysical Research Letters* **38**: L04809, doi: 10.1029/2010GL046451.

Böing SJ, Jonker JJ, Siebesma AP, Grabowski WW. 2012. Influence of the subcloud layer on the development of a deep convective ensemble. *Journal of the Atmospheric Sciences* **69**: 2682–2698, doi: 10.1175/JAS-D-11-0317.1.

Brooks HE. 2009. Proximity soundings for severe convection for Europe and the United States from reanalysis data. *Atmospheric Research* **93**: 546–553, doi: 10.1016/j.atmosres.2008.10.005.

Brooks HE, Lee JW, Craven JP. 2003. The spatial distribution of severe thunderstorms and tornado environments from global reanalysis data. *Atmospheric Research* **67–68**: 73–94, doi: 10.1016/S0169-8095(03)00045-0.

Costa S, Mezzasalma P, Levizzani V, Alberoni PP, Nanni S. 2001. Deep convection over Northern Italy: synoptic and thermodynamic analysis. *Atmospheric Research* **56**: 73–88, doi: 10.1016/S0169-8095(00)00091-0.

Davies HC. 1984. On the orographic retardation of a cold front. *Contributions to Atmospheric Physics (Beiträge zur Physik der Atmosphäre)* **57**: 409–418.

Davies JM. 2004. Estimations of CIN and LFC associated with tornadic and nontornadic supercells. *Weather and Forecasting* **19**: 714–726.

Delden AV. 1998. The synoptic setting of a thundery low and associated prefrontal squall line in Western Europe. *Meteorology and Atmospheric Physics* **65**: 113–131, doi: 10.1007/BF01030272.

Doswell CA III. 1987. The distinction between large-scale and mesoscale contribution to severe convection: A case study example. *Weather and Forecasting* **2**: 3–16.

Doswell CA III (ed). 2001. *Severe Convection Storms*, Vol. **28**. American Meteorological Society. American Meteorological Society: Boston, MA, 561 pp.

Doswell CA III, Evans JS. 2003. Proximity sounding analysis for derechos and supercells: an assessment of similarities and differences. *Atmospheric Research* **67–68**: 117–133, doi: 10.1016/S0169-8095(03)00047-4.

Droegemeier K, Wilhelmson R. 1985a. Three-dimensional numerical modeling of convection produced by interacting thunderstorm outflows. Part I: control simulation and low-level moisture variations. *Journal of the Atmospheric Sciences* **42**: 2381–2403, doi: 10.1175/1520-0469(1985)042<2381:TDNMOC>2.0.CO;2.

Droegemeier K, Wilhelmson RB. 1985b. Three-dimensional numerical modeling of convection produced by interacting thunderstorm outflows. Part II: variations in vertical wind shear. *Journal of the Atmospheric Sciences* **42**: 2404–2414, doi: 10.1175/1520-0469(1985)042<2404:TDNMOC>2.0.CO;2.

Droegemeier KK, Lazarus SM, Davies-Jones R. 1993. The influence of helicity on numerically-simulated convective storms. *Monthly Weather Review* **121**: 2005–2029.

Emanuel K. 1994. *Atmospheric Convection*. Oxford University Press, Inc.: New York, NY.

Fawbush EJ, Miller RC. 1954. The types of air masses in which North American tornadoes form. *Bulletin of the American Meteorological Society* **35**: 154–165.

Foote G, Krauss TW, Makitov V. 2005. *Hail metrics using convectional radar*. 16th Conference on Planned and Inadvertent Weather Modification. The 85th AMS Annual Meeting: San Diego, CA.

Gensini VA, Mote TL, Brooks HE. 2014. Severe thunderstorm reanalysis environments and collocated radiosonde observations. *Journal of Applied Meteorology and Climatology* **53**: 742–751, doi: 10.1175/JAMC-D-13-0263.1.

Germann U, Galli G, Boscacci M, Bolliger M. 2006. Radar precipitation measurement in a mountainous region. *Quarterly Journal of the Royal Meteorological Society* **132**: 1669–1692, doi: 10.1256/qj.05.190.

Giaiotti D, Nordio S, Stel F. 2003. The climatology of hail in the plain of Friuli Venezia Giulia. *Atmospheric Research* **67–68**: 247–259, doi: 10.1016/S0169-8095(03)00084-X.

Grams JS, Thompson RL, Snively DV, Prentice JA, Hodges GM, Reames LJ. 2012. A climatology and comparison of parameters for significant tornado events in the United States. *Weather and Forecasting* **27**: 106–123, doi: 10.1175/WAF-D-11-00008.1.

Groenemeijer P, Delden AV. 2007. Sounding-derived parameters associated with large hail and tornadoes in the Netherlands. *Atmospheric Research* **83**: 473–487, doi: 10.1016/j.atmosres.2005.08.006.

Haklander AJ, Delden AV. 2003. Thunderstorm predictors and their forecast skill for the Netherlands. *Atmospheric Research* **67–68**: 273–299, doi: 10.1016/S0169-8095(03)00056-5.

Hering AM, Morel C, Galli G, Sénési S, Ambrosetti P, Boscacci M. 2004. Nowcasting thunderstorms in the Alpine region using a radar based adaptive thresholding scheme. 3rd European Conference on Radar in Meteorology and Hydrology, 6–10 September 2004, Visby, Sweden.

Hewson DT. 1998. Objective fronts. *Meteorological Applications* **5**: 37–65, doi: 10.1017/S1350482798000553.

Heymsfield GM, Schotz S. 1985. Structure and evolution of a severe squall line over Oklahoma. *Monthly Weather Review* **113**: 1563–1589.

Holton JR. 2004. *An Introduction to Dynamic Meteorology*, 4 ed. Elsevier Academic Press: London, UK.

Houze RA Jr, Schmid W, Fovell RG, Schiesser HH. 1993. Hailstorms in Switzerland: left movers, right movers, and false hooks. *Monthly Weather Review* **121**: 3345–3370.

Huntrieser H, Schiesser HH, Schmid W, Waldvogel A. 1997. Comparison of traditional and newly developed thunderstorm indices for Switzerland. *Weather and Forecasting* **12**: 108–125, doi: 10.1175/1520-0434(1997)012<0108:COTAND>2.0.CO;2.

Jenkner J, Sprenger M, Schwenk I, Schwierz C, Dierer S, Leuenberger D. 2010. Detection and climatology of fronts in a high-resolution model

reanalysis over the Alps. *Meteorological Applications* **17**: 1–18, doi: 10.1002/met.142.

Joss J, Schädler B, Galli G, Cavalli R, Boscacci M, Held E, Bruna GD, Kappenberger G, Nespor V, Spiess R. 1998. *Operational Use of Radar for Precipitation Measurements in Switzerland*, Vol. **134**. Vdf Hochschulverlag AG ETH Zurich: Zuerich, Switzerland, http://www.meteoschweiz.admin.ch/content/dam/meteoswiss/fr/Mess-und-Prognosesysteme/doc/meteoswiss_operational_use_of_radar.pdf.

Kaltenboeck R, Steinheimer M. 2015. Radar-based severe storm climatology for Austrian complex orography related to vertical wind shear and atmospheric instability. *Atmospheric Research* **158–159**: 216–230, doi: 10.1016/j.atmosres.2014.08.006.

Khairoutdinov FM, Randall D. 2006. High-resolution simulation of shallow-to-deep convection transition over land. *Journal of the Atmospheric Sciences* **63**: 3421–3436, doi: 10.1175/Jas3810.1.

Klemp JB, Wilhelmson RB. 1978. The simulation of three-dimensional convective storm dynamics. *Journal of the Atmospheric Sciences* **35**: 1070–1096, doi: 10.1175/1520-0469(1978)035<1070:TSOTDC>2.0.CO;2.

Knupp K, Cotton W. 1982. An intense, quasi-steady thunderstorm over mountainous terrain. Part II: doppler radar observations of the storm morphological structure. *Journal of the Atmospheric Sciences* **39**: 343–358, doi: 10.1175/1520-0469(1982)039<0343:AIQSTO>2.0.CO;2.

Kunz M, Puskeiler M. 2010. High-resolution assessment of the hail hazard over complex terrain from radar and insurance data. *Meteorologische Zeitschrift* **10**: 427–439, doi: 10.1127/0941-2948/2010/0452.

Locatelli JD, Martin JE, Castle JA, Hobbs PV. 1995. Structure and evolution of winter cyclones in the Central United States and their effects on the distribution of precipitation. Part III: the development of a squall line associated with weak cold frontogenesis aloft. *Monthly Weather Review* **123**: 2641–2662.

Manzato A. 2012. Hail in northeast Italy: climatology and bivariate analysis with the sounding-derived indices. *Journal of Applied Meteorology and Climatology* **51**: 449–467, doi: 10.1175/JAMC-D-10-05012.1.

Manzato A. 2013. Hail in northeast Italy: a neural network ensemble forecast using sounding-derived indices. *Weather and Forecasting* **28**: 3–28, doi: 10.1175/WAF-D-12-00034.1.

Markowski P, Richardson Y. 2010. *Mesoscale Meteorology in Midlatitudes*. John Wiley & Sons: West Sussex, UK, doi: 10.1002/9780470682104.

Meischner PF, Bringi VN, Heimann D, Höller H. 1991. A squall line in southern Germany: kinematics and precipitation formation as deduced by advanced polarimetric and doppler radar measurements. *Monthly Weather Review* **119**: 678–701, doi: 10.1175/1520-0493(1991)119<0678:ASLISG>2.0.CO;2.

Merino A, Wu X, Gascón E, Berthet C, Garca-Ortega E, Dessens J. 2014. Hailstorms in southwestern France: incidence and atmospheric characterization. *Atmospheric Research* **140–141**: 61–75, doi: 10.1016/j.atmosres.2014.01.015.

Mohr S, Kunz M. 2013. Trend analysis of convective indices relevant for hail events in Germany and Central Europe. *Atmospheric Research* **123**: 211–228, doi: 10.1016/j.atmosres.2012.05.016.

Mohr S, Kunz M, Geyer B. 2015. Hail potential in Europe based on a regional climate model hindcast. *Geophysical Research Letters* **42**: 10,904–10,912, doi: 10.1002/2015GL067118.

Naylor J, Gilmore MS. 2012. Environmental factors influential to the duration and intensity of tornadoes in simulated supercells. *Geophysical Research Letters* **39**(L17): 802, doi: 10.1029/2012GL053041.

Neiman PJ, Wakimoto RM. 1999. The interaction of a pacific cold front with shallow air masses east of the Rocky Mountains. *Monthly Weather Review* **127**: 2102–2127.

Newton CW. 1950. Structure and mechanism of the prefrontal squall line. *Journal of Meteorology* **7**: 210–222, doi: 10.1175/1520-0469(1950)007<0210:SAMOTP>2.0.CO;2.

Nisi L, Martius O, Hering A, Germann U. 2016. Spatial and temporal distribution of hailstorm in the Alpine region: a long-term, high resolution, radar-based analysis. *Quarterly Journal of the Royal Meteorological Society*, doi: 10.1002/qj.2771.

Ogura Y, Portis D. 1982. Structure of the cold front observed in SESAME-AVE III and its comparison with the Hoskins-Bretherton frontogenesis model. *Journal of the Atmospheric Sciences* **39**: 2773–2792, doi: 10.1175/1520-0469(1982)039<2773:SOTCFO>2.0.CO;2.

Punge HJ, Bedka KM, Kunz M, Werner A. 2014. A new physically based stochastic event catalog for hail in Europe. *Natural Hazards* **73**: 1625–1645, doi: 10.1007/s11069-014-1161-0.

Rasmussen EN, Blanchard DO. 1998. A baseline climatology of sounding-derived supercell and tornado forecast parameters. *Weather and Forecasting* **13**: 1148–1164, doi: 10.1175/1520-0434(1998)013<1148:ABCOSD>2.0.CO;2.

Renard R, Clarke L. 1965. Experiments in numerical objective frontal analysis. *Monthly Weather Review* **93**: 547–556, doi: 10.1175/1520-0493(1965)093<0547:EINOFA>2.3.CO;2.

Rotunno R, Klemp J. 1985. On the rotation and propagation of simulated supercell thunderstorms. *Journal of the Atmospheric Sciences* **42**: 271–292, doi: 10.1175/1520-0469(1985)042<0271:OTRAPO>2.0.CO;2.

Rotunno R, Klemp JB, Weisman ML. 1988. A theory for strong, long-lived squall lines. *Journal of the Atmospheric Sciences* **45**: 463–485, doi: 10.1175/1520-0469(1988)045<0463:ATFSLL>2.0.CO;2.

Schemm S, Sprenger M. 2015. Frontal-wave cyclogenesis in the North Atlantic – a climatological characterisation. *Quarterly Journal of the Royal Meteorological Society* **141**: 2989–3005, doi: 10.1002/qj.2584.

Schemm S, Rudeva I, Simmonds I. 2015. Extratropical fronts in the lower troposphere – global perspectives obtained from two automated methods. *Quarterly Journal of the Royal Meteorological Society* **141**: 1686–1698, doi: 10.1002/qj.2471.

Schiesser HH, Waldvogel A, Schmid W, Willemse S. 1997. *Klimatologie der Stürme und Sturmsysteme anhand von Radar - und Schadendaten (in German)*, Vol. **132**. Vdf Hochschulverlag AG ETH Zurich: Zuerich, Switzerland.

Schlemmer L, Hohenegger C. 2014. The formation of wider and deeper clouds as a result of cold-pool dynamics. *Journal of the Atmospheric Sciences* **71**: 2842–2858, doi: 10.1175/JAS-D-13-0170.1.

Schultz DM. 2005. A review of cold fronts with prefrontal troughs and wind shifts. *Monthly Weather Review* **133**: 2449–2472, doi: 10.1175/MWR2987.1.

Schumann U. 1987. Influence of mesoscale orography on idealized cold fronts. *Journal of the Atmospheric Sciences* **44**: 3423–3441, doi: 10.1175/1520-0469(1987)044<3423:IOMOOI>2.0.CO;2.

Steppler JG, Doms G, Schättler U, Bitzer HW, Gassmann A, Damrath U, Gregoric G. 2003. Meso-gamma scale forecasts using the non-hydrostatic model LM. *Meteorology and Atmospheric Physics* **82**: 75–96, doi: 10.1007/s00703-001-0592-9.

Suwala K, Bednorz E. 2013. Climatology of hail in central Europe. *Quaestiones Geographicae* **32**: 99–110, doi: 10.2478/quageo-2013-0025.

Torris G, Kuang Z, Tian Y. 2015. Mechanisms for convection triggering by cold pools. *Geophysical Research Letters* **42**: 1943–1950, doi: 10.1002/2015GL063227.

Towery NG, Changnon SA. 1970. Characteristics of hail-producing radar echoes in Illinois. *Monthly Weather Review* **98**: 346–353, doi: 10.1175/1520-0493(1970)098<0346:COHPRE>2.3.CO;2.

Waldvogel A, Federer B, Grimm P. 1979. Criteria for the detection of hail cells. *Journal of Applied Meteorology and Climatology* **18**: 1521–1525, doi: 10.1175/1520-0450(1979)018<1521:CFTDOH>2.0.CO;2.

Weisman ML, Klemp JB. 1982. The dependence of numerically simulated convective storms on vertical wind shear and buoyancy. *Monthly Weather Review* **110**: 504–520, doi: 10.1175/1520-0493(1982)110<0504:TDONSC>2.0.CO;2.

Weisman M, Rotunno R. 2004. "A theory for strong, long-lived squall lines" revisited. *Journal of the Atmospheric Sciences* **61**: 361–382.

Weisman M, Klemp J, Rotunno R. 1988. Structure and evolution of numerically simulated squall lines. *Journal of the Atmospheric Sciences* **45**: 1990–2013, doi: 10.1175/1520-0469(1988)045<1990:SAEONS>2.0.CO;2.

Weissmann M, Braun F, Ganter L, Mayr G, Rahm S, Reitebuch O. 2005. The Alpine mountain-plain circulation: Airborne doppler lidar measurements and numerical simulations. *Monthly Weather Review* **133**: 3095–3109, doi: 10.1175/MWR3012.1.

Wilhelmson RB, Klemp JB. 1978. A numerical study of storm splitting that leads to long-lived storms. *Journal of the Atmospheric Sciences* **35**: 1974–1986, doi: 10.1175/1520-0469(1978)035<1974:ANSOSS>2.0.CO;2.

Willemse S. 1995. A statistical analysis and climatological interpretation of hailstorms in Switzerland. PhD thesis, ETH, Zurich, Switzerland, 10.3929/ethz-a-001486581.

21

Multi-parameter multi-physics ensemble (MPMPE): a new approach exploring the uncertainties of climate sensitivity

Hideo Shiogama,[1,2]* Masahiro Watanabe,[3] Tomoo Ogura,[1] Tokuta Yokohata[1] and Masahide Kimoto[3]

[1]Center for Global Environmental Research, National Institute for Environmental Studies, Tsukuba, Japan
[2]Environmental Change Institute, University of Oxford, Oxford, UK
[3]Atmosphere and Ocean Research Institute, The University of Tokyo, Kashiwa, Japan

*Correspondence to:
H. Shiogama, National Institute for Environmental Studies, Tsukuba, Ibaraki 305-8506, Japan.
E-mail:
shiogama.hideo@nies.go.jp

Abstract

To explore both the parametric and structural uncertainties of climate sensitivity (CS), we have proposed a new general circulation model (GCM) ensemble termed the multi-parameter multi-physics ensemble (MPMPE). We used eight multi-physics ensemble (MPE) models in which the MIROC5 physics schemes were replaced by those of MIROC3. MPMPE consisted of perturbed-physics ensembles in which the parameter values were swept for each MPE model. MPMPE resulted in a wide range of CS, which was related to the shortwave cloud feedback (SWcld). Coupling between low- and mid-level clouds controlled the differences in the parametric spread of SWcld among the MPE models.

Keywords: climate sensitivity; cloud feedback; ensemble

1. Introduction

Climate sensitivities (CSs, defined as the global mean surface air temperature responses to a doubling of the atmospheric CO_2 concentration) differ between general circulation models (GCMs). In the Fourth Assessment Report of the Intergovernmental Panel on Climate Change, the range of CS was 2.1–4.4 °C for the multi-model ensemble (MME) of GCMs (Randall et al., 2007). By analysing the outputs of MMEs, previous studies suggested that the spread of CSs was mainly caused by feedback (FB) and radiative forcing (RF) uncertainties according to cloud changes (Webb et al., 2006, 2013; Dufresne and Bony, 2008; Gregory and Webb, 2008). Different physical parameterization schemes (structural uncertainties) lead to the spreads of cloud changes. Although the MME is important, there are limitations to understanding the sources of the uncertainties in CS. Because each GCM in MME was developed by a different modelling centre, tracing the uncertainties of the climate simulations to particular differences in the physics scheme structures is difficult. Furthermore, because each GCM is tuned by adjusting uncertain parameter values in the physics schemes, it is impossible to perfectly distinguish between the structural uncertainties and the parametric uncertainties mentioned below by analysing only the MME.

Recently, Gettelman et al. (2012) and Watanabe et al. (2012) (hereafter W12) proposed the 'multi-physics ensemble' (MPE) approach to investigate the structural uncertainties of CS. In the MPEs, these authors replaced single or multiple physics schemes between two versions of a GCM developed in the same modelling centre. W12 developed seven hybrid models between the MIROC5 GCM (Watanabe et al., 2010) and the MIROC3 GCM (K-1 Model Developers, 2004). The eight MPE models (one MIROC5 and the seven hybrid models) have a wide range of CS, from 2.3 to 5.9 °C. This MPE approach can facilitate a better understanding of the structural uncertainties in CS. The obvious limitations of the MPE are that the results depend on the base models and that the MPE models have particular parameter value sets.

The parametric uncertainties of CS have also been investigated by examining a 'perturbed-physics ensemble' (PPE), in which uncertain parameter values of a single GCM were swept. The PPEs of some GCMs (Murphy et al., 2004; Stainforth et al., 2005; Klocke et al., 2011) indicated wide parametric uncertainties of CSs comparable with or greater than the structural uncertainty in the MME, whereas the PPEs of other GCMs have skewed CS distributions (Annan et al., 2005; Jackson et al., 2008; Sanderson, 2011; Shiogama et al., 2012 (hereafter S12)). The PPEs can provide information that is valuable for characterizing the parametric sensitivities of single GCMs. However, the properties of a climate system (such as the relationships between changes in clouds in a warming climate and their biases in the present climate) found in a PPE are not necessarily carried into other MME models or into the PPEs of different models (Yokohata et al., 2010; Klocke et al., 2011; Sanderson, 2011). Furthermore, the results of a PPE can be sensitive to the selection of the perturbed parameters, their ranges, and the parameter value sampling methods.

In this study, we propose a new approach termed the 'multi-parameter multi-physics ensemble' (MPMPE) to explore both the parametric and structural uncertainties of CS. Herein, we conducted PPEs with a

Table I. A list of hybrid model names, their ensemble sizes, and schemes of MIROC5 that were replaced by those of MIROC3.

Names	Ensemble sizes	Cloud	Cumulus convection	Turbulence
CLD + CNV + VDF	18	MIROC3	MIROC3	MIROC3
CLD + VDF	15	MIROC3		MIROC3
CLD + CNV	20	MIROC3	MIROC3	
CNV + VDF	12		MIROC3	MIROC3
VDF	11			MIROC3
CNV	20		MIROC3	
CLD	20	MIROC3		
MIROC5A	20			

common sampling strategy using each of the eight MPE models developed by W12. As shown below, although the PPEs of different MPE models indicated different parametric sensitivities for cloud FB, there was a common property across the eight MPE models. It should be noted that we attempted to develop a new ensemble technique, but not to derive any plausible, observationally constrained, range of CS in this paper.

2. Experimental design

We used the atmospheric general circulation model (AGCM) of MIROC5 (hereafter MIROC5A) (Watanabe *et al.*, 2010; the T42L40 resolution version) and seven hybrid AGCMs (W12) in which single or multiple physics schemes of clouds, cumulus convection and turbulence in MIROC5 were replaced by the corresponding scheme of the older version, MIROC3 (K-1 Model Developers, 2004). Table I shows the names of these MPE models and the schemes that were replaced. See W12 for the details and the references of the schemes. Applying the Latin-hypercube sampling method (McKay *et al.*, 1979) to each of the eight MPE models, we selected 20 sets of six physics parameters (two parameters for each scheme) swept within the

min-max ranges shown in Table II. Because there are large structural differences between the MIROC5 and MIROC3 schemes, we could not choose parameters common to both base models. Instead, we selected the perturbed parameters and their ranges from those used in the previous PPEs of MIROC5 (S12) and MIROC3 (Annan *et al.*, 2005), except for *aml0* in the MIROC3 turbulence scheme. See S12 and Yoshimori *et al.* (2011) for the details of the selected parameters.

Using these 20×8 MPMPE models, we performed the following three types of AGCM runs (6-year-long integrations; the mean values from years 2 to 6 were analysed) (cf S12 and W12): (1) AGCMs forced by the climatology of the sea surface temperature (SST) and sea ice (ICE) from the pre-industrial control run of the coupled atmosphere ocean general circulation model (CGCM) version of MIROC5 (MIROC5C) using the standard parameter settings and pre-industrial CO_2 concentrations (the control (CTL) runs); (2) AGCMs forced by the average of years 11–20 for the SST and ICE from the 20-year integrations of $4 \times CO_2$ runs of the standard MIROC5C and using pre-industrial CO_2 concentrations (the SST runs); (3) AGCMs forced by the SST and ICE from the MIROC5C control run and by the $4 \times CO_2$ concentrations (the CO_2 runs). Because the hybrid models have not been carefully configured and checked for performance, some runs stopped due to numerical problems, and these runs were removed from our analyses. Therefore, the PPE sizes varied among the MPE models (Table I).

Analysing the changes in radiation at the top of the atmosphere (R) and the global mean surface air temperature (T) from the above three types of AGCM runs, we can estimate the RF, FB, and effective climate sensitivity (ECS) of the models :

$$RF = [R(CO_2) - R(CTL)]/2 \qquad (1)$$

$$FB = [R(SST) - R(CTL)]/[T(SST) - T(CTL)] \qquad (2)$$

$$ECS = -RF/FB \qquad (3)$$

Table II. Lists of the perturbed physics parameters and their ranges.

MIROC5

Name	Category	Description	Min	Max
vicec	Cloud	Factor for ice falling speed [$m^{0.474}\,s^{-1}$]	25.0	40.0
b1_5	Cloud	Efficiency factor for liquid precipitation [$m^3\,kg^{-1}$]	0.07	0.11
wcbmax	Cumulus	Max. cumulus updraft velocity at cloud base [$m\,s^{-1}$]	0.70	2.80
clmd	Cumulus	Entrainment efficiency [ND]	0.40	0.60
faz1	Turbulence	Factor for PBL overshooting [ND]	1.00	3.00
alp1	Turbulence	Factor for length scale L_T [ND]	0.16	0.30

MIROC3

Name	Category	Description	Min	Max
prctau	Cloud	e-Folding time for ice precipitation [s]	4.02×10^3	3.05×10^4
b1_3	Cloud	Efficiency factor for liquid precipitation [$m^3\,kg^{-1}\,s^{-1}$]	6.77×10^{-3}	0.119
rhmcrt	Cumulus	Critical relative humidity for cumulus convection [ND]	0.683	0.893
elamin	Cumulus	Minimum entrainment factor of cumulus convection [m^{-1}]	0.00	5.46×10^{-4}
dfmmin	Turbulence	Minimum vertical diffusion coefficient [$m^2\,s^{-1}$]	0.0785	0.158
aml0	Turbulence	Maximum mixing length [m]	150	600

where the factor of 1/2 in Equation (1) converts the RF of $4 \times CO_2$ to that of $2 \times CO_2$. Because we use AGCM instead of CGCM, we can obtain statistically robust diagnoses of RF, FB, and ECS using short length simulations. We performed 20-year runs of CTL, SST, and CO_2 using the standard MIROC5 AGCM, and confirmed that 6-year simulations are enough for statistically robust results (not shown). Additionally, AGCM runs are free from climate drifts which could occur in MME, PPE, and MPE experiments. On the other hand, our method using AGCM has limitations. There is no interaction between atmosphere and ocean. While the atmosphere rapidly response to the increases of CO_2 concentration, the ocean including sea ice change slowly in decadal to several hundred year time scales. We cannot investigate these slow responses using our method. This disconnection between atmosphere and ocean is the reason why we used year 11–20 of the $4 \times CO_2$ runs of the standard MIROC5C to derive the SST conditions for the FB analysis rather than a longer average, for example, between 101 and 150 years, where ocean processes would be important. Therefore,

it should be noted that the ECS values calculated by our method can be taken as an estimate only.

3. Results

Figure 1(a) shows the RF, total FB, and ECS of the MPMPE. The MPMPE covered a very large range of ECS values: all but one ensemble model had ECS in the range 2.1–6 °C with a single very high member having 10.4 °C. Although the PPEs of five MPE models (MIROC5A, CLD, CNV, VDF, and CNV + VDF) indicate ECSs of less than 4 °C, the other three MPE models yielded larger ECSs (Figure 1(d)). As we move more closely towards MIROC3, we get higher ECS. This is consistent with that the ECS of standard MIROC3 model (3.6 °C) is higher than that of MIROC5C (2.85 °C).

The values of the global mean shortwave cloud feedback (SWcld) [shortwave cloud forcing changes (SST minus CTL) divided by surface air temperature changes] (hereafter SWcld) related well to the variations in the total FB and ECS (Figure 1(b) and

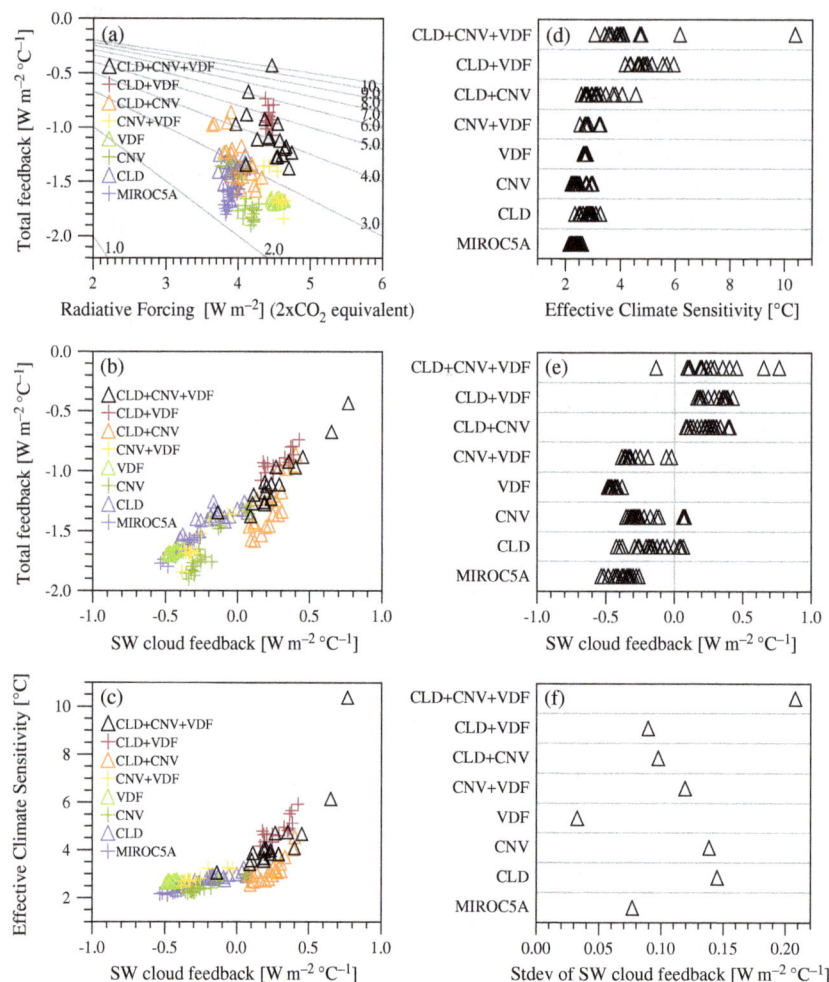

Figure 1. The left panels are scatter plots of the MPMPE runs between (a) the total FB [W m^{-2} °C^{-1}] and RF [W m^{-2}], (b) the total FB [W m^{-2} °C^{-1}] and SWcld [W m^{-2} °C^{-1}], and (c) the ECS [°C] and SWcld [W m^{-2} °C^{-1}]. The grey lines in panel (a) are the ECSs [°C]. (d) ECS [°C] for the MPMPE runs. (e) SWcld [W m^{-2} °C^{-1}] for the MPMPE runs. (f) σ(SWcld) [W m^{-2} °C^{-1}] for each MPE model.

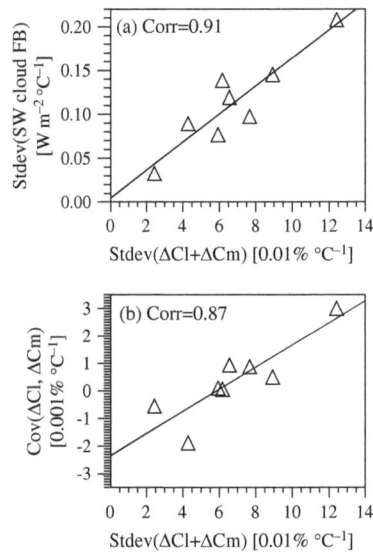

Figure 2. (a) Scatter plot between $\sigma(\mathrm{SWcld})$ [$\mathrm{W\,m^{-2}\,^{\circ}C^{-1}}$] and $\sigma(\triangle \mathrm{Cl} + \triangle \mathrm{Cm})$ [$0.01\%\,^{\circ}\mathrm{C}^{-1}$]. (b) Scatter plot between the covariance of $\triangle\mathrm{Cl}$ and $\triangle\mathrm{Cm}$ [$0.001\%\,^{\circ}\mathrm{C}^{-1}$] against $\sigma(\triangle \mathrm{Cl} + \triangle \mathrm{Cm})$ [$0.01\%\,^{\circ}\mathrm{C}^{-1}$].

(c)). Lower ECS runs had negative values of SWcld, and higher ECS runs had positive values of SWcld (Figure 1(c)–(e)). Here we define $\sigma(\mathrm{SWcld})$ as a standard deviation of SWcld across PPE members for a given MPE model. MIROC5A and VDF had negative values of SWcld and small $\sigma(\mathrm{SWcld})$ (Figure 1(f)). CLD, CNV, and CNV + VDF had negative values of SWcld and moderate values of $\sigma(\mathrm{SWcld})$. Although CLD + CNV + VDF had a positive SWcld and large $\sigma(\mathrm{SWcld})$, the other two MPE models with positive values of SWcld (CLD + CNV and CLD + VDF) did

not have large values of $\sigma(\mathrm{SWcld})$. The reasons for the differences in the values of SWcld among the different MPE models (for a particular parameter setting) have been reported by W12. Therefore, we focused on different $\sigma(\mathrm{SWcld})$ values of PPE among the various MPE models hereafter. We diagnosed SWcld by applying the conventional cloud RF method (Cess et al., 1990). Therefore, cloud masking would lead to systematic negative biases in the SWcld estimations (Soden et al., 2004). However, the systematic negative biases would not significantly affect the points of this paper: the correlation between ECS and SWcld, and differences in $\sigma(\mathrm{SWcld})$ between the MPE models.

The standard deviations of the global mean changes in the low-level cloud cover ($\triangle \mathrm{Cl}$) plus mid-level cloud cover ($\triangle \mathrm{Cm}$), $\sigma(\triangle \mathrm{Cl} + \triangle \mathrm{Cm})$, were well correlated with $\sigma(\mathrm{SWcld})$ (Figure 2(a)). $\sigma(\triangle \mathrm{Cl} + \triangle \mathrm{Cm})^2$ can be decomposed to $\sigma(\triangle \mathrm{Cl})^2 + \sigma(\triangle \mathrm{Cm})^2$ and the covariance between $\triangle \mathrm{Cl}$ and $\triangle \mathrm{Cm}$ ($2\mathrm{Cov}[\triangle \mathrm{Cl}, \triangle \mathrm{Cm}]$). Variations of $\sigma(\triangle \mathrm{Cl} + \triangle \mathrm{Cm})$ were mainly related to $\mathrm{Cov}(\triangle \mathrm{Cl},\triangle \mathrm{Cm})$ (Figure 2(b)) rather than to $\sigma(\triangle \mathrm{Cl})^2 + \sigma(\triangle \mathrm{Cm})^2$ (not shown). When anomalies of $\triangle \mathrm{Cl}$ and $\triangle \mathrm{Cm}$ relative to the PPE averages have the same sign, $\sigma(\mathrm{SWcld})$ is suggested to be enhanced. In contrast, when $\triangle \mathrm{Cl}$ and $\triangle \mathrm{Cm}$ fluctuate in opposite directions, the PPE spreads of the SWcld are decreased.

Figure 3 shows the correlations between SWcld and regional $\triangle \mathrm{Cl}$ and $\triangle \mathrm{Cm}$ in the tropics in CLD + CNV + VDF, CLD + VDF, and MIROC5A as examples. Negative correlations indicate that changes in the cloud cover enhanced values of $\sigma(\mathrm{SWcld})$ and vice versa. For CLD + CNV + VDF, both $\triangle \mathrm{Cl}$ and $\triangle \mathrm{Cm}$ led to large spreads of SWcld over large fractions

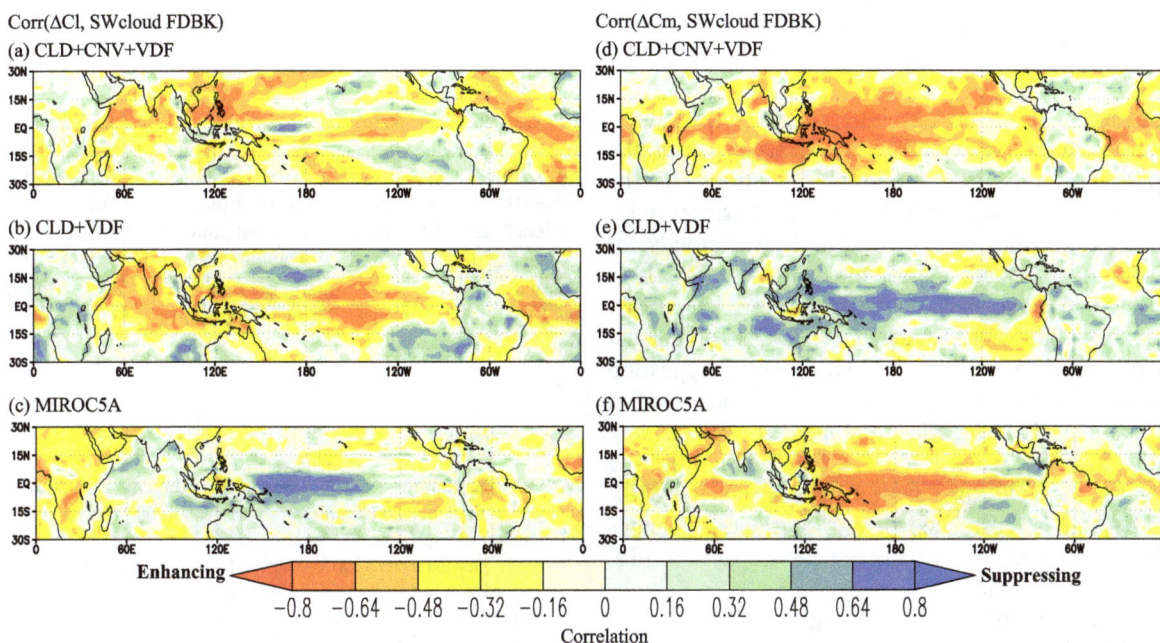

Figure 3. Intra-PPE correlations between the global mean SW cloud feedback and regional low-level cloud cover changes for (a) CLD+CNV+VDF, (b) CLD+VDF, and (c) MIROC5A. The right panels are the same as the left ones, except for the mid-level cloud cover changes.

of the tropical ocean. For CLD + VDF, whereas $\triangle$Cl over the equatorial ocean tended to enlarge the spread of SWcld, $\triangle$Cm counteracted $\triangle$Cl, causing a small value of σ(SWcld). In contrast, for MIROC5A, $\triangle$Cl and $\triangle$Cm over the equatorial ocean suppressed and enhanced the spread of SWcld, respectively.

Most MME studies and some PPE studies (e.g. Bony and Dufresne, 2005; Webb *et al.*, 2006, 2013; Yokohata *et al.*, 2010) have stressed the important role of low-level clouds in relation to the uncertainties of SW cloud feedback. S12 indicated that mid-level clouds also lead to the spread of SW cloud feedback in the MIROC5C PPE. If we had focused only on the low-level clouds or the mid-level clouds, we would have found only discrepancies in the role of clouds for the spread of SWcld across the MPE models, i.e. a cloud at a given level would enhance the intra-PPE spread of SWcld in some MPEs but suppress that intra-PPE spread in other MPE models. In addition to the individual roles of low- and mid-level clouds, we suggested that the coupling between low- and mid-level cloud covers can affect the spread of SW cloud feedback. Previous comparisons of PPEs were limited to two ensembles (Yokohata *et al.*, 2010; Sanderson, 2011). The production of PPEs from many models enabled us to explore common characteristics across the different PPEs.

4. Summary and discussion

To investigate both the parametric and structural uncertainties of ECS, we proposed a new approach termed the MPMPE. This MPMPE is a fusion of the PPE for exploring parametric uncertainties and the MPE, a type of MME for exploring structural uncertainties. Our MPMPE consisted of PPEs using eight MPE models developed by W12 based on the MIROC5 and MIROC3 models.

This MPMPE covered a wide range of ECS values. Variations in the ECS related well to those in the SWcld. The intra-PPE standard deviations of the SWcld varied among the MPE models. Discrepancies existed in the roles of low- and mid-level clouds for the spread of SWcld across the MPE models. It is suggested that the characteristics of cloud feedback found in a given PPE are not necessarily carried into other model PPEs or MMEs. This suggestion is consistent with the findings of previous studies (Yokohata *et al.*, 2010; Klocke *et al.*, 2011; Sanderson, 2011). However, we also found a common controlling factor for the PPE spread of SWcld across the MPE models; that is, correlated or anti-correlated changes in the different level cloud covers can impact the PPE spread of the SWcld.

It should be noted that we have not discussed about likelihood of ECS of each MPMPE member. Some evaluations of model's quality may discard outliers, for example, the model with ECS of 10.4 °C. However constraints of uncertainties in ECS and

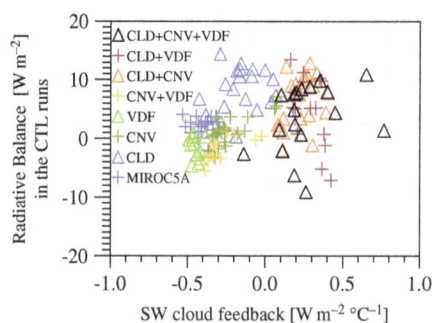

Figure 4. Scatter plot between the radiation balance at the top of atmosphere in the CTL runs [$W\,m^{-2}$] and the shortwave cloud feedback [$W\,m^{-2}\,°C^{-1}$].

SWcld are not easy. Figure 4 shows the scatter plot of the radiation balance at the top of atmosphere in the CTL runs (models with values close to zero are good) and SWcld. There is no relationship between these properties. Therefore, we cannot constrain the uncertainty of SWcld using the simple test of radiation balance. It may be necessary to apply process-oriented evaluations (e.g. Shiogama *et al.*, 2011; Fasullo and Trenberth, 2012).

The results of the MPMPE can depend on the base models, the replaced physics schemes, the perturbed parameters and their ranges, and the sampling method of the parameter values. Despite these issues, MPMPE is a new and useful approach. For a better understanding of the parametric and structural uncertainties of ECS, we hope that our finding of the common property across multiple PPEs will encourage the intercomparison of PPEs from a number of models, which have been referred to as "super-ensembles' by Murphy *et al.* (2004).

Acknowledgements

We appreciate useful comments from two reviewers, CD Jones, JD Annan, JC Hargreaves, M Yoshimori, and M Chikira. This work was supported by the Program for Risk Information on Climate Change, Grant-in-Aid 23310014 and Grant-in-Aid 23340137 from the Ministry of Education, Culture, Sports, Science and Technology of Japan, and by the Environment Research and Technology Development Fund (S-10) of the Ministry of the Environment of Japan. The Earth Simulator at JAMSTEC and NEC SX at NIES were used to perform the model simulations.

References

Annan JD, Hargreaves JC, Ohgaito R, Abe-Ouch A, Emori S. 2005. Efficiently constraining climate sensitivity with paleoclimate simulations. *SOLA* **1**: 181–184.

Bony S, Dufresne JL. 2005. Marine boundary layer clouds at the heart of tropical cloud feedback uncertainties in climate models. *Geophysical Research Letters* **32**: L20806.

Cess RD, Potter GL, Blanchet JP, Boer GJ, Delgenio AD, Dequé M, Dymnikov V, Galin V, Gates WL, Ghan SJ, Kiehl JT, Lacis AA, Le Treut H, Li ZX, Liang XZ, McAveney BJ, Meleshko VP, Mitchell JFB, Morcrette JJ, Randall DA, Rikus L, Roeckner E, Royer JF, Shlese U, Sheinin DA, Slingo A, Sokolov AP, Taylor KE,

Washington WM, Wetherald RT, Yagai I, Zhang MH. 1990. Intercomparison and interpretation of climate feedback processes in 19 atmospheric general circulation models. *Journal of Geophysical Research* **95**: 16601–16615.

Dufresne JL, Bony S. 2008. An assessment of the primary sources of spread of global warming estimates from coupled atmosphere–ocean models. *Journal of Climate* **21**: 5135–5144.

Fasullo JT, Trenberth KE. 2012. A less cloudy future: the role of subtropical subsidence in climate sensitivity. *Science* **338**: 792–794.

Gettelman A, Kay JE, Shell KM. 2012. The evolution of climate sensitivity and climate feedbacks in the Community Atmosphere Model. *Journal of Climate* **25**: 1453–1469.

Gregory JM, Webb MJ. 2008. Tropospheric adjustment induces a cloud component in CO2 forcing. *Journal of Climate* **21**: 58–71.

Jackson CS, Sen MK, Huerta G, Deng Y, Bowman KP. 2008. Error Reduction and Convergence in Climate Prediction. *Journal of Climate* **21**: 6698–6709.

K-1 Model Developers. 2004. K-1 coupled GCM (MIROC) description. K-1 Tech. Rep. 1, University of Tokyo, 1–34.

Klocke D, Pincus R, Quaas J. 2011. On constraining estimates of climate sensitivity with present-day observations through model weighting. *Journal of Climate* **24**: 6092–6099.

McKay MD, Beckman RJ, Conover WJ. 1979. A comparison of three methods for selecting values of input variables in the analysis of output from a computer code. *Technometrics* **21**: 239–245.

Murphy JM, Sexton DMH, Barnett DN, Jones GS, Webb MJ, Collins M. 2004. Quantification of modelling uncertainties in a large ensemble of climate change simulations. *Nature* **430**: 768–772.

Randall DA, Wood RA, Bony S, Colman R, Fichefet T, Fyfe J, Kattsov V, Pitman A, Shukla J, Srinivasan J, Stouffer RJ, Sumi A, Taylor KE. 2007. Cilmate models and their evaluation. In *Climate Change (2007) The Physical Science Basis. Contribution of Working Group I to the Fourth Assessment Report of the Intergovernmental Panel on Climate Change*, Solomon S, Qin D, Manning M, Chen Z, Marquis M, Averyt KB, Tignor M, Miller HL (eds). Cambridge University Press: Cambridge, UK and New York, NY.

Sanderson BM. 2011. A multimodel study of parametric uncertainty in predictions of climate response to rising greenhouse gas concentrations. *Journal of Climate* **24**: 1362–1377.

Shiogama H, Emori S, Hanasaki N, Abe M, Masutomi Y, Takahashi K, Nozawa T. 2011. Observational constraints indicate risk of drying in the Amazon basin. *Nature Communications* **2**: 253, DOI: 10.1038/ncomms1252

Shiogama H, Watanabe M, Yoshimori M, Yokohata T, Ogura T, Annan JD, Hargreaves JC, Abe M, Kamae Y, O'ishi R, Nobui R, Emori S, Nozawa T, Abe-Ouchi A, Kimoto M. 2012. Perturbed physics ensemble using the MIROC5 coupled atmosphere–ocean GCM without flux corrections: experimental design and results. *Climate Dynamics* **39**: 3041–3056.

Soden BJ, Broccoli AJ, Hemler RS. 2004. On the use of cloud forcing to estimate cloud feedback. *Journal of Climate* **17**: 3661–3665.

Stainforth DA, Aina T, Christensen C, Collins M, Faull N, Frame DJ, Kettleborough JA, Knight S, Martin A, Murphy JM, Piani C, Sexton D, Smith LA, Spicer RA, Thorpe AJ, Allen MR. 2005. Uncertainty in predictions of the climate response to rising levels of greenhouse gases. *Nature* **433**: 403–406.

Watanabe M, Suzuki T, O'ishi R, Komuro Y, Watanabe S, Emori S, Takemura T, Chikira M, Ogura T, Sekiguchi M, Takata K, Yamazaki D, Yokohata T, Nozawa T, Hasumi H, Tatebe H, Kimoto M. 2010. Improved climate simulation by MIROC5: mean states, variability, and climate sensitivity. *Journal of Climate* **23**: 6312–6335.

Watanabe M, Shiogama H, Yokohata T, Kamae Y, Yoshimori M, Ogura T, Annan JD, Hargreaves JC, Emori S, Kimoto M. 2012. Using a multi-physics ensemble for exploring diversity in cloud shortwave feedback in GCMs. *Journal of Climate* **25**: 5416–5431.

Webb MJ, Senior CA, Sexton DMH, Ingram WJ, Williams KD, Ringer MA, McAvaney BJ, Colman R, Soden BJ, Gudgel R, Knutson T, Emori S, Ogura T, Tsushima Y, Andronova NG, Li B, Musat I, Bony S, Taylor KE. 2006. On the contribution of local feedback mechanisms to the range of climate sensitivity in two GCM ensembles. *Climate Dynamics* **27**: 17–38.

Webb M, Lambert FH, Gregory JM. 2013. Origins of differences in climate sensitivity, forcing and feedback in climate models. *Climate Dynamics* **40**: 677–707.

Yokohata T, Webb MJ, Collins M, Williams KD, Yoshimori M, Hargreaves JC, Annan JD. 2010. Structural similarities and differences in climate responses to CO2 increase between two perturbed physics ensembles. *Journal of Climate* **23**: 1392–1410.

Yoshimori M, Hargreaves JC, Annan JD, Yokohata T, Abe-Ouchi A. 2011. Dependency of feedbacks on forcing and climate state in physics parameter ensembles. *Journal of Climate* **24**: 6440–6455.

An evaluation of CORDEX regional climate models in simulating precipitation over Southern Africa

Mxolisi E. Shongwe,[1,2]* Chris Lennard,[3] Brant Liebmann,[4] Evangelia-Anna Kalognomou,[5] Lucky Ntsangwane[1] and Izidine Pinto[2]

[1] South African Weather Service, Private Bag X097, Pretoria, South Africa
[2] University of Pretoria, South Africa
[3] Climate Systems Analysis Group, University of Cape Town, 7945, South Africa
[4] NOAA-CIRES Climate Diagnostics Center, Boulder, CO, USA
[5] Laboratory of Heat Transfer and Environmental Engineering, Aristotle University, Thessaloniki, Greece

*Correspondence to:
M. E. Shongwe, South African Weather Service, Private Bag X097, Pretoria, South Africa.
E-mail: mxolisi.shongwe @weathersa.co.za

Abstract

This article evaluates the ability of the Coordinated Regional Downscaling Experiment (CORDEX) regional climate models (RCMs) in simulating monthly rainfall variation during the austral summer half year (October to March) over southern Africa, the timing of the rainy season and the relative frequencies of rainfall events of varying intensities. The phasing and amplitude of monthly rainfall evolution and the spatial progression of the wet season onset are well simulated by the models. Notwithstanding some systematic biases in a few models, the simulated onset and end of the rainy season and their interannual variability are highly correlated with those computed from the reference data. The strongest agreements between the reference and modelled precipitation patterns are found north of about 20°S in the vicinity of the Inter Tropical Convergence Zone. A majority of the RCMs adequately capture the reference precipitation probability density functions, with a few showing a bias towards excessive light rainfall events.

Keywords: CORDEX – Africa; regional climate model evaluation; rainfall characteristics

1. Introduction

As the Earth's climate continues to change, the characteristics of regional precipitation and extreme events may change without necessarily being reflected in seasonal totals (Seneviratne et al., 2012; Trenberth et al., 2003). For example, changes in wet season timing may not affect calendar season totals, and changes in wet season duration may be balanced by changes in the intensity or frequency of daily precipitation during that season. Changes in the character of precipitation have important implications for a number of climate-sensitive sectors such as agriculture, forestry, water resources, ecosystem services and disaster risk management. For instance, steady, soaking, moderate rains are generally better for agriculture than the same amount of rainfall over a short period, which may result in rapid runoff and flash flooding, leaving the deeper soil layers dry. The timing and duration of the rainy season inter alia determine the planting dates and the selection of crop types. Despite their obvious importance, however, the changes in rainfall characteristics are seldom analysed in observations (e.g. Reason et al., 2005) or regional-scale climate simulations in Africa.

Changes in the scale and shape of the rainfall distribution, which may alter tail probabilities, affect the physical and natural systems more than changes in its central tendency (Easterling et al., 2000). An evaluation of the ability of climate models in simulating the entire precipitation probability distribution at regional scales (e.g. Perkins et al., 2007) while clearly warranted has not yet been done for southern Africa. The present study attempts to address these issues by assessing the ability of the Coordinated Regional Downscaling Experiment (CORDEX) regional climate models (RCMs) in capturing monthly rainfall evolution, selected rainfall characteristics and the observed rainfall probability density functions (PDFs) in their control simulations of daily rainfall events over predefined rainfall regions (Shongwe et al., 2009) in Africa south of 10°S.

Since the launch of the CORDEX Africa programme (Jones et al., 2011), a few studies have evaluated the RCM performance over parts of the continent, including southern Africa (Nikulin et al., 2012; Kalognomou et al., 2013). Monthly and seasonal data have been used to assess the RCM's ability to simulate the main features of seasonal mean rainfall distribution and the rainfall annual cycle. Notwithstanding biases in some regions and seasons, such as a wet bias (dry bias) close to Lesotho (over northern Mozambique) during austral summer months, the models were found to adequately simulate precipitation patterns (Kalognomou et al., 2013). However, monthly and seasonal averages or totals can conceal systematic biases in the simulated climate (e.g. Tadross et al., 2005). Also, given that cumulative effects of weather events on daily time scales have a direct impact on natural systems and human activities, an assessment of model ability to simulate the characteristics of daily rainfall events is clearly valuable.

Over southern Africa, seasonal rainfall characteristics have become a subject of interest in recent years/decades, particularly their relevance for agriculture (Tadross et al., 2005, 2009). Tadross et al. (2005), using observation-based data up to 1997, showed that the mean onset occurs earlier (September or October) over eastern South Africa and later (November to December) over northern Mozambique and Botswana. A trend towards a later onset was found over northeastern South Africa. Using seven Coupled Model Intercomparison Project Phase 3 (CMIP 3) global climate models statistically downscaled to station data, Tadross et al. (2009) projected a reduction in mid twenty-first century (2046–2065) austral spring precipitation, and an increase in the autumn, suggesting a shift of the rainy season to later dates over southern Africa north of about 20°S. No attempt was made to quantify the uncertainty associated with the downscaling procedure. The CORDEX Africa programme produces large volumes of data necessary to provide future projections of high-frequency precipitation statistics and a quantification of the inherent uncertainties. This paper provides the necessary foundation for such analyses by assessing the models' ability to replicate some rainfall statistics in their control simulations.

2. Data and methodology

Monthly and daily rainfall simulated by 10 CORDEX RCMs over the common 17-year period (1991–2007), driven by the ERA-Interim reanalysis (Simmons et al., 2006), are used. The RCMs include: (1) the Université du Québec à Montréal fifth-generation Canadian Regional Climate Model (CRCM5), (2) the Universidad de Cantabria Weather Research and Forecasting Model, version 3.1.1 (WRF3.1.1), (3) the Sveriges Meteorologiska och Hydrologiska institut (SMHI) Rossby Centre Regional Atmospheric Climate Model, version 3.5 (RCA3.5), (4) the Max Planck Institute Regional Model (REMO), (5) the Consortium for Small-scale Modelling (COSMO) Climate Limited-Area Model, version 4.8 (CCLM4.8), (6) the Centre National de Recherches Météorologiques Action de Recherche Petite Echelle Grande Echelle, version 5.1 (ARPEGE5.1), (7) the Abdus Salam International Centre for Theoretical Physics RCM, version 3 (RegCM3), (8) the University of Cape Town Providing Regional Climates for Impacts Studies (PRECIS), (9) the Danmarks Meteorologiske Institut HIRHAM, version 5 (HIRHAM5; the HIRHAM5 has days with missing data and has been omitted in the analysis of onset and withdrawal of the rainy season) and the Koninklijk Nederlands Meteorologisch Instituut Regional Atmospheric Climate Model, version 2.2b (RACMO2.2b). The RCM setup details and relevant

references are presented in Nikulin et al. (2012) and Kalognomou et al. (2013).

Owing to the dearth and/or inaccessibility of observational data over much of the study area, the ERA-Interim reanalysis is used as the reference data set to assess model performance. Throughout this paper, reference data refer to the ERA-Interim precipitation. Given the uncertainty associated with observation-based data sets (Kalognomou et al., 2013; Sylla et al., 2013), the Global Precipitation Climatology Project (GPCP; Huffman et al., 2009) data, which are available from 1997, are also used for comparison.

The homogeneous rainfall regions over southern Africa defined by Shongwe et al. (2009) are adopted in this study. CORDEX simulation of monthly spatially averaged precipitation is assessed using Taylor diagrams (Taylor, 2001), which graphically synthesize the degree of correspondence between RCMs and the reference data in terms of the phase and amplitude of their evolution, measured by Pearson correlation coefficients, the centred root-mean-square error (RMSE) and a comparison of their variances. Taylor diagrams are widely used to evaluate the multiple aspects of complex models and gauging the relative skill of many different models (e.g. Kalognomou et al., 2013).

Onset and cessation of the wet season is defined for each region in Figure 1, assuming spatial coherence of the timing across the individual regions, from anomalous precipitation accumulation in a given day [A (day)] as

$$A\,(\text{day}) = \sum_{n=1}^{\text{day}} R\,(n) - \overline{R} \qquad (1)$$

where $R(n)$ is the daily precipitation and $\overline{R}$ is the long-term annual daily mean (Liebmann et al., 2007; Rauscher et al., 2007). The calculation started on 1 July (climatologically, the driest month; see Figure 2 of Shongwe et al., 2009). The onset of the rainy season is defined as the date on which the curve reaches a minimum, since after that date precipitation exceeds the annual daily climatology and before that date precipitation is less than the annual climatology. Prior to onset, there are often brief periods of precipitation causing the curve to move upward, but these are considered 'false' onsets, because the curve ultimately falls to its absolute minimum for the year. Similarly, cessation is defined as the date on which the curve reaches the maximum, since after that date precipitation is less than climatology (e.g. Figure 2 of Liebmann et al., 2007). Pearson correlation coefficients and RMSE are computed between the simulated and reference onset and cessation dates. Statistical significance of the computed correlation is assessed using a parametric Student's t-test.

Regional PDFs are constructed for the reference data sets and CORDEX RCMs by considering all the grid points falling within a region, omitting daily precipitation values <0.1 mm day^{-1} at any given grid. The PDFs are compared visually.

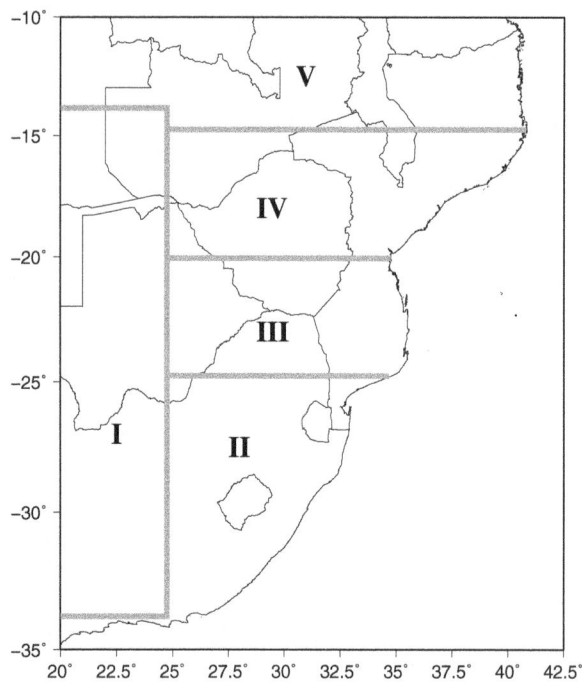

Figure 1. Location map of the Southern Africa homogeneous climate regions defined by Shongwe *et al.* (2009).

3. Simulation of monthly precipitation

There are notable spatial inhomogeneities and inter-model differences in simulating the phasing and amplitude of monthly rainfall evolution (Figure 2). Relative to the south (Figure 2(a)–(c)), the simulated monthly precipitation to the north of about 20°S (Regions IV and V) is in better agreement with the reference data (Figure 2(d) and (e)). Almost all RCMs attain correlation scores >0.5, centred RMS differences (the curved contours in each plot) ≤2 mm day^{-1} and have their intermonthly standard deviation close to the reference data in Regions IV and V. To the south (Regions I, II and III), the pattern correlation is lower (≤0.6) in most RCMs, but statistically significant (from re-randomization tests, not shown), indicating that the downscaled monthly precipitation can still provide useful information in these regions.

APERGE5.1 stands out as the RCM with the best agreement with the reference data. This model's pattern correlation exceeds 0.8 everywhere, and the RMS differences between this model's downscaled precipitation and the ERA-Interim reanalysis are the smallest. The WRF3.1.1 model, albeit comparable with the majority of the RCMs in its ability to simulate the phase of the monthly rainfall evolution, has the largest amplitude of precipitation variation.

4. Seasonal rainfall onset and withdrawal

The observed rainfall annual cycle plots for each climatic region are shown in Figure 2 of Shongwe *et al.* (2009). The rainy season begins first in the southeast

and progresses northward. Earlier onset over eastern South Africa (Region II) is associated with moisture advection from the warm Agulhas system and instabilities induced by mid-latitude disturbances (Tyson and Preston-Whyte, 2000). The spatial progression of the regional rainfall onset in the RCMs is broadly consistent with the reference and the GPCP data, which often present similar estimates of the rainy season timing for the common period 1997/1998–2007/2008, a notable exception being the 2002/2003 season. In ERA-Interim, the mean (median) rainfall onset date in Region II is 22 October (18 October), with an interannual standard deviation (σ) of 15 days. In the GPCP data, the mean onset date and σ for eastern South Africa (i.e. Region II) are 23 October and 17 days. Albeit not statistically significant, the trend towards later onset of 1 day year^{-1} in the GPCP data over eastern South Africa is consistent with Tadross *et al.* (2005) and the broadly held notion that in parts of southern Africa, the start of the rainy season has shifted to later dates in recent years/decades. In the ERA-Interim, a weaker delayed onset (4 days decade^{-1}) is only found in the late twentieth century.

Most of the CORDEX RCMs adequately capture the mean rainfall onset in Region II, falling within ±7 days of the reference data (Figure 3(b)). To the southwest (Region I), the mean start of the rainy season occurs around 8 November and 16 November in the ERA-Interim and GPCP reference data sets, respectively. A trend towards a delayed onset (3 ± 2 days year^{-1}) is found during the last decade of the twentieth century. With reference to ERA-Interim, RegCM3 has the overall early mean onset bias of ≈14 days, while the onset in the CCLM4.8 is delayed by about 11 days. Except these two and the RCA35, the mean onset dates in the rest of the RCMs fall within a 7 day margin from the ERA-Interim estimate in Region I (Figure 3(a)). To the north, where moisture recycling ratios are relatively higher during austral summer months (Trenberth, 1999), the seasonal rainfall onset occurs slightly later, around 12 (09), 19 (14) and 11 (12) November in Regions III, IV and V in the ERA-Interim (GPCP) data, respectively, with a relatively low interannual variability (Figure 3(c)–(e)). In Region III, all the RCMs except the RegCM3 have onset dates between 07 and 19 November (Figure 3(c)). In the RegCM3, the mean rainfall onset occurs around 25 October. Over northern Zimbabwe and central Mozambique (Region IV), the onset bias is low, with the CCLM4.8 showing a slight late bias (Figure 3(d)). The onset bias is similarly quite low over the northern-most region (Region V) will almost all RCMs simulating mean onset dates within ±7days of 11 November.

Correlations between reference and simulated onset dates are shown in Table 1. It is evident from the table that the RCMs capture the interannual variability of the rainfall onset reasonably well. In particular, the ARPEGE5.1 attains statistically significant correlations and lowest RMSE almost everywhere. There are a few cases when the RCM simulated onset dates are out of

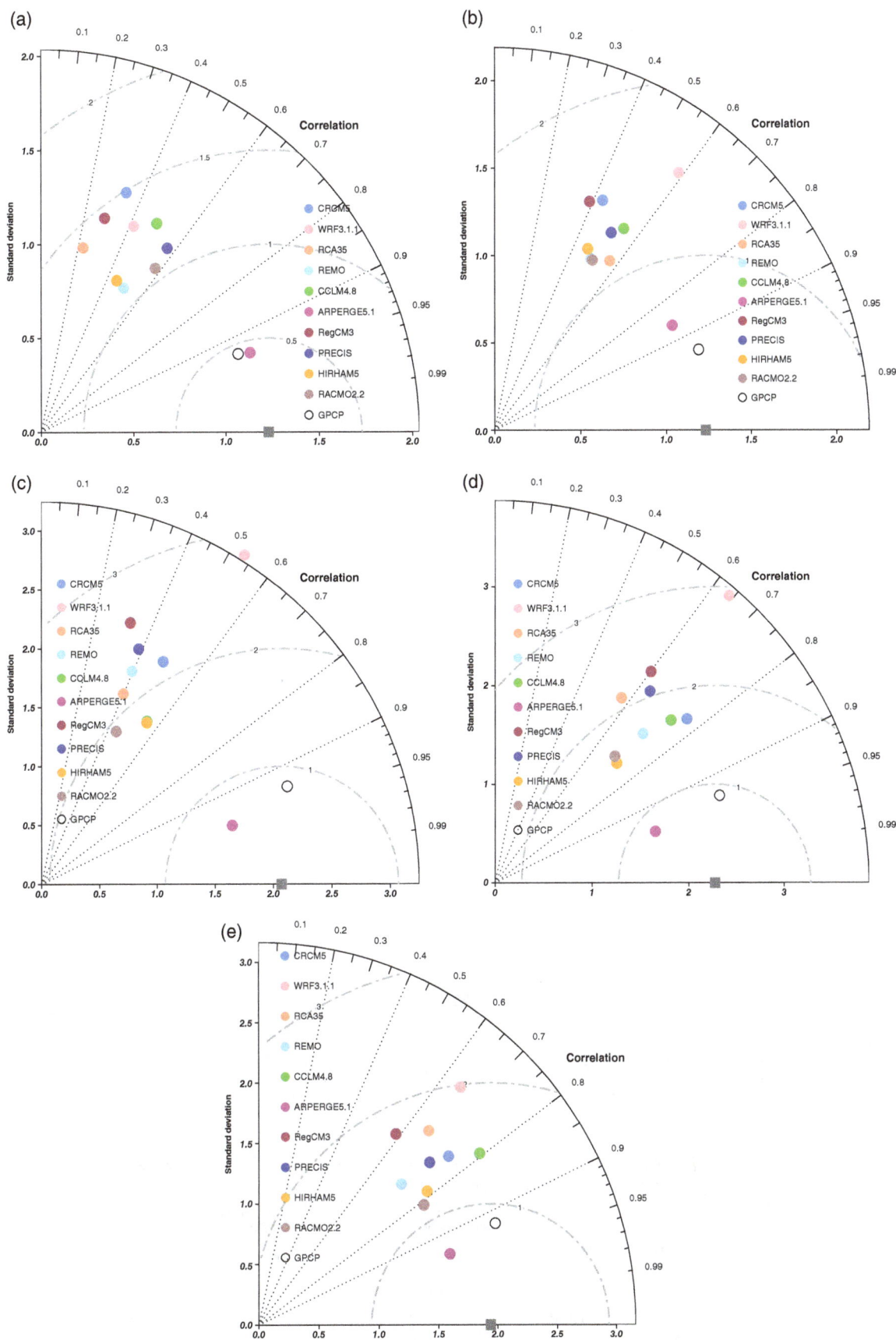

Figure 2. Taylor diagrams for area-averaged monthly precipitation in Regions (a) I, (b) II, (c) III, (d) IV and (e) V shown in Figure I. The reference (ERA-Interim) data are shown by the grey square along the horizontal axis. The individual RCMs are shown by the solid circles and the GPCP by the open circle. The radial coordinate shows the standard deviation. The azimuthal axis shows the correlation between the RCMs and the reference data. The centred root-mean square error is indicated by the dashed grey semi-circles about the reference point.

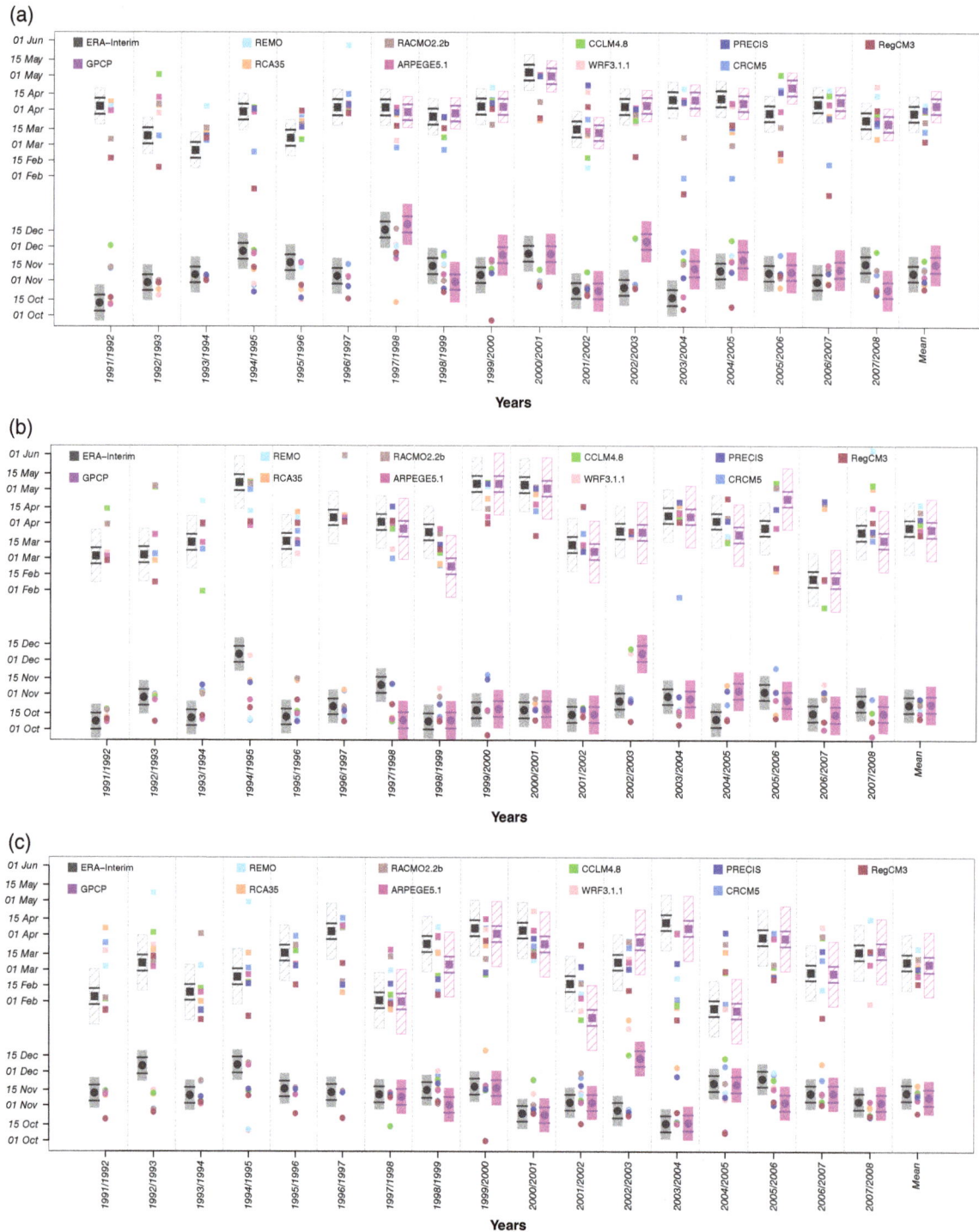

Figure 3. ERA-Interim (grey rectangles), GPCP from 1997/1998 (orchid rectangles) and CORDEX-simulated (individual points) interannual onset (solid circles) and cessation (solid squares) dates of the wet season for Regions (a) I, (b) II, (c) III, (d) IV and (e) V. In each figure, the onset (cessation) date is plotted at the lower (upper) ordinate running from 01 October to 15 December (01 February to 01 June). For the reference plots, the rectangles extend to $\pm 1\sigma$ of interannual variability from the estimated date, while the line segments within each rectangle extend to ± 7 days.

phase with the reference data (e.g. CRCM5 and REMO in Region II, RCA3.5 in Region III and WRF3.1.1 in Region IV). The weak correlation between the GPCP and ERA-Interim in Regions I and II demonstrates how different observational-based data sets may disagree on their representation of precipitation characteristics (e.g. Kalognomou *et al.*, 2013).

In the reference data, rainfall withdrawal does not show any notable spatial migration. Almost everywhere, except in Region III, the mean end date occurs between 28 March and 05 April (27 March and 06 April) in ERA-Interim (GPCP). In Region III, which has long been known to have strong ENSO teleconnections (Ropelewski and Halpert, 1987; Rocha

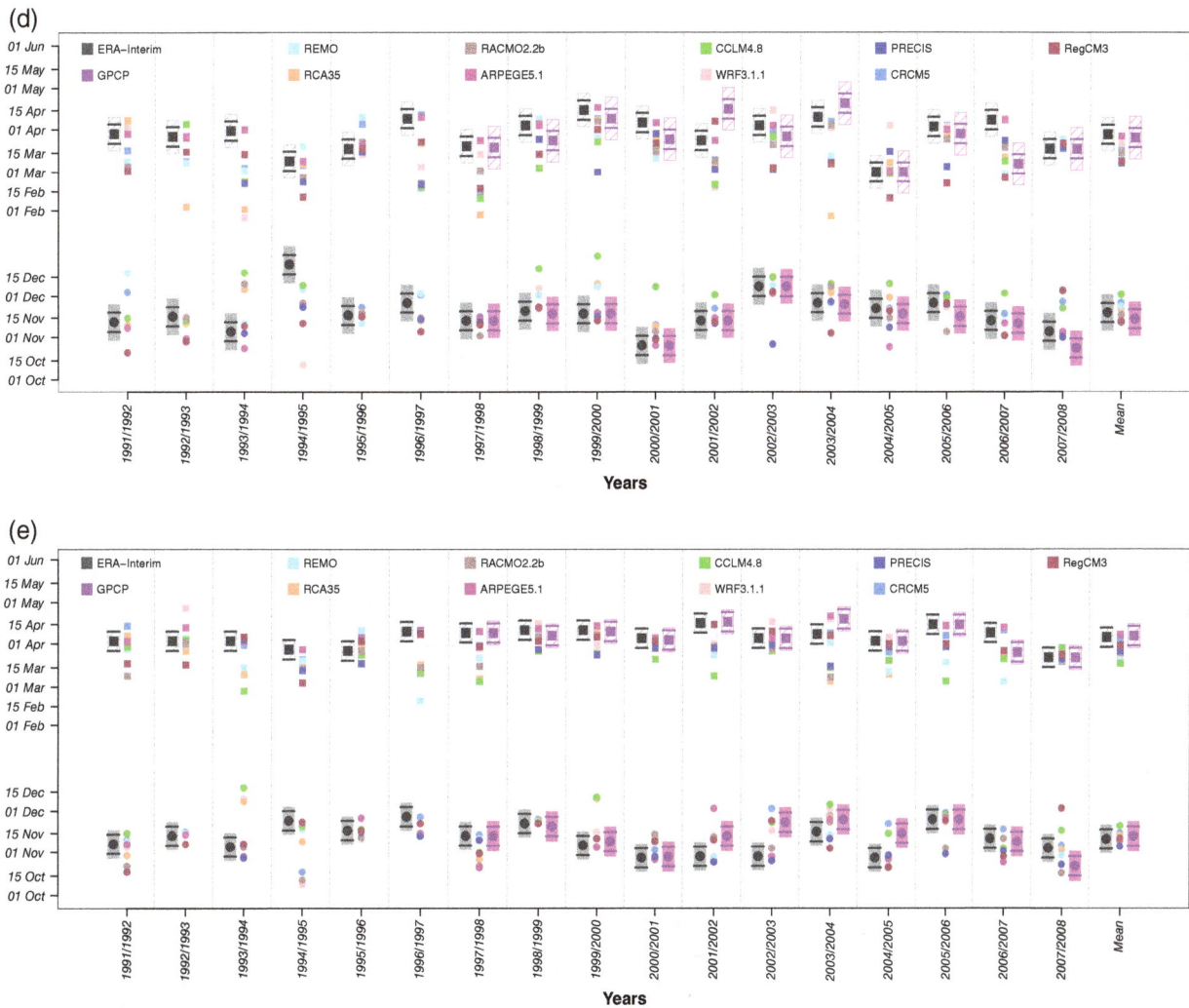

Figure 3. continued

Table 1. Correlation between the RCMs and the reference ERA-Interim onset dates.

RCM	Region I	Region II	Region III	Region IV	Region V
GPCP	0.48 (17)	0.13 (18)	**0.67**** (8)	**0.94**** (4)	**0.60**** (10)
ARPEGE5.1	**0.73**** (12)	**0.48** (13)	**0.59**** (12)	**0.69**** (10)	**0.49**** (13)
CCLM4.8	0.34 (16)	**0.45** (13)	0.29 (21)	0.16 (18)	0.15 (15)
CRCM5	0.16 (18)	−0.08 (21)	0.29 (23)	0.39 (13)	0.28 (14)
PRECIS	0.22 (16)	0.40 (15)	0.04 (15)	**0.46** (13)	**0.78**** (7)
RACMO2.2b	**0.56**** (13)	**0.46** (17)	0.34 (20)	**0.41** (13)	0.17 (15)
RCA3.5	0.18 (17)	0.08 (18)	−0.23 (31)	0.39 (14)	0.13 (16)
RegCM3	**0.63**** (13)	0.32 (14)	0.12 (23)	0.26 (15)	0.21 (15)
REMO	0.38 (17)	−0.09 (18)	**0.69**** (10)	**0.50**** (13)	**0.62**** (9)
WRF3.1.1	**0.67**** (12)	**0.46** (17)	**0.53**** (12)	−0.18 (22)	0.06 (18)

The RMSE (days) for each RCM is shown in brackets. Correlations that are significant at the 10% (5%) level are in bold (shown by two asterisks).

and Simmonds, 1997), the mean cessation, which occurs around 08 and 06 March in the reference data and GPCP, is influenced by six early cessation seasons (**1991/1992**, 1993/1994, **1994/1995**, **1997/1998**, 2001/2002 and **2004/2005**). In these extreme years, four (in bold font) of which coincided with a warm ENSO phase, the seasonal rainfall ends either late in

January (e.g. 2004/2005) or early to mid-February. Statistically significant correlations (at the 10% level) have been found between ENSO and early rainfall cessation over a number of stations in Zimbabwe, southern Mozambique and Malawi (Tadross *et al.*, 2009). Noteworthy, the southeastern regions (Regions II and III) have a high interannual variability ($\sigma \geq 25$ days) of the rainfall withdrawal (Figure 3(b) and (c)).

The pattern of rainfall cessation in the RCMs closely resembles that of the reference data, with some models showing an excellent agreement even during the unusual years. In Region I, all except two of the RCMs (RegCM3 and CRCM5) have end dates within ±7 days of the reference data. In RegCM3, the mean end date occurs quite early (04 March), while in CRCM5, the average cessation occurs around 13 March, 16 days earlier than in ERA-Interim. Over eastern South Africa (Region II), the largest biases are found in REMO and RACMO2.2 where the end of the rainy season is delayed by 21 and 14 days, respectively. In almost all the RCMs, the cessation dates vary greatly from year to year, in close agreement with the reference data (Figure 3(b)). In Region III, early withdrawals in the

Table 2. Correlation between the RCMs and the reference ERA-Interim cessation dates.

RCM	Region I	Region II	Region III	Region IV	Region V
GPCP	**0.84**** (7)	**0.88**** (12)	**0.93**** (11)	**0.54** (13)	**0.75**** (5)
ARPEGE5.1	**0.82**** (9)	**0.79**** (14)	**0.84**** (9)	**0.78**** (8)	**0.81**** (4)
CCLM4.8	**0.46** (19)	**0.60**** (9)	0.36 (28)	0.23 (16)	−0.29 (14)
CRCM5	−0.09 (29)	0.33 (31)	**0.43** (28)	**0.65**** (12)	**0.55**** (7)
PRECIS	**0.56**** (14)	**0.44** (21)	0.06 (31)	0.04 (17)	**0.48** (7)
RACMO2.2b	0.12 (18)	**0.72**** (15)	0.29 (30)	**0.47** (14)	0.25 (10)
RCA3.5	0.20 (18)	**0.46** (24)	0.07 (35)	0.17 (24)	0.17 (11)
RegCM3	0.36 (24)	**0.50**** (22)	**0.53**** (25)	**0.47** (15)	**0.55**** (8)
REMO	0.25 (27)	**0.61**** (18)	0.16 (38)	**0.64**** (19)	−0.19 (15)
WRF3.1.1	**0.47** (17)	0.38 (25)	**0.82**** (17)	0.37 (21)	0.32 (9)

The RMSE (days) for each RCM is shown in brackets. Correlations that are significant at the 10% (5%) level are in bold (shown by two asterisks).

RegCM3 occur often, which results in an average negative bias of 18 days. Further north (Region IV), where the ability to simulate monthly rainfall is highest, negative biases in RCM end dates in excess of 18 days are found in PRECIS, CCLM4.8, RCA3.5 and RegCM3. Over the north-most Region V, disagreements between average reference and models' end dates still persist (Figure 3(e)).

A much stronger agreement is found between the ERA-Interim and GPCP in terms of interannual variability of rainfall cessation (Table 2). The ARPEGE5.1 has the highest correlations in all the regions. With only a few exceptions (e.g. CCLM4.8 and REMO in Region V, and CRCM5 in Region I), the RCM simulated rainfall withdrawal is positively correlated with the reference dates indicating that the models are able to capture the interannual variation of the rainy season cessation.

5. Rainfall probability density functions

Daily precipitation PDFs for each region are shown in Figure 4(a)–(e). The x axes of the plots terminate at about 40 mm day^{-1} because at higher values exceeding 20 mm day^{-1}, the individual curves become almost indistinguishable. We show that very intense events ($\geq$40 mm day^{-1}) are simulated in both the reference data and in the RCMs, demonstrating that the models can provide useful information in relation to flood risks.

Although drizzle and light rainfall events $\leq$3 mm day^{-1} occur too often in a few of the RCMs (CCLM4.8, REMO and RCA3.5), in most regions, the majority have

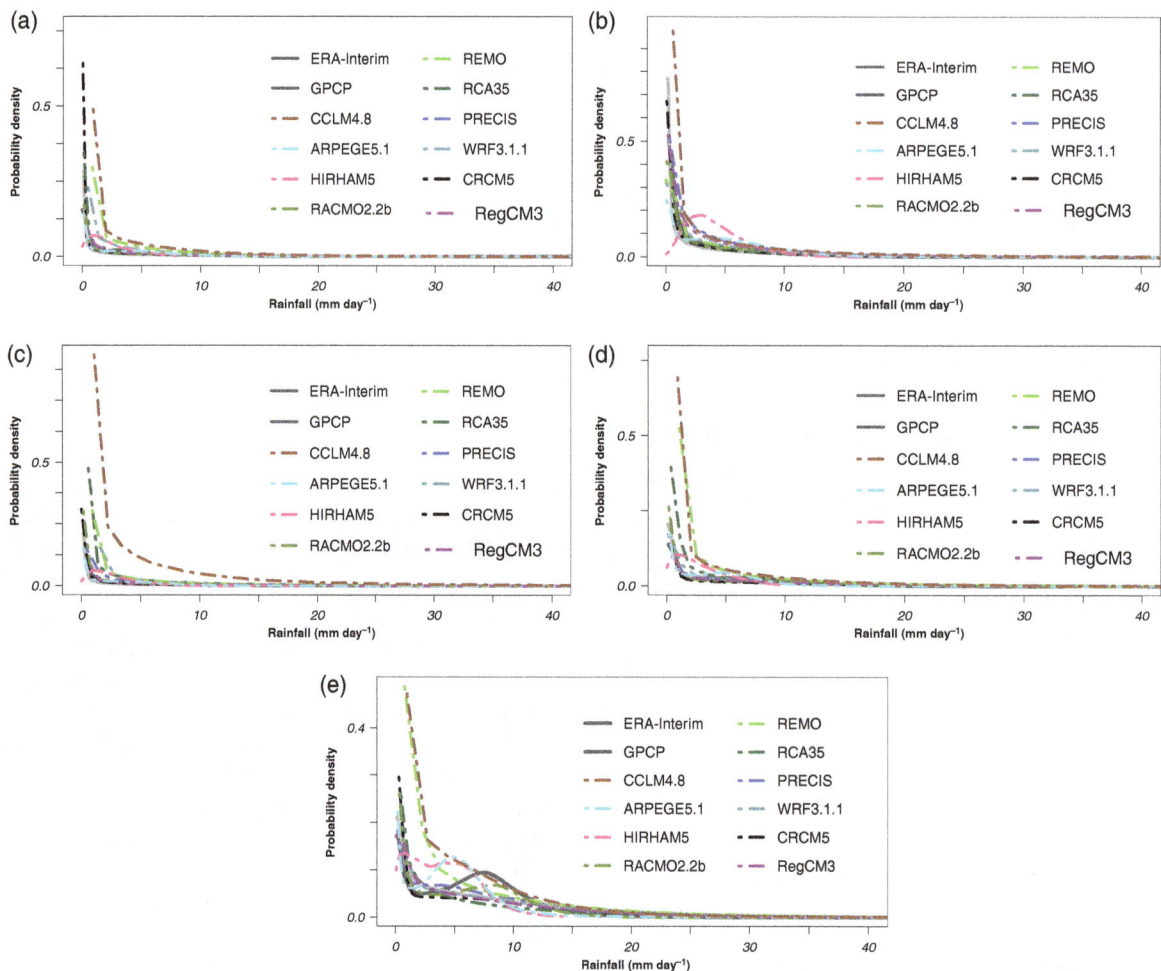

Figure 4. Reference (solid lines) and CORDEX (dot-dashed lines) precipitation probability density functions for Regions (a) I, (b) II, (c) III, (d) IV and (e) V.

low-tail probabilities comparable with the reference data. In most cases, the CCLM4.8, and in some cases, the REMO overestimate the probability of light rainfall by about 2–3 times. The RACMO2.2, the ARPEGE5.1 and the HIRHAM5 get very close to capturing the secondary peak probabilities located around 5–10 mm day^{-1} in the ERA-Interim data in Region V.

6. Summary and conclusions

The present study has enabled us to get to grips with how well the CORDEX RCMs simulate monthly rainfall amounts and some regional rainfall characteristics over southern Africa. The models have been evaluated against the ERA-Interim reanalysis, whilst the GPCP data have been used for comparison. In simulating monthly rainfall variability, the overall performance of the RCMs is good, particularly to the north of the study domain. The stretched grid ARPEGE5.1 appears to the best overall performer, in qualitative agreement with Kalognomou *et al.* (2013), where the individual RCM setups are presented. This RCM's correlation with the ERA-Interim data is highest almost everywhere in southern Africa and the biases are least. No attempt is made in this paper to identify the model physics and dynamics responsible for the differences in RCM performance. Other aspects of the simulated precipitation that are of practical significance such as rainy season timing and the relative frequencies of a spectrum of rainfall intensities have also been analysed. The spatial migration of the seasonal rainfall onset is well captured by the RCMs. The reference and simulated average onset and withdrawal dates are within a few days of each other, and the interannual variability is captured by the RCMs, a few notable biases notwithstanding. A visual comparison of the reference and simulated PDFs shows that a majority of the models do quite well in representing the probabilities of certain rainfall events.

There are prospects for the CORDEX RCMs to provide usable information on regional monthly precipitation, rainfall characteristics and probabilities of events of differing intensities over southern Africa, and their likely future changes, for climate change impacts assessments where impacts are related to precipitation. The uncertainties associated with the modelling process from the driving climate to impacts are yet to be quantified. One of the goals for future work is to investigate and compare the physical mechanisms underlying rainfall variability and changes in rainfall character in observations and in the models.

Acknowledgements

Review comments by Dr. Rachel James and two anonymous reviewers are greatly appreciated. This paper is a contribution of the Southern African Analysis group to the CORDEX-Africa programme. The CORDEX-Africa programme was supported by the Global Change System for Analysis, Research, and Training (START) through the Climate Systems Analysis Group of the University of Cape Town. Support from the World Climate Research Program (WCRP), the Climate and Development Knowledge Network (CDKN), the International Centre for Theoretical Physics (ICTP), the Swedish Meteorological and Hydrological Institute (SMHI) and the European Union Seventh Framework Programme is gratefully acknowledged. Special thanks to the modelling groups contributing to the CORDEX-Africa program and to Grigory Nikulin from SMHI for post-processing the data to a common grid, data format and domain size to enable direct comparison in the analyses.

References

Easterling DR, Meehl GA, Parmesan C, Changnon SA, Karl TR, Mearns LO. 2000. Climate extremes: observations, modeling, and impacts. *Science* **289**: 2068–2074.

Huffman GJ, Adler RF, Bolvin DT, Gu G. 2009. Improving the global precipitation record: GPCP version 2.1. *Geophysical Research Letters* **36**, doi: 10.1029/2009GL04000.

Jones C, Giorgi F, Asrar G. 2011. The Coordinated Regional Downscaling Experiment (CORDEX). An international downscaling link to CMIP5. *Clivar Exchanges* **16**: 34–40.

Kalognomou E-A, Lennard C, Shongwe M, Pinto I, Favre A, Kent M, Hewitson B, Dosio A, Nikulin G, Panitz H-J, Büchner M. 2013. A diagnostic evaluation of precipitation in CORDEX models over southern Africa. *Journal of Climate* **26**: 9477–9506.

Liebmann B, Camargo SJ, Seth A, Marengo JA, Carvalho LMV, Allured D, Fu R, Vera CS. 2007. Onset and end of the rainy season in South America in observations and the ECHAM 4.5 atmospheric general circulation model. *Journal of Climate* **20**: 2037–2050.

Nikulin G, Jones C, Giorgi F, Asrar G, Büchner M, Cerezo-Mota R, Christensen OB, Déqué M, Fernandez J, Hänsler A, van Meijgaard E, Samuelsson P, Sylla MB, Sushamak L. 2012. Precipitation climatology in an ensemble of CORDEX-Africa regional climate simulations. *Journal of Climate* **25**: 6057–6078.

Perkins SE, Pitman AJ, Holbrook NJ, Macaneney J. 2007. Evaluation of the AR4 climate models' simulated daily maximum temperature, minimum temperature, and precipitation over Australia using probability density functions. *Journal of Climate* **20**: 4356–4376.

Rauscher SA, Seth A, Liebmann B, Qian J-H, Camargo SJ. 2007. Regional climate model simulated timing and character of seasonal rains in South America. *Monthly Weather Review* **135**: 2642–2657.

Reason CJC, Hachigonta S, Phaladi RF. 2005. Interannual variability in rainy season characteristics over the Limpopo region of southern Africa. *International Journal of Climatology* **25**: 1835–1853.

Rocha A, Simmonds I. 1997. Interannual variability of south-eastern African summer rainfall. Part 1: relationships with air–seas interaction processes. *International Journal of Climatology* **17**: 235–265.

Ropelewski CF, Halpert MS. 1987. Global and regional scale precipitation patterns associated with the El Niño/Southern Oscillation. *Monthly Weather Review* **115**: 1606–1626.

Seneviratne S, Nicholls N, Easterling D, Goodess CM, Kanae S, Kossin J, Luo Y, Marengo J, McInnes K, Rahimi M, Reichstein M, Sorteberg A, Vera C, Zhan X. 2012. Changes in climate extremes and their impacts on the natural physical environment. In *Managing the Risks of Extreme Events and Disasters to Advance Climate Change Adaptation (SREX). Special Report of the Intergovernmental Panel on Climate Change (IPCC)*, Field CB, Barros V, Stocker TF, Dahe Q (eds). Cambridge University Press: Cambridge, UK and New York, NY; 109–230.

Shongwe ME, van Oldenborgh GJ, van den Hurk BJJM, de Boer B, Coelho CAS, van Aalst MK. 2009. Projected changes in mean and extreme precipitation in Africa under global warming. Part I: southern Africa. *Journal of Climate* **22**: 3819–3837.

Simmons A, Uppala S, Dee D, Kobayashi S. 2006. ERA-Interim: new ECMWF reanalysis products from 1989 onwards. *ECMWF Newsletter* **110**: 25–35.

Sylla MB, Giorgi F, Coppola E, Mariotti L. 2013. Uncertainties in daily rainfall over Africa: assessment of gridded observation products and evaluation of a regional climate model simulation. *International Journal of Climatology* **33**: 1805–1817.

Tadross MA, Hewitson BC, Usman MT. 2005. The interannual variability of the onset of the maize growing season over South Africa and Zimbabwe. *Journal of Climate* **18**: 3356–3372.

Tadross M, Suarez P, Lotsch A, Hachigonta S, Mdoka M, Unganai L, Lucio F, Kamdonyo D, Muchinda M. 2009. Growing-season rainfall and scenarios of future change in southeast Africa: implications for cultivating maize. *Climate Research* **40**: 147–161.

Taylor KE. 2001. Summarizing multiple aspects of model performance in a single diagram. *Journal of Geophysical Research* **106**: 7183–7192.

Trenberth KE. 1999. Atmospheric moisture recycling: role of advection and local evaporation. *Journal of Climate* **12**: 1368–1381.

Trenberth KE, Dai A, Rasmussen RM, Parsons DB. 2003. The changing character of precipitation. *Bulletin of the American Meteorological Society* **84**: 1205–1217.

Tyson PD, Preston-Whyte RA. 2000. *The Weather and Climate of Southern Africa*. Oxford University Press: Cape Town, South Africa; 396 pp.

23

Assessing the contribution of different factors in regional climate model projections using the factor separation method

Csaba Torma* and Filippo Giorgi

Earth System Physics Section, The Abdus Salam International Centre for Theoretical Physics, Trieste, Italy

*Correspondence to:
Cs. Torma, Earth System Physics
Section, The Abdus Salam
International Centre for
Theoretical Physics, Trieste
I-34151, Italy.
E-mail: ctorma@ictp.it*

Abstract

This study applies the factor separation (FS) method to investigate the contributions of different factors, along with their synergy, on a set of regional climate model (RCM) projections for the Mediterranean region. The FS method is applied to six projections for the period 1970–2100 performed with the regional model RegCM4 over the Med-CORDEX domain. Two different sets of factors are intercompared, namely the driving global climate model (GCM) boundary conditions against two model physics settings (convection scheme and irrigation). We demonstrate the usefulness of the FS method to assess different sources of uncertainty in RCM-based regional climate projections.

Keywords: factor separation; climate change; regional climate modeling

1. Introduction

Following the pioneering work of Stein and Alpert (1993), the factor separation (FS) method has been widely used for numerical studies in the last two decades in order to assess the linear and nonlinear interactions among factors which govern processes on different scales and in different spheres (Alpert *et al.*, 1996;Romero *et al.*, 1997; Claussen *et al.*, 2001; Alpert *et al.*, 2006; Jankov *et al.*, 2007; Teller and Levin, 2008; Healy *et al.*, 2009; Lynn *et al.*, 2009; Alpert and Sholokhman, 2011; Kumar *et al.*, 2013). An area where the FS method can be especially useful is in separating the contributions of different elements (or sources of uncertainty) involved in generating climate change projections and in assessing the importance of the synergies across these elements. Climate projections are indeed characterized by a variety of sources of uncertainties (Giorgi, 2005), and the identification of the relative importance of these sources may help their understanding.

As a contribution to the COordinated Regional Downscaling EXperiment (CORDEX, Giorgi *et al.*, 2009), a project aimed at producing large ensembles of regional model projections over regions worldwide, a set of 14 scenario projections was completed (extending from 1970 to 2100) for the Mediterranean CORDEX (Med-CORDEX) domain with the International Centre for Theoretical Physics (ICTP) regional climate model (RCM) RegCM4 (Giorgi *et al.*, 2012). The projections employ different combinations of seven model physics configurations and two driving global climate models (GCMs), the HadGEM2-ES (HadGEM) and MPI-ES-MR (MPI). This ensemble includes a combination of experiments in which different model components are changed individually and in combination, and thus lends itself optimally to the application of the FS method.

In this paper we apply the FS to a suitable subset of our Med-CORDEX ensemble. Specifically, we selected six simulations providing a combination of experiments that allowed us to separate the contributions of different RCM physics options (convection scheme and presence of irrigation) and GCM boundary conditions on temperature and precipitation over the Mediterranean basin. Boundary conditions and physics schemes are indeed two of the main sources of uncertainty in RCM-based regional projections (Giorgi, 2005). In addition, the analysis is carried out for different 20th and 21st century time slices, 1976–2005, 2010–2039, 2040–2069 and 2070–2099, in order to assess the temporal evolution of the different factor contributions.

2. Experiment design and analysis technique

2.1. Model simulations and study area

The region of interest in this work is the Med-CORDEX domain (Figure 1), which includes the Mediterranean region, North Africa, Middle East, and Central/Southern Europe. The model used in this study is the latest version of RegCM4 (Giorgi *et al.*, 2012), which is a hydrostatic and terrain following sigma vertical coordinate model with multiple physics options. Among those of interest in this study are: the Biosphere-Atmosphere Transfer Scheme (BATS) (Dickinson *et al.*, 1993) for the representation of land surface processes and the cumulus convection schemes of Grell (1993) and Emanuel and Zivkovic-Rothman

Table I. Experiments used for the factor separation analysis.

Experiment	Driving GCM	Convective scheme	Irrigation	Ensemble 1	Ensemble 2
I	HadGEM	Grell + Emanuel	no	f_0	f_0
2	MPI	Grell + Emanuel	no	f_1	f_1
3	HadGEM	Grell	no	f_2	–
4	MPI	Grell	no	f_{12}	–
5	HadGEM	Grell + Emanuel	yes	–	f_2
6	MPI	Grell + Emanuel	yes	–	f_{12}

The factors are as follow: different driving GCMs, different convective schemes and irrigation of crop fields. Two simulations (experiments I and 2) were used for both ensembles (Ens I and Ens2).

(1999). The RegCM modeling system has been used for more than two decades for a wide variety of applications, and the most up-to-date version of this community RCM is available at the following URL: http://gforge.ictp.it/gf/project/regcm/frs/.

The simulation period, 1970–2100, covers the late 20th century and the entire 21st century and the model grid spacing is 50 km, i.e. the coarse resolution setting of Med-CORDEX. The initial and lateral meteorological boundary conditions for the scenario runs were provided by two GCMs involved in the CMIP5 experiment (Taylor *et al.*, 2012): HadGEM (Collins *et al.*, 2011) and MPI (Jungclaus *et al.*, 2010). The selection of these GCMs was based on a preliminary analysis showing that they performed relatively well over the European region (see also Brands *et al.*, 2013). In addition, we carried out projections for two representative concentration pathways (RCPs), i.e. RCP8.5 and RCP4.5 (Moss *et al.*, 2010), corresponding to high-end and medium greenhouse gas concentration scenarios, respectively, but in this paper we only selected runs for the RCP8.5.

In total, 14 simulations were carried out with seven model physics configurations including various convection and land surface schemes. Of these 14 simulations, 6 fitted the needs of the FS method, as indicated in Table I. Our analysis focuses on the effects of applying different parameterization settings and driving GCMs on the simulation of surface air temperature and precipitation. We analyzed two sub-ensembles of experiments targeting different factors. Ensemble 1 (Ens1) includes four simulations using different driving GCMs (HadGEM and MPI) and convective schemes (Grell and Grell/Emanuel mixed convection, see Giorgi *et al.*, 2012), while Ensemble 2 (Ens2) considers the different driving GCMs along with the effect of irrigation over crop covered areas.

The irrigation experiments were originally carried out to investigate the effect of irrigation (as an adaptation tool) on the climate change signal, given that soil moisture has been proven to be an important element of climate projections over the Mediterranean (Seneviratne *et al.*, 2006). Following Kueppers *et al.* (2007), irrigation is applied within the BATS framework during the summer growing season by not allowing the soil moisture in crop covered areas to fall below the level of 60% of saturation. Figure 1 shows the model

domain/topography and distribution of crop covered areas.

Concerning the convection schemes analyzed, the Grell scheme considers a quasi-equilibrium between convective instability generated on the large scale and its dissipation at the cumulus scale by an updraft and downdraft pair, with no direct interactions between the environment and the convective air parcel except at the top and the bottom of the circulation. In the mixed convection set-up, the Grell scheme is used over land and the Emanuel scheme over ocean. In the latter, instead of a single detraining/entraining cloud, a spectrum of convective drafts inside the cloud system are considered depending on the buoyancy of the mixture within the environment. The two schemes also differ in terms of cloud microphysics representation within the updraft.

2.2. Factor separation method

The FS method aims to investigate the contributions of different factors (in our case driving GCM, convective scheme, irrigation), both individually and along with their synergy, to the temperature and precipitation change signals over the Mediterranean basin. In order to apply the method, a minimum of four simulations are needed. Following Stein and Alpert (1993) the subsequent equations describe the effects of the different factors:

$$\widehat{f_1} = f_1 - f_0 \tag{1}$$

$$\widehat{f_2} = f_2 - f_0 \tag{2}$$

$$\widehat{f_{12}} = f_{12} - (f_1 + f_2) + f_0 \tag{3}$$

where f_0 represents the control experiment, (f_1, f_2) two experiments in which one factor is changed individually compared to f_0, and f_{12} an experiment in which both factors are changed simultaneously. Then in Equations (1)–(3) $\widehat{f_1}$ represents the contribution of factor 1 (fact1), $\widehat{f_2}$ the contribution of factor 2 (fact2) and $\widehat{f_{12}}$ (fact12) the contribution owing to their interaction (or synergy).

For both our ensembles f_0 refers to a HadGEM-driven simulation, f_1 to a corresponding MPI-driven experiment, f_2 to an experiment in which either the convection scheme is changed or the irrigation option is activated, f_{12} an experiment in which both the driving GCM and convection (or irrigation option) are changed (Table I). In this way, Equations (1)–(3) allow us to separate the contributions of driving GCMs (HadGEM vs MPI) versus physics options (convection scheme or presence of irrigation) and to assess the importance of their synergy term.

The FS as expressed by Equations (1)–(3) was calculated for summer (JJA – June-July-August) and winter (DJF – December-January-February) temperature and precipitation averaged over the subregions shown in Figure 1 and over four 30-year long time slices, 1976–2005, 2010–2039, 2040–2069 and 2070–2099,

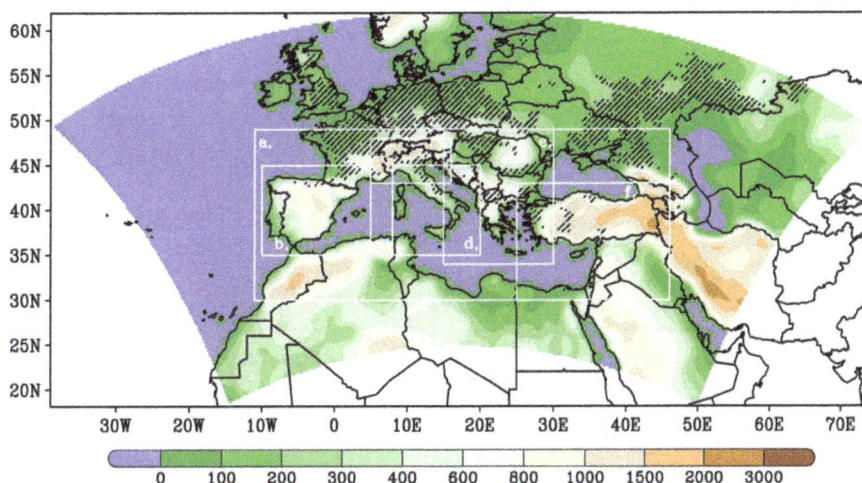

Figure 1. Model domain and topography (m) at a grid spacing of 50 km. Also shown are six subregions used in the analysis: Mediterranean (a), Spain (b), Alps (c), Adriatic (d), Balkans (e) and Turkey (f). Hatched regions indicate areas covered by crop (where the irrigation scheme is applied).

in order to assess the temporal evolution of the contributions. Only land grid points were included in the regional averages and in this paper we only report results for the summer season as those for the winter are of smaller magnitude.

3. Results

In general, the change patterns obtained in the present projections follow those found in previous generations of global and regional model projections (e.g. Giorgi and Lionello, 2008), namely a warming and a prevailing reduction in precipitation during the summer season. Figure 2 shows the summer warming and drying trends in the different simulations used here averaged over the entire Mediterranean land areas (Figure 1). Here we are specifically interested in the use of the FS method to separate different contributions associated with boundary conditions and model physics on the simulation of surface air temperature and precipitation. For temperature, Figure 3 summarizes the results of Equations (1)–(3) for both ensembles (Ens1 and Ens2), the summer season and all Mediterranean subregions of Figure 1. In Figure 3, fact1 measures the effect of the driving GCMs (common to both ensembles), fact2 the effect of convection scheme (black triangles) or irrigation (red triangles) and fact12 the synergy between the two contributions (black and red squares for Ens1 and Ens2, respectively).

For summer, over most subregions the difference arising from the different driving GCMs (fact1) increases until the mid-21st century and then it levels off. The only exception is the Alpine region, where the GCM effect keeps increasing until the end of the century. Maximum values of the driving GCM contribution are found over the Balkans, in excess of 4 °C. This contribution is essentially tied to the GCM climate sensitivity, as the HadGEM has a substantially higher sensitivity than

the MPI model. This is transferred to the RCM simulations through the lateral boundary conditions and leads to the negative values of fact1 in Figure 3. The fact that over the Alps fact1 continues to increase until the end of the century may be associated with the effect of snow melting and related snow albedo feedback mechanism (Giorgi *et al.*, 1997), which is greater in the HadGEM than MPI runs and keeps being active until the end of the simulations.

The contribution of different convection schemes on the summer temperature signal is smaller, <1 °C, while the associated convection/GCM synergy is essentially negligible. Conversely, the contribution of irrigation is significant, up to ~3 °C, especially over the regions where large irrigated crop areas are located (Balkan and Adriatic). Essentially, irrigation prevents the soil from drying, thereby reducing the associated feedback-induced warming (Seneviratne *et al.*, 2006). This results in negative values of fact2 (for irrigation) of up to 2–3 °C. The synergy contribution related to irrigation (red squares) is also considerable, order of 1 °C in the mid and end of century, and it is in the opposite direction compared to the individual effects. In other words, the effect of irrigation is more dominant in the HadGEM-driven than MPI-driven cases and thus it tends to reduce the differences across the two GCM-driven runs.

The results for summer precipitation are shown in Figure 4. First, we find that the variability in factor contributions across regions is greater than for temperature (Figure 3). This may be expected in view of the fact that precipitation generally shows a greater variability in space and time than temperature. In addition, the fraction of convective versus non-convective rain changes with regions, and this can also affect the different contributions. For precipitation the effect of changing driving GCM (fact1) appears to be generally less dominant than for temperature, indicating that for summer precipitation local processes within the RCM play an important

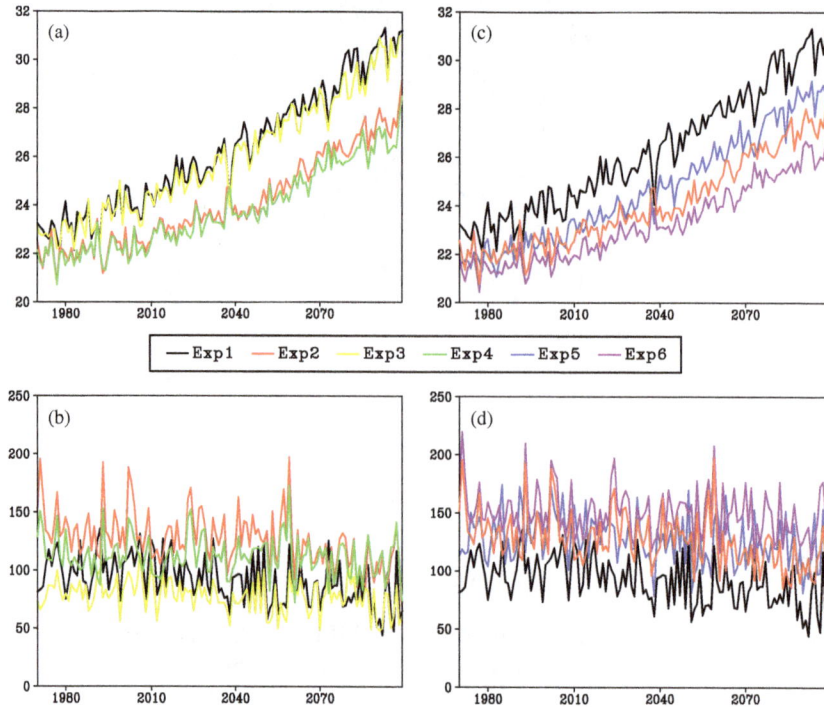

Figure 2. Summer temperature (a and c) (°C) and precipitation (b and d) (mm/season) throughout the period 1970–2100 in the six simulations used in the FS analysis averaged over all land points in the Mediterranean region. Both ensembles are depicted: Ens1 (a and b) and Ens2 (c and d).

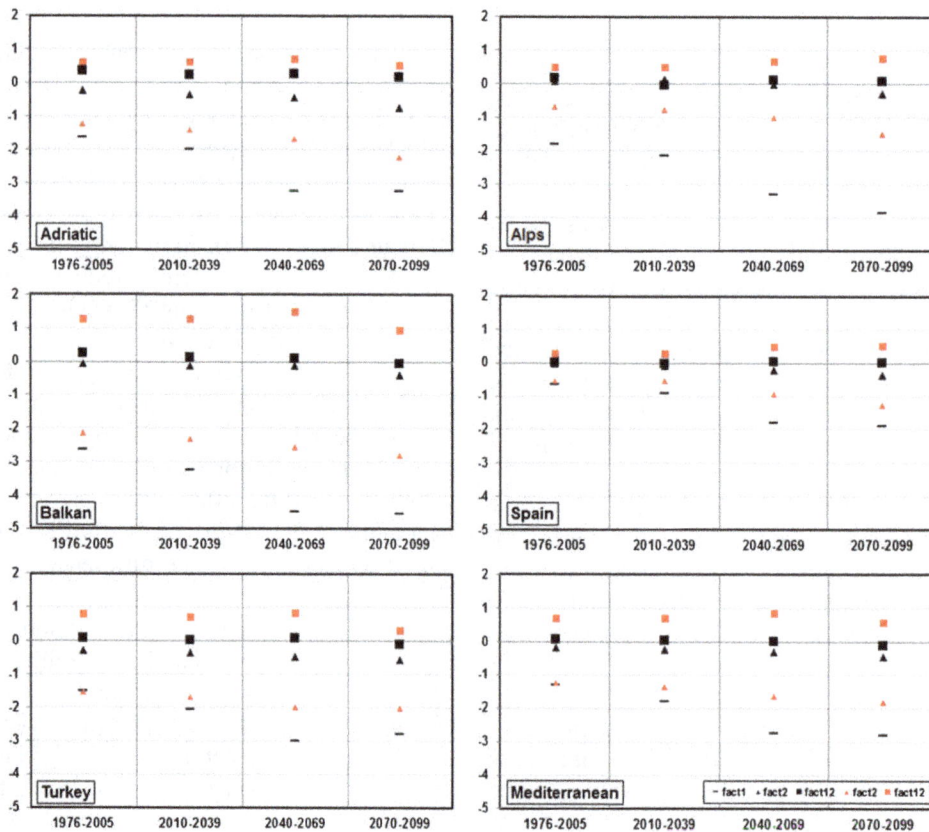

Figure 3. Individual contributions of different factors (fact1 and fact2) and their synergies (fact12) compared to the control simulation (f_0) over the six selected subregions for summer (JJA) temperature. Bars indicate fact1, triangles indicate fact2 and squares indicate the interaction term (fact12). Black colors are for Ens1 (GCM forcing vs convection scheme) while red color is for Ens2 (GCM forcing vs irrigation). Units are degrees Celsius (°C).

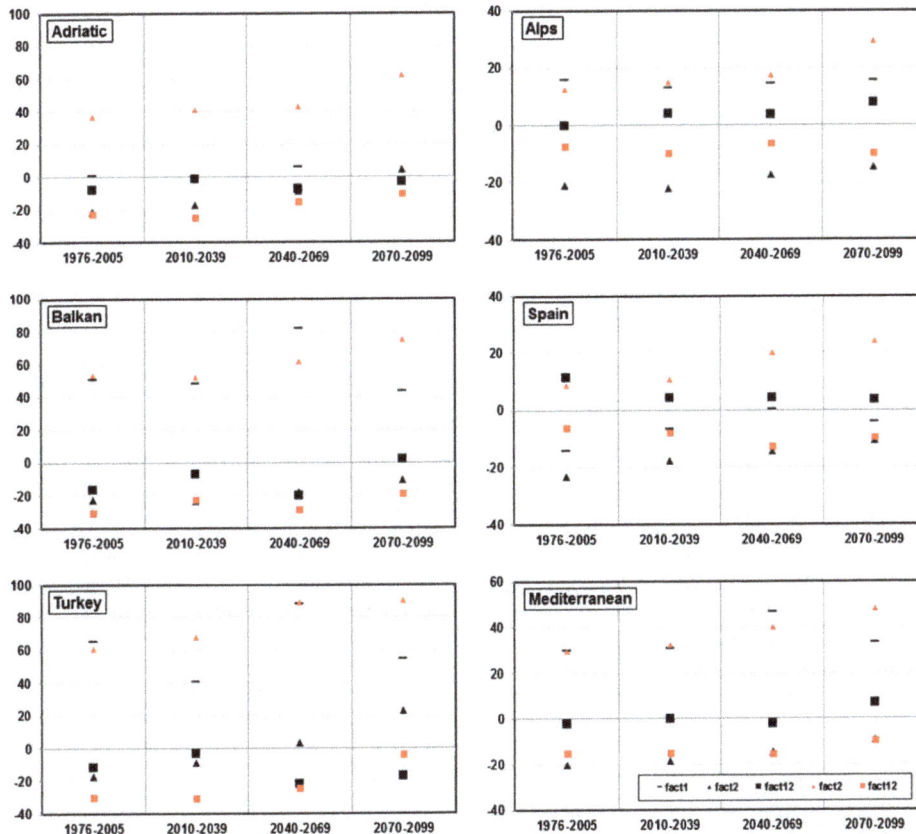

Figure 4. Same as Figure 3, but for summer precipitation [% of f_0 values]. [Correction added on 22 April 2014 after original online publication: Figure 4 has been amended to show results for JJA as described in the legend.]

role. Fact1 is largest and positive over the Alps, Balkan and Turkey regions, indicating greater precipitation in the MPI-driven than HadGEM-driven runs. In fact, as percentage of reference (f_0) summer precipitation, fact1 may reach up to 80% over the Balkan region and over 40% for the whole Mediterranean, although this value is inflated considerably by the fact that little precipitation occurs in summer over the Mediterranean. Irrigation has a comparable or stronger effect compared to the GCM forcing over all regions and the Mediterranean as a whole, with enhanced summer precipitation leading to positive values of fact2 (red triangles). Indeed, irrigation considerably reduces the summer drying over the region in the future climate projections (Figure 2). Again, the synergy contribution associated with irrigation versus driving GCM is in the opposite direction compared to the irrigation-alone effect, suggesting a negative feedback response.

The effect of convection scheme is mostly negative (black triangles), indicating lower precipitation amounts produced by the Grell-only compared to the mixed scheme over land, it is maximum over the Alps and Spain (where it is of comparable magnitude, up to 20%, as the irrigation and GCM contributions), and it tends to decrease for time slices farther into the century. This may be due to the general decrease of precipitation found in the projections over the region. The only exception is Turkey, where the effect of convection scheme shows a continuous upward trend throughout

the 21st century. Also for this GCM-versus-convection ensemble the synergy term tends to be in the opposite direction compared to the individual components.

4. Discussion and summary considerations

In this paper we applied the FS method of Stein and Alpert (1993) to isolate the effects of different GCM boundary conditions and regional model physics, along with their synergy term, in modulating 21st century temperature and precipitation projections over the Mediterranean region. Toward this purpose, we used simulations from an ensemble of RegCM4 projections over the Med-CORDEX domain. Specifically, we studied the contributions of the HadGEM and MPI driving GCMs versus the irrigation and convection schemes.

For temperature, the largest contributions to differences across model projections were due to the driving GCMs and the presence of irrigation, while for precipitation the use of different convection schemes had also a comparable contribution. We also found that the contributions of different factors changed across geographical regions and showed temporal trends throughout the 21st century. The synergy term, when not negligible, mostly acted in counterbalancing differences across the individual factor contributions, suggesting the presence of negative feedback responses in the model (Alpert and Sholokhman, 2011).

Although limited to an available 'ensemble of opportunity', our study illustrates how the FS method can provide a useful tool to assess the contributions of different simulation aspects to the uncertainties underlying climate change projections at the regional scale. In our case for example, the internal model physics provided a contribution to uncertainty often of the same order of magnitude as the driving GCMs. Being able to identify different sources of uncertainty is especially important within the context of the CORDEX international framework in view of the availability of large multi-model ensembles of RCM projections.

Acknowledgements

This work was partially supported by the FP7 EU project ATOPICA and by the Project NextData of the Italian Ministry for Education, University and Research, and the Italian Ministry of Environment, Land and Sea. We acknowledge the World Climate Research Programme's Working Group on Coupled Modeling, which is responsible for CMIP5, and we thank the climate modeling groups for producing and making available their model output.

References

Alpert P, Sholokhman T. 2011. In *Factor Separation in the Atmosphere. Applications and Future Prospects*, Alpert P, Sholokhman T (eds). Cambridge University Press: New York, NY.

Alpert P, Krichak SO, Krishnamurti TN, Stein U, Tsidulko M. 1996. The relative roles of lateral boundaries, initial conditions and topography in mesoscale simulation of lee cyclogenesis. *Journal of Applied Meteorology* **35**: 1091–1099.

Alpert P, Niyogi D, Pielke RA, Eastman JL, Xue YK, Raman S. 2006. Evidence for carbon dioxide and moisture interactions from the leaf cell up to global scales: perspective on human-caused climate change. *Global Planetary Change* **54**: 202–208.

Brands S, Herrera S, Fernandez J, Gutierrez JM. 2013. How well do CMIP5 Earth System Models simulate present climate conditions in Europe and Africa? *Climate Dynamics* **41**: 803–817.

Claussen M, Brovkin V, Ganopolski A. 2001. Biogeophysical versus biogeochemical feedbacks of large-scale land cover change. *Geophysical Research Letters* **28**(6): 1011.

Collins WJ, Bellouin N, Doutriaux-Boucher M, Gedney N, Halloran P, Hinton T, Hughes J, Jones CD, Joshi M, Liddicoat S, Martin G, O'Connor F, Rae J, Senior C, Sitch S, Totterdell I, Wiltshire A, Woodward S. 2011. Development and evaluation of an Earth-system model HadGEM2. *Geoscientific Model Development Discussions* **4**: 997–1062.

Dickinson R, Henderson-Sellers A, Kennedy P. 1993. Biosphere-Atmosphere Transfer Scheme (BATS) version 1 as coupled to the NCAR community climate model. NCAR technical note NCAR/TN-387+ STR, NCAR, Boulder, CO.

Emanuel KA, Zivkovic-Rothman M. 1999. Development and evaluation of a convection scheme for use in climate models. *Journal of Atmospheric Sciences* **56**: 1766–1782.

Giorgi F. 2005. Climate change prediction. *Climatic Change* **73**: 239–265.

Giorgi F, Lionello P. 2008. Climate change projections for the Mediterranean region. *Global Planet Change* **63**: 90–104.

Giorgi F, Hurrell JW, Marinucci MR, Beniston M. 1997. Elevation signal in surface climate change: a model study. *Journal of Climate* **10**: 288–296.

Giorgi F, Jones C, Asrar G. 2009. Addressing climate information needs at the regional level, the CORDEX framework. *WMO Bulletin* **58**(3): 175–183.

Giorgi F, Coppola E, Solmon F, Mariotti L, Sylla MB, Bi X, Elguindi N, Diro GT, Nair V, Giuliani G, Turuncoglu UU, Cozzini S, Güttler I, O'Brien TA, Tawfik AB, Shalaby A, Zakey AS, Steiner AL, Stordal F, Sloan LC, Brankovic C. 2012. RegCM4: Model description and preliminary tests over multiple CORDEX domains. *Climate Research* **52**: 7–29.

Grell GA. 1993. Prognostic evaluation of assumptions used by cumulus parameterizations. *Monthly Weather Review* **121**: 764–787.

Healy R, Druyan ML, Lynn BH. 2009. Quantifying the sensitivity of simulated climate change to model configuration. *Climatic Change* **92**: 275.

Jankov I, Schultz PJ, Anderson CJ, Koch SE. 2007. The impact of different physical parameterizations and their interactions on cold season QPF in the American River Basin. *Journal of Hydrometeorology* **8**(5): 1141–1151.

Jungclaus JH, Lorenz SJ, Timmreck C, Reick CH, Brovkin V, Six K, Segschneider J, Giorgetta MA, Crowley TJ, Pongratz J, Krivova NA, Vieira LE, Solanki SK, Klocke D, Botzet M, Esch M, Gayler V, Haak H, Raddatz TJ, Roeckner E, Schnur R, Widmann H, Claussen M, Stevens B, Marotzke J. 2010. Climate and carbon-cycle variability over the last millennium. *Climate of the Past* **6**(5): 723.

Kueppers LM, Snyder MA, Sloan LC. 2007. Irrigation cooling effect: regional climate forcing by land-use change. *Geophysical Research Letters* **34**: L03703.

Kumar SV, Peters-Lidard CD, Mocko D, Tian Y. 2013. Multiscale evaluation of the improvements in surface snow simulation through terrain adjustments to radiation. *Journal of Hydrometeorology* **14**(1): 220–232.

Lynn BH, Healy R, Druyan LM. 2009. Quantifying the sensitivity of simulated climate change to model configuration. *Climatic Change* **92**: 275–298.

Moss RH, Edmonds JA, Hibbard KA, Manning MR, Rose SK, van Vuuren DP, Carter TE, Emori S, Kainuma M, Kram T. 2010. The next generation of scenarios for climate change research and assessment. *Nature* **463**: 747–756.

Romero R, Ramis C, Alonso S. 1997. Numerical simulation of an extreme rainfall event in Catalonia: role of orography and evaporation from the sea. *The Quarterly Journal of the Royal Meteorological Society* **123**: 537–559.

Seneviratne SI, Luthi D, Litschi M, Schar C. 2006. Land-atmosphere coupling and climate change in Europe. *Nature* **443**: 205–209.

Stein U, Alpert P. 1993. Factor separation in numerical simulations. *Journal of the Atmospheric Sciences* **50**: 2107–2115.

Taylor KE, Stouffer RJ, Meehl GA. 2012. An overview of CMIP5 and the experiment design. *Bulletin of the American Meteorological Society* **93**: 485–498.

Teller A, Levin Z. 2008. Factorial method as a tool for estimating the relative contribution to precipitation of cloud microphysical processes and environmental conditions: method and application. *Journal of Geophysical Researches* **113**: D02202.

24

Trend and pattern classification of surface air temperature change in the Arctic region

Wandee Wanishsakpong,[1,2] Nittaya McNeil[1,2,*] and Khairil A. Notodiputro[3]

[1]Department of Mathematics and Computer Science, Faculty of Science and Technology, Prince of Songkla University, Pattani, Thailand
[2]Centre of Excellence in Mathematics, Commission on Higher Education, Bangkok, Thailand
[3]Department of Statistics, Faculty of Mathematics and Science, Bogor Agricultural University, Indonesia

*Correspondence to:
N. McNeil, Department of Mathematics and Computer Science, Faculty of Science and Technology, Prince of Songkla University, Rusamilae, Pattani Campus, Pattani 94000, Thailand.
E-mail: nchirtki@gmail.com

Abstract

Monthly seasonally adjusted temperatures above latitude 45°N were investigated from January 1973 to November 2013. The study area was divided into 69 sub-regions of similar size each in the shape of an igloo brick. The data were filtered with a second-order autoregressive process to remove autocorrelation. Two sub-regions did not have sufficient data due to substantial numbers of missing values. Factor analysis was then applied to the remaining 67 sub-regions and was used to classify regions with similar temperature changes. As a result, 63 sub-regions could be classified based on 12 factors but 4 sub-regions could not be grouped due to uniqueness. The temperatures for each group of sub-regions were found to increase during 1973–2013. The largest temperature increases of 0.19 °C/decade were found in northern and southern Siberia and part of the Arctic Ocean. In northern Canada, Alaska, the northern Pacific Ocean and eastern Siberia the temperatures increased by at least 0.16 °C/decade. In Iceland, Norway, Sweden and part of the Pacific and Arctic Oceans the temperature increased by around 0.15 °C/decade. In northeastern Canada, Greenland and its surrounding Atlantic Ocean and the Arctic Ocean the temperature increased by about 0.15 °C/decade.

Keywords: Arctic; climate change; Time series analysis; autocorrelation; linear regression model; factor analysis

1. Introduction

Variability in climate across the Arctic has attracted the interest of many researchers during the last two decades. The Arctic area is located in the northern hemisphere and consists of the Arctic Ocean and parts of Canada, Russia, Alaska, Denmark, Norway, Sweden, Finland and Iceland. The warmest temperatures of the northern hemisphere were found in 1990 (Houghton et al., 2001), with temperatures rising on average by 0.37 °C from 1925 to 1944 and 0.32 °C from 1978 to 1997 (Jones et al., 1999). Since the beginning of the 20th century, average Arctic temperatures have increased by approximately 0.6 °C, whereas maximum temperatures, recorded in 1945, were 1.2 °C higher than those in 1910 (Overpeck et al., 1997). The eastern Arctic Ocean had maximum temperature change of approximately +1 °C/decade while temperatures during the winter season have decreased by about −1 °C/decade (Rigor et al., 2000). In 2008, the National Academies, non-government organization that were set up to provide independent and technological advice to the US government and nation, reported that the warming trend was associated with an increase in land surface temperature, sea level rise, and a decrease in snow cover. According to Houghton et al. (2001) the warming trend was also related to changes in atmospheric and ocean circulation patterns. Other researches have mentioned

that warming trend was reflected in the Arctic Oscillation and impacted the climate system in the late 1970s and early 1980s (Wang et al., 2012), in permafrost warming and in melting of glaciers and ice caps (Opel et al., 2003). Significant warming has been detected in most of the Arctic region (Rigor et al., 2000).

Natural Arctic temperature changes and pressure variations are complicated to understand due to the range of possible causes of the variability (Polyakov et al., 2002). This, in turn, also complicates the process of making reliable predictions for future climatic change (Opel et al., 2003). However, there are many studies using various methodologies including computer simulation models (Johannessen et al., 2003), empirical orthogonal functions (Semenov, 2007) and statistical techniques that have assessed the trend and pattern of temperature change. These techniques include cluster analysis (Mahlstein and Knutti, 2009), principal component and hierarchical cluster analysis using the Ward method (Rebetez and Reinhard, 2007) and T-mode principle component analysis by convergent K-mean clustering (Jiang et al., 2012). Factor analysis is another method for examining the spatial and temporal patterns of temperatures. Factor analysis is used to analyze the structure of interrelationships among a large number of variables.

In this article, the trend and pattern of monthly temperatures in the Arctic region above 45°N from January

1973 to November 2013 were investigated using statistical methods, including time series analysis, linear regression and factor analysis for classification of the trends and patterns of temperature change.

2. Data and methods

2.1. Data

Monthly temperature data were downloaded from the Climatic Research Unit at the University of East Anglia UK (CRU, 2013) and the detailed description of the data set was made by Brohan *et al.* (2006). The area in this study was divided into 648 grid-boxes of $5° \times 5°$ latitude–longitude. Since there were missing data in some grid-boxes, the grid-boxes were combined into 69 sub-regions of similar size using the average temperature between adjoining regions in the same latitude band. The 69 sub-regions used in the analysis are shown in Figure 1. These 69 sub-regions include one covering up to 5° below the North Pole (85°–90°N), 8 in the 75–85°N band, 12 in the 65°–75°N band, 24 in the 55°–65°N band and another 24 in the 45°–55°N band, covering the Arctic Ocean, the northern areas of the Atlantic and Pacific Oceans, and the North American, Asian and European continents. The study areas included both land and sea, and there were 491 monthly temperatures recorded over this 41-year period for each sub-region. Four sub-regions had missing data. Figure 1 shows the number of missing data (in month) in brackets. For example, in sub-region 1 the data from 389 months were missing, whereas in sub-region 14 the data from the first month was missing.

2.2. Statistical methods

Temperatures were seasonally adjusted since temperature series have seasonal components. Linear regression models were fit to the data to obtain temperature trends of each sub-region. A factor analysis was then conducted to produce clusters of sub-regions which were further grouped based on the slopes or the rates of increase in temperatures within the clusters. More detailed explanation of these models and techniques are provided in the following paragraph.

The seasonal adjustment was carried out using the backward difference operator Δ as well as the backshift operator Δ (Montgomery *et al.*, 2008). If x_{it} is the monthly temperature data in sub-region i for time t then the backward difference is given by:

$$\Delta^d = \left(1 - \dot{B}\right)^d$$

where $\dot{B}^d x_{it} = x_{i(t-d)}$ then the lag-d seasonal operator is defined as

$$\Delta^d x_{it} = \left(1 - \dot{B}\right)^d x_{it} = x_{it} - x_{i(t-d)}$$

Hence, y_{it} which denotes the seasonally adjusted temperature in sub-region i for month t can be found from

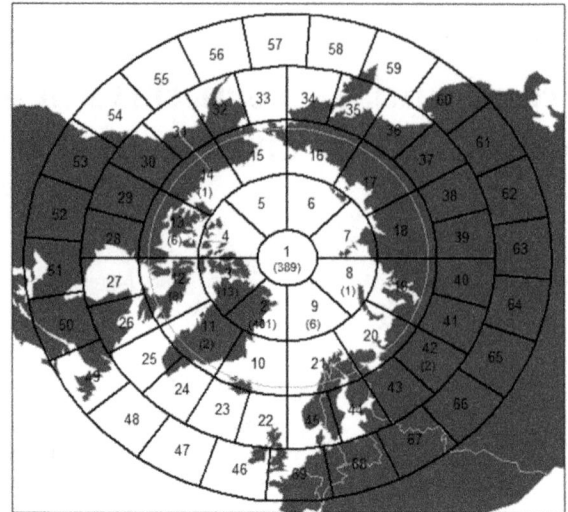

Figure 1. Map of the 69 sub-regions above latitude 45°N. Numbers in brackets are the number of months with missing data for 1973–2013.

the above equation by choosing $d = 12$ as follows

$$y_{it} = \Delta^{12} x_{it} = \left(1 - \dot{B}\right)^{12} x_{it} = x_{it} - x_{i(t-12)}$$

Separate linear regression models were fit to the seasonally adjusted temperature for each of the 69 sub-regions. The model takes the form

$$y_{it} = b_{0i} + b_{1i}t + e_{it} \text{ for } i = 1, 2, \ldots, 69$$
$$\text{and } t = 1, 2, \ldots, 491 \quad (1)$$

where y_{it} denotes the seasonally adjusted temperature in sub-region i for month t; b_{1i} corresponds to the temperature change per month in sub-region i; b_{0i} is the average temperature in sub-region i over the whole period; and e_{it} is the error assumed to follow a normal distribution with constant variance.

In order to analyze the time series data, an autoregressive (AR) model was used to account for auto-correlation among residuals from the fitted linear regression model. Residuals (e) from the linear model (1) were analyzed using a second order auto-regressive model AR(2) which takes the form

$$e_{it} = a_1 e_{i(t-1)} + a_2 e_{i(t-2)} \text{ for } i = 1, 2, \ldots, 69$$
$$\text{and } t = 1, 2, \ldots, 491 \quad (2)$$

where e_{it} is the difference between temperatures (y) and the corresponding fitted values $(\hat{y})$; $e_{i(t-1)}$ are residuals at lag 1; $e_{i(t-2)}$ are residuals at lag 2; and a_1, a_2 are the coefficients of the fitted second-order auto-regressive model AR(2).

Filtered seasonally adjusted temperatures were then obtained by

$$z_{it} = e_{it} - \left(a_1 e_{i(t-1)} + a_2 e_{i(t-2)}\right) \text{ for } i = 1, 2, \ldots, 69$$
$$\text{and } t = 1, 2, \ldots, 491 \quad (3)$$

where z_{it} is the filtered temperatures in sub-region i at month t; and a_1, a_2 are the average coefficients of the fitted second order auto-regressive models AR(2).

The inter-correlation between sub-regions among the 67 sub-regions were also considered using factor analysis (Johnson, 1998). Similar patterns of temperature changes were identified and regrouped to form a larger region (factor) by maximizing the likelihood of the covariance matrix and minimizing the correlation between the factors for a specified number of factors. These factors were identified using factor loadings. The loadings were controlled by rotating the factors using the Promax method (Venables and Riply, 2002). The cut off value for factor loadings was 0.3 (Hair *et al.*, 1998). Sub-regions with high uniqueness, which have variations in temperature that are not similar to the other sub-regions, were also considered in this study.

The factor model formulation with m factors ($m < 67$), denoted by $f_1, f_2, \ldots, f_m$, takes the form

$$z_{it} = \mu_i + \sum_{k=1}^{m} \lambda_{ik} f_k \text{ for } i = 1, 2, \ldots, 67,$$

$$= 3, 4, \ldots, 491 \text{ and } k = 1, 2, \ldots, m \qquad (4)$$

where z_{it} is the temperatures in sub-region i at month t; μ_i is the average across 491 months for sub-region i; λ_{ik} is the factor loading at the ith sub-region on the kth factor; and f_k is the kth common factor.

For each of the large regions, a linear model was fit to estimate temperature changes. All analyses and graphical displays were carried out using R (R Development Core Team, 2009).

3. Results

A separate linear model (1) was fit to the temperatures for each of the 67 sub-regions. The two sub-regions (North Pole and latitude 75°–85°N longitude 180°–135°W) with insufficient data were omitted from further analysis. Coefficients (b_{1i}), which represent the temperature change per decade, and p-values obtained from the linear regression model are shown in Table 1. Of the 67 sub-regions, the temperature changes increased in 64 sub-regions ($p < 0.05$). Exceptions were for three sub-regions which showed no evidence of significant temperature change. These three sub-regions were latitude 55°–65°N longitude 180°–165°W (sub-region 22, $p = 0.117$), latitude 55°–65°N longitude 75°–60°E (sub-region 39, $p = 0.057$), and latitude 45°–55°N longitude 145°–130°W (sub-region 48, $p = 0.072$).

From a fitted linear model in 67 sub-regions, Figure 2 shows a graph of the auto-correlation function (ACF) of the residuals for a sample of two sub-regions, where temperatures have increased. Correlations in residuals from this fitted model in time series are assumed to be stationary. The average coefficients (a_1 and a_2) were 0.335 and 0.042, respectively. Figure 3 shows the ACF of the filtered residuals for the two sub-regions illustrated in Figure 2. This filtering reduced auto-correlation substantially.

Table 1. Regression coefficients of 67 sub-regions.

Sub-regions	b_{1i}	P-value	Sub-regions	b_{1i}	P-value
3	0.083	<0.05	37	0.746	0.005
4	0.134	<0.05	38	0.259	0.014
5	0.416	<0.05	39	0.220	0.057
6	0.151	<0.05	40	0.261	0.021
7	0.166	<0.05	41	0.287	0.004
8	0.124	<0.05	42	0.393	<0.05
9	0.034	<0.05	43	0.259	<0.05
10	0.253	<0.05	44	0.269	<0.05
11	0.357	<0.05	45	0.258	<0.05
12	0.223	<0.05	46	0.103	<0.05
13	0.422	<0.05	47	0.066	0.008
14	0.154	<0.05	48	0.048	0.072
15	0.249	<0.05	49	0.179	<0.05
16	0.309	<0.05	50	0.197	0.039
17	0.273	<0.05	51	0.274	0.001
18	0.316	0.001	52	0.244	<0.05
19	0.266	<0.05	53	0.406	<0.05
20	0.312	<0.05	54	0.381	<0.05
21	0.307	<0.05	55	0.257	<0.05
22	0.058	0.117	56	0.251	<0.05
23	0.167	0.021	57	0.226	<0.05
24	0.141	0.019	58	0.331	<0.05
25	0.204	0.007	59	0.401	<0.05
26	0.283	<0.05	60	0.402	<0.05
27	0.393	<0.05	61	0.399	<0.05
28	0.316	<0.05	62	0.257	0.006
29	0.297	<0.05	63	0.228	0.003
30	0.337	<0.05	64	0.283	<0.05
31	0.375	<0.05	65	0.274	<0.05
32	0.338	<0.05	66	0.307	<0.05
33	0.251	<0.05	67	0.247	<0.05
34	0.267	<0.05	68	0.155	<0.05
35	0.337	<0.05	69	0.139	<0.05
36	0.347	<0.05			

The spatial correlations of temperature in 67 sub-regions were investigated. The correlations of temperatures ranged from −0.485 to 0.81 with a median of 0.43. A factor model was fit to reduce these correlations and as a result the 67 sub-regions could be factored into 16 groups based on the values of the factor loadings. These 16 groups consisted of 12 groups containing sub-regions and four individual sub-regions as shown in Table 2.

Since the rates of temperature increase in each large region were different then it is of interest to further group these 12 large regions as well as the 4 individual sub-regions based on their regression slopes. For this purpose, the average temperature changes per decade were estimated for each of the 12 large regions. This estimation was also carried out for the four sub-regions with high uniqueness. Table 3 provides estimates of these temperature increases in °C/decade with the corresponding 95% confidence intervals. As shown in the table, the confidence intervals do not cover zero implying that the temperatures have significantly increased during 1973–2013 for each of the large 12 regions as well as the four individual sub-regions. The gradient coefficients vary from a minimum of 0.063 °C/decade to a maximum of 0.255 °C/decade. Each such region

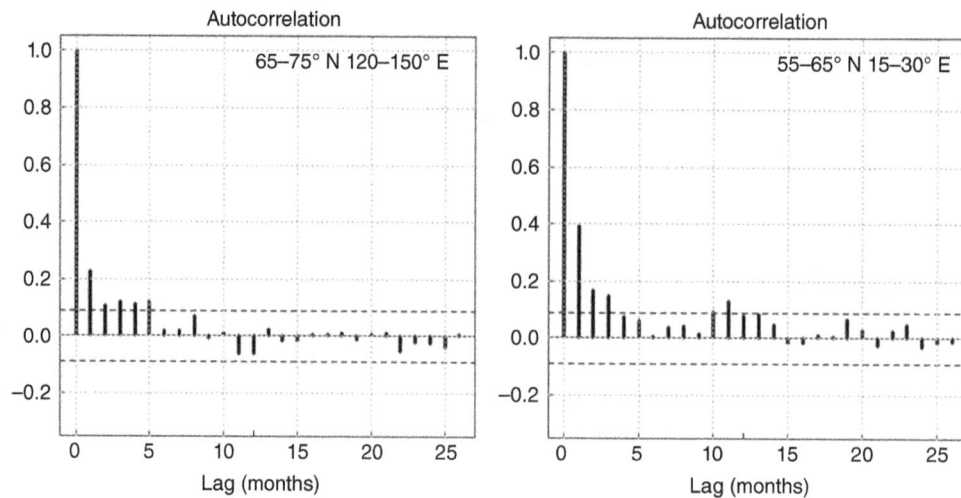

Figure 2. Auto-correlation function plots of the residuals for 2 of the 67 sub-regions. The dotted line represents the 95% confidence interval of a zero correlation.

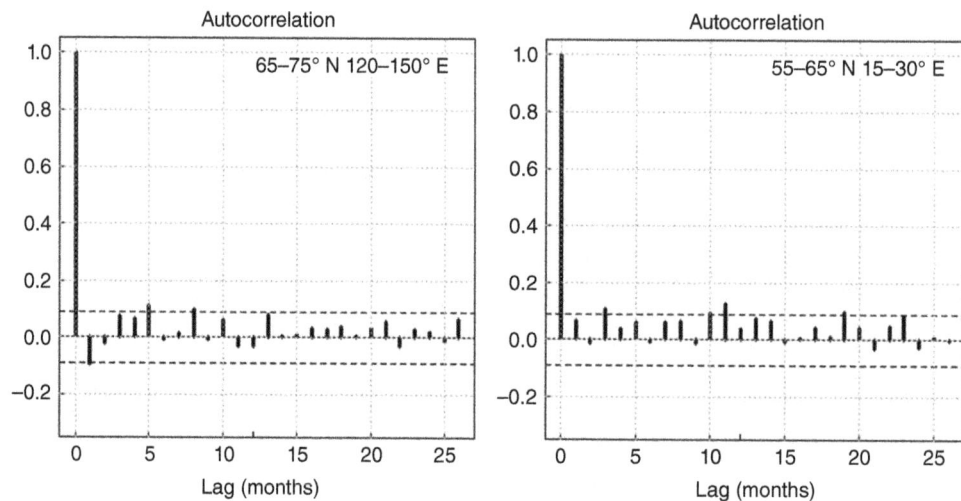

Figure 3. Auto-correlation function plots of the filtered residuals for the two sub-regions. The dotted line represents the 95% confidence interval for a zero correlation.

was re-classified into four levels according to the inter-quartile range of average increase. The overall quartiles of the combined sample of these 16 increases are: up to 0.15 °C/decade (lower quartile); from 0.15 to 16 °C/decade (second quartile); from 0.16 to 0.19 °C/decade (upper quartile); and 0.19 °C/decade and above.

A thematic map of the temperature change per decade for each large region is shown in Figure 4. This figure also provides the levels of increase in temperature which were classified into four different levels, namely <0.15 °C/decade, 0.15–0.16 °C/decade, 0.16–0.19 °C/decade, and ≥0.190 °C/decade. The temperatures in large regions IV and IX as well as sub-region 5 have increased by at least 0.19 °C/decade in northern Siberia, southern Siberia and part of the Arctic Ocean. The temperatures in large regions II, III, VIII, X and XI have increased by at least 0.16 °C/decade and less than 0.19 °C/decade in northern Canada, Alaska, the northern Pacific Ocean and eastern Siberia. The increase in large region VII and sub-regions 55 and 67

were also evident at least 0.15 °C/decade and less than 0.16 °C/decade in Iceland, Norway, Sweden and part of the Pacific and Arctic Oceans, while the temperatures in large regions I, V, VI, XII and sub-region 6 have increased below 0.15 °C/decade in north-eastern Canada, Greenland and its surrounding Atlantic Ocean and the Arctic Ocean. The two sub-regions with more than 80% missing data are shown in black shaded.

It is important to note that the conclusion about the increases of temperature was based on ordinary regression models assuming independent errors. The existence of auto-correlation among these errors may result in underestimated variances which, in turn, could affect the significance levels of the test.

4. Conclusions and discussion

The findings of the present study provide the information about temperature changes in the Arctic region (land and sea). It was also shown that an AR(2)

Table 2. Factor loading of 67 sub-regions from Promax-rotated factor analysis.

Factors	Sub-regions	F1	F2	F3	F4	F5	F6	Uniqueness
	50	**0.95**	−0.17					0.15
	26	**0.87**	0.16				0.27	0.10
	51	**0.82**	0.15				−0.27	0.22
I	25	**0.75**					0.25	0.12
	49	**0.73**	−0.24				0.12	0.49
	27	**0.67**	0.24					0.19
	52	**0.43**	0.22				−0.29	0.47
	28		**0.89**				−0.11	0.26
	29		**0.84**					0.33
	13	−0.12	**0.79**				0.11	0.35
II	53	0.12	**0.63**				−0.18	0.56
	12	0.25	**0.60**	−0.10			0.23	0.37
	4	−0.13	**0.53**			0.21		0.52
	3		**0.40**	−0.12			0.21	0.63
	62			**0.85**	0.21			0.17
	62			**0.58**		−0.15		0.22
III	39			**0.79**		0.27		0.05
	38			**0.68**	0.23	0.11		0.07
	64			**0.66**		−0.31		0.31
	40			**0.62**		0.12		0.31
	60			0.16	**0.88**	−0.20		0.19
	36			−0.16	**0.87**			0.06
IV	37			0.12	**0.86**	0.21		0.07
	59		0.11		**0.68**	−0.20		0.28
	61			0.21	**0.65**			0.19
	18				0.23	**0.94**		0.17
V	19		0.11		−0.15	**0.70**		0.21
	71			−0.15		**0.63**		0.52
	8				−0.12	**0.59**		0.52
	23						**0.92**	0.19
	24		0.18				**0.84**	0.09
VI	10		−0.15				**0.73**	0.23
	11		0.14	0.22		0.11	**0.61**	0.41
	22		−0.14			−0.11	**0.59**	0.43

Factors	Sub-regions	F7	F8	F9	F10	F11	F12	Uniqueness
	44	**0.95**						0.21
	21	**0.91**		−0.14				0.18
VII	45	**0.77**		−0.17		0.12		0.24
	43	**0.72**		0.19	−0.20			0.36
	20	**0.49**		0.28	−0.28			0.25
	9	**0.46**			−0.13			0.65
	34		**0.90**			-.21		0.15
	58	−0.11	**0.77**			−0.13		0.41
VIII	35		**0.75**			0.26		0.06
	57		**0.67**		0.11			0.52
	33		**0.56**		0.22			0.28
IX	42			**0.89**		−0.20		0.13
	41	−0.13		**0.85**				0.07
	66			**0.77**	0.11			0.37
	65			**0.68**				0.27
	32		0.13		**0.94**			0.17
	31	−0.10			**0.80**			0.31
	30	−0.12	−0.18		**0.56**			0.43
X	15				**0.47**	0.18		0.68
	14		−0.16		**0.45**			0.62
	56		0.13		**0.40**			0.73
	54				**0.35**			0.72
XI	17					**0.70**		0.21
	16		0.24		0.13	**0.58**		0.36
	46	−0.11					**0.96**	0.17
	47	−0.14			−0.14		**0.79**	0.37
XII	69	0.14					**0.67**	0.44
	48		−0.10	−0.10			**0.49**	0.54
	68	0.26	−0.11		0.12	0.12	**0.39**	0.65
	5				0.16			0.83
	6			−0.11	0.14	0.27		0.82
	55		−0.13		0.29		0.10	0.79
	67	0.13	−0.11		0.21	0.23	0.11	0.79

The factor loading scores using the cut-off at 0.3 are shown in bold.

Table 3. Estimated average temperature increase per decade (with 95% CI).

Variables	Increase	95% CI
Factor I	0.143	(0.030−0.282)
Factor II	0.160	(0.112−0.206)
Factor III	0.160	(0.004−0.315)
Factor IV	0.235	(0.090−0.381)
Factor V	0.142	(0.058−0.226)
Factor VI	0.117	(0.032−0.204)
Factor VII	0.156	(0.066−0.226)
Factor VIII	0.181	(0.120−0.244)
Factor IX	0.204	(0.076−0.332)
Factor X	0.185	(0.165−0.206)
Factor XI	0.187	(0.118−0.256)
Factor XII	0.063	(0.040−0.087)
Sub-region 5	0.255	(0.130−0.381)
Sub-region 6	0.096	(0.065−0.127)
Sub-region 55	0.156	(0.121−0.194)
Sub-region 67	0.157	(0.118−0.194)

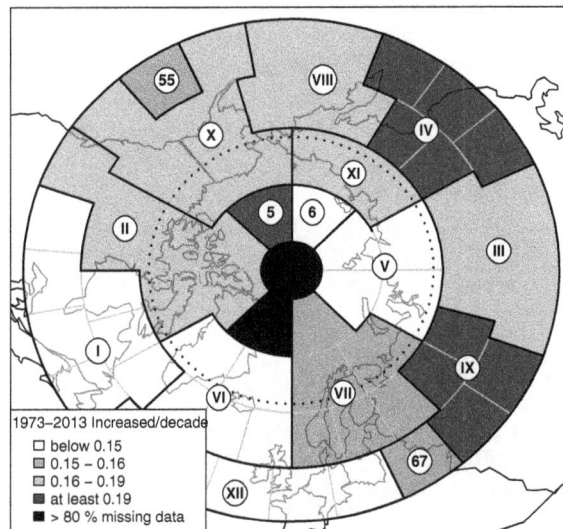

Figure 4. Thematic map of Arctic regions showing temperature increases per decade in regions.

process could successfully remove the correlations between consecutive temperatures. By using factor analysis in combination with regression models we have shown that almost all regions experienced some level of temperature increase. The large region I, V, VI and XII have experienced increases in temperature below 0.15 °C/decade. Most of those regions cover the North Atlantic Ocean and Arctic Ocean which are influenced by the slow change in oceanic thermohaline circulation (Delworth and Knutson, 2000). The variability between the Arctic and North Atlantic regions may occur due to surface air temperature variation which is stronger in the polar region than in lower latitudes (Polyakov *et al.*, 2002). In addition, the Arctic Ocean current and the ice drifting from the Arctic Ocean effect temperature changes in and around East Greenland (Woodgate *et al.*, 1999; Rudels *et al.*, 1999).

The temperatures in other large regions were also found to have a significant increase more than 0.15 °C/decade. The increase of temperature in Russia was found to be different in early and late 20th century. The maximum of temperature increase was found in high latitudes of the northern hemisphere but the recent warming in Russia is shifted inside the continent and mostly occurred in Central Siberia. The causes of warming might be due to natural factors as well as increased atmospheric transparency attributable to a weakening of volcanic activity (Anisimov *et al.*, 2007). The Pacific climate shift, wind, precipitation and the difference in sea level pressure have influenced the Pacific variability and responded to the warming trend in the different regions of Alaska (Hartmann and Wendler, 2005).

This study faced missing data problems so that several sub-regions could not be included in the analysis. Since recent statistical approaches have provided ways to deal with missing data, estimation of these missing values prior to analysis could be considered in future studies to obtain better estimates.

Finally, this classification of temperature changes in the Arctic region could explain the climate variability and could also be evident in other climate parameters such as surface air temperature and atmospheric circulation (Wang and Key, 2005; Rigor *et al.*, 2000). For future studies it would be interesting to consider further breakdown of the data so that different trends between summer and winter could be identified.

Acknowledgements

This research is supported by the Center of Excellence in Mathematics, Commission on Higher Education, Thailand. The authors are grateful to Emeritus Prof Don McNeil, Macquarie University, Australia for supervising our research.

References

Anisimov OA, Lobanov VA, Reneva SA. 2007. Analysis of change in air temperature in Russia and empirical forecast for the first Quarter of the 21st century. *Russian Meteorology and Hydrology* **32**(10): 620–626.

Brohan P, Kennedy IJ, Harris I, Tett SFB, Jones PD. 2006. Uncertainty estimate in regional and global observed temperature changes: a new data set from 1850. *Journal of Geophysical Research: Atmospheres* **11**: D12106, doi: 10.1029.

CRU. 2013. Climatic research unit. http://www.cru.uea.ac.uk/cru/data/temperture (accessed 1 March 2013).

Delworth TL, Knutson TR. 2000. Simulation of the early 20th century global warming. *Science* **287**: 2246–2250.

Hair JF, Anderson RE, Tatham RL, Black WC. 1998. *Multivariate Data Analysis*. Prince Hall International (UK) Limited: London.

Hartmann B, Wendler G. 2005. The significance of the 1976 Pacific climate shift in the climatology of Alaska. *Journal of Climate* **18**: 4824–4839.

Houghton JT, Ding Y, Griggs DJ, Noguer PJ, Linden VD, Xiaosu D. 2001. *Climate change 2001: The Scientific Basic*. Cambridge University Press: Cambridge.

Jiang N, Cheung K, Luo K, Beggs PJ, Zhou W. 2012. On two different objective procedures for classifying synoptic wether types over east Australia. *International Journal of Climatology* **32**: 1475–1494, doi: 10.1002/joc.2373.

Johannessen OM, Bengtsson L, Miles MW, Kuzmina SI, Semenov VA, Alekseev GV, Nagurnyi AP, Zakharovn VF, Bobylev LP, Pettersson LH, Hasselmann K, Cattle HP. 2003. Arctic climate change: observe and modeled temperature and sea-ice variability. *Tellus* **56**: 328–341.

Johnson DE. 1998. *Applied Multivariate Methods for Data Analysts*. Duxbury: Pacific Grove, CA.

Jones PD, New M, Parker DE, Martin S, Rijor IG. 1999. Surface air temperature and its changes over the past 150 years. *Reviews of Geophysics* **37**: 173–199.

Mahlstein I, Knutti R. 2009. Regional climate change patterns identified by cluster analysis. *Climate Dynamics* **35**: 587–600.

Montgomery CM, Jennings CL, Kulahci M. 2008. *Introduction to Time Series Analysis and Forecasting*. John Wiley and Sons: Hoboken, NJ.

Opel T, Fritzsche D, Meyer H. 2003. Eurasian Arctic Climate over the past millennium as recorded in the Akademii Nauk ice core (Severnaya Zemlya). *Climate of the Past* **9**: 2379–2389, doi: 10.5194/cp-9-2379-2013.

Overpeck J, Hughen K, Hardy D, Bradley R, Case R, Douglas M, Finney B, Gajewski K, Jacoby G, Jennings A, Lamoureux S, Lasca A, MacDonale G, Moore J, Retelle M, Smith S, Wolfe A, Zielinski G. 1997. Arctic environmental change of the last four centuries. *Science* **278**: 1251–1256.

Polyakov VI, Brkryaev RV, Bhatt GVUS, Colony RL, Johnson MA, Maskshtas AP, Walash D. 2002. Variability and trends of air temperature and pressure in the maritime Arctic, 1875–2000. *American Meteorological Society* **16**: 2067–2076.

R Development Core Team. 2009. *R: A language and environment for statistical computing*. R Foundation for Statistical computing: Vienna, Austria.

Rebetez M, Reinhard M. 2007. Monthly air temperature trends in Switzerland 1901–2000 and 1975–2004. *Theoretical and Applied Climatology* **91**: 27–34, doi: 10.1007/s00704-007-0296-2.

Rigor IG, Colony RL, Martin S. 2000. Variation in surface air temperature observations in the Arctic, 1979–97. *Journal of Climate* **13**: 896–914.

Rudels B, Friedrich HJ, Quadfasel D. 1999. The Arctic circumpolar boundary current. *Deep Sea Research* **46**: 1023–1062.

Semenov VA. 2007. Structure of temperature variability in the high latitudes of the northern hemisphere. *Izvestiya, Atmospheric and Oceanic Physics* **43**: 744–753.

Venables WN, Riply BD. 2002. *Modern Applied Statistics with S*. Springer: New York.

Wang X, Key J. 2005. Arctic surface, cloud, and radiation properties based on the AVHRR polar pathfinder dataset. Part II: recent trends. *American Meteorological Society* **18**: 2575–2593.

Wang X, Key J, Fowler C, Maslanik J, Tschudi M. 2012. Arctic climate variability and trends from satellite observations. *Advances in Meteorology* **2012**: 505613, doi: 10.1155/2012/505613.

Woodgate RA, Fahrbach E, Ruhardt G. 1999. Structure and transport of the East Greenland current at 75°N from moored current meters. *Journal of Geophysical Research* **104**: 18059–18072.

Long-term changes in the relationship between stratospheric circulation and East Asian winter monsoon

Ke Wei,[1,2]* Masaaki Takahashi[2] and Wen Chen[1]

[1]Center for Monsoon System Research, Institute of Atmospheric Physics, Chinese Academy of Sciences, Beijing, China
[2]Atmosphere and Ocean Research Institute, University of Tokyo, Kashiwa, Japan

*Correspondence to:
K. Wei, Center for Monsoon System Research, Institute of Atmospheric Physics, Chinese Academy of Sciences, P.O. Box 2718, Zhong-Guan-Cun, Haidian District, Beijing 100190, China. E-mail: weike@mail.iap.ac.cn

Abstract

Using two generations of reanalysis datasets from the National Center for Atmospheric Research (NCAR), European Centre for Medium-Range Weather Forecasts (ECMWF), and Japan Meteorological Agency (JMA), we showed that on the interannual timescale the two leading modes of the East Asian winter monsoon (EAWM) are associated with tropospheric annular mode (AM) and stratospheric polar vortex (SPV), respectively. The relationship between AM and the first EAWM mode remained stable during 1958–2013, whereas that between SPV and the second EAWM mode increased suddenly since the late 1980s. The SPV-related circulation and planetary wave activities are intensified in the latter period. We suggested that this change might be caused by the global warming and ozone depletion.

Keywords: stratospheric circulation; East Asian winter monsoon; climate change

1. Introduction

In the boreal winter months, the East Asian winter monsoon (EAWM) can have tremendous influences on the East Asian countries, which usually causes substantial damages and large economic losses in its extreme state. For example, extreme cold events, blizzards, and freezing rain in southern China during January and February 2008 caused 129 casualties and an economic loss of ~25 billion USD (Gu *et al.*, 2008; Tao and Wei, 2008; Zhou *et al.*, 2009; Bao *et al.*, 2010). However, the variation of the EAWM is complicated and is influenced by several factors like El Niño Southern Oscillation (ENSO), the thermal state of the tropical western Pacific warm pool, Eurasian snow cover, and atmospheric internal processes such as Arctic Oscillation (AO), North Atlantic Oscillation, and planetary wave activities (e.g. Huang *et al.*, 2012 and references therein). These complexities therefore lead to the difficulty in seasonal prediction and issuance of disaster precautions. Furthermore, the EAWM can also be influenced by downward propagation of stratospheric anomalies associated with the polar vortex, providing a possible predictor for short-term East Asian winter climate anomalies of the East Asian winter because stratospheric anomalies generally have longer time scales (Chen and Wei, 2009; Chen *et al.*, 2013).

Usually, the influence of the stratospheric polar vortex on the EAWM is coupled with the influence of AO or northern annular mode (NAM), because of its near barotropic nature extending from the Earth's surface to the lower stratosphere (Thompson and Wallace, 1998, 2000). Using reanalysis data and a coupled model (IAP_FGOALS) simulation, Wei and Bao (2012) showed that the first and second empirical orthogonal

function modes (EOF1 and EOF2) of EAWM are related to AO and with the strength of the stratospheric polar vortex, respectively. Therefore, the influence of the tropospheric AO and stratospheric polar vortex can be orthogonal in the East Asian monsoon region despite the high correlation between AO and the polar jet. Most of the above-mentioned studies are focused on the interannual variation of the EAWM and there is very few dealing in longer timescale. Earlier studies revealed that the EAWM–ENSO relationship has weakened since the mid-1980s (Zhou *et al.*, 2007; Wang *et al.*, 2008; Wang and He, 2012; He, 2013); however, the stability of the relationship between atmospheric internal processes and EAWM has not been assessed. In this paper, an attempt has been made to study the long-term changes in the relationship among EAWM, AM, and the stratospheric polar vortex, using the reanalysis datasets. We offer preliminary evidence of a stronger correlation between EAWM and the stratospheric polar vortex since the late 1980s.

2. Data and methodology

2.1. Datasets

The primary datasets used in this study include two generations of reanalysis datasets from the National Centers for Environmental Prediction (NCEP), European Centre for Medium-Range Weather Forecasts (ECMWF), and Japan Meteorological Agency (JMA). These datasets include the NCEP/National Center for Atmospheric Research (NCAR) 40-year Reanalysis Project (NCEP1) (Kalnay *et al.*, 1996), the ECMWF 40-year Reanalysis dataset (ERA40) (Uppala *et al.*, 2005), the Japanese 25-year Reanalysis Project

Table 1. Reanalysis datasets employed in this study.

Dataset	Horizontal resolution	Vertical levels	Data period	Source
NCEP1[a]	2.5° × 2.5°	17	1948–2013	Kalnay et al. (1996)
JRA55*	1.125° × 1.125°	37	1958–2013	Ebita et al. (2011)
ERA40*	2.5° × 2.5°	23	1957–2002	Uppala et al. (2005)
NCEP2	2.5° × 2.5°	17	1979–2013	Kanamitsu et al. (2002)
JRA25	1.125° × 1.125°	23	1979–2013	Onogi et al. (2007)
ERAINT	2.5° × 2.5°	37	1979–2013	Dee and Uppala (2009)

[a]Data were used in long-term analysis.

(JRA-25) (Onogi et al., 2007), the NCEP–Department of Energy (DOE) Reanalysis Project (NCEP2) (Kanamitsu et al., 2002), the ECMWF Interim Reanalysis (ERAINT) (Dee and Uppala, 2009), and the Japanese 55-year Reanalysis (JRA-55) (Ebita et al., 2011). Table 1 shows the brief descriptions of these six reanalysis datasets. Additionally, we used monthly surface temperature data from 160 meteorological stations across China that were collected and compiled by the China Meteorological Administration from January 1951 to the present. Since the durations of NCEP2, JRA25, and ERAINT are less than 35 years, the changes in the long-term relationship between the EAWM and stratospheric circulation is studied based on the longer datasets, i.e. NCEP1, JRA55, and ERA40. Our analyses are mainly focused on the boreal winter months and therefore, the wintertime mean is computed by averaging the 3-month periods from December to February (DJF) in each year. The climatological wintertime means are then calculated by averaging all the DJF over the entire data period.

2.2. Methodology

We adopt the partial correlation and regression analyses to exclude the possible influence of one factor from another and a two-tailed Student's t-test is employed to test the significance of the results. To obtain the basic modes of EAWM, EOF analysis is applied in this study by constructing an area-weighted covariance matrix. We have applied a square root of cosine latitude weighting on the wintertime mean gridded surface (2 m) air temperature (SAT) data prior to the analyses.

In this study, a modified zonal index (ZI), defined by the zonal-mean sea level pressure (SLP) anomalies between 35 and 65°N (Li and Wang, 2003), is employed to represent the tropospheric annular oscillation. In contrast to the AO index using the EOF analysis by Thompson and Wallace (1998), the ZI showed more pronounced annular ring structure than that in the EOF AO index. However, the significant correlations (Li and Wang, 2003) (>99.9% confidence limit) between ZI and the EOF AO index are 0.92, 0.99, 0.91, 0.92, 0.92, and 0.91 in NCEP1, NCEP2, ERA40, ERAINT, JRA25,

and JRA55, respectively. The results from EOF AO index are quite consistent with our present study. We also use the stratospheric polar-night jet index (SPJI), which is defined by the zonal-mean wind at 60°N at 30 hPa, an index measuring the strength of the circumpolar wind and the polar vortex in the stratosphere.

To examine the propagation conditions of stationary planetary waves, the zonal-mean quasi-geostrophic refractive index (Hu and Tung, 2002; Lorenz and Hartmann, 2003; Perlwitz and Harnik, 2003) is calculated, and is defined as $n_k^2 = [q]_\phi / [u] - (k/a\cos\phi)^2 - (f/2NH)^2$. Here $[q]_\phi$, a, ϕ, f, k, N, and H denote the meridional gradient of the zonal-mean potential vorticity, Earth radius, latitude, Coriolis parameter, zonal wave number, buoyancy frequency, and scale height respectively. Following Li et al. (2007), we define a parameter to denote the winter frequency distribution of days with positive n_0^2 ($f\left(n_0^2 > 0\right)$, FDPN2 hereafter). Our analysis differs from that of Li et al. (2007) that they used $f\left(n_0^2 < 0\right)$. In each winter, the FDPN2 is expressed by the ratio of the number of days with positive n_0^2 to the total number of days at individual latitudes and heights. Previous studies (Matsuno, 1970; Andrews et al., 1987) showed that planetary waves can propagate in regions of positive refractive index and are evanescent in regions of negative refractive index. Moreover, planetary waves have the tendency to refract toward large positive index, i.e. toward larger FDPN2. Furthermore, the Eliassen-Palm (EP) flux in spherical geometry (Edmon et al., 1980; Andrews et al., 1987) is also employed to investigate the propagation of stationary planetary waves in two dimensional planes.

3. Results

The EAWM is dominated by two distinct EOF modes, which together account for ~70% of the total SAT variance. The leading mode, or the northern mode by Wang et al. (2010), is characterized by the positive SAT anomalies in the northern part of East Asia, representing the westerly advection and is associated with the positive phase of AO. While, the second mode features an SAT seesaw pattern north and south of 40°N over EA, representing the combined effect of westerly advection over higher latitude and planetary effect in the mid-latitudes, which possibly associated with the variation in strength of stratospheric polar vortex (Wei and Bao, 2012). The correlation coefficients (CCs) of the two leading PCs, the ZI, and the SPJI in the six reanalysis datasets are presented in Table 2. All the CCs between ZI and PC1 are significant above the 99% statistical confidence level (0.71 in NCEP1, 0.71 in JRA55, 0.67 in ERA40, 0.72 in NCEP2, 0.71 in JRA25, and 0.72 in ERAINT), accounting for approximately 50% of the PC1 variance. Considering the high correlation between the ZI and the SPJI (0.37 in NCEP1, 0.43 in JRA55, 0.44 in ERA40, 0.36 in NCEP2, 0.36 in JRA25, and 0.46 in ERAINT), pCCs are also calculated. After excluding the possible influence of the stratospheric

Table 2. Correlation coefficients of the two leading PCs and zonal index (ZI) and stratospheric polar jet index (SPJI) in the employed reanalysis datasets.

	Years	ZI		SPJI	
		PC1	**PC2**	**PC1**	**PC2**
NCEPR1	65	0.71** (0.66**)	−0.08 (−0.23)	0.34** (0.12)	0.33** (0.39**)
JRA55	55	0.71** (0.65**)	−0.10 (−0.33*)	0.43** (0.20)	0.40** (0.50**)
ERA40	45	0.67** (0.57**)	−0.30* (−0.51**)	0.54** (0.37*)	0.31* (0.52**)
NCEP2	34	0.72** (0.70**)	0.04 (−0.19)	0.23 (−0.04)	0.53** (0.55**)
JRA25	34	0.71** (0.67**)	−0.04 (−0.36)	0.33 (−0.11)	0.62** (0.68**)
ERAINT	34	0.72** (0.68**)	−0.09 (−0.39*)	0.37* (0.18)	0.58** (0.66**)

The numbers in the brackets are the partial correlation coefficients (pCCs) between leading PCs and ZI (jet) index, excluding the influence of the jet (ZI) index.
Numbers with ** and * indicate the value above the 99% and 95% statistical confidence level, respectively.

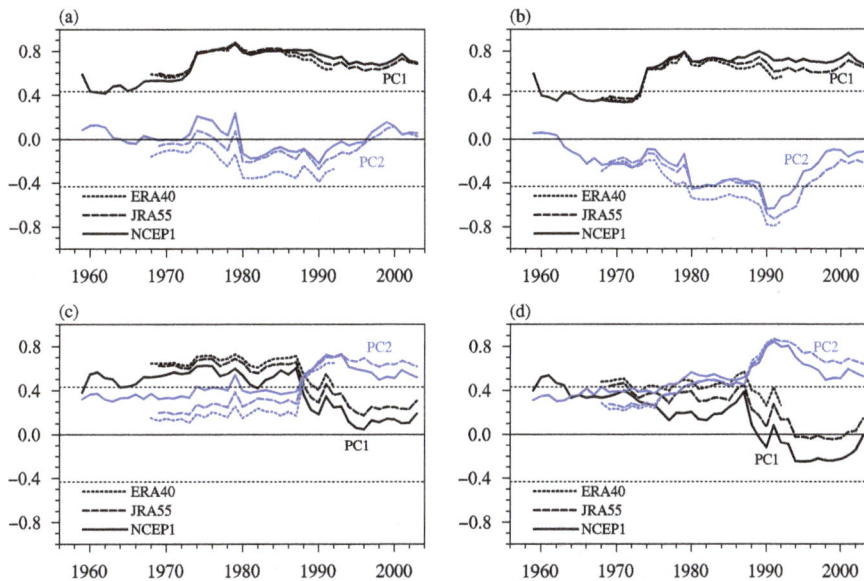

Figure 1. Sliding correlation coefficients (CCs, left panels) and partial correlation coefficients (pCCs, right panels) of the two East Asian winter monsoon (EAWM) principal components (PCs) and the zonal index (ZI, a and b) and stratospheric polar vortex jet index (SPJI, c and d) with a 21-year window. Red lines denote PC1, and blue lines denote PC2. The correlation shown in the figure is for the center year of the 21-year window. Solid lines indicates National Centers for Environmental Prediction/National Center for Atmospheric Research (NCEP1) data, dashed lines indicate Japanese 55-year Reanalysis Project (JRA55) data, and dotted lines indicate European Centre for Medium-Range Weather Forecasts 40-year reanalysis (ERA40) data. The dotted straight lines represent the 95% confidence level for the sliding correlations.

polar vortex, and the pCCs in all the reanalysis datasets are 0.66, 0.65, 0.57, 0.70, 0.67, and 0.68 in NCEP1, JRA55, ERA40, NCEP2, JRA25, and ERAINT, respectively. The differences among datasets are negligible, particularly NCEP2, JRA25, and ERAINT. On the other hand, the CCs between the ZI and the PC2 are small in all datasets and are nonsignificant except for ERA40 (−0.30, above the 95% statistical confidence level).

In all datasets, the CCs between PC2 and SPJI are significant above the 95% statistical confidence level (0.33 in NCEP1, 0.40 in JRA55, 0.31 in ERA40, 0.53 in NCEP2, 0.62 in JRA25, and 0.58 in ERAINT). After removing the linear influence from the low-level ZI, the pCCs between PC2 and SPJI are 0.39, 0.50, 0.52, 0.55, 0.68, and 0.66 in NCEP1, JRA55, ERA40, NCEP2, JRA25, and ERAINT, respectively. The three long-term datasets, i.e. the NCEP1, JRA55, and ERA40, have larger CCs and pCCs between PC1 and SPJI than the three short-term datasets, i.e. the NCEP2, JRA25, and

ERAINT. Using multiple reanalysis, our findings confirmed the results of Wei and Bao (2012) and Chen *et al.* (2013), i.e. the EAWM EOF1 is related to the NH annular oscillation, whereas the EAWM EOF2 is associated with the strength of the stratospheric polar vortex.

To investigate the variability of EAWM−SPJI relationships, sliding correlation is calculated with a 21-year sliding window. It is worth mentioning that similar results can be obtained by using a different sliding window such as 31 or 15 years, or using only interannual values by filtering out the low frequency decadal variability. If the individual 21 winter's data are independent of each other, the CC corresponding to the 95% significance levels is 0.433, as shown by the dotted straight lines in Figure 1. For all of the datasets, the running CCs between ZI and PC1 (red lines in Figure 1(a)) is statistically significant, whereas those between ZI and PC2 are always nonsignificant (blue lines in Figure 1(c)).

The running CCs between SPJI and PCs exhibit a clear interdecadal change with significant correlation between PC1 and SPJI before the late 1980s, whereas it is nonsignificant afterward. In contrast, the correlation between PC2 and SPJI exhibits opposite trend from nonsignificant to more than significant after the late 1980s. During the overlap period in the three reanalysis datasets from the late 1960s to the early 1990s, the running CCs between PC2 and SPJI have consistent long-term trends and interannual variation. Because the ZI correlate highly with SPJI, the running pCCs are calculated to remove the linear trend from the ZI. Similarly, after the late 1980s, as shown in Figure 1(b) and (d), the weakened and strengthened correlations between PC1 and SPJI and between PC2 and SPJI are noticed, respectively.

The stronger correlation between SPJI and PC2 can also be tested by using station observation data over China. After removing the linear trend from the ZI, the partial regression coefficient of the station air temperature across China with the standardized SPJI is shown in Figure 2, using the NCEP1 dataset during two periods from 1951/1952 to 1986/1987 and from 1988/1989 to 2012/2013. In the earlier period (Figure 2(a)), significant positive values appear across the northern China, and significant negative values at several stations in southwestern China and the southeastern coastal region, and this pattern is identical to that of EAWM EOF1 (Wang et al., 2010; Wei and Bao, 2012). In the latter period (Figure 2(b)), overall negative regression is observed at most of the stations in China stations except some warming in the northeastern China. The increased influence of the stratospheric polar vortex is significant at most of the central and southern stations, which resembles closely to the EAWM EOF2 pattern. The Siberian high has strengthened and extended southward over central and eastern China (Figure 2(b)), and the northerly winds along the coastal region of China has intensified which leads to the cooling in central and southern parts of China. Studies (Ding and Krishnamurti, 1987; Zhang et al., 1997) show that the intensified Siberian high provides favorable condition for triggering the cold surges, which may lead to more cooling in southern EA.

Zonal wind anomalies due to thermal advection play an essential role in the influence of ZI/NAM on the surface temperature, particularly over Siberia (Thompson and Wallace, 2000). Therefore, using NCEP1 data the regression/correlation between the averaged zonal wind and the SPJI in the earlier (1951/1952 to 1986/1987) and later (1988/1989 to 2010/2011) periods are presented in Figure 3. Moreover, the features are consistent with the results from ERA40 and JRA55 (figures not shown). On the interannual timescales, an intensification of the polar vortex is associated with anomalous downward and equatorward propagation of planetary waves, which causes planetary wave flux to converge in the mid-latitudes, leading to the deceleration of zonal wind through wave-mean flow interaction. The features are quite opposite (zonal wind

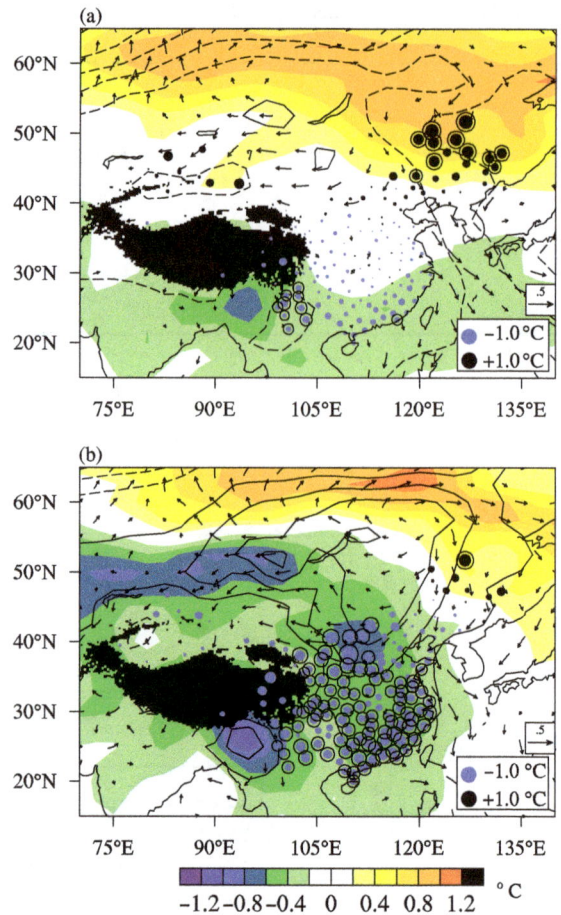

Figure 2. Partial regression (pReg) of NCEP1 surface air temperature (shading, °C), sea level pressure (contours, CI = 0.2 hPa), surface 10 m wind (vectors), and station air temperature (dots) across China upon the standardized stratospheric polar jet time series during the periods of (a) 1951/1952–1986/1987 and (b) 1988/1989–2012/2013. The diameters of the dots are proportional to the regression coefficient; with red dots denote positive values, and blue dots denote negative values. The outer circle shows the stations with significant correlations above the 95% confidence level.

acceleration in the mid-latitudes) when the polar vortex is weaker. Therefore, the westerly anomaly seesaw pattern between high and mid-latitudes has formed with the variation of stratospheric polar vortex. The near barotropic structure of the pattern extends downward from upper stratosphere to the Earth's surface, in the two zones centering around 37 and 60°N. It is worth mentioning that these two zones are approximately 5°N of the two centers of the ZI/AO, which are close to 30 and 55°N (Thompson and Wallace, 2000; Li and Wang, 2003). This is indicative to the fact that the influences of the annular mode from the stratosphere and troposphere need to distinguish explicitly.

During the later period, the SPJI-associated EP flux anomalies are intensified, with stronger downward and equatorward propagation of planetary waves around the subpolar region, leading to stronger westerly anomalies northward of 60°N. The difference in EP flux between the two periods indicates stronger planetary

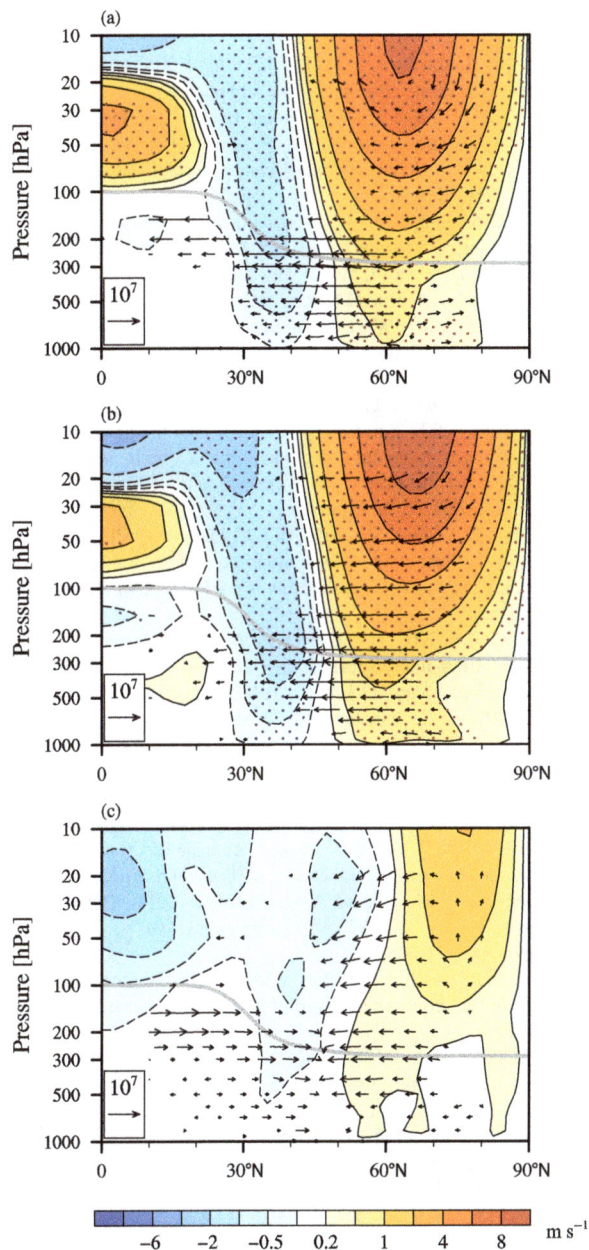

Figure 3. Zonal-mean zonal wind (shadings and contours, units are m s^{-1} per std) and EP flux regressed on the standardized SPJI during the periods of (a) 1951/1952–1986/1987 and (b) 1988/1989–2012/2013. (c) The differences between the two time periods (b minus a). Stippling shows the significant region above the 95% confidence level for the zonal-mean zonal wind regression. Gray thick lines denote the climatology position of the tropopause.

wave convergence by meridional EP flux dominantly in the mid-latitudes and is associated with stronger easterly anomalies there. Boyle (1986) shows that stronger westerly winds are unfavorable for the occurrence of cold air outbreak over eastern Asia because the synoptic scale waves have the tendency to pass rapidly without causing any appreciable disturbances. Therefore, the easterly anomalies around 37°N may favor larger wave amplitude, leading to more cold air intrusions over eastern Asia.

3.1. Discussions and conclusion

The running correlation between SPJI and EAWM PCs reveals that the influence of stratosphere on the first EAWM mode has decreased around the late 1980s, whereas its influence on the second mode has increased significantly. The results using three long-term reanalysis datasets, e.g. NCEP1, ERA40, and JRA55 are unequivocal despite the various assimilation schemes and models used in the reanalysis. This result is further confirmed by the correlation between SPJI and Chinese station temperatures during two time periods, in which the correlation is strongly negative over central and southern China in the later period. The current short-term climate prediction model depends heavily on the external forcing such as the ENSO SST (sea surface temperature). Our findings on the increased EAWM–SPJI relationship will help to better construct the new statistical/numerical model including the stratospheric signals in the seasonal climate prediction for the East Asian region.

Further studies are needed to understand the mechanisms that change the relationship between SPJI and basic EAWM modes, particularly the strengthened relationship between SPJI and PC2. The possible external forcing factor may be the anthropogenic global warming and ozone depletion, which caused the tropospheric warming and stratospheric cooling. Since the tropopause has an average height of ~ 17 km in the tropics and only ~ 9 km at the poles, both the temperature and pressure gradients increase near the upper troposphere and lower stratosphere in the extratropical mid-latitude regions (Figure 4(a)), leading to zonal wind acceleration (Figure 4(b)) from upper troposphere to middle stratosphere in the mid-latitudes. Because the transmission of the propagating planetary waves in the stratosphere is bounded within the polar waveguide (Dickinson, 1968; Huang and Gambo, 1983) along with the polar-night jet, the mid-latitude zonal wind acceleration in the later period has extended the waveguide further equatorward. As shown in Figure 4(b), the increased frequency of positive refractive index ($f\left(n_0^2 > 0\right)$) can be observed from upper troposphere to middle stratosphere in the mid-latitudes, favoring the propagation of planetary waves in this region, and the extension of the polar waveguide equatorward to the subpolar region, south of the EP flux climatology. Meanwhile, the tropospheric equatorward wave flux got stronger in the later period, help to extend the SPJ's influence to the lower latitudes via the linkage of planetary activity anomalies.

It seems that the zonal wind speed in the equatorial stratosphere has increased during the later period, leading to the bias of the quasi-biennial oscillation (QBO) to more westerly phase. During the later period, the ratio of QBO westerly/easterly is 15/10, while that in the earlier period is 13/15 in the NCEP datasets. In the mid-latitudes, westerly QBO modulation of the

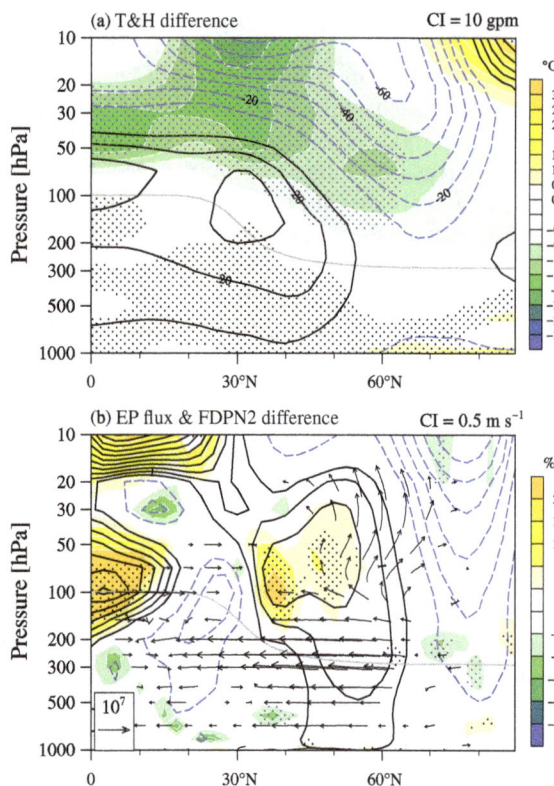

Figure 4. Circulation differences between the two time periods, 1988/1989–2010/2011 minus 1951/1952–1986/1987: (a) zonal-mean geopotential height (contours) and temperature (shading); (b) zonal-mean zonal wind (contours) and FDPN2 (frequency distribution of days with positive n_0^2, $f\left(n_0^2 > 0\right)$, shading). Stippling indicates significantly different regions above the 95% confidence level by using the two-sided t-test for zonal-mean temperature in (a) and FDPN2 in (b). Gray thick lines denote the climatology position of the tropopause.

wintertime EP fluxes is marked by downward and equatorward EP flux anomalies from the stratosphere into the troposphere (Figure 9c of Chen and Li (2007)). This helps to increase the EP flux convergence in the mid-latitudes and contributes to the weaker westerly in the mid-latitudes. However, further studies are needed to understand the exact mechanism; such studies may help better understand the interaction between stratosphere and troposphere under the background of global warming and ozone recovery.

Acknowledgements

The first author would like to acknowledge the hospitality of Atmosphere and Ocean Research Institute, University of Tokyo, for providing the visiting researcher program. In this study, the NCEP1 and NCEP2 reanalysis data are provided by the NOAA/OAR/ESRL PSD, and are available online at ftp://ftp.cdc.noaa.gov/Datasets/, the ERA40 and ERAINT data are provided by the ECMWF and are available online at http://apps.ecmwf.int/datasets/data/, and the JRA25 and JRA55 data are provided by JMA and are available online at http://jra.kishou.go.jp/. This research was supported by the National Basic Research Program of China (973 Program) (Grant No. 2010CB428603) and the National Natural Science Foundation of China (Grant Nos 41175041 and 41375046).

References

Andrews DG, Holton JR, Leovy CB. 1987. *Middle Atmosphere Dynamics*. Academic Press: San Diego, CA; 489 pp.

Bao Q, Yang J, Liu Y, Wu G, Wang B. 2010. Roles of anomalous Tibetan Plateau warming on the Severe 2008 winter storm in Central-Southern China. *Monthly Weather Review* **138**: 2375–2384, doi: 10.1175/2009mwr2950.1.

Boyle JS. 1986. Comparison of the synoptic conditions in midlatitudes accompanying cold surges over eastern Asia for the months of December 1974 and 1978. Part I: Monthly mean fields and individual events. *Monthly Weather Review* **114**: 903–918.

Chen W, Li T. 2007. Modulation of northern hemisphere wintertime stationary planetary wave activity: East Asian climate relationships by the Quasi-Biennial Oscillation. *Journal of Geophysical Research-Atmospheres* **112**: D20120, doi: 10.1029/2007jd008611.

Chen W, Wei K. 2009. Anomalous propagation of the quasi-stationary planetary waves in the atmosphere and its roles in the impact of the stratosphere on the East Asian winter climate. *Advances in Earth Science* **24**: 272–285 (in Chinese).

Chen W, Wei K, Wang L, Zhou Q. 2013. Climate variability and mechanisms of the East Asian winter monsoon and the impact from the stratosphere. *Chinese Journal of Atmospheric Sciences* **37**: 425–438 (in Chinese).

Dee DP, Uppala S. 2009. Variational bias correction of satellite radiance data in the ERA-Interim reanalysis. *Quarterly Journal of the Royal Meteorological Society* **135**: 1830–1841, doi: 10.1002/qj.493.

Dickinson RE. 1968. Planetary Rossby waves propagating vertically through weak westerly wind wave guides. *Journal of Atmospheric Sciences* **25**: 984–1002.

Ding Y, Krishnamurti TN. 1987. Heat budget of the Siberian high and the winter monsoon. *Monthly Weather Review* **115**: 2428–2449.

Ebita A, Kobayashi S, Ota Y, Moriya M, Kumabe R, Onogi K, Harada Y, Yasui S, Miyaoka K, Takahashi K, Kamahori H, Kobayashi C, Endo H, Soma M, Oikawa Y, Ishimizu T. 2011. The Japanese 55-year reanalysis "JRA-55": an interim report. *Sola* **7**: 149–152, doi: 10.2151/sola.2011-038.

Edmon HJ Jr, Hoskins BJ, McIntyre ME. 1980. Eliassen-Palm cross sections for the troposphere. *Journal of the Atmospheric Sciences* **37**: 2600–2616.

Gu L, Wei K, Huang R. 2008. Severe disaster of Blizzard, freezing rain and low temperature in January 2008 in China and its association with the anomalies of East Asian monsoon system. *Climatic and Environmental Research* **13**: 405–418 (in Chinese).

He S. 2013. Reduction of the East Asian winter monsoon interannual variability after the mid-1980s and possible cause. *Chinese Science Bulletin* **58**: 1331–1338, doi: 10.1007/s11434-012-5468-5.

Hu YY, Tung KK. 2002. Interannual and decadal variations of planetary wave activity, stratospheric cooling, and Northern Hemisphere Annular mode. *Journal of Climate* **15**: 1659–1673.

Huang R, Gambo K. 1983. On other wave guide in stationary planetary wave propagations in the winter Northern Hemisphere. *Science in China* **26**: 940–950.

Huang R, Chen J, Wang L, Lin Z. 2012. Characteristics, processes, and causes of the spatio-temporal variabilities of the East Asian monsoon system. *Advances in Atmospheric Sciences* **29**: 910–942, doi: 10.1007/s00376-012-2015-x.

Kalnay E, Kanamitsu M, Kistler R, Collins W, Deaven D, Gandin L, Iredell M, Saha S, White G, Woollen J, Zhu Y, Chelliah M, Ebisuzaki W, Higgins W, Janowiak J, Mo KC, Ropelewski C, Wang J, Leetmaa A, Reynolds R, Jenne R, Joseph D. 1996. The NCEP/NCAR 40-year reanalysis project. *Bulletin of the American Meteorological Society* **77**: 437–471.

Kanamitsu M, Ebisuzaki W, Woollen J, Yang SK, Hnilo JJ, Fiorino M, Potter GL. 2002. NCEP-DOE AMIP-II reanalysis (R-2). *Bulletin of the American Meteorological Society* **83**: 1631–1643, doi: 10.1175/bams-83-11-1631.

Li JP, Wang JXL. 2003. A modified zonal index and its physical sense. *Geophysical Research Letters* **30**: 1632, doi: 10.1029/2003gl017441.

Li Q, Graf HF, Giorgetta MA. 2007. Stationary planetary wave propagation in Northern Hemisphere winter - climatological analysis of the refractive index. *Atmospheric Chemistry and Physics* **7**: 183–200.

Lorenz DJ, Hartmann DL. 2003. Eddy-Zonal flow feedback in the Northern Hemisphere winter. *Journal of Climate* **16**: 1212–1227.

Matsuno T. 1970. Vertical propagation of stationary planetary waves in the winter Northern Hemisphere. *Journal of Atmospheric Sciences* **27**: 871–883.

Onogi K, Tslttsui J, Koide H, Sakamoto M, Kobayashi S, Hatsushika H, Matsumoto T, Yamazaki N, Kaalhori H, Takahashi K, Kadokura S, Wada K, Kato K, Oyama R, Ose T, Mannoji N, Taira R. 2007. The JRA-25 reanalysis. *Journal of the Meteorological Society of Japan* **85**: 369–432.

Perlwitz J, Harnik N. 2003. Observational evidence of a stratospheric influence on the troposphere by planetary wave reflection. *Journal of Climate* **16**: 3011–3026.

Tao SY, Wei J. 2008. Severe snow and freezing-rain in January 2008 in the Southern China. *Climatic and Environmental Research* **13**: 337–350 (in Chinese).

Thompson DWJ, Wallace JM. 1998. The Arctic Oscillation signature in the wintertime geopotential height and temperature fields. *Geophysical Research Letters* **25**: 1297–1300.

Thompson DWJ, Wallace JM. 2000. Annular modes in the extratropical circulation. Part I: Month-to-month variability. *Journal of Climate* **13**: 1000–1016.

Uppala SM, KÅllberg PW, Simmons AJ, Andrae U, Da Costa Bechtold V, Fiorino M, Gibson JK, Haseler J, Hernandez A, Kelly GA, Li X, Onogi K, Saarinen S, Sokka N, Allan RP, Andersson E, Arpe K, Balmaseda MA, Beljaars ACM, Van De Berg L, Bidlot J, Bormann N, Caires S, Chevallier F, Dethof A, Dragosavac M, Fuentes M, Hagemann S, Holm E, Hoskins BJ, Isaksen L, Janssen PAEM, Jenne R, McNally AP, Mahfouf JF, Morcrette JJ, Rayner NA, Saunders RW, Simon P, Sterl A, Trenberth KE, Untch A, Vasiljevic D, Viterbo P, Woollen J. 2005. The ERA-40 re-analysis. *Quarterly Journal of the Royal Meteorological Society* **131**: 2961–3012.

Wang H, He S. 2012. Weakening relationship between East Asian winter monsoon and ENSO after mid-1970s. *Chinese Science Bulletin* **57**: 3535–3540, doi: 10.1007/s11434-012-5285-x.

Wang L, Chen W, Huang R. 2008. Interdecadal modulation of PDO on the impact of ENSO on the East Asian winter monsoon. *Geophysical Research Letters* **35**: L20702, doi: 10.1029/2008gl035287.

Wang B, Wu Z, Chang CP, Liu J, Li J, Zhou T. 2010. Another look at interannual-to-interdecadal variations of the East Asian winter monsoon: the Northern and Southern temperature modes. *Journal of Climate* **23**: 1495–1512.

Wei K, Bao Q. 2012. Projections of the East Asian winter monsoon under the IPCC AR5 scenarios using a coupled model: IAP_FGOALS. *Advances in Atmospheric Sciences* **29**: 1200–1214, doi: 10.1007/s00376-012-1226-5.

Zhang Y, Sperber KR, Boyle JS. 1997. Climatology and interannual variation of the East Asian winter monsoon: results from the 1979–95 NCEP/NCAR reanalysis. *Monthly Weather Review* **125**: 2605–2619.

Zhou W, Wang X, Zhou TJ, Chan JCL. 2007. Interdecadal variability of the relationship between the East Asian winter monsoon and ENSO. *Meteorology and Atmospheric Physics* **98**: 283–293, doi: 10.1007/s00703-007-0263-6.

Zhou W, Chan JCL, Chen W, Ling J, Pinto JG, Shao YP. 2009. Synoptic-scale controls of persistent low temperature and icy weather over Southern China in January 2008. *Monthly Weather Review* **137**: 3978–3991, doi: 10.1175/2009mwr2952.1.

26

Variations in temperature-related extreme events (1975–2014) in Ny-Ålesund, Svalbard

Ting Wei,[1] Minghu Ding,[1,2]* Bingyi Wu,[1] Changgui Lu[1] and Shujie Wang[3]

[1]Climate System Institute, Chinese Academy of Meteorological Sciences, Beijing, China
[2]State Key Laboratory of Cryospheric Sciences, CAREERI, Lanzhou, China
[3]College of Geography and Environment, Shandong Normal University, Jinan, China

*Correspondence to:
M. Ding, Chinese Academy of Meteorological Sciences, 46 Zhongguancun South Street, Haidian District, Beijing 100081, China.
E-mail: dingminghu@cams.cma.gov.cn

Abstract

We present a comprehensive analysis of temperature-related extreme events in Ny-Ålesund (78.9°N, 11.9°E) using data from three meteorological stations. The results show that annual mean temperatures in Ny-Ålesund increase at a rate that is four times faster than the global mean from 1975 to 2014 with no 'hiatus' in recent decades. The annual diurnal temperature range shows a negative trend as minimum daily temperature increases at a faster rate than maximum daily temperature. A negative trend in cold extremes and a positive trend in warm extremes are observed. This asymmetry hints at potential changes in the probability distribution of temperatures in Ny-Ålesund.

Keywords: Arctic; climate change; extremes

1. Introduction

Extreme events, such as heat waves, droughts, floods, and hurricanes, interact with exposed and vulnerable human and natural systems to produce disasters (IPCC, 2013). Overwhelming evidence indicates that climate extremes have frequently occurred on both regional and global scales in recent years (Osborn et al., 2000; Alexander et al., 2006; Donat et al., 2013; Chen and Sun, 2015). The intensification of climate extremes is causing dangerous changes in the economy and ecosystems, and to society and human health (Handmer et al., 2012). Enhanced warming is observed at high latitudes in response to increased CO_2 relative to the warming trends at mid- and low-latitudes. Polar amplification may cause an increased probability of extreme weather events in mid-latitudes (Francis and Vavrus, 2012). Therefore, the variations in mean climate and extreme events at high latitudes should be investigated.

Recent studies indicate that climate extremes have changed at mid-high latitudes. Cold/warm extreme events show a positive trend in North America since the late 1960s (CCSP, 2008a; Peterson et al., 2008; Zhang et al., 2010). In European and Mediterranean countries, a decrease in large-scale occurrences of cold extremes and an increase in warm/hot extremes are generally consistent with global trends of temperatures and their extremes (Moberg et al., 2006; Della-Marta et al., 2007). A significant positive trend in the frequency of heat waves and a negative trend in the number of cold spells have also been observed in Australia and the majority of Asia since the middle of the 20th century (Chambers and Griffiths, 2008; Choi et al., 2009; You et al., 2010; Wang et al., 2013; Chen and Sun, 2014).

However, few studies have examined extreme events in polar regions. Tuomenvirta et al. (2000) indicated that mean maximum and minimum temperatures in western coastal Greenland show a decreasing trend from 1950 to 1995 and opposite trends over the Nordic Seas and Fenno-Scandia. These results indicate considerable regional differences for extreme events over the Arctic. A high frequency in warm days is observed at Svalbard Airport from 1975 to 2010 with a negative trend in cold nights (Bednorz, 2011; Bednorz and Kolendowicz, 2013; Tomczyk and Bednorz, 2014).

Ny-Ålesund (78.9°N, 11.9°E), which is located on the west coast of Svalbard (Spitsbergen), is one of the northernmost archipelagos in the Arctic. Although Ny-Ålesund has its own weather features, it provides valuable evidence for climate change in the general Arctic. Several studies have analysed climate change in Ny-Ålesund and identified a significant warming in recent decades (Førland and Hanssen-Bauer, 2000; Førland et al., 2011; Maturilli et al., 2013, 2015). However, extreme events and their changes have been given minimal attention. In this study, we investigate changes in temperature and its related extremes in Ny-Ålesund based on data from three meteorological stations. This investigation may advance our understanding of intensity and frequency of climate extremes in the Arctic and the influence of global warming on extreme events.

2. Methods

2.1. Data

We obtained data from three meteorological stations in Ny-Ålesund, Svalbard (Figure S1, Supporting

Table 1. List of the 14 ETCCDI temperature extreme indices (TX for maximum temperature and TN for minimum temperature).

Index	Index name	Definition	Units
TXx	Max TX	Warmest daily maximum temperature	°C
TXn	Min TX	Coldest daily maximum temperature	°C
TNn	Min TN	Coldest daily minimum temperature	°C
TNx	Max TN	Warmest daily minimum temperature	°C
DTR	Diurnal temperature range	Mean difference between daily maximum and minimum temperature	°C
FD	Frost days	Annual number of days when $TN < 0\,°C$	days
ID	Ice days	Annual number of days when $TX < 0\,°C$	days
GSL	Growing season length	Annual number of days between the first occurrence of 6 consecutive days with $((TX + TN)/2) > 5\,°C$ and first occurrence of consecutive 6 days with $[(TX + TN)/2] < 5\,°C$. For the Northern Hemisphere this is calculated from 1 January to 31 December while for the Southern Hemisphere it is calculated from 1 July to 30 June	days
CSDI	Cold spell duration	Annual number of days with at least 6 consecutive days when $TN < 10th$ percentile	days
WSDI	Warm spell duration	Annual number of days with at least 6 consecutive days when $TX > 90th$ percentile	days
TX90p	Warn days	Share of days when $TX > 90th$ percentile	%
TX10p	Cold days	Share of days when $TX < 10th$ percentile	%
TN90p	Warn nights	Share of days when $TN > 90th$ percentile	%
TN10p	Cold nights	Share of days when $TN < 10th$ percentile	%

Information). Sverdrup Research station, which was established by the Norwegian Polar Institute (NPI), provides the greatest amount of daily meteorological observation data (1975 to present) for the vicinity of Ny-Ålesund. The joint French-German Arctic Research Base by Alfred Wegener Institute for Polar and Marine Research and Polar Institute Paul Emile Victor (AWIPEV) operates a 10 m meteorological tower, which is mounted on a measurement field with soft tundra ground. It provides data with 5-min intervals from 1994 to 1998 and data with 1-min intervals from 1998 to 2011. China established the Yellow River (YR) station in 2003. This station operates a meteorological tower on the eastside of the township near the Bird Sanctuary. The height of the tower was changed from 6 to 10 m in August 2007. Continuous observation data from 2005 are available with a 1-h resolution at a level of 2 m. The observations of the three stations are obtained from the World Meteorological Organization (WMO) standard; the data are extensively applied (Førland and Hanssen-Bauer, 2000; Førland *et al.*, 2011; Maturilli *et al.*, 2013, 2015). In the following sections, the three stations are denoted by NPI, AWI, and YR.

2.2. Climate extreme indices

The 14 temperature-related extreme indices proposed by the Expert Team on Climate Change Detection and Indices (ETCCDI) are analysed in this study (Table 1). The ETCCDI climate indices are very popular and extensively utilized in climate research and related fields (e.g. Zhang *et al.*, 2011; Sillmann *et al.*, 2013) due to their robustness and fairly straightforward calculation and interpretation. The indices are primarily selected for the assessment of changing climates, including changes in intensity, frequency and

duration of temperature-related events. The indices are calculated using a FORTRAN package, as documented at the ETCCDI climate change indices website (http://etccdi.pacificclimate.org/software.shtml). All raw data from the three stations are resampled to a common daily resolution using an arithmetic mean. Note that we use the Sen's slope method (Sen, 1968) to compute the linear trend of the considered variables and the non-parametric Mann–Kendall approach for the significance test. Regarding probability density distribution (PDF), we employ the Kolmogorov–Smirnov two-sample test to determine if two samples are from the same distribution.

3. Results

The annual mean air temperature in Ny-Ålesund shows a significant warming trend from 1975 to 2014 ($\sim 0.79\,°C\,decade^{-1}$; Table S1), which is four times larger than the trend for global mean temperature ($\sim 0.17\,°C\,decade^{-1}$) (Figure 1(a)). Temperatures during the winter season exhibit an enhanced warming ($1.75\,°C\,decade^{-1}$) with greater inter-annual variability ($3.34\,°C\,year^{-1}$). The observations from the three stations show reasonable consistency (Figure 1). Based on an 11-year sliding linear trend, the magnitude of the warming trend in Ny-Ålesund is gradually decreased (Figure 1(b)), which is consistent with the global mean. The results show a continuous warming trend in the winter temperatures in Ny-Ålesund since the mid-1990s, whereas the trend of global mean temperature is weakened and turns to a negative phase in recent years (Figure 1(d)). The global mean summer temperatures exhibit a 'hiatus' since the late 1990s, whereas summer temperatures in Ny-Ålesund exhibit a negative trend in recent years (Figure 1(f)). These results

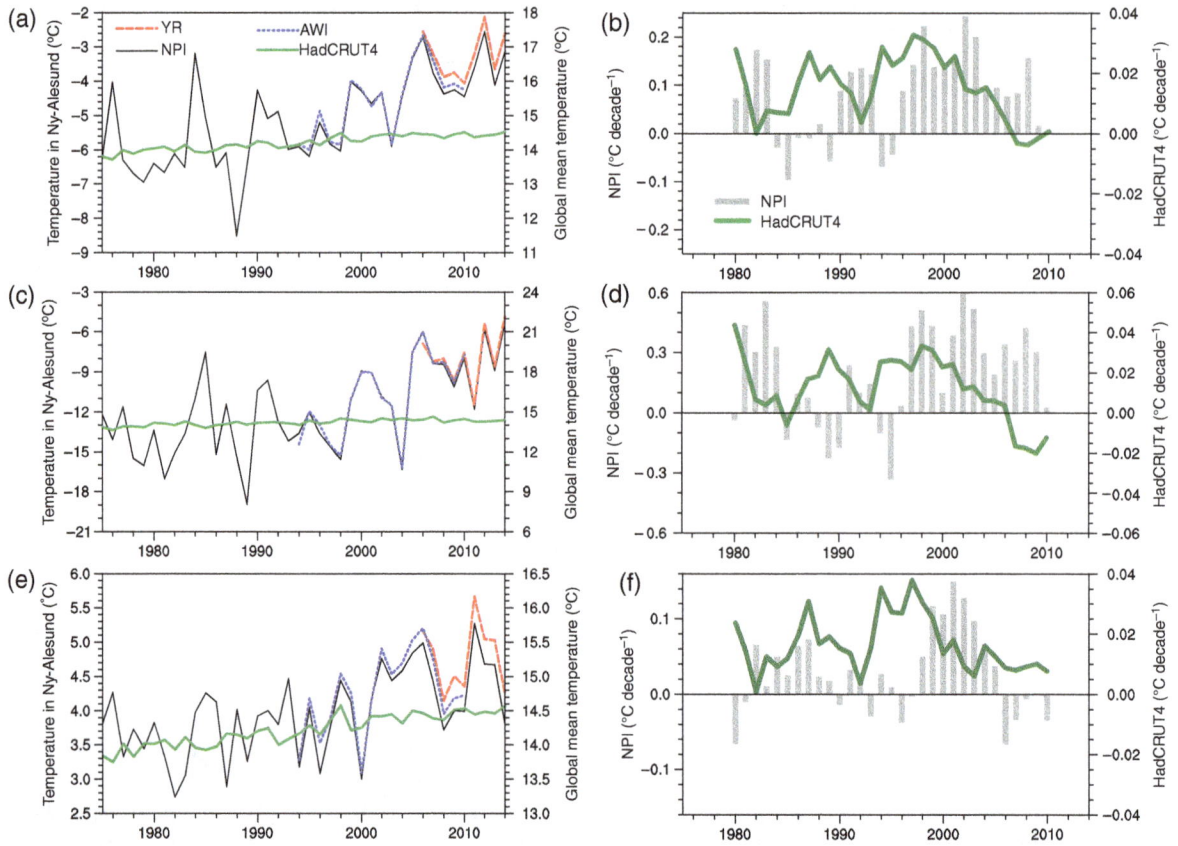

Figure 1. Left: the time series of air temperature for the global mean and the three stations in Ny-Ålesund for the annual mean (a), the winter season (c) and the summer season (e). Right: the 11-yr sliding linear trends of air temperature for the global mean and the NPI station in Ny-Ålesund for the annual mean (b), the winter season (d) and the summer season (f). The global averaged data are obtained from HadCRUT4.

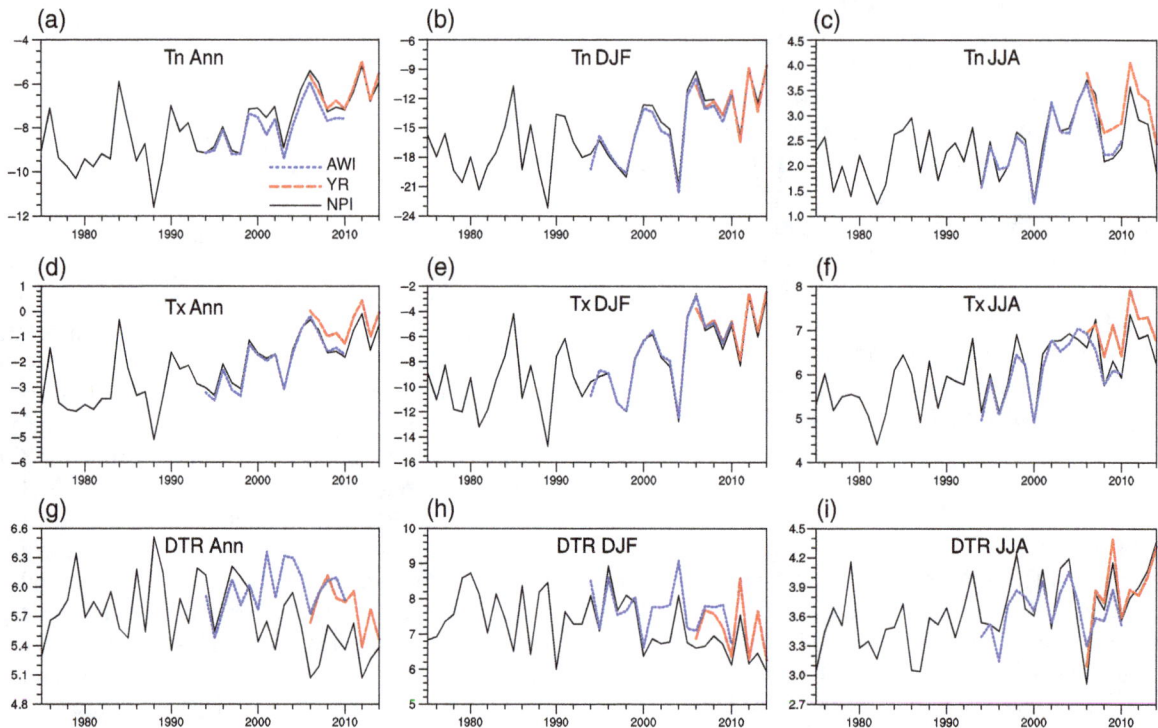

Figure 2. Time series of minimum daily temperature (a–c), maximum daily temperature (d–f) and diurnal temperature range (g–i) from the three stations in Ny-Ålesund for the annual mean (a, d, g), winter season (b, e, h), and summer season (c, f, i).

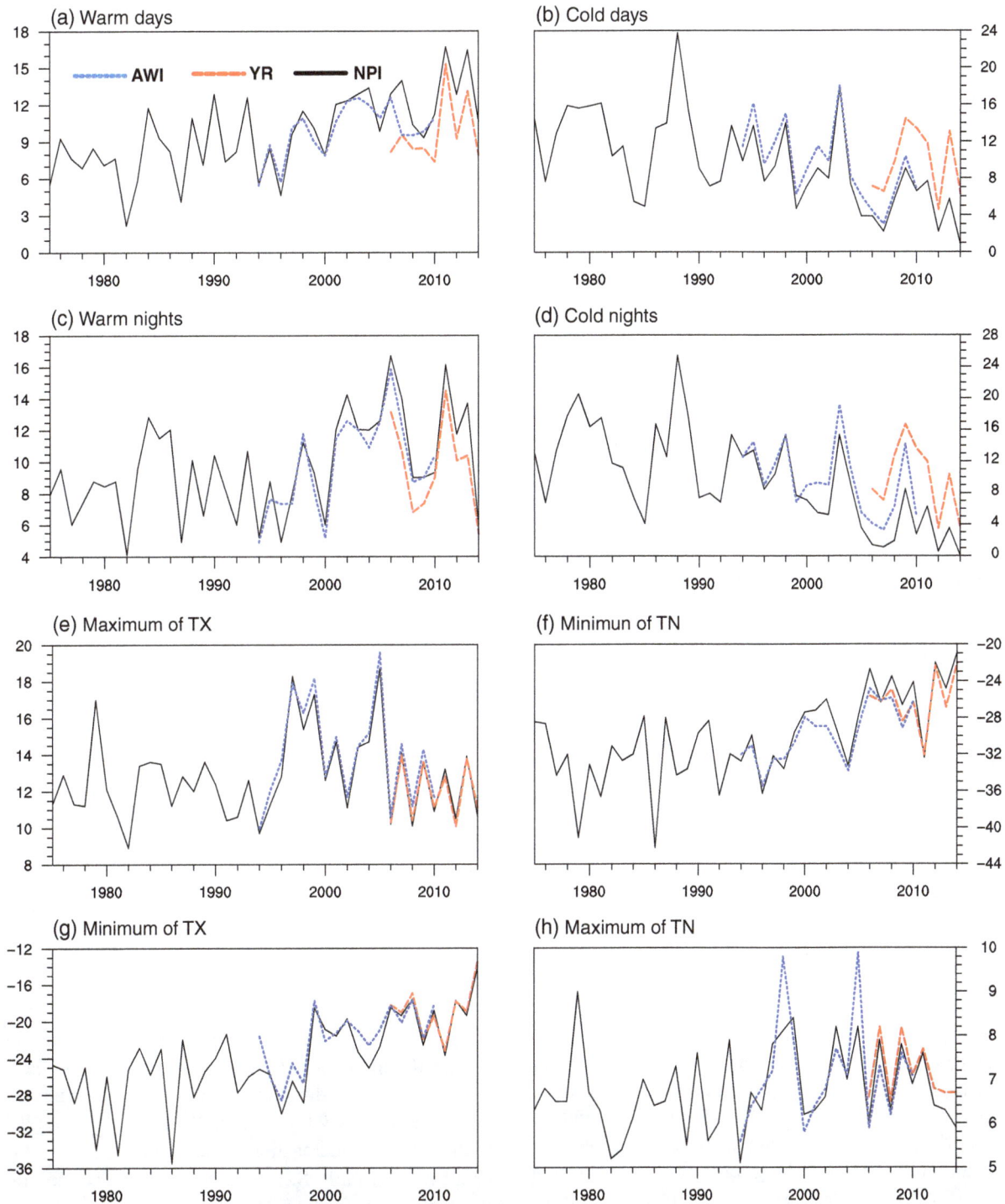

Figure 3. Time series of the annual mean maximum of TX (TXx), minimum of TX (TXn), maximum of TN (TNx), minimum of TN (TNn), frost days (FD), ice days (ID), warm spell duration (WSDI), cold spell duration (CSDI), warm days (TX90p), cold days (TX10p), warm nights (TN90p), cold nights (TN10p), and growing season length. Please refer to Table I for detailed definitions.

indicate different temperature behaviours between the Ny-Ålesund and the globe. Thus, the variations in temperature-related extremes in Ny-Ålesund should be examined.

Figure 2 illustrates the time series of mean maximum and minimum daily temperatures in Ny-Ålesund. Both time series show a significant positive trend in the past 40 years (Figure 2(a)–(c)), especially for the winter season since mid-1990s (Table S1). The annual diurnal

temperature range (DTR) exhibits a decreasing trend (0.012 °C year^{-1}) during 1975–2014 (Figure 1(g)) as the minimum daily temperature increases at a faster rate than the maximum daily temperature (Table S1). The negative trend in annual DTR since the mid-1980s is contradictory to the DTR behaviour over global land surfaces (Rohde *et al.*, 2013). Note that the summer DTR shows a positive trend (0.02 °C year^{-1}) due to the rapid increase in the maximum daily temperature.

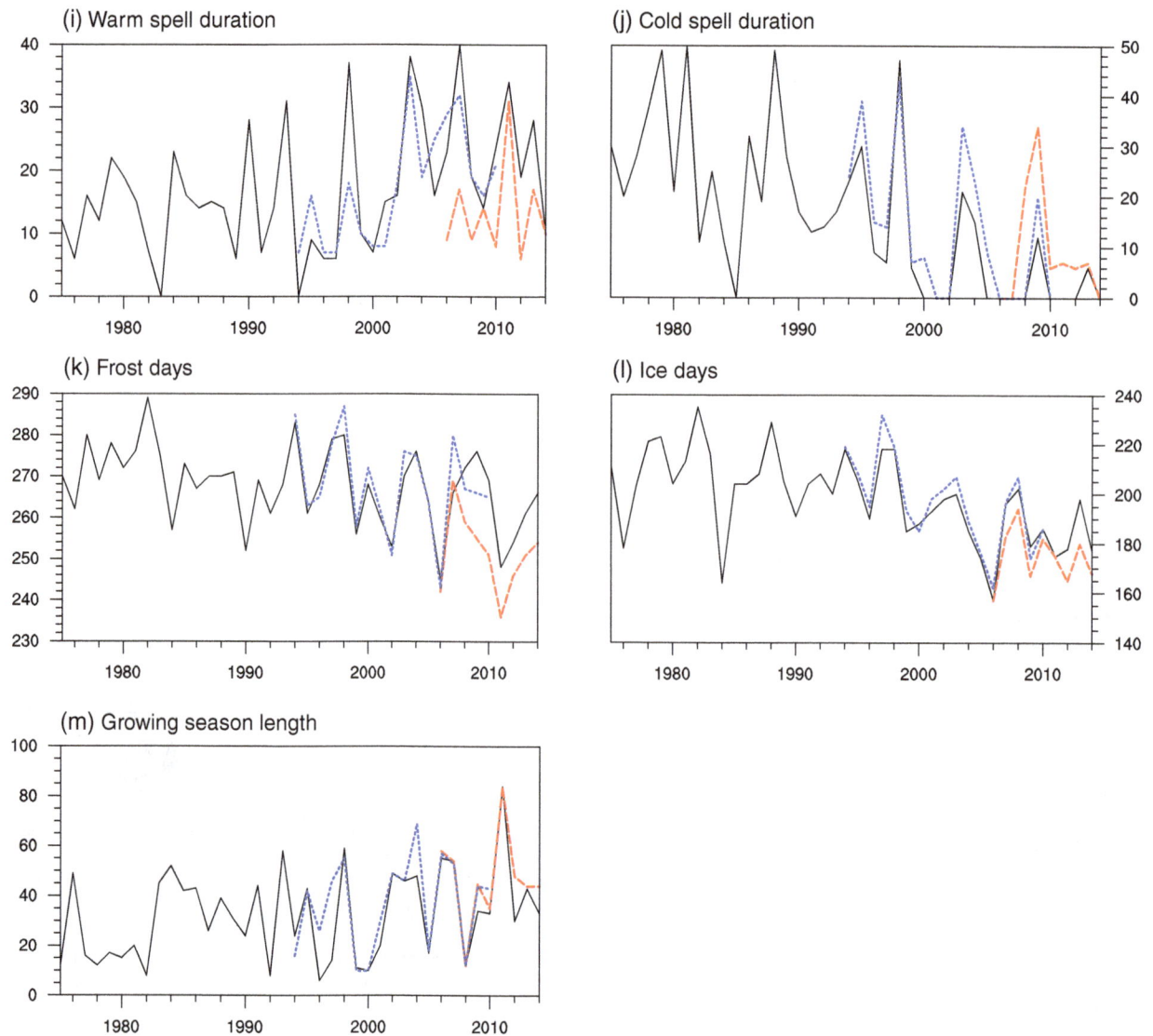

Figure 3. continued

In response to the increased temperatures, a significant negative trend in the annual occurrence of cold days (~2.7% decade^{-1}) and nights (~3.5% decade^{-1}) is observed from 1975 to 2014 (Figure 3(b) and (d); Table S2). Although the annual occurrence of warm days and nights shows a long-term upward trend in the past 40 years (Figure 3(a) and (c)), their magnitudes (1.3–1.8% decade^{-1}) are relatively small. The annual occurrence of warm nights does not exhibit a significant trend from 1975 to 2000 and shows a negative trend in recent years (Figure 3(c)). The temperatures on the coldest days (nights) significantly increase during 1975–2014 (Figure 3(f) and (g)) with a significant positive trend of 2.3 (2.5) °C decade^{-1} (Table S2). However, the temperatures on the hottest days and nights reveal a large magnitude of inter-annual variations with an insignificant long-term upward trend (Figure 3(e) and (h)). This result may be attributed to a small warming trend in summer temperatures. In addition, the local SST and large-scale circulation patterns are likely to influence the absolute temperature indices (e.g.

temperatures on the hottest days) based on previous studies (Scaife *et al.*, 2008; Alexander *et al.*, 2009).

The cold spell duration exhibits a significant negative trend of ~8 days decade^{-1} from 1975 to 2014 with a larger decreasing trend from 1994 to 2010 (Figure 3(j); Table S2). The significant decline since the early 1990s is similar to the trend in cold spells at middle and high latitudes (Alexander *et al.*, 2006; Chen and Sun, 2015). Although the warm spell duration increased over the past 40 years, minimal change and a downward trend since the early 2000s is observed (Figure 3(i)). The annual occurrence of frost days and ice days in Ny-Ålesund shows a significant negative trend (3–9 days decade^{-1}) during the study period, especially since the early 1990s (Figure 3(k) and (l)). The growing season length, which affects hydrologic factors and biologic territories (Logan *et al.*, 2003; CCSP, 2008b), increases in Ny-Ålesund from 1975 to 2014 (Figure 3(m)) with a long-term trend of ~4 days decade^{-1}. This result is similar to the change

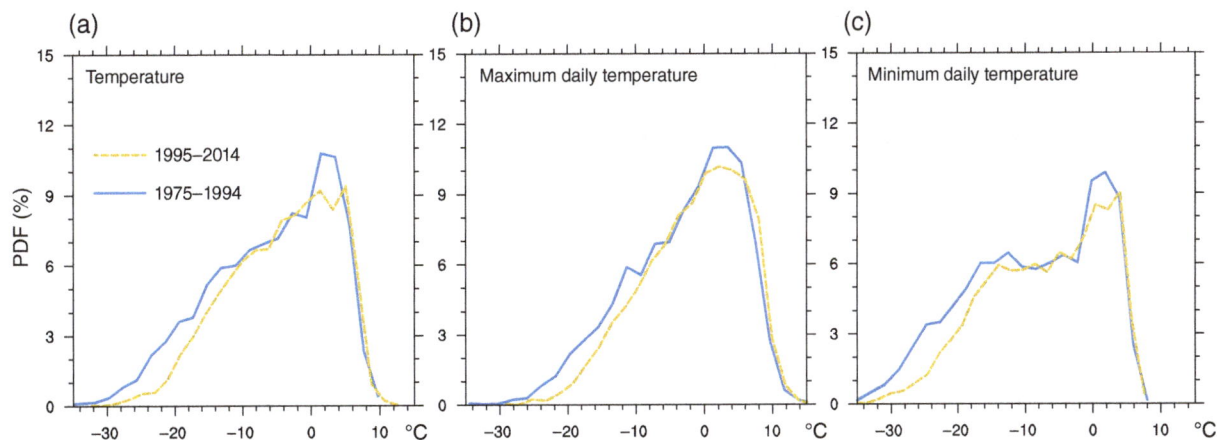

Figure 4. Probability distribution functions for (a) daily temperature, (b) maximum daily temperature, and (c) minimum daily temperature at the NPI station.

in the growing season length in the eastern and central Tibetan Plateau (Liu *et al.*, 2006).

The temperature-related extremes in Ny-Ålesund show a negative trend in cold extremes and a positive trend in warm extremes. The negative trend in extreme minimum temperatures is larger than the positive trend in extreme maximum temperatures. This asymmetry hints at potential changes in the shape and/or scale of the distribution of temperatures in Ny-Ålesund. According to the probability distribution function, it is likely (at 99% confidence level) that there is a decreased frequency in the cold tails of temperatures after mid-1990s and a decreased frequency of temperature above freezing point (Figure 4). The maximum and minimum temperatures indicate a similar shift. However, additional observations are required to accurately assess potential changes in the shape of the distribution of temperature.

4. Summary

In this study, we present a comprehensive picture of changes in temperature-related extremes from 1975 to 2014 based on the station data in Ny-Ålesund. The annual warming trend in Ny-Ålesund during the past 40 years is four times larger than the global mean. The continuous warming since the mid-1990s is inconsistent with the 'hiatus' in global warming trend in terms of annual mean temperature. The annual DTR exhibits a negative trend from 1975 to 2014, whereas the summer DTR shows a positive trend. In response to the increased temperature, the temperature-related extremes in Ny-Ålesund show a negative (positive) trend in cold (warm) extremes. Specifically, Ny-Ålesund experiences a significant increase (decrease) in the annual occurrence of warm days (cold nights). The temperatures on the coldest days (nights) exhibit a significant positive trend of 2.3 (2.5) $°C\,decade^{-1}$, whereas the trend of the temperatures on the hottest days (nights) is moderate. The annual occurrence of cold spells, frost days, and ice days shows a long-term negative trend from 1975 to 2014, whereas the annual occurrence of warm spells shows a positive trend. In addition, the growing season length is prolonged in Ny-Ålesund with a trend of ~4 days $decade^{-1}$.

Acknowledgements

We thank Marion Maturilli for her helpful discussions. We thank the two anonymous reviewers for their constructive comments, which significantly improved this paper. This study was jointly funded by the National Basic Research and Development Program of China (2015CB453202), National Science Fund for Distinguished Young Scholars (41425003), National Natural Science Foundation of China (41505068), and Fundamental Research Funds of CAMS (2015Y004).

Supporting information

The following supporting information is available:

Figure S1. (a) Map of Svalbard in the Arctic region showing the location of Ny-Ålesund. (b) The location of three meteorological stations (NPI, AWI, YR).

Table S1. Linear trends ($°C\,year^{-1}$) of the maximum temperature, minimum temperature, diurnal temperature range and mean temperature at the three stations.

Table S2. Linear trends of temperature-related extremes (Table 1) at the three stations. Units are % $year^{-1}$ for TX90p, TX10p, TN90p, and TN10p, $°C\,year^{-1}$ for TXx, TXn, TNx, and TNn, and days $year^{-1}$ for the remaining indices.

References

Alexander LV, Zhang X, Peterson TC, Caesar J, Gleason B, Klein Tank AMG, Haylock M, Collins D, Trewin B, Rahimzadeh F, Tagipour A, Rupa Kumar K, Revadekar J, Griffiths G, Vincent L, Stephenson DB, Burn J, Aguilar E, Brunet M, Taylor M, New M, Zhai P, Rusticucci M, Vazquez-Aguirre JL. 2006. Global observed changes in daily climate extremes of temperature and precipitation. *Journal of Geophysical Research* **111**: D05109, doi: 10.1029/2005JD006290.

Alexander LV, Uotila P, Nicholls N. 2009. Influence of sea surface temperature variability on global temperature and precipitation extremes. *Journal of Geophysical Research* **114**: D18116, doi: 10.1029/2009JD012301.

Bednorz E. 2011. Occurrence of winter air temperature extremes in Central Spitsbergen. *Theoretical and Applied Climatology* **106**: 547–556.

Bednorz E, Kolendowicz L. 2013. Summer mean daily air temperature extremes in Central Spitsbergen. *Theoretical and Applied Climatology* **113**: 471–479.

CCSP. 2008a. The effects of climate change on agriculture, land resources, water resources, and biodiversity in the United States. A Report by the U.S. Climate Change Science Program and the Subcommittee on Global Change Research, U.S. Department of Agriculture, Washington, DC; 362 pp.

CCSP. 2008b. Weather and climate extremes in a changing climate: regions of focus: North America, Hawaii, Caribbean, and U.S. Pacific Islands. A Report by the U.S. Climate Change Science Program and the Subcommittee on Global Change Research, Department of Commerce, NOAA's National Climatic Data Center, Washington, DC; 164 pp.

Chambers L, Griffiths G. 2008. The changing nature of temperature extremes in Australia and New Zealand. *Australian Meteorological Magazine* **57**: 13–35.

Chen H, Sun J. 2014. Changes in climate extreme events in China associated with warming. *International Journal of Climatology* **35**: 2735–2751, doi: 10.1002/joc.4168.

Chen H, Sun J. 2015. Assessing model performance of climate extremes in China: an intercomparison between CMIP5 and CMIP3. *Climatic Change* **129**: 197–211.

Choi G, Collins D, Ren G, Trewin B, Baldi M, Fukuda Y, Afzaal M, Pianmana T, Gomboluudev P, Huong PTT, Lias N, Kwon WT, Boo KO, Cha YM, Zhou Y. 2009. Changes in means and extreme events of temperature and precipitation in the Asia Pacific Network region, 1955–2007. *International Journal of Climatology* **29**: 1906–1925.

Della-Marta PM, Haylock MR, Luterbacher J, Wanner H. 2007. Doubled length of western European summer heat waves since 1880. *Journal of Geophysical Research* **112**: D15103, doi: 10.1029/2007JD008510.

Donat MG, Alexander LV, Yang H, Durre L, Vose R, Dunn RJH, Willett KM, Aguilar E, Brunet M, Caesar J, Hewitson B, Jack C, Klein Tank AMG, Kruger AC, Marengo J, Peterson TC, Renom M, Oria Rojas C, Rusticucci M, Salinger J, Elrayah AS, Sekele SS, Srivastava AK, Trewin B, Villarroel C, Vincent LA, Zhai P, Zhang X, Kitching S. 2013. Updated analyses of temperature and precipitation extreme indices since the beginning of the twentieth century: the HadEX2 dataset. *Journal of Geophysical Research* **118**: 2098–2118.

Førland EJ, Hanssen-Bauer I. 2000. Increased precipitation in the Norwegian Arctic: true or false? *Climatic Change* **46**: 485–509.

Førland EJ, Benestad R, Hanssen-Bauer I, Haugen JE, Skaugen TE. 2011. Temperature and precipitation development at Svalbard 1900–2100. *Advances in Meteorology*, doi: 10.1155/2011/893790.

Francis JA, Vavrus SJ. 2012. Evidence linking Arctic amplification to extreme weather in mid-latitudes. *Geophysical Research Letters* **39**: L06801, doi: 10.1029/2012GL051000.

Handmer J, Honda Y, Kundzewicz ZW, Arnell N, Benito G, Hatfield J, Mohamed IF, Peduzzi P, Wu S, Sherstyukov B, Takahashi K, Yan Z. 2012. Changes in impacts of climate extremes: human systems and ecosystems. In Managing the Risks of Extreme Events and Disasters to Advance Climate Change Adaptation. Field CB, Barros V, Stocker TF, Qin D, Dokken DJ, Ebi KL, Mastrandrea MD, Mach KJ, Plattner GK, Allen SK, Tignor M, Midgley PM (eds). A Special Report of Working Groups I and II of the Intergovernmental Panel on Climate Change (IPCC). Cambridge University Press: Cambridge, UK, and New York, NY, USA, 231–290.

IPCC. 2013. *Climate Change 2013: The Physical Science Basis. Contribution of Working Group I to the Fifth Assessment Report of the Intergovernmental Panel on Climate Change.* Stocker TF, Qin D, Plattner GK, Tignor M, Allen SK, Boschung J, Nauels A, Xia Y, Bex V, Midgley PM (eds). Cambridge University Press: Cambridge, UK and New York, NY; 1535 pp.

Liu X, Yin ZY, Shao X, Qin N. 2006. Temporal trends and variability of daily maximum and minimum, extreme temperature events, and growing season length over the eastern and central Tibetan Plateau during 1961–2003. *Journal of Geophysical Research* **111**: D19109, doi: 10.1029/2005JD006915.

Logan JA, Regnière J, Powell JA. 2003. Assessing the impacts of global warming on forest pest dynamics. *Frontiers in Ecology and the Environment* **1**: 130–137.

Maturilli M, Herber A, König-Langlo G. 2013. Climatology and time series of surface meteorology in Ny-Ålesund, Svalbard. *Earth System Science Data* **5**: 155–163.

Maturilli M, Herber A, König-Langlo G. 2015. Surface radiation climatology for Ny-Ålesund, Svalbard (78.9°N), basic observations for trend detection. *Theoretical and Applied Climatology* **120**: 331–339.

Moberg A, Jones PD, Lister D, Walther A, Brunet M, Jacobeit J, Alexander LV, Della-Marta PM, Luterbacher J, Yiou P, Chen DL, Klein Tank AMG, Saladié O, Sigró J, Aguilar E, Alexandersson H, Almarza C, Auer I, Barriendos M, Begert M, Bergström H, Böhm R, Butler CJ, Caesar J, Drebs A, Founda D, Gerstengarbe F, Micela G, Maugeri M, Österle H, Pandzic K, Petrakis M, Srnec L, Tolasz R, Tuomenvirta H, Werner PC, Linderholm H, Philipp A, Wanner H, Xoplaki E. 2006. Indices for daily temperature and precipitation extremes in Europe analyzed for the period 1901–2000. *Journal of Geophysical Research* **111**: D22106, doi: 10.1029/2006JD007103.

Osborn TJ, Hulme M, Jones PD, Basnett TA. 2000. Observed trends in the daily intensity of United Kingdom precipitation. *International Journal of Climatology* **20**: 347–364.

Peterson TC, Zhang XB, Brunet-India M, Vazquez-Aguirre JL. 2008. Changes in North American extremes derived from daily weather data. *Journal of Geophysical Research* **113**: D07113, doi: 10.1029/2007JD009453.

Rohde R, Muller RA, Jacobsen R, Muller E, Perlmutter S, Rosenfeld A, Wurtele J, Groom D, Wickham C. 2013. A new estimate of the average earth surface land temperature spanning 1753 to 2011. *Geoinformatics & Geostatistics: An Overview* **1**: 1, doi: 10.4172/2327-4581.1000101.

Scaife A, Folland C, Alexander L, Moberg A, Knight J. 2008. European climate extremes and the North Atlantic Oscillation. *Journal of Climate* **21**: 72–83.

Sen PK. 1968. Estimates of the regression coefficient based on Kendall's tau. *Journal of the American Statistical Association* **63**: 1379–1389.

Sillmann J, Kharin VV, Zhang X, Zwiers FW, Bronaugh D. 2013. Climate extremes indices in the CMIP5 multimodel ensemble: part 1. Model evaluation in the present climate. *Journal of Geophysical Research* **118**: 1716–1733.

Tomczyk AM and Bednorz E. 2014. Warm waves in north-western Spitsbergen. *Polish Polar Research* **35** : 497–511.

Tuomenvirta H, Alexandersson H, Drebs A, Frich P, Nordli PO. 2000. Trends in Nordic and Arctic temperature extremes and ranges. *Journal of Climate* **13**: 977–990.

Wang X, Trewin B, Feng Y, Jones D. 2013. Historical changes in Australian temperature extremes as inferred from extreme value distribution analysis. *Geophysical Research Letters* **40**: 573–578.

You QL, Kang SC, Aguilar E, Pepin N, Flügel W, Yan YP, Xu YW, Zhang YJ, Huang J. 2010. Changes in daily climate extremes in China and their connection to the large scale atmospheric circulation during 1961–2003. *Climate Dynamics* **36**: 2399–2417.

Zhang XB, Wang JF, Zwiers FW, Groisman PY. 2010. The influence of large-scale climate variability on winter maximum daily precipitation over North America. *Journal of Climate* **23**: 2902–2915.

Zhang XB, Alexander L, Hegerl GC, Jones P, Tank AK, Peterson TC, Trewin B, Zwiers FW. 2011. Indices for monitoring changes in extremes based on daily temperature and precipitation data. *Wiley Interdisciplinary Reviews: Climate Change* **2**: 851–870.

27

Strong subsurface soil temperature feedbacks on summer climate variability over the arid/semi-arid regions of East Asia

Lingyun Wu[1] and Jingyong Zhang[2]*

[1] State Key Laboratory of Numerical Modeling for Atmospheric Sciences and Geophysical Fluid Dynamics (LASG), Institute of Atmospheric Physics, Chinese Academy of Sciences, Beijing 100029, China
[2] Center for Monsoon System Research, Institute of Atmospheric Physics, Chinese Academy of Sciences, Beijing 100029, China

*Correspondence to:
J. Zhang, Center for Monsoon System Research, Institute of Atmospheric Physics, Chinese Academy of Sciences, Beijing 100029, China.
E-mail: zjy@mail.iap.ac.cn

Abstract

Two long-term regional climate model simulations with and without subsurface soil temperature feedbacks are performed to investigate the role of soil temperature–atmosphere coupling in influencing interannual variability of summer climate over East Asia. Results indicate that soil temperature–atmosphere coupling depends on climate regimes, mainly affecting summer climate variability over the arid/semi-arid regions. Over these areas, subsurface soil temperature feedbacks play an important role in amplifying summer surface air temperature variability, accounting for about 30–70% of total variance. The feedbacks on precipitation variability are weaker than those on surface air temperature variability over the arid/semi-arid regions, but are still significant over many areas of western part.

Keywords: soil temperature feedbacks; surface air temperature variability; precipitation variability; arid/semi-arid regions; regional climate model simulations

1. Introduction

Over recent decades, land–atmosphere interactions have received increasing attention in the scientific literature (e.g. Namias, 1962; Manabe, 1975; Walker and Rowntree, 1977; Kurbatkin et al., 1979). In 1975, Charney's pioneering work on the effects of albedo on the African climate emphasized the importance of land surface feedbacks on climate (Charney, 1975). Since then, many researchers addressed studies on land effects on climate at a variety of time and space scales (Shukla and Mintz, 1982; Pielke et al., 1998; Koster et al., 2004; Bonan, 2008; Zhang et al., 2008a, 2011a). They found that soil moisture, vegetation, and snow cover play important roles in influencing climate variations (e.g. Yeh et al., 1984; Yang and Huang, 1992; Dirmeyer, 2000; Wang et al., 2008; Ma et al., 2009; Wu et al., 2009; Yang et al., 2009; Dutra et al., 2011; Xia et al., 2011; Chen and Zhou, 2013). For example, soil moisture–atmosphere coupling has been found to make an important contribution to summer precipitation and surface air temperature variability over particular areas, depending on climate regimes (e.g. Koster et al., 2004; Seneviratne et al., 2006; Zhang et al., 2008b).

Soil temperature, another important component of land surface, can influence the climate through its effects on surface energy and water budgets and resulted changes in regional atmospheric circulation (e.g. Tang and Reiter, 1986; Retnakumari et al., 2000; Hu and Feng, 2004a). Only a few published studies have explored the effects of soil temperature on climate variations (e.g. Hu and Feng, 2004a, 2004b; Mahanama

et al., 2008; Xue et al., 2012). In particular, the role of soil temperature feedbacks for interannual climate variability has been less discussed. Mahanama et al. (2008) provided global estimates of subsurface soil temperature impacts on interannual variability of surface air temperature using three long-term atmospheric general circulation model (AGCM) experiments for the first time. However, there are large uncertainties in AGCM studies of land–atmosphere interactions on local to regional scales (e.g. Dirmeyer et al., 2006; Zhang et al., 2008a; Pitman et al., 2009). Compared with AGCMs, regional climate models (RCMs) can better reproduce many small-scale processes involved in land–atmosphere interactions, and thus have been intensively applied in studies on the role of land surface feedbacks for climate variability (e.g. Giorgi, 1990; Fu and Yan, 1996; Lü and Chen, 1999; Lee and Suh, 2000; Gao et al., 2003; Zhang, 2004; Seneviratne et al., 2006; Diffenbaugh et al., 2007; Fischer et al., 2007; Zhang et al., 2008a, 2011b; Leung et al, 2011; Notaro and Zarrin, 2011; Jerez et al., 2012).

Increased climate variability potentially associated with more climate extremes can have severe impacts on human society and ecosystems (IPCC, 2012). The goal of this study is to investigate the role of soil temperature–atmosphere coupling in influencing interannual variability of summer climate over East Asia with a RCM. This is achieved by carrying out two long-term Weather Research and Forecasting (WRF) model simulations, with and without subsurface soil temperature feedbacks. To our knowledge, this study provides the first RCM-based subsurface soil temperature feedbacks on summer climate variability over East Asia.

2. Model and methods

The RCM employed in the study is National Center for Atmospheric Research WRF version 3.3 with the advanced research WRF (ARW) dynamics solver (Skamarock *et al.*, 2008). Model setup includes 50-km horizontal grid resolution [9250 km (west–east) × 6350 km (south–north), centered at 35°N and 115°E], 28 vertical layers. In this study, we use the unified Noah land model for surface physics. The land model consists four soil layers, which are 0–10, 10–40, 40–100 and 100–200 cm, respectively (Chen and Dudhia, 2001). We use the WRF single-moment six-class microphysics scheme (Hong and Lim, 2006), the Kain–Fritsch convective parameterizations (Kain, 2004), the Yonsei University planetary boundary layer scheme (Hong *et al.*, 2006) and the NCAR Community Atmosphere Model (CAM 3.0) spectral-band shortwave and longwave radiation schemes (Collins *et al.*, 2006). The initial and boundary conditions and sea surface temperature (SST)

are obtained from the National Centers for Environmental Prediction (NCEP)-Department of Energy (DOE) reanalysis (Kanamitsu *et al.*, 2002).

This study performs two simulations to explore subsurface soil temperature feedbacks over East Asia. A 27-year-long control simulation (CTL) allows subsurface soil temperature to interact with the atmosphere, covering the period of 1 January 1979 to 31 December 2005. An additional soil temperature experiment (SoilT) disables interactive subsurface soil temperature by replacing soil temperature at three layers below 10 cm with the climatology of CTL at each time step, including 25 summers (1981–2005). In SoilT, we restart CTL on 1 June of each year and integrate to 31 August. The differences between CTL and SoilT reflect the role of subsurface soil temperature feedbacks for summer climate variability over East Asia.

In this study, we use a percentage parameter PV_v to measure the relative contribution of subsurface soil temperature to summer surface air temperature and precipitation variability:

$$PV_v = \frac{\sigma_V^2 \,(\text{CTL}) - \sigma_V^2 \,(\text{SoilT})}{\sigma_V^2 \,(\text{CTL})} \times 100\% \qquad (1)$$

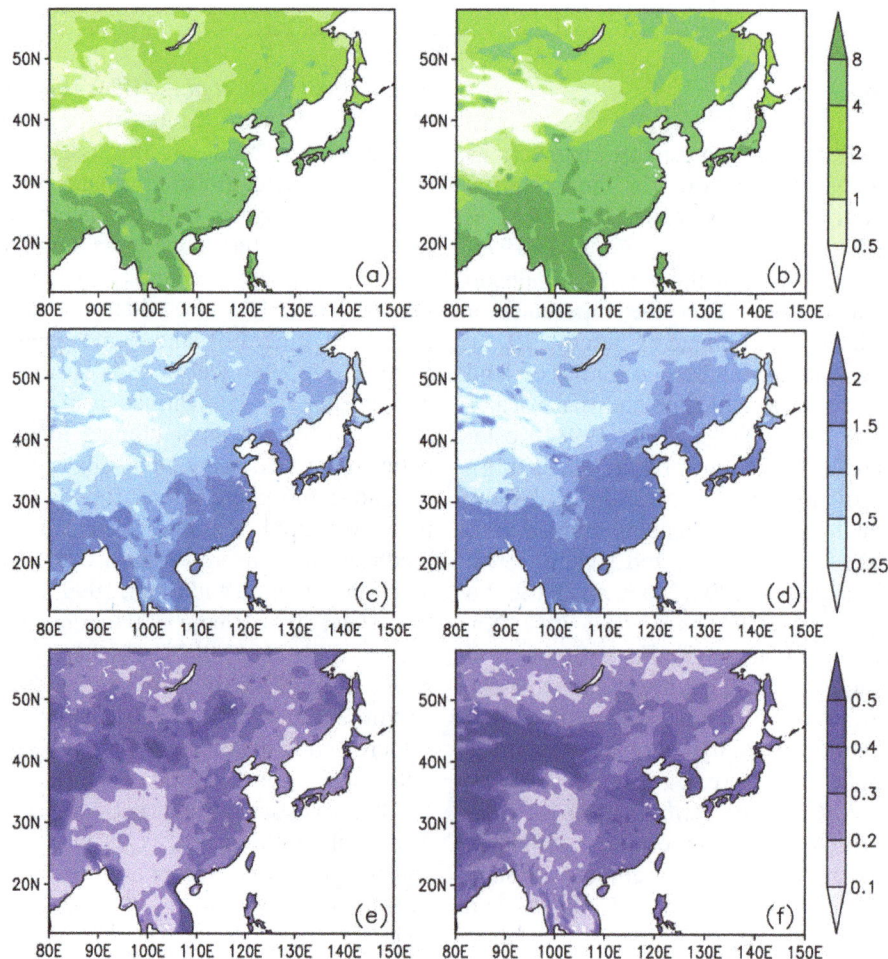

Figure 1. The 1981–2005 summer mean precipitation, standard deviation of summer precipitation, and the coefficient of variation (standard deviation divided by the mean) in (left) observations (Xie *et al.*, 2007) and (right) CTL: (a, b) mean (in mm day^{-1}), (c, d) standard deviation (in mm day^{-1}), and (e, f) coefficient of variation.

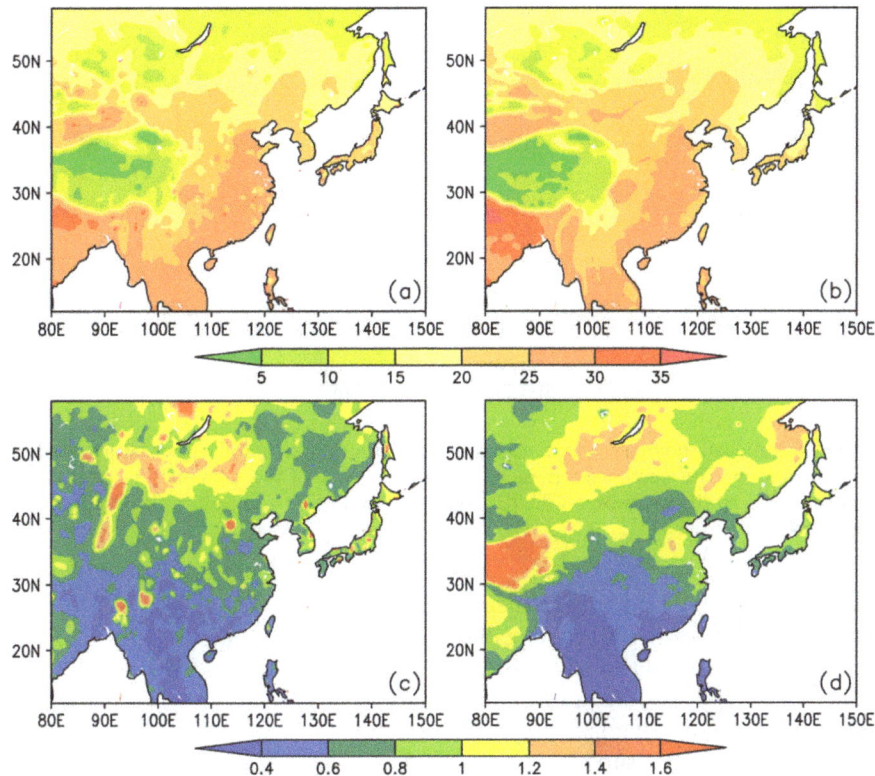

Figure 2. The 1981–2005 summer mean surface air temperature and standard deviation of summer surface air temperature in (left) observations (Fan and van den Dool, 2008) and (right) CTL: (a, b) mean and (c, d) standard deviation.

Where σ_V^2 (CTL) and σ_V^2 (SoilT) represent the interannual variance of summer mean V in the CTL and SoilT simulations, respectively.

3. Results

Figure 1 compares geographic distributions of summer mean precipitation, standard deviation of summer precipitation, and the coefficient of variation over East Asia during 1981–2005 between observations (Xie *et al.*, 2007) and WRF model simulations. Spatially, the observed summer mean precipitation exhibits a clear southeast to northwest gradient, with relatively high values over the tropics and southern China, and relatively low values over the Northwest. The WRF model captures the observed summer mean precipitation rather well both in the geographic distributions and magnitude, but produces too much precipitation over some humid/semi-humid areas. The observed interannual variability expressed by its standard deviation of summer precipitation shows a similar pattern to that of summer mean precipitation. The WRF model simulates relatively well geographic distributions of interannual variability of summer precipitation over most areas of East Asia. In this study, we also adopt the coefficient of variation (standard deviation divided by the mean) to evaluate the performance of the WRF model. The index is a more objective measure of interannual variability as the dependency of the standard deviation on the mean precipitation has been removed (Giorgi *et al.*, 2004).

Compared with the observations, the WRF model performs well in simulating the coefficient of variation both the geographic distributions and magnitude.

Figure 2 compares geographic distributions of summer mean surface air temperature and standard deviation of summer surface air temperature over East Asia during 1981–2005 between the observations (Fan and van den Dool, 2008) and WRF model simulations. The WRF model generally simulates mean surface air temperature well, both in geographic distributions and magnitude. The observed surface air temperature variability exhibits a large south-to-north gradient. The WRF model successfully simulates the observed pattern, but overestimates interannual variability of surface air temperature over some areas of Tibetan Plateau and Northeast Asia.

Figure 3 shows percent contribution of subsurface soil temperature feedbacks to the variance of interannual summer surface air temperature and precipitation variability during the period of 1981–2005. Subsurface air temperature feedbacks significantly increase interannual variability of summer surface air temperature mainly over the arid/semi-arid regions extending from northwestern part of Tibetan Plateau across northwestern China to southern Mongolia. Over these areas, subsurface soil temperature feedbacks make a large contribution to the interannual variability of summer surface air temperature, accounting for about 30–70% of the total variance. This study shows that soil temperature feedbacks dominate summer surface air temperature variability over many areas of the arid/semi-arid

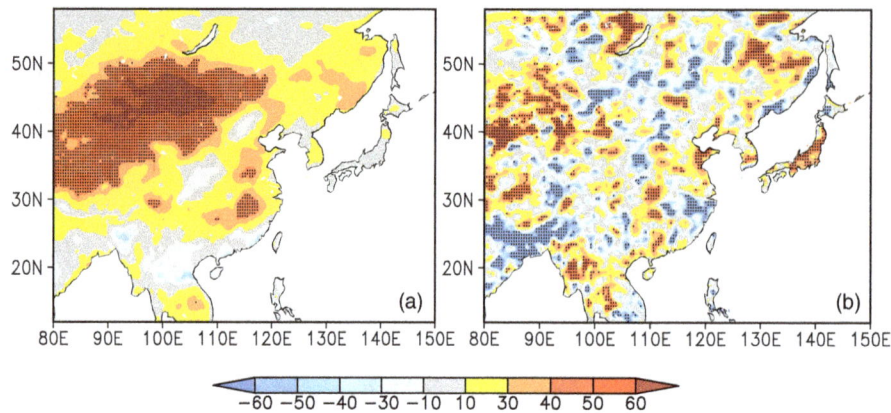

Figure 3. Percentage of interannual summer climate variance due to subsurface soil temperature feedbacks: (a) surface air temperature and (b) precipitation. The percentage is calculated as the ratio of the difference in interannual variance between CTL and SoilT (CTL minus SoilT) to the variance in CTL [Equation (1)]. Grid cells with values significant at the 90% level by *F*-test are marked by the solid circles.

regions of East Asia. Over most land areas outside the arid/semi-arid regions, Zhang *et al.* (2011b) demonstrated that soil moisture feedbacks play a leading role in influencing summer surface air temperature variability. Together, the results suggest that the land plays an important role in influencing summer surface air temperature variability over East Asia.

In contrast, the subsurface soil temperature feedbacks on precipitation variability are weaker than those on surface air temperature variability over the arid/semi-arid regions. Significant positive subsurface soil moisture feedbacks mainly occur over western part of arid/semi-arid regions, explaining about half of the total variance of interannual precipitation variability over many areas of this region. Over the climatic and ecological transition zones of the southern Siberia-northern Mongolia region and northern China and many areas of western China, soil moisture is found to make a dominant contribution to summer precipitation variability (Zhang *et al.*, 2011b). These results suggest that precipitation variability outside humid regions of East Asia is strongly affected by soil moisture and soil temperature. It needs to mention that, in addition to soil moisture and soil temperature, other factors such as El Niño-Southern Oscillation (ENSO) and Indian Ocean SST, also can influence summer climate variability over East Asia (e.g. Chang, 2004; Yang *et al.*, 2007).

Figure 4 shows how subsurface soil temperature feedbacks on summer surface air temperature and precipitation variability depend on soil moisture. PV_T and PV_P represent relative contribution of subsurface soil temperature feedbacks to summer surface air temperature and precipitation variability, respectively. All PV_T values are positive, implying subsurface soil temperature feedbacks increase interannual variability of surface air temperature. It is clear that PV_T exhibits strong soil moisture dependence. We find that averaged PV_T is 30–50% when soil moisture is lower than 50 cm. This indicates that over the arid/semi-arid regions, subsurface soil temperature feedback make an important contribution to the interannual variability of surface air

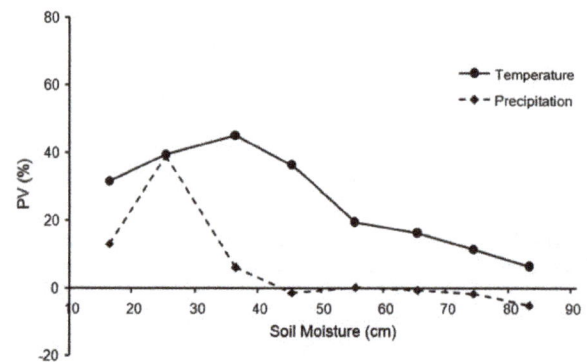

Figure 4. Relationship between soil moisture and percentage of interannual summer climate variance (*PV* binned by 10-cm soil moisture intervals) due to subsurface soil temperature feedbacks on surface air temperature (*PV_T*) and precipitation (*PV_P*).

temperature. PV_P has relatively high values when soil moisture is low, and is nearly zero when soil moisture is high (soil moisture >40 cm). It indicates that the subsurface soil temperature feedbacks on precipitation variability are strongest over the arid/semi-arid regions. Comparatively, the feedback strength on precipitation variability is weaker than that on surface air temperature variability over the arid/semi-arid regions.

We further examine changes in surface energy balance components to explain our findings. It is found that subsurface soil temperature strongly influences summer surface air temperature variability over the arid/semi-arid regions mainly through its effects on ground heat flux (Figure 5(a)). Sensible heat flux is less sensitive to subsurface soil temperature than ground heat flux over arid/semi-arid areas. The pattern of changes in standard deviation of latent heat flux generally resembles that of changes in summer precipitation variability over northern part of East Asia (Figure 5(b)). This suggests that changes in latent heat flux induced by subsurface soil temperature make an important contribution to summer precipitation variability over these areas. In the meanwhile, it should be noted that indirect

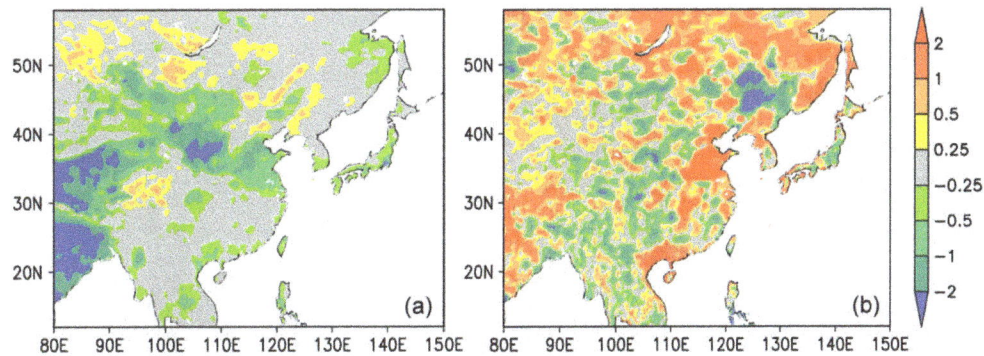

Figure 5. Differences in standard deviation of (a) ground heat flux and (b) latent heat flux (in w m^{-2}) between CTL and SoilT.

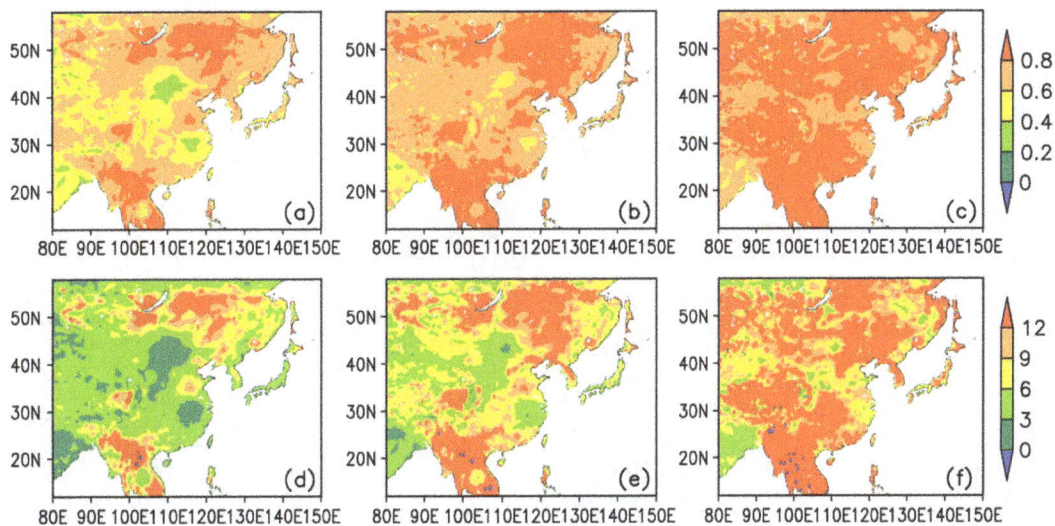

Figure 6. One-month autocorrelation (top) and decorrelation time (bottom) of soil temperature at soil layers of 10−40 cm (a, d), 40−100 cm (b, e), and 100−200 cm (c, f).

processes such as water vapor and atmospheric circulation also play a role.

Over the regions where land surface plays an important role in influencing climate, land memory can contribute to improve seasonal to interannual climate prediction skills. We further estimate subsurface soil temperature memory using decorrelation time ($t_d = (1 + \alpha) / (1 - \alpha)$, where α is 1-month autocorrelation) (von Storch and Zwiers, 1999). Over the arid/semi-arid regions where subsurface soil temperature makes important contributions to summer climate variability, subsurface soil temperature generally has a long climate memory spanning 3−15 months and soil temperature memory increases with soil layer depth (Figure 6). This suggests that over these areas, subsurface soil temperature can be a useful predictor of summer temperature and precipitation. Over regions outside of the arid/semi-arid regions, subsurface soil temperature also shows a long climate memory (>3 months in general), however, subsurface soil temperature has small effects on summer climate variability. Thus, over these regions soil temperature memory can be less transferred into the atmospheric persistence, and may have small effects on summer climate prediction.

4. Conclusions

This study investigates the role of subsurface soil temperature feedbacks in influencing summer surface air temperature and precipitation variability over East Asia by means of RCM simulations. For this aim, two long-term simulations with and without subsurface soil temperature feedbacks are performed with the WRF model. From our investigation, it is evident that subsurface soil temperature feedbacks make an important contribution to amplifying summer surface air temperature variability over the arid/semi-arid regions. The feedbacks on precipitation variability are weaker than those on surface air temperature variability over the arid/semi-arid regions, but are still significant over many areas of western part of the arid/semi-arid regions.

This study provides the first RCM-based demonstration that subsurface soil temperature feedbacks play an important role in influencing climate variability over East Asia. In the meanwhile, the model bias should be recognized. The results achieved by this study thus need to be further confirmed in a multi-model framework to eliminate the model dependence. This study highlights the importance of soil temperature feedbacks to summer surface air temperature variability over the arid/semi-arid regions. Outside the arid/semi-arid regions, the important role of soil moisture over land areas for summer surface air temperature variability is previously identified (Zhang *et al.*, 2011b). These findings together suggest that the land makes an important contribution to summer surface air temperature variability over land areas of East Asia. This study also finds that subsurface air temperature makes a large contribution to precipitation variability over many areas of western part of arid/semi-arid regions. Previous study demonstrated that soil moisture makes a dominant contribution to precipitation variability over the climatic and ecological transition zones of the southern Siberia-northern Mongolia region and northern China and many areas of western China. Together, these studies suggest that land strongly influences precipitation variability outside humid regions of East Asia. Knowledge of land-climate relationships could aid in improving summer climate prediction over East Asia.

Acknowledgements

We would like to thank Pingping Xie for providing the East Asia gauge-based precipitation analysis. NCEP/DOE reanalysis which is used to drive the WRF was produced with the support of the U.S. National Weather Service and of PCMDI (U.S. Department of Energy). This work was supported by the National Natural Science Foundation of China (Grant No. 41275089 and 41305071), the National Basic Research Program of China (2012CB955604), and "100-talent program" of the Chinese Academy of Sciences. This work was also supported by the Jiangsu Collaborative Innovation Center for Climate Change.

References

Bonan GB. 2008. *Ecological Climatology*, 2nd ed. Cambridge University Press: Cambridge, UK; 550.

Chang CP. 2004. In *Preface, in East Asian Monsoon*, Chang CP (ed). World Sci: Singapore; v–vi.

Charney JG. 1975. Dynamics of deserts and drought in Sahel. *Quarterly Journal of the Royal Meteorological Society* **101**: 193–202.

Chen F, Dudhia J. 2001. Coupling and advanced land surface-hydrology model with the Penn State-NCAR MM5 modeling system. Part I: Model implementation and sensitivity. *Monthly Weather Reviewer* **129**: 569–585.

Chen H, Zhou J. 2013. Impact of interannual soil moisture anomaly on simulation of extreme climate in China Part II: sensitivity experiment analysis (in Chinese). *Chinese Journal of Atmospheric Sciences* **37**: 1–13.

Collins WD, Rasch PJ, Boville BA, Hack JJ, McCaa JR, Williamson DL, Briegleb BP, Bitz CM, Lin S, Zhang M. 2006. The formulation and atmospheric simulation of the Community Atmosphere Model version 3 (CAM3). *Journal of Climate* **19**: 2144–2161.

Diffenbaugh NS, Pal JS, Giorgi F, Gao X. 2007. Heat stress intensification in the Mediterranean climate change hotspot. *Geophysical Research Letters* **34**: L11706, DOI: 10.1029/2007GL030000.

Dirmeyer PA. 2000. Using a global soil wetness dataset to improve seasonal climate simulation. *Journal of Climate* **13**: 2900–2922.

Dirmeyer PA, Koster RD, Guo Z. 2006. Do global models properly represent the feedback between land and atmosphere? *Journal Hydrometeorology* **7**: 1177–1198.

Dutra E, Schär C, Viterbo P, Miranda PMA. 2011. Land-atmosphere coupling associated with snow cover. *Geophysical Research Letters* **38**: L15707, DOI: 10.1029/2011GL048435.

Fan Y, van den Dool H. 2008. A global monthly land surface air temperature analysis for 1948-present. *Journal of Geophysical Research* **113**: D01103, DOI: 10.1029/2007JD008470.

Fischer EM, Seneviratne SI, Lüthi D, Schär C. 2007. Contribution of land-atmosphere coupling to recent European summer heat waves. *Geophysical Research Letters* **34**: L06707, DOI: 10.1029/2006GL029068.

Fu C, Yan Z. 1996. *Global Change and the Future Life-supporting Environment of China* (in Chinese). Meteorology Press: Beijing.

Gao X, Luo Y, Lin W, Zhao Z, Giorgi F. 2003. Simulation of effects of land use change on climate in China by a regional climate model. *Advances in Atmospheric Sciences* **20**: 583–592.

Giorgi F. 1990. Simulation of regional climate using a limited area model nested in general circulation model. *Journal of Climate* **3**: 941–963.

Giorgi F, Bi X, Pal JS. 2004. Mean, interannual variability and trends in a regional climate change experiment over Europe. I: Presentday climate (1961–1990). *Climate Dynamics* **22**: 733–756.

Hong SY, Lim JOJ. 2006. The WRF single-moment 6-class microphysics scheme (WSM6). *Journal of the Korean Meteorological Society* **42**: 129–151.

Hong SY, Noh Y, Dudhia J. 2006. A new vertical diffusion package with an explicit treatment of entrainment processes. *Monthly Weather Review* **134**: 2318–2341.

Hu Q, Feng S. 2004a. A role of the soil enthalpy in land memory. *Journal of Climate* **17**: 3633–3643.

Hu Q, Feng S. 2004b. Why has the land memory changed? *Journal of Climate* **17**: 3236–3243.

IPCC. 2012. Managing the risks of extreme events and disasters to advance climate change adaptation. In *A Special Report of Working Groups I and II of the Intergovernmental Panel on Climate Change*, Field, CB, Barros V, Stocker TF, Qin D, Dokken DJ, Ebi KL, Mastrandrea MD, Mach KJ, Plattner G-K, Allen SK, Tignor M, Midgleg PM (eds). Cambridge University Press: Cambridge, NY; 582.

Jerez S, Montavez JP, Gomez-Navarro JJ, Jimenez PA, Jimenez-Guerrero P, Lorente R, Gonzalez-Rouco JF. 2012. The role of the land-surface model for climate change projections over the Iberian Peninsula. *Journal of Geophysical Research* **117**: D01109, DOI: 10.1029/2011JD016576.

Kain JS. 2004. The Kain-Fritsch convective parameterization: an update. *Journal of Applied Meteorology* **43**: 170–181.

Kanamitsu M, Ebisuzaki W, Woollen J, Yang SK, Hnilo J, Fiorino M, Potter GL. 2002. NCEP-DOE AMIP-II reanalysis (R-2). *Bulletin of the American Meteorological Society* **83**: 1631–1643.

Koster RD, Dirmeyer PA, Guo Z, Bonan G, Chan E, Cox P, Gordon CT, Kanae S, Kowalczyk E, Lawrence D, Liu P, Lu C-H, Malyshev S, McAvaney B, Mitchell K, Mocko D, Oki T, Oleson K, Pitman A, Sud YC, Taylor CM, Verseghy D, Vasic R, Xue Y, Yamada T. 2004. Regions of strong coupling between soil moisture and precipitation. *Science* **305**: 1138–1140.

Kurbatkin GP, Manabe S, Hahn DG. 1979. The moisture content of the continents and the intensity of summer monsoon circulation. *Meteorologiya i Gidrologiya* **11**: 5–11.

Lee D, Suh M. 2000. Ten-year east Asian summer monsoon simulation using a regional climate model (RegCM2). *Journal of Geophysical Research* **105**: 29565–59577.

Leung LR, Huang M, Qian Y, Liang X. 2011. Climate-soil-vegetation control on groundwater table dynamics and its feedbacks in a climate model. *Climate Dynamics* **36**: 57–81.

Lü S, Chen Y. 1999. The influence of Northwest China afforestation on regional climate in China (in Chinese). *Plateau Meteorology* **18**: 416–424.

Ma Y, Wang Y, Wu R, Hu Z, Yang K, Li M, Ma W, Zhong L, Sun F, Chen X, Zhu Z, Wang S, Ishikawa H. 2009. Recent advances on the study of atmosphere-land interaction observations on the Tibetan Plateau. *Hydrology and Earth System Sciences* **13**: 1103–1111.

Mahanama SPP, Koster RD, Reichle RH, Suarez MJ. 2008. Impact of subsurface temperature variability on surface air temperature variability: an AGCM study. *Journal of Hydrometeorology* **9**: 804–815.

Manabe S. 1975. A study of the interaction between the hydrological cycle and climate using a mathematical model of the atmosphere.

Report on meeting on weather-food interactions, Endicott House, Massachusetts Institute of Technology: Cambridge, MA, 9–11 May 1975; 21–45.

Namias J. 1962. Influences of abnormal surface heat sources and sinks on atmospheric behavior. In Proceedings of the International Symposium on Numerical Weather Prediction, Tokyo, 7–13 November 1960. Meteorological Society of Japan, c/o Japan Meteorological Agency, Tokyo; 615–627.

Notaro M, Zarrin A. 2011. Sensitivity of the North American monsoon to antecedent Rocky Mountain snowpack. *Geophysical Research Letters* **38**: L17403, DOI: 10.1029/2011GL048803.

Pielke RA, Avissar SR, Raupach M, Dolman AJ, Zeng X, Denning AS. 1998. Interactions between the atmosphere and terrestrial ecosystems: Influence on weather and climate. *Global Change Biology* **4**: 461–475.

Pitman AJ, Noblet-Ducoudré Nd, Cruz FT, Davin EL, Bonan GB, Brovkin V, Clussen M, Delire C, Ganzeveld L, Gayler V, van den Hurk BJJM, Lawrence PJ, van der Molen MK, Müller C, Reick CH, Seneviratne SI, Strengers BJ, Voldoire A. 2009. Uncertainties in climate responses to past land cover change: first results from the LUCID intercomparison study. *Geophysical Research Letters* **36**: L14814, DOI: 10.1029/2009GL039076.

Retnakumari K, Renuka G, Rao G. 2000. Relation between pre-monsoon soil temperature and monsoon rainfall in a tropical station. *Mausam* **51**: 365–366.

Seneviratne SI, Lüthi D, Litschi M, Schär C. 2006. Land-atmosphere coupling and climate change in Europe. *Nature* **443**: 205–209.

Shukla J, Mintz Y. 1982. Influence of land-surface evapotranspiration on the Earth's climate. *Science* **215**: 1498–1501.

Skamarock WC, Klemp JB, Dudhia J, Gill DO, Barker DM, Duda MG, Huang XY, Wang W, Powers JG. 2008. *A Description of the Advanced Research WRF Version 3, Rep. NCAR/TN-475+STR*. National Centre for Atmospheric Research: Boulder, CO; 125.

von Storch H, Zwiers FW. 1999. *Statistical Analysis in Climate Research*. Cambridge University Press: Cambridge, UK; 484.

Tang M, Reiter ER. 1986. The similarity between the maps of soil temperature in U.S. and precipitation anomaly of the subsequent season. *Plateau Meteorology* **5**: 293–307.

Walker J, Rowntree PR. 1977. The effect of soil moisture on circulation and rainfall in a tropical model. *Quarterly Journal of the Royal Meteorological Society* **103**: 29–46.

Wang C, Cheng G, Deng A, Dong W. 2008. Numerical simulation on climate effects of freezing-thawing processes using CCM3. *Sciences in Cold and Arid Regions* **1**: 68–79.

Wu B, Yang K, Zhang R. 2009. Eurasian snow cover variability and its association with summer rainfall in China. *Advances in Atmospheric Sciences* **26**: 31–44.

Xia K, Luo Y, Li W. 2011. Simulation of freezing and melting of soil on northeast Tibetan Plateau. *Chinese Science Bulletin* **56**: 2145–2155.

Xie P, Yatagai A, Chen M, Hayasaka T, Fukushima Y, Liu C, Yang S. 2007. A gauge-based analysis of daily precipitation over East Asia. *Journal of Hydrometeorology* **8**: 607–626.

Xue Y, Vasic R, Janjic Z, Liu YM, Chu PC. 2012. The impact of spring subsurface soil temperature anomaly in the western U.S. on North American summer precipitation: A case study using regional climate model downscaling. *Journal of Geophysical Research* **117**: D11103, DOI: 10.1029/2012JD017692.

Yang X, Huang S. 1992. Climatic effects of Eurasian snow cover and their impact on the formation of winter monsoon circulation (in Chinese). *Journal of Nanjing University* **28**: 326–335.

Yang J, Liu Q, Xie SP, Liu Z, Wu L. 2007. Impact of the Indian Ocean SST basin mode on the Asian summer monsoon. *Geophysical Research Letters* **34**: L02708, DOI: 10.1029/2006GL028571.

Yang K, Chen Y, Qin J. 2009. Some practical notes on the land surface modeling in the Tibetan Plateau. *Hydrology and Earth System Sciences* **13**: 687–701.

Yeh TC, Wetherald RT, Manabe S. 1984. The effect of soil moisture on the short-term climate and hydrology change – A numerical experiment. *Monthly Weather Review* **112**: 474–490.

Zhang Y. 2004. Virtual numerical experiments on the climatic effects of vegetation type changes over Northern China (in Chinese). *Journal of Nanjing University* **40**: 684–691.

Zhang J, Wang WC, Leung LR. 2008a. Contribution of land-atmosphere coupling to summer climate variability over the contiguous United States. *Journal of Geophysical Research* **113**: D22109, DOI: 10.1029/2008JD010136.

Zhang J, Wang WC, Wei J. 2008b. Assessing land-atmosphere coupling using soil moisture from the Global Land Data Assimilation system and observational precipitation. *Journal of Geophysical Research* **113**: D17119, DOI: 10.1029/2008JD009807.

Zhang J, Wu L, Huang G, Zhu W, Zhang Y. 2011a. The role of May vegetation greenness on the southeastern Tibetan Plateau for East Asian summer monsoon prediction. *Journal of Geophysical Research* **116**: D05106, DOI: 1029/2010JD015095.

Zhang J, Wu L, Dong W. 2011b. Land-atmosphere coupling and summer climate variability over East Asia. *Journal of Geophysical Research* **116**: D05117, DOI: 10.1029/2010JD014714.

28

How does model development affect climate projections?

Jussi S. Ylhäisi,[1]* Jouni Räisänen,[2] David Masson,[3] Olle Räty[2] and Heikki Järvinen[2]

[1]*Weather and Safety, Finnish Meteorological Institute, Finland*
[2]*Department of Physics, University of Helsinki, Finland*
[3]*Federal Office of Meteorology and Climatology, MeteoSwiss, Switzerland*

Correspondence to:
J. S. Ylhäisi, Weather and Safety,
Finnish Meteorological Institute,
P.O. Box 503 (Erik Palménin
aukio 1), FIN-00101 Helsinki,
Finland.
E-mail: jussi.ylhaisi@fmi.fi

Abstract

We apply analysis of variance to assess the effect of climate model development on temperature, precipitation and sea level pressure projections using three consecutive climate model generations provided by the Coupled Model Intercomparison Project. The introduction of a new model version mostly randomly affects individual model projections, but models also show some generation-independent characteristics which can potentially serve a basis for model weighting. Mean values of temperature and pressure change differ significantly between the model generations only over limited regions, mainly near the sea-ice edge. If multi-model ensembles are uniformly weighted, the efforts of model development are not completely exploited.

Keywords: CMIP; ANOVA; climate change; model development

1. Introduction

An overarching objective of climate science is to estimate future anthropogenic climate change. This is most plausibly done with the help of climate models, which are constantly being developed through the inclusion of new scientific understanding. As the new models simulate observed climate better (Reichler and Kim, 2008; Knutti *et al.*, 2013), it is natural for users of climate model data to assume that the latest generation of models also provides the best climate change projections. However, out-of-sample performance of climate models in the projected future is notoriously hard to verify in absence of future observations (Räisänen *et al.*, 2010). Therefore, the effect of model development of climate change projections must be evaluated separately.

One reason for the users to prefer the latest model generation is that the emissions scenarios used as the basis of future climate projections have considerably evolved over time. This also prevents the direct comparison of older and newer model versions to each other. However, idealized CO_2-induced climate change experiments, which still are relevant for real-world climate change analysis if adjusted for the scenario-dependency of global mean temperature increase (Räisänen and Ylhäisi, 2014), have remained a model benchmark for several years. Correspondingly, they exist for several model generations and provide a means to compare different model versions with each other without the complication from changing scenarios.

If systematic differences between model generations are small compared to the variations within them, then the benefit of focusing on the latest models comes into question. If this happens, one can further ask whether this is mainly because (1) subsequent versions of the same basic model tend to give similar results (as found for the present-day climate by Masson and Knutti (2011) and Knutti *et al.* (2013) and for 21st century projections by Knutti and Sedláček (2012)), or (2) model development changes the projections of different models in seemingly random ways. Case (1) would make it potentially easier to pinpoint the model-specific characteristics in the climate projections, after first identifying those differences in model formulation that remain the same from generation to generation. Model weights defined through evaluation of key processes in individual models would not only have a chance of improving the climate projections, but they would also have effects to model 'ranking': If robust, version-independent, characteristics of a climate model could be identified, they could provide background information for evaluating the performance of the improved version of the model.

Model weighting is controversial in the scientific literature (Weigel *et al.*, 2010; DelSole *et al.*, 2013), but can substantially impact the multi-model mean climate change projection in some regions (e.g. Räisänen *et al.*, 2010, Figure 10). In the absence of unambiguously better alternatives, uniform weights are often conservatively used as e.g. in the IPCC reports (Cubasch *et al.*, 2001; Meehl *et al.*, 2007a; Collins *et al.*, 2013). How much can we expect further model development efforts to change the upcoming multi-model mean (MMM) estimates? Should more attention be paid to postprocessing and analysis methods of climate model output (such as methodologies used to define model weights), or model development itself? In this paper, we study these questions focussing on CO_2-induced change in time-mean climate in three generations of the Coupled Model Intercomparison Project, [Covey, *et al.*, 2000 (CMIP2), Meehl *et al.*, 2007b (CMIP3), Taylor *et al.*, 2012 (CMIP5)]. We use analysis of variance (ANOVA) to dissect the sources of variability in this multi-generation data set.

2. Methods and data sets

We analyze three of the most commonly used surface variables: temperature (tas), precipitation (pr), sea level pressure (psl) in CMIP2, CMIP3 and CMIP5. Each of these model generations reflects the state of scientific knowledge at the time of its release. Data from simulations with CO_2 increasing 1% per year were used in this study and were first smoothed using 20-year running mean values. We identified those model institutes that had provided data to each of the three phases of CMIP and therefore had three consecutive generations of data from the same model available. Using this criteria, 13 models were available for temperature and precipitation and 10 for sea level pressure (Table S1, Supporting Information for a list of used models). In CMIP2 and CMIP5, the increase in CO_2 continues throughout the 80-year period, whereas it stops after doubling in the year 70 in CMIP3.

To illustrate the behavior of the 13 models in each of CMIP2, CMIP3 and CMIP5, we show in Figure 1 the changes in global mean temperature straddling the doubling of CO_2 at year 70 in the idealized simulations (i.e. the Transient Climate Response, TCR). The 13 different climate models and their MMM are shown at the individual columns. Each column consists from three colored and one gray bar, corresponding to model values for the three CMIP generations and their multi-generation means (MGM). In common with studies based on more comprehensive evaluation metrics (Masson and Knutti, 2011), the TCR estimates seem to have semi-persistent model-dependent characteristics. Models simulating low or high TCR

to some extent preserve this characteristic in relation to the other models for all model generations, but the development of models has also had model-dependent effects. Proceeding from one model generation to the next, TCR is increased in some models and decreased in others. None of the pairwise MMM differences between the three CMIPs are statistically significant (two-sided *t*-test on a 5% level) and the inter-model variance of TCR is also statistically indistinguishable (two-sided *F*-test on a 5% level) between the model generations.

We apply an ANOVA framework for separating the total variance in climate projections into three additive components: the *model institute* component, the *generation* component, and the *interaction* component (Appendix S1, Section 3). Figure 1 can be used to illustrate these three variance components:

1. *Model institute* component: Variance in TCR induced by differences of 'generation-independent model characteristics' of each model, i.e. variance between the 13 MGMs. The relative contribution of this component to the total variance is 40% $(0.058\,^\circ C^2)$.

2. *Generation* component (model-independent part of model development): variance due to differences of collective information from each of the model generations used, i.e. variance of the three MMM estimates; 5% $(0.007\,^\circ C^2)$ of the total.

3. *Interaction* component (model-dependent part of model development): as transition from one model generation to subsequent one has unique effects for each climate model, this component describes 'how typical model characteristics are obscured by the introduction of a new model generation'. The relative contribution of this component to the total variance is 55% $(0.080\,^\circ C^2)$.

The variances diagnosed with the used ANOVA framework are potentially sensitive to the sample sizes. Compared with a hypothetical infinite sample from the same statistical population, our framework tends to underestimate the *generation* variance relative to the *model institute* variance, because the number of model generations is smaller than the number of models (Eqs. (9) and (10) in the Appendix S1). However, our general conclusions are robust to this bias. Further information on the ANOVA methodology and the significance testing of the three components (as done in Figure 3) is provided in the Appendix S1.

For any model, differences in climate change between CMIP2, CMIP3 and CMIP5 are partly due to internal variability rather than model development *per se* (Ylhäisi *et al.*, 2014), and to a lesser extent the same also applies to the inter-model and inter-generation differences. Therefore, our three variance components are also affected by internal variability, *interaction* component the most and *generation* component the least. In Section 4 of the Appendix S1, we use a previously published method for estimating the internal variability in the 20-year mean values (Räisänen, 2001), and

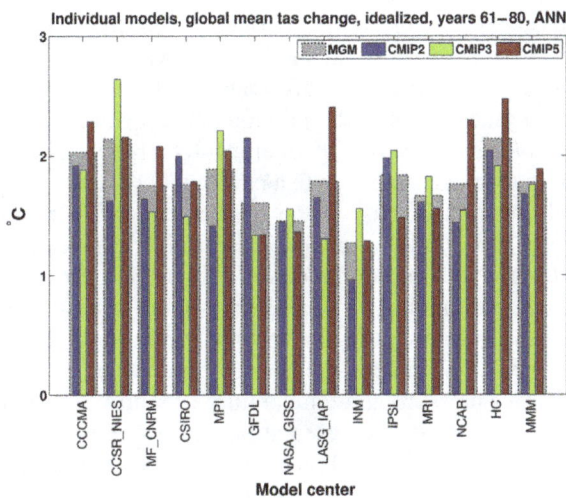

Figure 1. Global mean temperature change. Transient climate responses for CMIP2, CMIP3 and CMIP5 models in together with their multi-generation mean values (MGM) are averaged over the years 61–80. Multi-model mean (MMM) in the last column.

Annual averages, years 61–80
Absolute ANOVA components

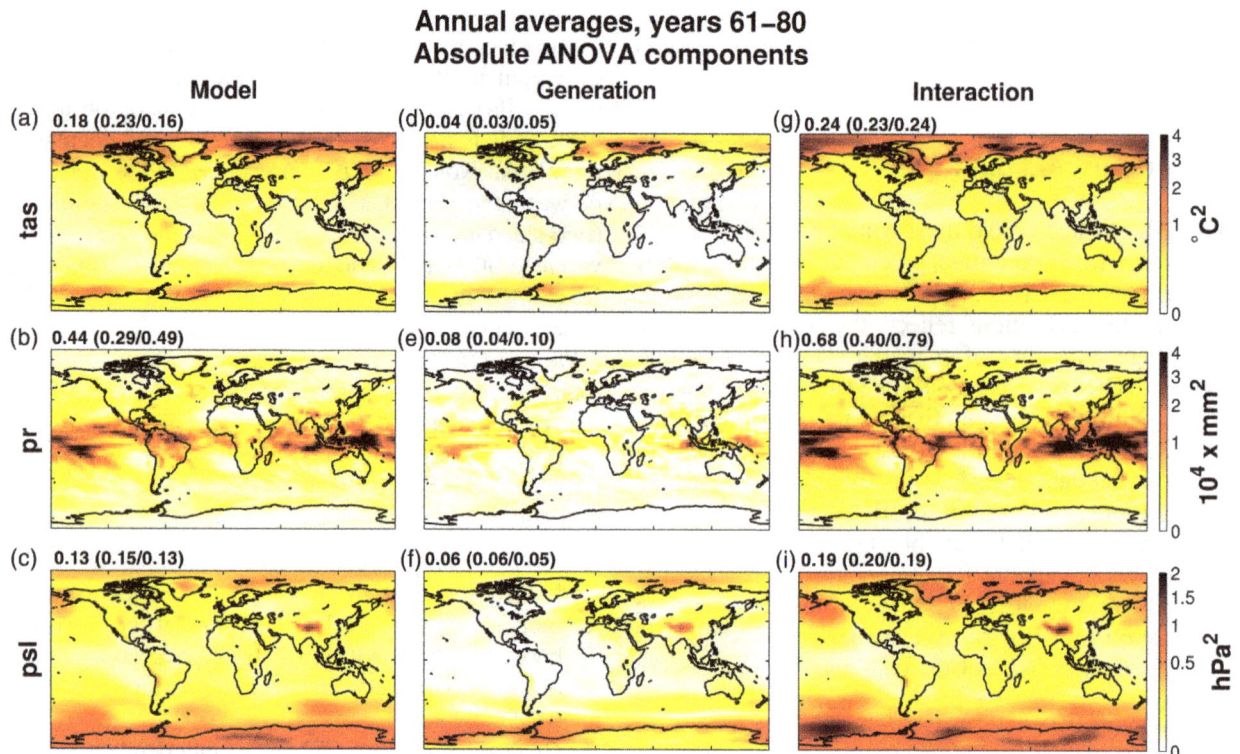

Figure 2. Absolute variances. Variance of 20-year annual mean change of temperature (upper row), precipitation (middle row), and sea level pressure (bottom row) at the time of doubling of CO_2 as divided into three additive components: model differences (left column), systematic inter-generation differences (middle column) and model-specific inter-generation differences (right column). The global (land area/sea area) means of the variances are given above each panel. Note the non-linear color scale.

apply the result to estimate the variance components that would have been obtained in absence of internal variability.

3. Results

Maps of the variance components are provided in Figures 2 (absolute values) and 3 (relative contributions to total variance), for the 20-year annual mean changes of temperature, precipitation, and sea level pressure at the doubling of CO_2. All three variance components of temperature and pressure change are the highest over high latitudes (for temperature in particular over the high-latitude oceans near the sea ice edge), whereas the highest values for precipitation are seen in the tropics (Figure 2). Over those areas with large variances, large absolute uncertainty is implied in the future change of the variable considered.

Many of the characteristics are shared by each of the variance components, but the maps are not linearly scalable as shown by the relative variances in Figure 3. The thick contours crop the areas where the corresponding relative variances are significantly smaller (gray) or larger (black) than would be expected by using independent and normally distributed random data (see Appendix S1). The different versions of the same climate model are not independent from each other as can be clearly seen from the strong *model institute* component of temperature over land areas.

Over land, an individual climate model is likely to preserve its model-specific characteristics whenever a new model version is introduced. These areas with significantly high *model institute* variance widely overlap with those of significantly low *interaction* variance. The maps for precipitation are very noisy, giving little evidence of persistent model-specific features. For sea level pressure, model-specific characteristics mostly occur over oceans.

The *generation* component represents the variance between the three multi-model mean values and summarizes the collective information which the model development adds to the previous model generation as a whole. Over areas with significantly large *generation* component the transition to a new model generation has had coherent effects on the projections of 13 (tas/pr) or 10 (psl) CMIP models. This could be potentially traceable to some common set of physical processes which are included in the more sophisticated model versions but absent in the previous ones. Such effects are weakly seen near the sea-ice borderline for mean sea level pressure and temperature (possibly reflecting improved sea-ice representation in new models) and over mid-latitudes of both hemispheres for sea level pressure (the physical attribution of which is unclear). Elsewhere, the three MMM estimates are statistically indistinguishable.

The order of importance for the variance components remains very similar for all three variables throughout the 80-year simulation period, with the *generation*

Annual averages, years 61–80
Relative ANOVA components

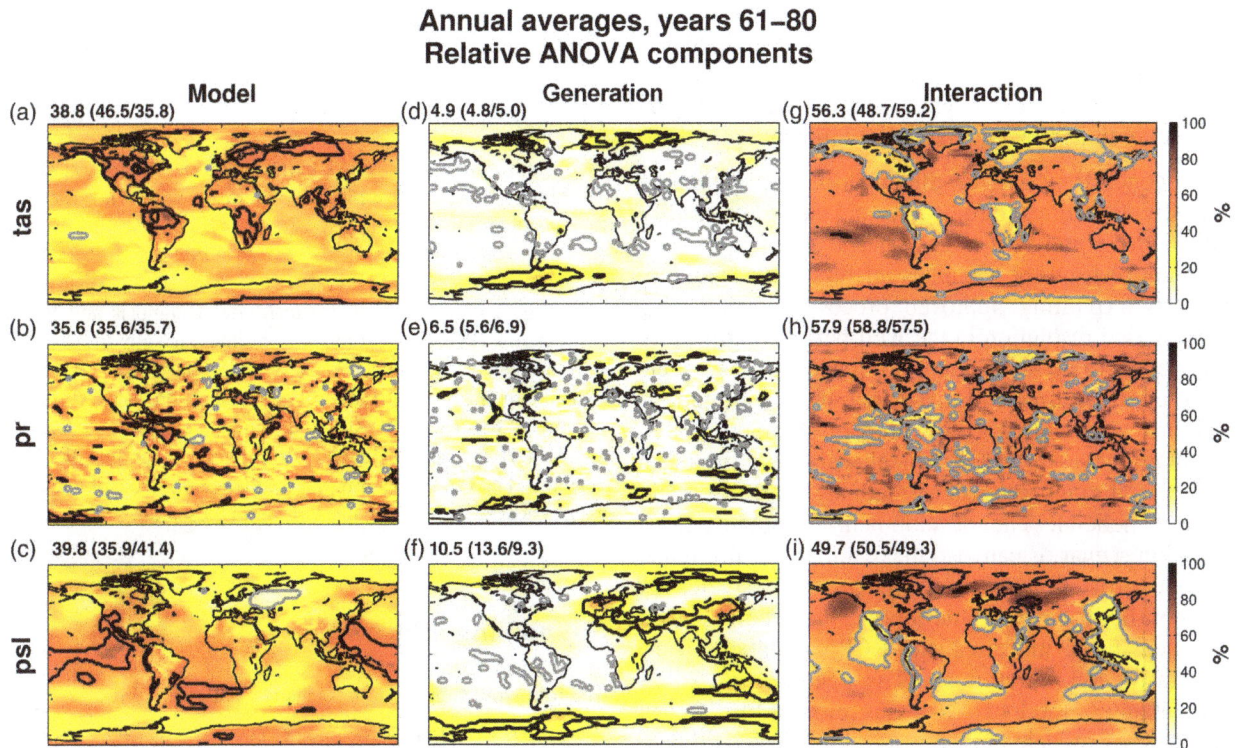

Figure 3. Relative variances. Fractional contributions of the three components to total variance as in Figure 2. Areas where the relative variance is at the 2.5% confidence level larger (smaller) than expected from chance are contoured in black (gray).

(*interaction*) variance being the smallest (largest) (Figs. S1–S3). All three components increase with time, implying an increased uncertainty as the CO_2 forcing increases. The relative rate of increase is, however, slightly faster for the *model institute* and *generation* components than the *interaction* component. This is probably because (1) short-term projections have a low signal-to-noise ratio and are thus heavily affected by internal variability (Deser *et al.*, 2012; Räisänen, 2001), and (2) the *interaction* component is more strongly amplified by internal variability than the others. However, in a globally averaged sense, internal variability does not appear to dominate any of the three variance components during the period considered in Figures. 1–3 (Figs. S1–S3 and Table S2). At regional scales, the availability of multiple parallel runs would still make it easier to separate model-related differences in climate change from the effects of internal variability.

4. Conclusions and discussion

The main conclusions of our ANOVA analysis and the recommendations following from it are summarized as:

1. Despite clustering of individual climate models (Knutti *et al.*, 2013), inter-generation differences in individual climate model runs can be substantial and seemingly stochastic. This is partly indicative of undersampling of internal variability: several parallel runs from the same climate model instead of just one should be used. This would increase the potential for robust model weighting (making the model

institute component of variance relatively larger), even though this cannot be tested using our dataset.

2. Owing to persistent inter-generation characteristics of individual climate models over certain regions (areas with significant values of *model institute* component), model weighting has the potential to robustly change MMM projections. This persistence suggests that model weights defined for one particular CMIP generation can serve as a first-order estimate even for the next climate model generation.

3. The uniform weighting approach in a multi-model framework can result in an inadequate exploitation of advances in climate model development. Should an individual climate model (in relation to other climate models) perform considerably better whenever a new model version is introduced, it should be rewarded for this. However, if uniform weighting is used, the more improved (better-performing) models are put to the same line with the less improved (worse-performing) models and the effects of the improvements in the projections is greatly diluted (or: the fruits of intensive model development work are largely wasted!) in combined MMM projections. This may at least partly explain why regional patterns of MMM climate change have remained largely the same from model generation to model generation (Collins *et al.*, 2013, Räisänen and Ylhäisi, 2014) when scenario-related differences in the global mean temperature change are accounted for.

4. The relative importance of generation and model institute components clearly indicate the secondary

importance of inter-generation differences compared to inter-model differences (as in Räisänen and Ylhäisi, 2014). More emphasis should be put on improved climate model analysis techniques in parallel with continued model development.

Our conclusions and recommendations are applicable to best-guess climate projections and cannot be readily extended to probabilistic estimates. Their relevance is conditional on the variables analyzed: for example, properly exploiting improved process understanding or increased resolution in the well-performing models is expected to be more important for many less understood climate variables (such as extreme precipitation) that are important for IAV studies (Maraun, 2013; Swart and Avelar, 2011). Model weighting for these variables through consistent methods might be even more beneficial. The development of sophisticated emission scenarios can also make a marked difference on local projections of these higher-order statistics, as the non-homogeneous anthropogenic forcing agents included in the more realistic RCP-simulations (representative concentration pathways, Moss *et al.*, 2010) allow many of the local processes relevant for them to be explicitly modeled in the more sophisticated climate models. Additional regions with significant effects from improved process understanding might emerge (e.g. improved aerosol representation over polluted areas), should RCP simulations from several model versions be available.

Persistent model-specific characteristics in climate projections suggest that model differences could, in principle, be traced back to physical behavior of the climate models, which could be utilized for model weighting. The best known example from performance-based weighting concerns the areas near the sea-ice edge, where misrepresentation of present-day conditions distorts the sea-ice feedback in a warming world (Räisänen *et al.*, 2010, Bracegirdle and Stephenson, 2012). Depending on the variable analyzed, comprehensive evaluation for the reasons behind the model differences might be extremely difficult to achieve, but this attribution would provide consistent criterias for model weighting. Causal attribution is possible for the *model institute* and *generation* components, while this is highly controversial for the interaction component, it being stochastic by nature. An effort to identify causalities for these persistent model-dependent features should be a research priority. On the other hand, research on non-uniform weighting of climate model projections is still at an early stage (Collins *et al.*, 2013), and our results indicate that this technique is unlikely to be equally efficient in all parts of the world.

The findings of this study are far from conclusive. We primarily argue that combining apples and oranges in a multi-model framework (where the sophistication level of individual climate models is not comparable and the inter-generation changes in different models partially counteract each other) most likely will need proper model weighting. Otherwise, it is logical to anticipate that the long-term mean MMM projections from forthcoming climate model generations will not differ much from those that are already available.

Acknowledgements

We acknowledge the World Climate Research Programme's Working Group on Coupled Modelling (WGCM), which is responsible for CMIP, and we thank the climate modeling groups (listed in Table S1 of this paper) for producing and making available their model output. For CMIP the U.S. Department of Energy's Program for Climate Model Diagnosis and Inter-comparison (PCMDI) provides coordinating support and led development of software infrastructure in partnership with the Global Organization for Earth System Science Portals. Academy of Finland (decision 140801) and Finnish Academy of Science and Letters are acknowledged for supporting funding for this research. The Finnish Centre of Excellence in Atmospheric Science is acknowledged for supporting this research.

Supporting information

The following supporting information is available:

Appendix S1. How does model development affect climate projections?

Figure S1. Globally averaged variance components for running 20-year mean changes in temperature (Eqs. (2)–(5)) and the contributions of internal variability to these as estimated with Eqs. (8)–(11).

Figure S2. Globally averaged variance components for running 20-year mean changes in precipitation.

Figure S3. Globally averaged variance components for running 20-year mean changes in mean sea level pressure.

Table S1. The models used in the study. Mean sea level pressure simulations of those model centers in *italics* were not used.

Table S2. The contributions of the model institute variance, generation variance and interaction variance to the globally averaged total variance in the years 61–80, as obtained directly from Eqs. (2)–(5) (the first values in each cell) and after removing the contribution of internal variability using Eqs. (8)–(11) (in parentheses).

References

Bracegirdle TJ, Stephenson DB. 2012. More precise predictions of future polar winter warming estimated by multi-model ensemble regression. *Climate Dynamics* **39**: 2805–2821, doi: 10.1007/s00382-012-1330-3.

Collins M, Knutti R, Arblaster J, Dufresne J-L, Fichefet T, Friedlingstein P, Gao X, Gutowski WJ, Johns T, Krinner G, Shongwe M, Tebaldi C, Weaver AJ, Wehner M. 2013. Long-term climate change: projections, commitments and irreversibility. In *Climate Change 2013: The Physical Science Basis. Contribution of Working Group I to the Fifth Assessment Report of the Intergovernmental Panel on Climate Change*, Stocker TF, Qin D, Plattner G-K, Tignor M, Allen SK, Boschung J, Nauels A, Xia Y, Bex V, Midgley PM (eds). Cambridge University Press: Cambridge, UK and New York, NY.

Covey C, AchutaRao KM, Lambert SJ, Taylor KE. 2000. Intercomparison of Present and Future Climates Simulated by Coupled Ocean-Atmosphere GCMs. PCMDI Report No. 66, Program for Climate Model Diagnosis and Intercomparison, Lawrence Livermore National Laboratory, University of California, Livermore, CA.

Cubash U, Meehl GA, Boer GJ, Stouffer RJ, Dix M, Noda A, Senior CA, Raper S, Yap KS. 2001. Projections of future climate change. In *Climate Change 2001: The Scientific Basis. Contribution of Working Group I to the Third assessment Report of the Intergovernmental Panel on Climate Change.* Houghton JT, Ding Y, Griggs DJ, Noguer M, van der Linden PJ, Dai X, Maskell K, Johnson CA (ed). Cambridge University Press: Cambridge, UK and New York, NY; 881 pp.

DelSole T, Xiaosong Y, Tippett MK. 2013. Is unequal weighting significantly better than equal weighting for multimodel forecasting? *Quarterly Journal of the Royal Meteorological Society* **139**(670): 176–183.

Deser C, Knutti R, Solomon S, Phillips AS. 2012. Communication of the role of natural variability in future North American climate. *Nature Climate Change* **2**: 775–779, doi: 10.1038/nclimate1562.

Knutti R, Sedláček J. 2012. Robustness and uncertainties in the new CMIP5 climate model projections. *Nature Climate Change* **3**: 369–373, doi: 10.1038/nclimate1716.

Knutti R, Masson D, Gettelman A. 2013. Climate model genealogy: Generation CMIP5 and how we got there. *Geophysical Research Letters* **40**: 1194–1199, doi: 10.1002/grl.50256.

Maraun D. 2013. Bias correction, quantile mapping, and downscaling: Revisiting the inflation issue. *Journal of Climate* **26**: 2137–2143, doi: 10.1175/JCLI-D-12-00821.1.

Masson D, Knutti R. 2011. Climate model genealogy. *Geophysical Research Letters* **38**: L08703, doi: 10.1029/2011GL046864.

Meehl GA, Stocker TF, Collins WD, Friedlingstein P, Gaye AT, Gregory JM, Kitoh A, Knutti R, Murphy JM, Noda A, Raper SCB, Watterson IG, Weaver AJ, Zhao Z-C. 2007a. Global climate projections. In *Climate Change 2007: The Physical Science Basis. Contribution of Working Group I to the Fourth Assessment Report of the Intergovernmental Panel on Climate Change.* Solomon S, Qin D, Manning M, Chen Z, Marquis M, Averyt KB, Tignor M, Miller HL (ed). Cambridge University Press: Cambridge, UK and New York, NY.

Meehl GA, Covey C, Delworth T, Latif M, McAvaney B, Mitchell JFB, Stouffer RJ, Taylor KE. 2007b. The WCRP CMIP3 multimodel dataset: a new era in climate change research. *Bulletin of the American Meteorological Society* **88**: 1383–1394.

Moss RH, Edmonds JA, Hibbard KA, Manning MR, Rose SK, van Vuuren DP, Carter TR, Emori S, Kainuma M, Kram T, Meehl GA, Mitchell JFB, Nakicenovic N, Riahi K, Smith SJ, Stouffer RJ, Thomson AM, Weyant JP, Wilbanks TJ. 2010. The next generation of scenarios for climate change research and assessment. *Nature* **463**: 747–756, doi: 10.1038/nature08823.

Reichler T, Kim J. 2008. How well do coupled models simulate today's climate? *Bulletin of the American Meteorological Society* **89**: 303–311, doi: 10.1175/BAMS-89-3-303.

Räisänen J. 2001. CO_2-induced climate change in CMIP2 experiments: quantification of agreement and role of internal variability. *Journal of Climate* **14**: 2088–2104, doi: 10.1175/1520-0442(2001)014<2088:CICCIC>2.0.CO;2.

Räisänen J, Ruokolainen L, Ylhäisi J. 2010. Weighting of model results for improving best estimates of climate change. *Climate Dynamics* **35**: 407–422.

Räisänen J, Ylhäisi JS. 2014. CO_2-induced climate change in northern Europe: CMIP2 versus CMIP3 versus CMIP5. *Climate Dynamics* , doi: 10.1007/s00382-014-2440-x.

Swart R, Avelar D. 2011. Bridging climate research data and the needs of the impact community. *Proceeding of IS-ENES/EEA/CIRCLE-2 workshop*, 11–12 January 2011. EEA, Copenhagen.

Taylor KE, Stouffer RJ, Meehl GA. 2012. An overview of CMIP5 and the experiment design. *Bulletin of the American Meteorological Society* **93**: 485–498, doi: 10.1175/BAMS-D-11-00094.1.

Weigel AP, Knutti R, Liniger MA, Appenzeller C. 2010. Risks of model-weighting in multimodel climate projections. *Journal of Climate* **23**: 4175–4191.

Ylhäisi JS, Garrè L, Daron J, Räisänen J. 2014. Quantifying sources of climate uncertainty to inform risk analysis for climate change decision-making. *Local Environment: The International Journal of Justice and Sustainability*, doi: 10.1080/13549839.2013.874987.

Pathways between soil moisture and precipitation in southeastern South America

Romina C. Ruscica,[1]* Anna A. Sörensson[1] and Claudio G. Menéndez[1,2]

[1]Centro de Investigaciones del Mar y la Atmósfera, CONICET/UBA, UMI IFAECI/CNRS, Buenos Aires, Argentina
[2]Departamento de Ciencias de la Atmósfera y los Océanos, FCEN, UBA, Buenos Aires, Argentina

*Correspondence to:
R. C. Ruscica, CIMA
(CONICET-UBA), Ciudad
Universitaria, Int. Guiraldes
2160, Pabellón 2, Piso 2, Ciudad
Autónoma de Buenos Aires
C1428EGA, Argentina.
E-mail: ruscica@cima.fcen.uba.ar

Abstract

Southeastern South America (SESA) is found to be the main hot spot of soil moisture–evapotranspiration coupling of South America during a dry summer. However, only its eastern part is a soil moisture–precipitation hot spot. Pathways between soil moisture and precipitation are evaluated through studying the coupling of soil moisture with surface and boundary layer variables. The outcome suggests that both the moist static energy and its vertical gradient are important for the development of precipitation, as a result of the total surface heat fluxes that are affected by soil moisture only in the eastern part of SESA.

Keywords: soil moisture; precipitation; coupling strength; southeastern South America; moist static energy; regional climate modelling

1. Introduction

Land–atmosphere interactions over southeastern South America (SESA) have been recognized as an important issue for a correct representation of regional climate during the austral summer season. A major improvement in the simulated low-level winds, sensible heat flux and Bowen ratio (BR) was achieved when both soil and vegetation processes were included in a land surface scheme coupled to a climate model (Ma et al., 2011). Surface temperature and precipitation were better represented when soil moisture–atmosphere interactions were taken into account (Barreiro and Díaz, 2011). SESA has also been identified as a hot spot region, where both soil moisture – evapotranspiration and soil moisture – precipitation coupling are strong (Sörensson and Menéndez, 2011). The concept of coupling refers to the influence of soil moisture on some variable, isolating it from the reverse influence of the variable on soil moisture. In particular, this is important when studying precipitation, which exerts a strong control on soil moisture. Ruscica et al. (2014) found that the deep soil moisture memory is lower inside the soil moisture – evapotranspiration hot spot in SESA, than in other regions of southern South America, i.e. the soil moisture memory and the strength of the coupling are anticorrelated within the hot spot.

The coupling strength (CS) index permits isolating and quantifying how much the soil moisture influences on the atmosphere using a methodology based on model experiments (Koster et al., 2004). Coupling strength studies have diagnosed the coupling between soil moisture and precipitation (e.g. Koster et al., 2006), evapotranspiration (e.g. Guo et al., 2006), 2 m temperature (e.g. Seneviratne et al., 2006) and surface sensible and latent heat fluxes (e.g. Guo and Dirmeyer, 2013). The coupling between soil moisture and precipitation tends to be less robust than the coupling between other variables because there are many processes that influence precipitation ranging from local to regional to large scale. It is therefore important to assess the different variables that connect soil moisture to precipitation. Analysis range from the simplest 'soil moisture–evapotranspiration-precipitation chain' (e.g. Sörensson and Menéndez, 2011; Wei and Dirmeyer, 2012) to more complex systems which can include several pathways (e.g. Santanello et al., 2011).

This study is designed with two purposes: first to determine whether the SESA region is a hot spot of soil moisture and precipitation coupling during anomalously dry and wet soil moisture conditions for different seasons of the year, and second to study the processes involved in creating favourable conditions for high soil moisture–precipitation coupling strength.

2. Model and methodology

In order to study the coupling index we follow an approach similar to Koster et al. (2006), which require performing ensembles of simulations with a climate model. The Rossby Centre Atmospheric regional model (RCA4, Samuelsson et al., 2011) is used for this study. This model has been used for climate studies in South America in the context of the CLARIS and CLARIS LPB projects (http://www.claris-eu.org/, Menéndez et al., 2010; Solman et al., 2013). The geographical domain covers South America with a horizontal resolution of 0.44° and 40 vertical levels. ERA-Interim reanalysis data are used at 0.75° horizontal resolution for the initial and boundary conditions (Dee *et al.*, 2011). For soil moisture (SM) prognostic variables, the soil column is divided into three layers where the two upper layers have a depth of 7 and 21 cm respectively, and the depth of the lowest layer is defined from the rooting depth of the Ecoclimap database (Masson *et al.*, 2003). RCA4 employs three land surface cover tiles for the separate calculation of fluxes of momentum and latent and sensible heat fluxes: open land, coniferous forest and broadleaved forest. The open land tile is subdivided into a vegetated and a bare soil part, and the two forest tiles include the canopy and the forest floor. The vegetation and soil parameters are taken from Ecoclimap.

A favourable condition for soil moisture to induce a response in precipitation (PP) is that the evapotranspiration (ET) variability is high (Guo *et al.*, 2006). As this condition is not met in SESA during austral winter (JJA), the study focuses on spring (SON), summer (DJF) and autumn (MAM). For each one of these seasons, one dry and one wet year from a simulation spanning 1980–1999, are identified through examining the seasonal soil water availability anomaly for a rectangle identified over the SESA region (see Figure 1). The dry periods selected are SON of 1988, DJF of 1988–1989 and MAM of 1989 and the wet periods are SON of 1985, DJF of 1997–1998 and MAM of 1998. For each of these six cases, the seasonal coupling strength index CS is calculated.

The methodology consists in the comparison of a similarity index (Ω) for two ensembles, one with prescribed soil moisture conditions (ensemble S) and the other with free interaction between surface and atmosphere (ensemble W). Each ensemble consists of 15 members, which were initialized on different dates so that each simulation has at least 45 days of atmospheric spin up. In order to have the soil in equilibrium with the atmosphere, the soil moisture initial conditions were taken from a multi-year integration so that each simulation has at least 2 years of soil moisture spin up. The Ω_X index measures the similarity of the amplitude, phase and mean of ensemble members of variable X (Yamada *et al.*, 2007), and quantifies the signal variance respect to the total variance (signal + noise) through the equation:

$$\Omega_X = \frac{15\,\sigma_{\hat{X}}^2 - \sigma_X^2}{14\,\sigma_X^2} \qquad (1)$$

where $\sigma_{\hat{X}}^2$ is the variance of the mean time series of all members of the ensemble and σ_X^2 is the ensemble inter member variance which was obtained by calculating the variance among all time steps and ensemble members. The synoptic scale was filtered by using time series of 6-day mean values. Then, the coupling strength of the variable X (hereafter CS[SM,X]) is calculated as the difference between the Ω_X of the ensembles:

$$CS[SM, X] = \Omega_X(S) - \Omega_X(W) \qquad (2)$$

As soil moisture is a boundary condition only for the ensemble S, CS[SM,X] is positive over regions where soil moisture explains some of the variance of X. In the following, coupling strength of different variables that connect soil moisture with precipitation is analysed.

3. Results

High values of both the coupling strength of evapotranspiration (CS[SM,ET]) and its daily variability (σ_{ET}) have been proposed to be necessary conditions for soil moisture to have a controlling effect over precipitation (Guo *et al.*, 2006). To identify when this occurs for the six cases, the product of these two statistics CS[SM,ET] $\times \sigma_{ET}$ (defined by Ruscica *et al.* (2014) as the 'coupling efficiency') is shown over South America in Figure 1. In general, the CS[SM,ET] $\times \sigma_{ET}$ hot spots appear over SESA and neighbouring regions and over the eastern coast of Brazil. Over the SESA region, CS[SM,ET] is higher for dry conditions than for wet conditions while the σ_{ET} is highest over SESA in DJF (not shown). The strongest coupling efficiency occurs when the land surface is dry during DJF (Figure 1(e)). Figure 2 shows that the coupling strength of precipitation (CS[SM,PP]) is weaker and less spatially coherent than CS[SM,ET] $\times \sigma_{ET}$, and that during the dry DJF the eastern SESA region is one of the main CS[SM,PP] hot spots while northwestern SESA has a CS[SM,PP] close to zero. In the following, we will therefore focus on the dry DJF season to understand why soil moisture controls precipitation only in the eastern SESA although the coupling efficiency is high over the entire region.

Eltahir (1998) proposed a theory of pathways between soil moisture conditions and subsequent rainfall, based mainly on considerations of the energy balance, in order to dissect the soil moisture influence on precipitation. To study the processes leading to high CS[SM,PP] in eastern SESA, some of these pathways were examined by analysing the coupling strength between soil moisture and the variables in Figure 3.

Most of the total energy in the boundary layer can be described by the moist static energy (MSE):

$$MSE = c_p T + Lq + gZ \qquad (3)$$

where the terms on the right hand side represent the internal, latent and potential energies (with c_p the specific heat of dry air at constant pressure, T the air temperature, L the latent heat of vaporization, q the specific humidity, g the gravitational acceleration and Z the geopotential height). MSE is sometimes used as an alternative to the equivalent potential temperature in studies of convection. A high MSE in the boundary layer plays an important role in the development of precipitation (Eltahir, 1998). When the MSE increases, the vertical gradient of MSE between the boundary layer and the free atmosphere also increases, favouring unstable conditions which can trigger precipitation. Furthermore, horizontal gradients of MSE at a range of scales induce thermally direct circulations which redistribute energy towards a flatter horizontal distribution of MSE.

$$CS[SM,ET]^x\sigma_{ET} \text{ (mm day}^{-1})$$

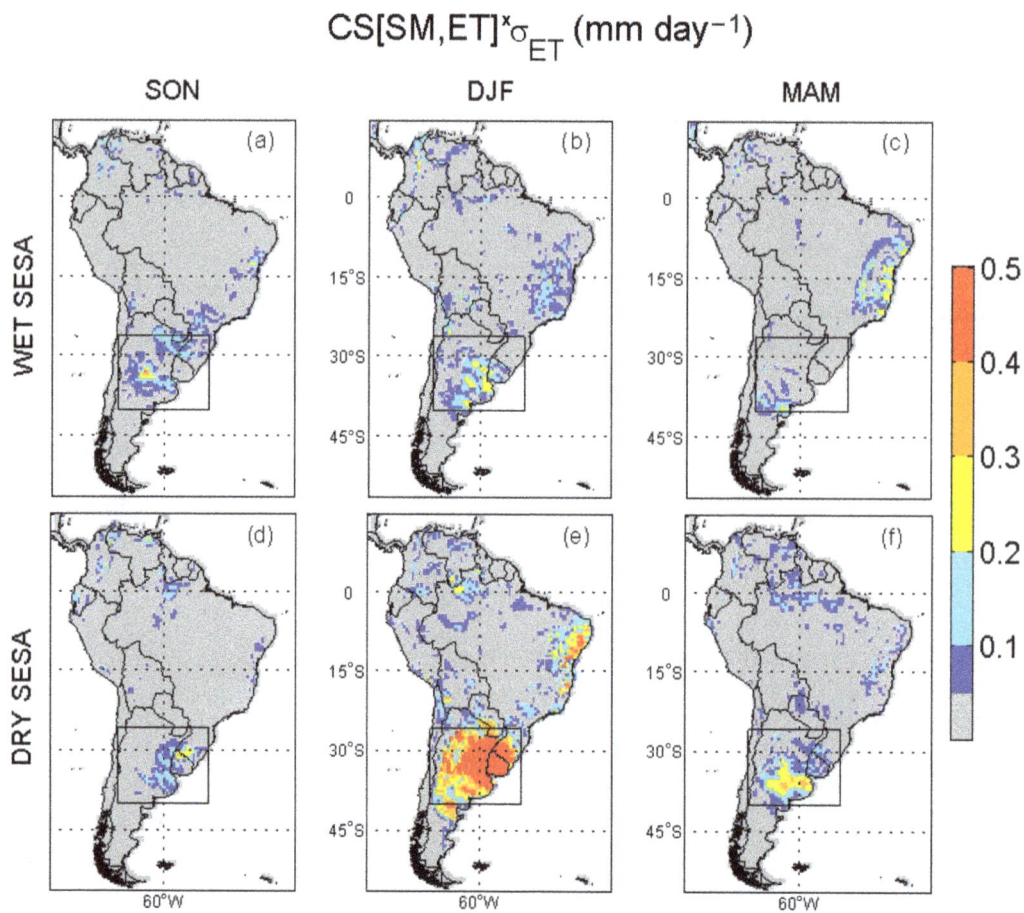

Figure 1. The product between the coupling strength of evapotranspiration (CS[SM,ET]) and its daily variability (σ_{ET}) in South America, for the three austral seasons: spring (SON), summer (DJF) and autumn (MAM) with wet (a–c) and dry (d–f) soil conditions in southeastern South America (inside rectangle).

Previous studies over subtropical South America support the statement that low level circulation and precipitation are sensitive to enhanced soil moisture/surface temperature gradients (Saulo *et al.*, 2010 and references therein).

Over large regions where the horizontal heat advection is small compared with the vertical heat flux, the MSE of the boundary layer is supplied by the sum of the sensible and latent surface heat fluxes (SHF + LHF). The total heat fluxes (SHF + LHF) directly affect the first two terms on the right hand side of Equation (3) and SHF affects the boundary layer depth (BLD). The soil moisture conditions play an important role in the partitioning of surface fluxes into SHF and LHF, represented by the Bowen ratio (BR = SHF/LHF). For example, in the case of dry soils, the BR, SHF and surface temperatures are higher than for wet conditions, resulting in an increase of BLD. However, the sum SHF + LHF does not have such a direct connection with soil moisture as BR has, as SHF + LHF depends on the surface net radiation which is also dependent on cloud cover (Findell and Eltahir, 1999).

In Figure 4, the coupling strength between soil moisture and the variables of Figure 3 for the dry DJF season are shown for SESA. It can be seen that soil moisture is highly coupled to the BR and to the BLD in the entire SESA (Figure 4(a) and (b)). It should be noted that the coupling strength between the consecutive variables in Figure 3 have not been calculated, and that for example high CS[SM,BLD] is interpreted as a result of high CS[SM,BR] through increased/decreased temperatures for dry/wet surface anomalies. The CS[SM,SHF + LHF] (Figure 4(c)) hot spot, however, does not cover the whole SESA but is concentrated over the eastern SESA, consistent with the hypothesis that the relationship between SM and SHF + LHF is not as direct as the relation SM–BR. The coupling strength of MSE (CS[SM,MSE], Figure 4(d), MSE is here calculated at 925 hPa) also shows its maximum value in the eastern SESA area. The coupling strength of the vertical gradient of MSE (CS[SM,Δ(MSE)/Δz]), Figure 4(e), calculated between 925 and 850 hPa) is quite similar to the coupling strength of the vertical humidity flux (CS[SM,wq]), showing the regions where soil moisture influences on the vertical humidity flux through atmospheric instability. Both patterns are similar to the CS[SM,SHF + LHF] pattern, suggesting that this influence is exerted through the coupling of the fluxes. Figure 4(g) indicates that soil moisture could also affect the local to regional scale circulation, here approximated by sea level pressure (SLP).The CS[SM,SLP] pattern has a maximum over southern

CS[SM,PP]

SON   DJF   MAM

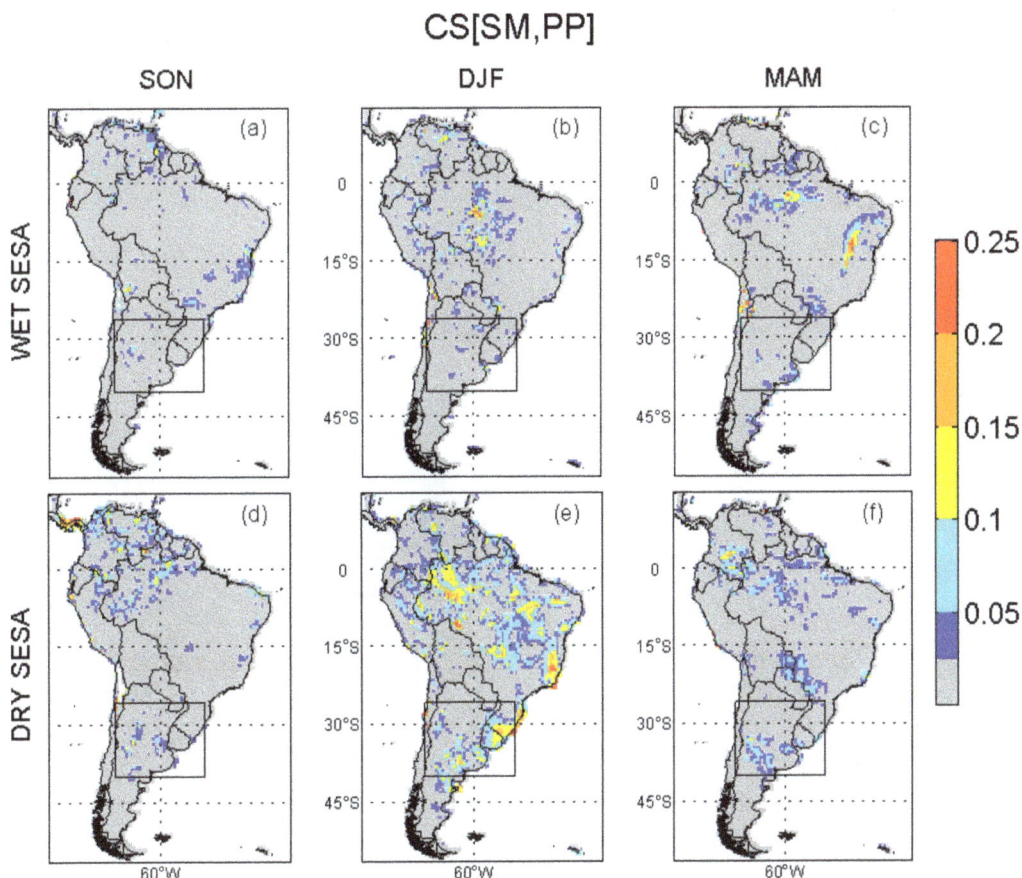

Figure 2. Coupling strength of precipitation (CS[SM,PP]) for the austral seasons: spring (SON), summer (DJF) and autumn (MAM) with wet (a–c) and dry (d–f) soil conditions in southeastern South America (inside rectangle).

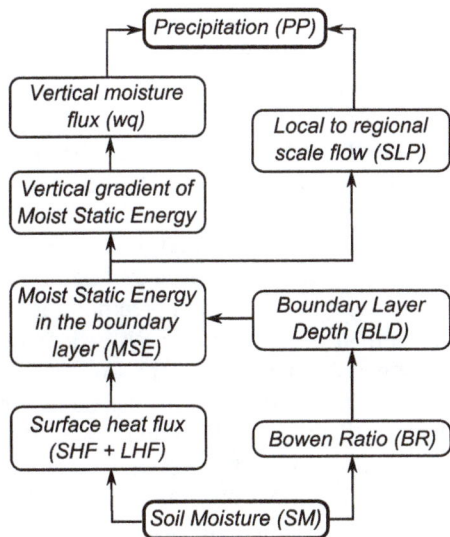

Figure 3. Pathways between soil moisture and precipitation processes (adapted from Eltahir (1998)).

Brazil where the horizontal contrast of CS[SM,MSE] is highest (see Figure 4(d)). Finally, the CS[SM,PP] pattern is shown in Figure 4(h). Considering the analysis of the Figure 4(a)–(g), the hot spot of soil moisture and precipitation coupling over Uruguay and southern Brazil (Figure 2(e)) is probably associated with changes

in MSE and in the circulation, induced by soil moisture anomalies.

4. Discussion and conclusions

SESA has a high coupling efficiency when the soil conditions are dry during the summer months. Weaker coupling efficiencies are found in the other seasons and for the wet summer case. This result is a combination of a high variability of the evapotranspiration during summer together with higher soil moisture – evapotranspiration coupling over dry than wet soil conditions. The latter can be understood as a major sensitivity of the evapotranspiration to changes in soil moisture content when the soil is dry, which has already been documented in previous studies (e.g. Wei and Dirmeyer, 2012). With respect to precipitation, high coupling values appear during the same dry DJF period but only over the eastern part of the SESA region. To understand this difference between eastern and western SESA, the coupling between soil moisture and some of the variables that are included in the pathways between soil moisture and precipitation were analysed.

While the couplings of soil moisture with the BR and of soil moisture with the BLD are high over all SESA, the coupling of soil moisture with the surface heat flux

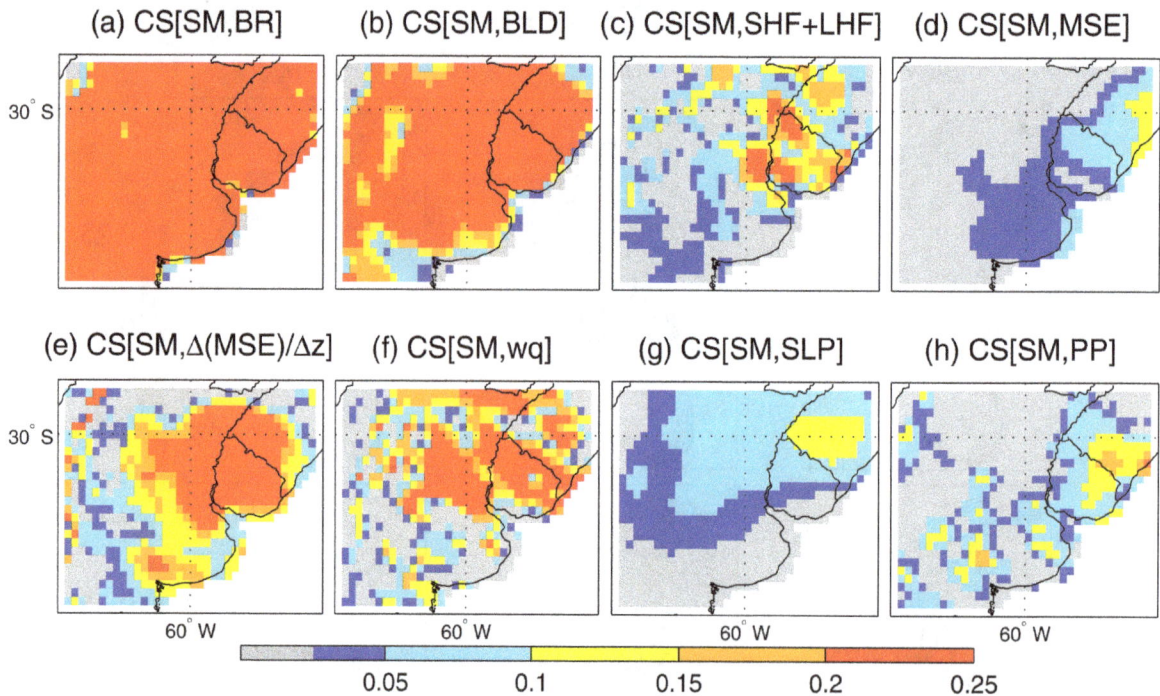

Figure 4. Coupling strength of the variables: (a) Bowen ratio (BR), (b) boundary layer depth (BLD), (c) surface heat flux (SHF + LHF), (d) moist static energy at 925 hPa (MSE), (e) vertical gradient of MSE between 925 and 850 hPa (Δ(MSE)/Δz), (f) vertical humidity flux (wq), (g) sea level pressure (SLP) and (h) precipitation in southeastern South America.

has a hot spot over eastern SESA. This generates coupling between soil moisture and MSE in the boundary layer as well as a higher coupling of soil moisture with the vertical gradient of MSE over eastern SESA. The vertical gradient of MSE seems to affect precipitation through coupling of soil moisture with the vertical moisture flux out of the boundary layer (we assumed that this vertical flow is enhanced in unstable conditions, thus favouring convection). It is also seen that the coupling of soil moisture with MSE is high over Uruguay and southern Brazil near the region of high coupling between soil moisture and SLP, suggesting that in this area soil moisture could affect the circulation in the lower levels of the atmosphere. This leads to the conclusion that precipitation is linked with soil moisture through the MSE in SESA during this dry summer.

The CS index embeds all aspects of the soil moisture field influence on precipitation, such as local water recycling (e.g. Trenberth, 1999), indirect local processes such as atmospheric thermodynamic properties (e.g. Beljaars *et al.*, 1996) and non-local processes such as advection of external moisture into the region (e.g. Spracklen *et al.*, 2012). It is worth mentioning that although a high coupling efficiency is commonly considered to be a necessary condition for a high coupling of precipitation, we find regions where this condition does not apply, such as in central and northern Brazil during the SESA-dry DJF season. This is an indication of non-local impact of soil moisture on precipitation such as advection of moisture from a remote source as suggested by e.g. Goessling and Reick (2011). Wei and Dirmeyer (2012) estimated the remote coupling between soil moisture and precipitation to around 20%

of total global coupling and van der Ent and Savenije (2011) found that 70% of the water resources in the SESA region comes from evapotranspiration from the Amazon rainforest.

Contrasting hypothesis about the sign of soil moisture –precipitation coupling, have been presented by different authors. In relation to the vertical gradient of MSE, while Eltahir (1998) suggested a positive influence on precipitation, Cook *et al.* (2006) proposed that an increased in the vertical gradient of MSE would lead to reduced precipitation through higher atmospheric stability. The results presented here are based on the CS index, which does not distinguish between positive and negative influences. The serial correlation between the vertical gradient of MSE and precipitation was calculated for all grid points of SESA where the coupling strength is higher than 0.025. At 83% of these grid points the correlation is positive, indicating that the vertical gradient of MSE influences positively on precipitation. The hot spot of both precipitation and vertical gradient of MSE and the positive correlation between the two variables are evidence for a positive feedback mechanism for eastern SESA during a dry summer.

Acknowledgements

Simulations were made with the high-performance computing clusters available at CIMA/UBA-CONICET, Argentina. We acknowledge the project PIP 112-200801-01788 (CONICET, Argentina). We also thank Patrick Samuelsson, Marco Kupiainen and Gabriel Vieytes for technical support and the anonymous reviewers for their valuable comments on the manuscript.

References

Barreiro M, Díaz N. 2011. Land–atmosphere coupling in El Niño influence over South America. *Atmospheric Science Letters* **12**: 351–355, doi: 10.1002/asl.348.

Beljaars ACM, Viterbo P, Miller MJ, Betts AK. 1996. The anomalous rainfall over the United States during July 1993: sensitivity to land surface parameterization and soil moisture anomalies. *Monthly Weather Review* **124**: 362–383, doi: 10.1175/1520-0493(1996)124<0362:tarotu>2.0.co;2.

Cook BI, Bonan GB, Levis S. 2006. Soil moisture feedbacks to precipitation in Southern Africa. *Journal of Climate* **19**: 4198–4206, doi: 10.1175/JCLI3856.1.

Dee DP et al. 2011. The ERA-Interim reanalysis: configuration and performance of the data assimilation system. *Quarterly Journal of the Royal Meteorological Society* **137**: 553–597, doi: 10.1002/qj.828.

Eltahir EAB. 1998. A soil moisture–rainfall feedback mechanism: 1. Theory and observations. *Water Resources Research* **34**(4): 765–776, doi: 10.1029/97WR03499.

van der Ent RJ, Savenije HHG. 2011. Length and time scales of atmospheric moisture recycling. *Atmospheric Chemistry and Physics* **11**: 1853–1863, doi: 10.5194/acp-11-1853-2011.

Findell KL, Eltahir EAB. 1999. Analysis of the pathways relating soil moisture and subsequent rainfall in Illinois. *Journal of Geophysical Research* **104**(D24): 31565–31574, doi: 10.1029/1999JD900757.

Goessling HF, Reick CH. 2011. What do moisture recycling estimates tell us? Exploring the extreme case of non-evaporating continents. *Hydrology and Earth System Sciences* **15**: 3217–3235, doi: 10.5194/hess-15-3217-2011.

Guo Z, Dirmeyer PA. 2013. Interannual variability of land–atmosphere coupling strength. *Journal of Hydrometeorology* **14**: 1636–1646, doi: 10.1175/JHM-D-12-0171.1.

Guo Z et al. 2006. GLACE: the global land-atmosphere coupling experiment. Part II: Analysis. *Journal of Hydrometeorology* **7**: 611–625, doi: 10.1175/JHM511.1.

Koster RD et al. 2004. Regions of strong coupling between soil moisture and precipitation. *Science* **305**: 1138–1140, doi: 10.1126/science.1100217.

Koster RD et al. 2006. GLACE: the global land-atmosphere coupling experiment. Part I: Overview. *Journal of Hydrometeorology* **7**: 590–610, doi: 10.1175/jhm510.1.

Ma HY, Mechoso CR, Xue Y, Xiao H, Wu CM, Li JL, de Sales F. 2011. Impact of land surface processes on the South American warm season climate. *Climate Dynamics* **37**: 187–203, doi: 10.1007/s00382-010-0813-3.

Masson V, Champeaux JL, Chauvin F, Meriguet C, Lacaze R. 2003. A global database of land surface parameters at 1-km resolution in meteorological and climate models. *Journal of Climate* **16**: 1261–1282, doi: 10.1175/1520-0442-16.9.1261.

Menéndez CG, De Castro M, Sörensson AA, Boulanger JP. 2010. CLARIS project: towards climate downscaling in South America. *Meteorologische Zeitschrift* **19**: 357–362, doi: 10.1127/0941-2948/2010/0459.

Ruscica RC, Sörensson AA, Menéndez CG. 2014. Hydrological links in Southeastern South America: soil moisture memory and coupling within a hot spot. *International Journal of Climatology* **34**: 3641–3653, doi: 10.1002/joc.3930.

Samuelsson P et al. 2011. The Rossby Centre Regional Climate model RCA3: model description and performance. *Tellus Series A: Dynamic Meteorology and Oceanography* **63**: 4–23, doi: 10.1111/j.1600-0870.2010.00478.x.

Santanello JA, Peters-Lidard CD, Kumar SV. 2011. Diagnosing the sensitivity of local land–atmosphere coupling via the soil moisture–boundary layer interaction. *Journal of Hydrometeorology* **12**: 766–786, doi: 10.1175/JHM-D-10-05014.1.

Saulo C, Ferreira L, Nogués-Paegle J, Seluchi M, Ruiz J. 2010. Land atmosphere interactions during a northwestern Argentina low event. *Monthly Weather Review* **138**: 2481–2498, doi: 10.1175/2010mwr3227.1.

Seneviratne SI, Lüthi D, Litschi M, Schär C. 2006. Land–atmosphere coupling and climate change in Europe. *Nature* **443**: 205–209, doi: 10.1038/nature05095.

Solman S et al. 2013. Evaluation of an ensemble of regional climate model simulations over South America driven by the ERA-Interim reanalysis: model performance and uncertainties. *Climate Dynamics* **41**: 1139–1157, doi: 10.1007/s00382-013-1667-2.

Sörensson AA, Menéndez CG. 2011. Summer soil-precipitation coupling in South America. *Tellus Series A: Dynamic Meteorology and Oceanography* **63**: 56–68, doi: 10.1111/j.1600-0870.2010.00468.x.

Spracklen DV, Arnold SR, Taylor CM. 2012. Observations of increased tropical rainfall preceded by air passage over forests. *Nature* **489**: 282–285, doi: 10.1038/nature11390.

Trenberth KE. 1999. Atmospheric moisture recycling: role of advection and local evaporation. *Journal of Climate* **12**: 1368–1381, doi: 10.1175/1520-0442(1999)012<1368:AMRROA>2.0.CO;2.

Wei J, Dirmeyer PA. 2012. Dissecting soil moisture-precipitation coupling. *Geophysical Research Letters* **39**: L19711, doi: 10.1029/2012GL053038.

Yamada TJ, Koster RD, Kanae S, Oki T. 2007. Estimation of predictability with a newly derived index to quantify similarity among ensemble members. *Monthly Weather Review* **135**(7): 2674–2687, doi: 10.1175/MWR3418.

Permissions

List of Contributors

C. Andrade
Mathematics and Physics Department, Polytechnic Institute of Tomar, Tomar 2300-313, Portugal
Centre for the Research and Technology of Agro-Environmental and Biological Sciences, Universidade de Trás-os-Montes e Alto Douro, Vila Real 5001-801, Portugal

H. Fraga and J. A. Santos
Centre for the Research and Technology of Agro-Environmental and Biological Sciences, Universidade de Trás-os-Montes e Alto Douro, Vila Real 5001-801, Portugal

Roger W. Bodman and Roger N. Jones
Victoria Institute of Strategic Economic Studies, Victoria University, Melbourne, Australia

Peter J. Rayner
School of Earth Sciences, The University of Melbourne, Melbourne, Australia

Carmen Sánchez de Cos, Jose M. Sánchez-Laulhé and Carlos Jiménez-Alonso
Centro Meteorológico de Málaga, AEMET, Málaga, Spain

Ernesto Rodríguez-Camino
Servicios Centrales, AEMET, Madrid, Spain

Ana Paula M. A. Cunha and Regina C. S. Alvalá
Brazilian Centre for Monitoring and Warning of Natural Disasters, São José dos Campos, SP, Brazil

Paulo Y. Kubota and Rita M. S. P. Vieira
National Institute for Space Research, Cachoeira Paulista, SP, Brazil

Richard Davy and Igor Esau
Nansen Environmental and Remote Sensing Centre, Centre for Climate Dynamics (SKD), 5006, Bergen, Norway

Andrew J. Dowdy
The Centre for Australian Weather and Climate Research, Bureau of Meteorology, Docklands, 3007, Australia

Sonya Louise Fiddes, Alexandre Bernardes Pezza and Vaughan Barras
School of Earth Sciences, University of Melbourne, Parkville, VIC, 3010, Australia

D. Founda, C. Giannakopoulos, F. Pierros and M. Petrakis
National Observatory of Athens, Institute for Environmental Research and Sustainable Development, Athens, Greece

A. Kalimeris
Department of Environmental Technology and Ecology, Technological and Educational Institute of the Ionian Islands, Zakynthos, Ionian Islands, Greece

Fumiaki Fujibe
Meteorological Research Institute, JMA, Tsukuba, Japan

Chao He
Guangzhou Institute of Tropical and Marine Meteorology, China Meteorological Administration, Guangzhou, China
LASG, Institute of Atmospheric Physics, Chinese Academy of Sciences, Beijing, China

Tianjun Zhou
LASG, Institute of Atmospheric Physics, Chinese Academy of Sciences, Beijing, China
CCRC, Institute of Atmospheric Physics, Chinese Academy of Sciences, Beijing, China

Elke Hertig and Jucundus Jacobeit
Institute of Geography, University of Augsburg, 86159 Augsburg, Germany

Ping Huang
Center for Monsoon System Research, Institute of Atmospheric Physics, Chinese Academy of Sciences, Beijing 100190, China

Masaru Inatsu
Faculty of Science, Hokkaido University, Sapporo, Japan

Tomonori Sato
Faculty of Earth Environmental Sciences, Hokkaido University, Sapporo, Japan

Tomohito J. Yamada and Murad A. Farukh
Faculty of Engineering, Hokkaido University, Sapporo, Japan

Ryusuke Kuno
ARK Information Systems, Yokohama, Japan

Shiori Sugimoto
Graduate School of Urban Environmental Sciences, Tokyo Metropolitan University, Japan

Yadu N. Pokhrel and Shuichi Kure
Department of Civil and Environmental Engineering, Michigan State University, East Lansing, MI, USA
International Research Institute of Disaster Science, Tohoku University, Sendai, Japan

Joan Kleypas
NCAR, Boulder, CO, USA

John Latham
NCAR, Boulder, CO, USA
SEAS, University of Manchester, Manchester, UK

Rachel Hauser
NCAR, Boulder, CO, USA
Environmental Studies Program and Center for Science and Technology Policy Research, University of Colorado, Boulder, UK

Ben Parkes
ICAS, University of Leeds, West Yorkshire, UK

Alan Gadian
NCAS, ICAS, University of Leeds, West Yorkshire, UK

Cyrille Meukaleuni, André Lenouo and David Monkam
Department of Physics, Faculty of Science, University of Douala, Cameroon

G. W. K. Moore
Department of Physics, University of Toronto, Toronto, Ontario, Canada

R. Nogherotto, E. Coppola, F. Giorgi and L. Mariotti
Earth System Physics Section, The Abdus Salam International Centre for Theoretical Physics, Trieste, Italy

Victor Privalsky
Space Dynamics Laboratory (ret.), Logan, UT, USA

Vladislav Yushkov
Physics Department, Moscow State University, Russia

R. C. Ruscica and A. A. Sörensson
Centro de Investigaciones del Mar y la Atmósfera, Consejo Nacional de Investigaciones Científicas y Técnicas, Universidad de Buenos Aires, Argentina

C. G. Menéndez
Centro de Investigaciones del Mar y la Atmósfera, Consejo Nacional de Investigaciones Científicas y Técnicas, Universidad de Buenos Aires, Argentina

Departamento de Ciencias de la Atmósfera y los Océanos, FCEN, Universidad de Buenos Aires, Argentina

Sebastian Schemm
Geophysical Institute and Bjerknes Centre for Climate Research, University of Bergen, Bergen, Norway
Institute of Geography, University of Bern, Bern, Switzerland
Oeschger Centre for Climate Change Research, University of Bern, Bern, Switzerland

Luca Nisi, Andrey Martinov and Olivia Martius
Institute of Geography, University of Bern, Bern, Switzerland
Oeschger Centre for Climate Change Research, University of Bern, Bern, Switzerland
Mobiliar Lab for Natural Risks, University of Bern, Bern, Switzerland

Daniel Leuenberger
Federal Office of Meteorology and Climatology MeteoSwiss, Zurich, Switzerland

Tomoo Ogura and Tokuta Yokohata
Center for Global Environmental Research, National Institute for Environmental Studies, Tsukuba, Japan

Hideo Shiogama
Center for Global Environmental Research, National Institute for Environmental Studies, Tsukuba, Japan
Environmental Change Institute, University of Oxford, Oxford, UK

Masahiro Watanabe and Masahide Kimoto
Atmosphere and Ocean Research Institute, The University of Tokyo, Kashiwa, Japan

Lucky Ntsangwane
South African Weather Service, Private Bag X097, Pretoria, South Africa

Izidine Pinto
University of Pretoria, South Africa

Mxolisi E. Shongwe
South African Weather Service, Private Bag X097, Pretoria, South Africa
University of Pretoria, South Africa

Chris Lennard
Climate Systems Analysis Group, University of Cape Town, 7945, South Africa

Brant Liebmann
NOAA-CIRES Climate Diagnostics Center, Boulder, CO, USA

Evangelia-Anna Kalognomou
Laboratory of Heat Transfer and Environmental
Engineering, Aristotle University, Thessaloniki, Greece

Csaba Torma and Filippo Giorgi
Earth System Physics Section, The Abdus Salam
International Centre for Theoretical Physics, Trieste,
Italy

Wandee Wanishsakpong and Nittaya McNeil
Department of Mathematics and Computer Science,
Faculty of Science and Technology, Prince of Songkla
University, Pattani, Thailand
Centre of Excellence in Mathematics, Commission on
Higher Education, Bangkok, Thailand

Khairil A. Notodiputro
Department of Statistics, Faculty of Mathematics and
Science, Bogor Agricultural University, Indonesia

Wen Chen
Center for Monsoon System Research, Institute of
Atmospheric Physics, Chinese Academy of Sciences,
Beijing, China

Ke Wei
Center for Monsoon System Research, Institute of
Atmospheric Physics, Chinese Academy of Sciences,
Beijing, China
Atmosphere and Ocean Research Institute, University
of Tokyo, Kashiwa, Japan

Masaaki Takahashi
Atmosphere and Ocean Research Institute, University
of Tokyo, Kashiwa, Japan

Ting Wei, Bingyi Wu and Changgui Lu
Climate System Institute, Chinese Academy of
Meteorological Sciences, Beijing, China

Minghu Ding
Climate System Institute, Chinese Academy of
Meteorological Sciences, Beijing, China
State Key Laboratory of Cryospheric Sciences,
CAREERI, Lanzhou, China

Shujie Wang
College of Geography and Environment, Shandong
Normal University, Jinan, China

Lingyun Wu
State Key Laboratory of Numerical Modeling for
Atmospheric Sciences and Geophysical Fluid Dynamics
(LASG), Institute of Atmospheric Physics, Chinese
Academy of Sciences, Beijing 100029, China

Jingyong Zhang
Center for Monsoon System Research, Institute of
Atmospheric Physics, Chinese Academy of Sciences,
Beijing 100029, China

Jussi S. Ylhäisi
Weather and Safety, Finnish Meteorological Institute,
Finland

Jouni Räisänen, Olle Räty and Heikki Järvinen
Department of Physics, University of Helsinki, Finland

David Masson
Federal Office of Meteorology and Climatology,
MeteoSwiss, Switzerland

Romina C. Ruscica and Anna A. Sörensson
Centro de Investigaciones del Mar y la Atmósfera,
CONICET/UBA, UMI IFAECI/CNRS, Buenos Aires,
Argentina

Claudio G. Menéndez
Centro de Investigaciones del Mar y la Atmósfera,
CONICET/UBA, UMI IFAECI/CNRS, Buenos 3Aires,
Argentina
Departamento de Ciencias de la Atmósfera y los
Océanos, FCEN, UBA, Buenos Aires, Argentina

Index

A

Africa, 16, 19, 30, 73, 82, 95, 101-106, 115-117, 119-121, 131, 150-152, 155, 157-159, 164, 203

Agro-forestry Systems, 7

Area-means, 3, 7

Athens, 53, 55, 59

Azores High, 16-19

B

Bias Corrections, 1-2

Biosphere-atmosphere Transfer Scheme, 116, 159, 164

C

Carbon Cycle Uncertainties, 9-10, 13-14

Central Tendency, 1, 102, 150

Change, 1, 5, 7-15, 21-26, 28-31, 39, 42, 45, 48, 51, 57-61, 67, 73, 77-89, 95, 99, 103, 107, 115, 117, 121, 127, 134, 144, 146, 150, 161, 164-171, 179, 182, 184, 197, 203

Clausius-clapeyron Relation, 60

Climate, 1-2, 4-9, 11-17, 19-23, 28-32, 37-40, 44-55, 60, 66, 70, 75, 82, 89, 97, 103, 108, 112-115, 119, 122, 133, 142, 144, 146, 152, 161, 165, 170, 179, 185, 188-198, 203

Climate Change, 1, 5, 7-9, 14-15, 21-23, 31, 37, 45, 54, 60, 65, 73, 78, 85-88, 94, 99, 107, 114, 121, 127, 129, 133, 144, 149, 157, 160, 165, 171, 179, 184, 190-193, 197, 203

Climate Change Projections, 1, 7, 53, 59, 164, 190, 192

Climate Model Evaluation, 16, 150

Climate Models, 1, 8-9, 12, 14, 16-17, 19, 21, 31-32, 37, 45, 49-50, 53, 58-59, 65, 80, 94, 107, 115, 121-122, 126, 130, 148-150, 157, 159, 164, 185, 192-193, 195-196, 203

Climate Projections, 15, 48, 52, 55, 94, 159-160, 163, 192-193, 196-197

Climatology, 8, 30-33, 37, 39-40, 44-45, 51-52, 55, 58-59, 65, 79-81, 83, 85, 89-90, 93-94, 101, 106, 120, 133, 137, 141-143, 145, 151, 157, 170, 177, 184, 190, 192, 203

Cloud Seeding, 95, 100

Cmip5, 8-19, 21, 31-33, 35-39, 45, 52, 67, 72, 74-75, 79-82, 84-86, 122, 124-125, 131, 157, 160, 164, 184, 193, 197

Conus, 122, 124-125

Convective Available Potential Energy, 101, 106, 133

Coral Bleaching, 95-100

Coupling Strength, 132, 198-202

Cyclone, 8, 39-40, 44-45

D

Decadal Change, 66-67, 70-72

Deforestation, 24, 30, 115-117, 119-121

Dynamical Downscaling, 87, 91, 132

E

East Asian Jet Stream, 107

Extreme Precipitation, 51, 53, 55, 57-61, 63, 65, 157, 196

F

Feedback, 9, 14, 16-19, 21, 31, 37, 85, 117, 131-132, 144, 146, 148-149, 161, 163, 177, 188, 190, 196, 202-203

Feedback-induced Warming, 161

G

Global Climate Modeling, 95

Global Warming, 15, 32, 44-46, 49, 51, 53, 80-81, 84-87, 91-94, 99-100, 113, 149, 157, 170-171, 175, 178, 183-184

H

Hierarchical Cluster Analysis, 165

Hurricane, 39, 45, 97

I

Ice Core, 107-109, 111-114, 170

Indices of Extremes, 1

Inter-tropical Convergence Zone, 101-102

Internal Climate Variability, 66, 70

J

Japan, 32, 37, 40, 60-65, 69, 72, 87-89, 93-94, 107, 109, 144, 148, 171, 177, 191

K

Kurtosis, 5

L

Land Use Change, 22, 30, 190

Linear Trend Rates, 122, 125

M

Marine Cloud Brightening, 95, 100

Mean Values And Standard Deviations, 122, 124

Moist Static Energy, 115, 198-199, 202

Monsoon, 21, 45, 72, 80, 85, 101-106, 113-117, 119-120, 131, 171, 173, 176-177, 185, 190-191

Multi-gcm By Multi-ram Experiments, 87, 93

Multi-model Ensemble, 1, 10, 15, 55, 87, 94, 126, 144, 196

N

Non-stationarities, 73, 76, 78

O

Observational Dataset, 1-2, 7

P

Paleoclimate, 107, 148

Planetary Boundary Layer, 31, 37, 137

Portugal, 1-3, 6-8

Precipitation, 8, 22-25, 28, 46-51, 53-65, 72-74, 81, 83, 95, 101, 108, 112, 115-121, 126, 129, 132, 135, 142, 145, 150-154, 156-164, 170, 185, 187-194, 196, 199, 201-203

R

Regional Climate Change, 22, 73, 87, 170, 190

Regional Climate Model, 1, 8, 30, 115, 132, 142, 150-151, 157, 159, 185, 190, 203

Regional Climate Modeling, 94, 115, 121, 126, 159

Regional Response, 80

S

Sea Surface Temperature, 21, 30, 44, 66, 72, 80, 85-86, 90, 95, 97-98, 145, 175

Seasonal Mean (left Panels) Tx, 3

Seasonal Study, 101

Semiarid, 22-24, 29-30, 127

Simple Climate Model, 9, 12

Snowfall, 46-51, 61, 79, 109

Soil Moisture, 21, 23, 31, 37, 53, 116-117, 121, 126-127, 131-132, 160, 185, 188, 190-191, 198-203

Southeastern South America, 126, 132, 198, 200-203

Spectral Densities, 122, 124

Statistical Properties, 122-123

Synergy, 159-161, 163

T

Temperature, 1-3, 5-16, 18, 21-24, 28, 33, 35, 38, 44-50, 53, 57, 59-66, 69, 73, 80, 87, 91, 98, 100, 106, 109, 115, 117, 120, 125, 132, 140, 145, 157, 163, 170, 172, 185, 196, 200

Temperature Extremes, 1, 5, 8, 184

Temperature-change Projections, 9

Trends, 6-8, 31-34, 38, 43, 45-51, 53-55, 59-60, 62, 64-66, 85-86, 94, 101-103, 105-107, 114, 124, 161, 163, 166, 170, 174, 178, 180, 183-184, 190

Tropical, 21, 24, 30, 39, 44-45, 52-53, 66, 70-72, 80-86, 89, 93, 95-97, 99, 101-102, 104, 106, 113-115, 117, 120-121, 131, 148, 150, 171, 191, 203

Tropical Rainfall, 80, 82, 85, 203

Type Of Probability Density Functions, 122

Typhoon, 39, 45

V

Variability, 1-2, 5-8, 23, 31-33, 35-45, 49, 51-55, 59, 66, 74, 79, 94, 96, 102, 108, 112, 122, 127, 130, 139, 141, 150, 157, 161, 165, 170, 173, 177, 179, 185, 191, 197, 203

Verification, 32, 122

W

Weather Regimes, 73-76, 78-79

www.ingramcontent.com/pod-product-compliance
Lightning Source LLC
Chambersburg PA
CBHW080647200326
41458CB00013B/4758